Advances in Nonlinear Dynamos

The fluid mechanics of astrophysics and geophysics
A series edited by Andrew Soward
University of Exeter, UK and

Michael Ghil
University of California, Los Angeles, USA

Founding Editor: Paul Roberts,
University of California, Los Angeles, USA

Advances in Nonlinear Dynamos

Edited by
Antonio Ferriz-Mas
and
Manuel Núñez

CRC Press
Taylor & Francis Group
Boca Raton London New York

CRC Press is an imprint of the
Taylor & Francis Group, an **informa** business
A TAYLOR & FRANCIS BOOK

First published 2003 by Taylor & Francis

Published 2019 by CRC Press
Taylor & Francis Group
6000 Broken Sound Parkway NW, Suite 300
Boca Raton, FL 33487-2742

First issued in paperback 2019

No claim to original U.S. Government works

ISBN 13: 978-0-367-45450-0 (pbk)
ISBN 13: 978-0-415-28788-3 (hbk)

Visit the Taylor & Francis Web site at
http://www.taylorandfrancis.com

and the CRC Press Web site at
http://www.crcpress.com

Typeset in Times New Roman by
Newgen Imaging Systems (P) Ltd, Chennai. India

British Library Cataloguing in Publication Data
A catalogue record for this book is available from the British Library

Library of Congress Cataloging in Publication Data
A catalog record for this book has been requested

Cover image: A snapshot of the 3D magnetic field structure simulated with the
Glatzmaier-Roberts geodynamo model. Reproduced with permission of Gary A Glatzmaier,
University of California, Santa Cruz and Paul H Roberts, University of California, Los Angeles.

Contents

Contributors

Rainer Arlt
Astrophysikalisches Institut Potsdam
An der Sternwarte 16
D-14482 Potsdam
Germany
E-mail: rarlt@aip.de

Mitchell A. Berger
Department of Mathematics
University College London
Gower Street
London WC1E 6BT, UK
E-mail: m.berger@ucl.ac.uk

Stanislav Braginsky
Institute of Geophysics and Planetary
  Physics
University of California
Los Angeles
CA 90095, USA
E-mail: sbragins@igpp.ucla.edu

Axel Brandenburg
Nordic Institute for Theoretical Physics
(NORDITA)
Blegdamsvej 17
DK-2100 Copenhagen Ø
Denmark
E-mail: brandenb@nordita.dk

Antonio Ferriz-Mas
Department of Physical Sciences
Astronomy Division
P.O. Box 3000
FIN-90014 University of Oulu
Finland
E-mail: antonio.ferriz@oulu.fi

David Galloway
School of Mathematics and Statistics
University of Sydney
NSW 2006 Sydney
Australia
E-mail: dave@maths.usyd.edu.au

Peter Hoyng
SRON Laboratory for Space Research
Sorbonnelaan 2
3584 CA Utrecht
The Netherlands
E-mail: p.hoyng@sron.nl

Keith Julien
Department of Applied Mathematics
University of Colorado
Boulder
CO 80309-0526, USA
E-mail: julien@colorado.edu

Edgar Knobloch
Department of Physics
University of California
Berkeley
CA 94720, USA
E-mail: knobloch@physics.berkeley.edu

Paul H. Roberts
Institute of Geophysics and Planetary
  Physics
University of California
Los Angeles
CA 90095, USA
E-mail: roberts@math.ucla.edu

Günther Rüdiger
Astrophysikalisches Institut Potsdam
An der Sternwarte 16
D-14482 Potsdam
Germany
E-mail: gruediger@aip.de

Dieter Schmitt
Max-Planck-Institut für Aeronomie
Max-Planck-Str. 2
D-37191 Katlenburg-Lindau
Germany
E-mail: schmitt@linmpi.mpg.de

Manfred Schüssler
Max-Planck-Institut für Aeronomie
Max-Planck-Str. 2
D-37191 Katlenburg-Lindau
Germany
E-mail: msch@linmpi.mpg.de

Andrew M. Soward
School of Mathematical Sciences
University of Exeter
Laver Building
North Park Road
Exeter, EX4 4QE, UK
E-mail: A.M.Soward@ex.ac.uk

Steve Tobias
Department of Applied Mathematics
University of Leeds
Leeds, LS2 9JT, UK
E-mail: smt@amsta.leeds.ac.uk

Preface

This book presents an up-to-date survey on nonlinear topics in dynamo theory. The ten chapters, starting with a review of the fundamentals of mean-field theory, cover aspects of planetary, solar/stellar and galactic dynamos, as well as recent developments in fast dynamos, convection in strong magnetic fields and topological techniques in magnetohydrodynamics.

Astrophysical dynamos are governed by nonlinear partial differential equations. This renders impossible to find analytical solutions in all but the simplest cases and highlights the necessity of performing numerical simulations for the study of the growth and maintenance of astrophysical magnetic fields (chapter by Brandenburg). Massive computations (see chapter by Braginsky and Roberts) reproduce succesfully important aspects of the geodynamo.

The situation is different for the Sun: no numerical simulations reproducing the behaviour of the solar cycle in detail have been performed so far. The large spatial dimension of the system and the small value of the molecular viscosity cause strong nonlinear interactions in the flows, leading to a state of turbulent convection, for which no generally accepted theory exists. The situation is further complicated by the effects of stratification, the existence of penetrative boundaries and the interaction of convection with differential rotation: neither the spatial and temporal structure of the convection nor the profile of differential rotation can be directly derived from the basic equations of hydrodynamics.

Although sophisticated computer modelling of the highly nonlinear processes will continue to provide new insights, computational resources are nowadays still insufficient to model a complete system that includes both small- and large-scale effects. Therefore, a complete theory encompassing in a self-consistent way the generation, structure, transport and evolution of the magnetic fields in stars does not exist yet.

One way around the difficulties of tackling the dynamo problem as a whole consists of studying a number of underlying processes in artificial isolation – usually in a simplified setting – in order to understand their fundamental physics and evaluate their relevance for the global problem. These individual processes are the building blocks of the full dynamo mechanism and must eventually be fitted together to allow for a reasonable description of the system. Some of the individual processes treated in this book are the storage and the transport of the large-scale magnetic field (in the form of flux tubes) from the bottom of the convection zone to the surface (see chapter by Schüssler and Ferriz-Mas), the role of magnetic buoyancy in producing an α-effect (see chapters by Schmitt and by Schüssler and Ferriz-Mas) and the intermittency in long-term solar activity (chapter by Rüdiger and Arlt). Also, the study of the rapid evolution of magnetic fields in many astrophysical phenomena has created the field of research of 'fast dynamos' (chapter by Galloway). An approach for the parametrization of turbulence is the mean-field magnetohydrodynamic model, which has been rather successful in predicting some observed features in astrophysical dynamos. The foundations of this model are reviewed and critically examined in the chapter by Hoyng.

Another theoretical approach with the ability to produce specific solutions is asymptotic analysis: when some of the parameters of the problem are small with respect to others, an asymptotic expansion may often be used whose first terms provide an excellent approximation to the real solution (see chapters by Julien, Knobloch and Tobias and by Soward). Also, the topological constraints of ideal magnetohydrodynamics, notably the conservation of magnetic helicity, yield relevant insights into the dynamo behaviour (chapter by Berger).

This monograph represents not only a study of specific problems by leading researchers in the field of dynamo theory, but also an exposition of some of the most important methods used in this exciting field of astrophysical science.

The book is intended for graduate students and researchers in theoretical astrophysics and applied mathematics interested in cosmic magnetism and related topics such as turbulence, convection and, more general, nonlinear physics.

Antonio Ferriz-Mas
University of Oulu and University of Vigo

Manuel Núñez
University of Valladolid

1 The field, the mean and the meaning

Peter Hoyng

SRON Laboratory for Space Research, Sorbonnelaan 2, 3584 CA Utrecht, The Netherlands, E-mail: p.hoyng@sron.nl

This chapter is about the fundamentals of mean field dynamo theory, with an emphasis on the statistical aspects of the theory. The equations for passive vectorial transport (dynamo equation) and passive scalar transport are derived on a par, as applications of the theory of stochastic equations with multiplicative noise. Only approximate transport equations for mean quantities exist, and the approximations are scrutinized in relation to the ensemble average and the azimuthal average. A summary of the elementary physics of the dynamo equation is followed by an analysis of the influence of mean shear flows and resistive effects on the transport coefficients α and β. It is argued that the dynamo equation contains in practical situations no higher than second-order spatial derivatives. The physics of turbulent diffusion and the information content of the dynamo equation is analysed, and it is concluded that transport equations for the mean can only make probabilistic statements about the physical systems to which they apply. They cannot predict the evolution for all times. This is elucidated for ensemble and azimuthal averages at the hand of examples, referring to the phase memory and magnetic energy losses of the solar dynamo, and to the variability and reversals of the geodynamo. The accent is on clarity of presentation, and on linear theory. A few remarks on nonlinear theory are made.

1.1. Introduction

Mean field dynamo theory seems to have a controversial status. To some it is their natural habitat, while others shy away from it because they maintain that the basic tenets are not understood or do not apply to real dynamos. In spite of this, much work in the literature on global magnetic fields is based on mean field theory. There are two main reasons for that. The first is that there is often no viable alternative. It is true that the magnetic Reynolds number R_m of the geodynamo is sufficiently small that several groups have succesfully performed fully resolved three-dimensional simulations (Glatzmaier and Roberts, 1995; Kageyama and Sato, 1997; Kuang and Bloxham, 1997; Christensen *et al.*, 1998). While these computations have demonstrated that self-consistent dynamo action is able to overcome resistive decay of the currents, the demands on computing resources are extreme. A routine study of many different cases, and runs extending over a long time and in the correct parameter regime are impossible. And stellar dynamos have such high Reynolds numbers that fully resolved MHD simulations of the whole dynamo will be impossible for the forseeable future. Only a small section of the dynamo can be addressed, and that is helpful for an understanding of the physical mechanisms at work (Nordlund *et al.*, 1992, 1994; Brandenburg *et al.*, 1996).

A second important reason for the popularity of mean field theory is that it works rather well. It produces results that 'look good', and provides a simple physical picture of what is going on. Given this state of affairs, it seems that mean field theory will stay in the niche it presently occupies for a long time to come, and that numerical simulations and mean field theory will go hand in hand in quest of the holy grail. A good example of this cross-fertilisation is the interface wave dynamo, a mean field model for the solar dynamo proposed by Parker (1993) on the basis of helioseismological observations and numerical simulations of the type cited above.

But this is not to say that all is well. There are many justified doubts about the approximations on which the dynamo equation is based, on the nature of the average and the role of nonlinear effects, etc. And there is the question why the theory works apparently so well. All this underlines the importance of a good understanding of mean field theory. Against this background I shall review the basics of the linear theory, from the perspective of stochastic equations with multiplicative noise. The advantage is that it permits treatment of scalar and vector transport on equal footing, and to jump back and forth between them for instructive analogies and differences. I shall present mean field theory as a *physical* theory (as opposed to a mathematical construction), with all its imperfections, and point out where the major problems are.

An important point to realise is that transport theory for mean quantities is a statistical theory. The equations for the mean are obtained by taking an average, and this has the unavoidable consequence that the predictive character of the equations disappears. Even when solved exactly, they can make only *probabilistic* statements about the scalars and vectors (magnetic fields) in the systems to which they apply. I shall restrict myself mainly to linear theory, with only occasional remarks on nonlinear theory, the problem of a long correlation time, and spectral theory (e.g. Pouquet *et al.*, 1976). A review of these topics would take too much space.

The organisation of this article is as follows. The transport equation for the mean is derived in Section 1.2. The attention then shifts to the ensemble and azimuthal averaging procedures in Section 1.3. We analyse to what extent they possess the required properties, and the traditional derivation based on the First Order Smoothing Approximation is reviewed. In Section 1.4 the transport equation is applied to passive scalar and vectorial transport for isotropic turbulence, and the elementary properties of the dynamo equation are discussed. Section 1.5 gives a general treatment of the influence of mean shear flows and resistive effects on the transport coefficients α and β, stressing the issue of Galilean invariance. The meaning and the information content of the mean field $\langle \mathbf{B} \rangle$ is explained in Sections 1.6 and 1.7 for the two popular averages in use, the ensemble average and the azimuthal average, and a few examples are given. A few comments on nonlinear theory are made in Section 1.8.

1.2. Turbulent transport

Passive advection of a scalar or a vector such as the magnetic field in random flows is governed by the continuity equation or the induction equation, respectively. Accordingly, we consider in this section linear equations of the type

$$\frac{\partial f}{\partial t} = A(t)f; \quad A = R + C(t). \tag{1.1}$$

The quantity f may be a scalar, a vector such as $\mathbf{B}$, anything really. The operator A can be split in a time-independent part R and a time-dependent part $C(t)$ that fluctuates randomly, with a finite correlation time τ_c. For example, for magnetic field transport

$$R = \nabla \times \mathbf{u} \times +\eta \nabla^2; \quad C = \nabla \times \mathbf{v} \times, \tag{1.2}$$

where **u** is the systematic flow ('the differential rotation'), and **v** the turbulent convection. For scalar transport we have

$$R = -\nabla \cdot \mathbf{u} + \eta \nabla^2; \quad C = -\nabla \cdot \mathbf{v}. \tag{1.3}$$

A dissipative term has been added to R, representing molecular diffusion or heat conduction, when necessary; R and C are of course operators with ∇ operating on everything to the right, $Cg = -\nabla \cdot (\mathbf{v}g)$, etc. We shall not make use of any specific property of f, R and C. The analysis of this section is therefore generally applicable, and only occasional references are made to scalar or vectorial transport.

Note that a random element is essential. Transport in media with purely systematic flows is not covered here. These include some very complicated flows such as ABC flows, which have chaotic streamlines but no random element in the sense that the flow can be predicted for all future times (Childress and Gilbert, 1995). The crucial point is that **v** must have a finite memory. After a correlation time τ_c the flow must have forgotten its past. The relevant number is the Strouhal number $v\tau_c/\lambda_c$, the ratio of τ_c and the eddy turnover time (λ_c = eddy or cell size, or correlation length). The theory that follows is restricted to $v\tau_c/\lambda_c \lesssim 1$. Flows with a long memory, $v\tau_c/\lambda_c \gg 1$ (so called frozen turbulence) require a separate treatment.

1.2.1. Stochastic equations with multiplicative noise

The multiplicative noise term $C(t)f$ renders (1.1) as a rule impossible to solve, and we seek an equation for $\langle f \rangle$, i.e. f averaged over the fluctuations.[1] The solution to this problem is long since known, and I shall follow a method originally due to Bourret (1962), as outlined by Van Kampen (1976), which has the merit that it offers the right balance between rigour and readability. In the literature on this topic, rigour has a tendency to turn stochastic equations into scholastic equations. The connection with the traditional derivation is given in Section 1.3.3. There exists an alternative and potentially more powerful method based on path integral techniques (see e.g. Dittrich *et al.*, 1984; Rogachevskii and Kleeorin, 1997).

The average is required to obey certain rules but the precise nature of the average is left open. In the end, any operation $\langle \cdot \rangle$ satisfying the rules will be acceptable. The first step is to transform to the interaction representation:

$$f \equiv \exp(Rt)u, \tag{1.4}$$

upon which (1.1) transforms into

$$\frac{\partial u}{\partial t} = \tilde{C}(t)u; \quad \tilde{C}(t) = \exp(-Rt)C(t)\exp(Rt). \tag{1.5}$$

For the moment we assume that the exponential operators $\exp(\pm Rt)$ are well defined entities, and defer a discussion of their meaning and existence to Sections 1.5.1 and 1.5.3. We write $u = u(0) + \int_0^t d\tau \tilde{C}(\tau)u(\tau)$, and iterate once more:

$$u = u(0) + \int_0^t d\tau \tilde{C}(\tau) \left\{ u(0) + \int_0^\tau d\sigma \tilde{C}(\sigma)u(\sigma) \right\}. \tag{1.6}$$

Next the average is taken, supposing that $\langle \cdot \rangle$ and $\int dt$ commute, and that $\langle \tilde{C}(t)u(0) \rangle = \langle \tilde{C}(t) \rangle u(0) = 0$ because $\langle \tilde{C}(t) \rangle = 0$:

$$\langle u \rangle = u(0) + \int_0^t d\tau \int_0^\tau d\sigma \langle \tilde{C}(\tau)\tilde{C}(\sigma)u(\sigma) \rangle. \tag{1.7}$$

It follows that the time derivative of $\langle u \rangle$ obeys

$$\frac{\partial}{\partial t} \langle u \rangle = \int_0^t d\sigma \, \langle \tilde{C}(t)\tilde{C}(\sigma)u(\sigma) \rangle. \tag{1.8}$$

The initial condition $u(0)$ has disappeared. As soon as t is more than a few τ_c removed from zero, we may push the lower integration limit to $-\infty$, because for these values of σ there is no correlation between $\tilde{C}(t)$ and $\tilde{C}(\sigma)u(\sigma)$ so that $\langle \tilde{C}(t)\tilde{C}(\sigma)u(\sigma) \rangle = \langle \tilde{C}(t) \rangle \langle \tilde{C}(\sigma)u(\sigma) \rangle = 0$. We also transform the integration variable to $\tau = t - \sigma$:

$$\frac{\partial}{\partial t} \langle u \rangle = \int_0^\infty d\tau \, \langle \tilde{C}(t)\tilde{C}(t - \tau)u(t - \tau) \rangle. \tag{1.9}$$

This is an exact result provided that $\langle \tilde{C}(t) \rangle = 0$. A few words on the rather compact notation may be helpful. Eulerian co-ordinates are employed, and the arguments, insofar not specified, are always $\mathbf{r}$ and t. For example, $u \equiv u(t) \equiv u(\mathbf{r}, t)$, and with (1.2): $\tilde{C}(\sigma) = \exp(-R\sigma) \, \nabla \times \mathbf{v}(\mathbf{r}, \sigma) \times \exp(R\sigma)$ with $R = \nabla \times \mathbf{u}(\mathbf{r}) \times + \eta\nabla^2$.

It is now time for a major sin, and we break the average in two parts:

$$\langle \tilde{C}(t)\tilde{C}(t - \tau)u(t - \tau) \rangle \simeq \langle \tilde{C}(t)\tilde{C}(t - \tau) \rangle \langle u(t - \tau) \rangle. \tag{1.10}$$

This approximation may be justified if $\tau_c \ll |C|^{-1}$, see Fig. 1.1.[2] Consider the average $X = \langle \tilde{C}^t \tilde{C}^{t-\tau} u^{t-\tau} \rangle$ in (1.9) – for brevity time arguments are written as upper indices. We need only worry about values of $\tau \lesssim \tau_c$ because for larger τ the factor $\tilde{C}^{t-\tau} u^{t-\tau}$ in X becomes uncorrelated with $\tilde{C}^t$, and $X = \langle \tilde{C}^t \rangle \langle \tilde{C}^{t-\tau} u^{t-\tau} \rangle = 0$. We substitute $u(t-\tau) = \langle u(t-\tau) \rangle + \delta u$ in X and estimate δu from (1.5) as $\delta u \simeq |C|\tau_c \langle u(t - \tau) \rangle$, see Fig. 1.1. The relative error involved in (1.10) will therefore be of order $|C|\tau_c$. It follows that

$$\frac{\partial}{\partial t} \langle u \rangle = \int_0^\infty d\tau \, \langle \tilde{C}(t)\tilde{C}(t - \tau) \rangle \langle u(t - \tau) \rangle + O(|C|\tau_c). \tag{1.11}$$

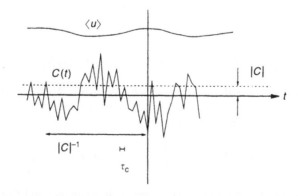

Figure 1.1 The function $\langle u \rangle$ evolves on a time scale $|C|^{-1}$, much slower than the fast time scale τ_c on which C changes itself (C is an operator and cannot be drawn, but the idea will be clear). This assumption of a short correlation time, $|C|\tau_c \ll 1$, is the key element in the derivation of an equation for the averages $\langle u \rangle$ and $\langle f \rangle$.

Equation (1.11) is an integro-differential equation for the average $\langle u \rangle$, but it is easy to see that it is equivalent to a much simpler differential equation. We set

$$\langle u(t - \tau) \rangle \simeq \langle u \rangle - \tau \frac{\partial}{\partial t} \langle u \rangle. \tag{1.12}$$

Since τ in (1.11) is restricted to $\tau \lesssim \tau_c$ we may estimate the magnitude of the second term with the help of (1.11): $\tau \partial_t \langle u \rangle \sim \tau_c \cdot \tau_c |C|^2 \langle u \rangle$. Without penalty we may therefore replace $\langle u(t - \tau) \rangle$ by $\langle u(t) \rangle$,

$$\frac{\partial}{\partial t} \langle u \rangle = \left\{ \int_0^\infty d\tau \langle \tilde{C}(t) \tilde{C}(t - \tau) \rangle \right\} \langle u \rangle + O(|C|\tau_c), \tag{1.13}$$

since the associated relative error is $\propto (|C|\tau_c)^2$, of higher order than the error already made. Note that the ∇'s in $\tilde{C}(t)$ and $\tilde{C}(t - \tau)$ operate on everything to their right, also on $\langle u \rangle$.

1.2.2. The transport equation for the mean

We now transform back to the original representation, and insert $u = \exp(-Rt)f$ and $\tilde{C}$ from (1.5), and suppress the O-term:

$$\left(\frac{\partial}{\partial t} - R \right) \langle f \rangle \simeq \left\{ \int_0^\infty d\tau \langle C(t) \exp(R\tau) C(t - \tau) \rangle \exp(-R\tau) \right\} \langle f \rangle. \tag{1.14}$$

This is the required equation for the mean $\langle f \rangle$ which we shall apply below to various problems, such as passive scalar and vector transport in a turbulent medium. It is no longer a stochastic equation, and (1.14) is for example the dynamo equation in disguise. The correlation function $\langle C(t) \exp(R\tau) C(t - \tau) \rangle$ does not depend on t because the fluctuations are supposed to be stationary in time. The physical reason for the emergence of a differential equation is that the evolution of $\langle f \rangle$ is slow relative to τ_c for $|C|\tau_c \ll 1$. Many realisations of $C(t)$ materialise while $\langle f \rangle$ changes little. Under these circumstances $\langle f \rangle$ evolves approximately as a Markov process, and only the last value of $\langle f \rangle$ matters rather than all previous values, as they formally still do in (1.11). In the limit of a continuous time axis a differential equation ensues.

The derivation is general since no use has been made of (1.2) or (1.3). The conditions for validity are that $|C|\tau_c \ll 1$ and $\langle C(t) \rangle = 0$, and the average has to obey certain rules, which we discuss in Section 1.3. There is no requirement that the 'fluctuating component' of f, viz. $f_1 \equiv f - \langle f \rangle$ be somehow small, nor that C be small with respect to R. The basic result (1.14) has established in different ways and has been used in many applications, see Van Kampen (1976, 1992). It is an expansion in the parameter $|C|\tau_c$; the right-hand side of (1.13) and (1.14) is actually the first term of a power series expansion in the small parameter $|C|\tau_c$. The starting equation (1.1) must be linear, but nonlinear equations may always be recast in an equivalent linear form, see Section 1.8.

Two special forms of (1.14) deserve to be mentioned. The integral in (1.14) is in practice restricted to $\tau \lesssim \tau_c$, otherwise the correlation function $\langle \cdot \rangle$ vanishes. In many applications not only $|C|\tau_c \ll 1$ but also $|R|\tau_c \ll 1$ and then a much simpler equation follows since $\exp(\pm R t) \simeq 1$:[3]

$$\left(\frac{\partial}{\partial t} - R \right) \langle f \rangle \simeq \left\{ \int_0^\infty d\tau \langle C(t) C(t - \tau) \rangle \right\} \langle f \rangle. \tag{1.15}$$

In the second place, if we transform back to the original representation starting from (1.11) we obtain:

$$\left(\frac{\partial}{\partial t} - R\right)\langle f\rangle \simeq \int_0^\infty \mathrm{d}\tau \langle C(t)\exp(R t)C(t-\tau)\rangle\langle f(t-\tau)\rangle. \tag{1.16}$$

This equation, known as Bourret's integral equation, is a perfectly valid alternative to (1.14). Equations (1.14) and (1.16) are quite different mathematically but physically equivalent in the sense that they are equally inaccurate. An exact and closed transport equation for the mean does not exist. There is also no point in striving for such an equation, since, loosely speaking, the mean determines the value of a quantity only to within a standard deviation from the mean.

1.2.3. Proper definition of R and C

Some averages may not obey the condition $\langle C(t)\rangle = 0$. The problem arises when R and C are defined on the basis of an intuitive splitting in mean and fluctuating part, that does not correspond to the definition of the average. The problem can always be handled by redefining R and C in (1.1):

$$A = \mathcal{R} + \mathcal{C}: \tag{1.17}$$

with

$$\mathcal{R} \equiv \langle A\rangle = R + \langle C\rangle; \quad \mathcal{C} = C - \langle C\rangle, \tag{1.18}$$

and $\langle \mathcal{C}\rangle = 0$, but the price is that $\mathcal{R}$ may no longer be time-independent. An example is the azimuthal average in a planetary or stellar convection zone, when (1.1) is the induction equation. In this case, we have

$$\mathcal{R} = \nabla \times (\mathbf{u} + \langle \mathbf{v}\rangle) \times + \eta\nabla^2; \quad \text{and} \quad \mathcal{C} = \nabla \times (\mathbf{v} - \langle \mathbf{v}\rangle) \times. \tag{1.19}$$

The mean flow is *defined* as the result of the azimuthal average of the total velocity field $\mathbf{u} + \mathbf{v}$ (and is therefore time-dependent), and the remainder is *defined* as the fluctuating component of the flow.

The exponential operator $\exp\{R(t - t_0)\}$ specifying the evolution of the system from t_0 to t must now replaced by a more general evolution operator $\mathcal{E}(t|t_0)$, defined as

$$\frac{\partial}{\partial t}\mathcal{E}(t|t_0) = \mathcal{R}(t)\mathcal{E}(t|t_0). \tag{1.20}$$

The solution is readily seen to be

$$\mathcal{E}(t|t_0) = 1 + \int_{t_0}^t \mathrm{d}\tau_1 \mathcal{R}(\tau_1) + \int_{t_0}^t \mathrm{d}\tau_1 \int_{t_0}^{\tau_1} \mathrm{d}\tau_2 \mathcal{R}(\tau_1)\mathcal{R}(\tau_2) + \cdots \tag{1.21}$$

$$= \left\lceil \exp\left\{\int_{t_0}^t \mathcal{R}(\tau)\,\mathrm{d}\tau\right\}\right\rceil. \tag{1.22}$$

The second expression is given here only to show that it is still possible to write $\mathcal{E}$ as a special kind of exponential operator. The time ordering symbol $\lceil\cdot\rceil$ indicates that after the exponent has been expanded in a series, the operators $\mathcal{R}(\tau_1)\mathcal{R}(\tau_2)\cdots\mathcal{R}(\tau_n)$ in each term are to be rearranged so that their time arguments appear in decreasing order. This complication arises because $\mathcal{R}(\tau_1)$ and $\mathcal{R}(\tau_2)$ do in general not commute. Some properties: $\mathcal{E}(t|t_1)\mathcal{E}(t_1|t_2) = \mathcal{E}(t|t_2)$; $\mathcal{E}(t|t_1)^{-1} = \mathcal{E}(t_1|t)$; $\mathcal{E}(t|t) = 1$. The interaction representation is now given by

$$f = \mathcal{E}(t|0)u; \quad \frac{\partial u}{\partial t} = \tilde{\mathcal{C}}(t)u; \quad \tilde{\mathcal{C}}(t) = \mathcal{E}(0|t)\mathcal{C}(t)\mathcal{E}(t|0). \tag{1.23}$$

The analysis then proceeds to (1.13) exactly as before, and upon transforming back to the original representation we obtain (1.14) with R, C replaced by $\mathcal{R}$, $\mathcal{C}$, and $\exp(Rt)$ by $\mathcal{E}(t|t-\tau)$, and $\exp(-Rt)$ by $\mathcal{E}(t-\tau|t)$. Of practical interest is the case that the correlation time is so small that both evolution operators $\mathcal{E}(*|*)$ become unity, and then we retrieve the equivalent of (1.15):

$$\left(\frac{\partial}{\partial t} - \mathcal{R}\right)\langle f \rangle \simeq \left\{\int_0^\infty d\tau \langle \mathcal{C}(t)\mathcal{C}(t-\tau)\rangle\right\}\langle f \rangle. \tag{1.24}$$

One may surmise that there is often little difference between $\mathcal{R}$ and R, and between $\mathcal{C}$ and C, so that (1.24) is practically equivalent to (1.15). The purpose of this section is to provide a formal basis for this idea, which is usually taken for granted. However, for simplicity only the interaction representation (1.4) is used below.

1.3. Averages

It is now time to take a closer look at the averaging procedures. This has always been a somewhat confused issue (e.g. Moffatt, 1978; Krause and Rädler, 1980), and the situation is still far from clear.

Retracing the steps of the analysis in Section 1.2.1, we see that the following properties are required:

$$\langle \lambda P + \mu Q \rangle = \lambda \langle P \rangle + \mu \langle Q \rangle; \tag{1.25}$$

$$\langle \cdot \rangle \text{ commutes with } \partial_t \text{ and } \nabla; \tag{1.26}$$

$$\langle PQ \rangle = \langle P \rangle \langle Q \rangle, \tag{1.27}$$

the last property (1.27) should hold only if the fluctuating elements in P and Q are statistically independent. The average must be linear, and λ and μ in (1.25) are arbitary numbers, not functions or anything more complicated. Note that if $\langle \cdot \rangle$ commutes with ∂_t, it also commutes with $\int dt$. Relation (1.27) cannot be missed in the analysis as presented, but it is stronger than

$$\langle P \langle Q \rangle \rangle = \langle P \rangle \langle Q \rangle, \tag{1.28}$$

that is (1.27) implies (1.28) but not vice versa. Relations (1.25), (1.26), (1.28), plus the trivial relation $\langle \mu \rangle = \mu$, constitute the Reynolds relations (Monin and Yaglom, 1973, section 3.1; Krause and Rädler, 1980), and the message is that the Reynolds relations and a short correlation time, $|C|\tau_c \ll 1$, do not suffice to derive the equation for the mean – property (1.27) is also needed.

1.3.1. Ensemble averages

The ensemble average is an average over a probability distribution (see Monin and Yaglom, 1973, ch. 2). The system is divided in small cells labeled i, as in three-dimensional numerical grid, and, assuming that the relevant quantities are velocities and magnetic fields, let i stand for the velocity and magnetic field at position i. There are various kinds of distribution functions, e.g. $\mathcal{P}(1, 2, 3, \ldots, t)$ would be the joint distribution of the velocities and fields at time t. The mean field at position i equals

$$\langle \mathbf{B} \rangle_i = \int \mathbf{B}_i \mathcal{P} \prod_k \mathrm{d}^3 \mathbf{v}_k \mathrm{d}^3 \mathbf{B}_k, \tag{1.29}$$

where $\mathcal{P}$ is supposed to be normalised to unity, $\int \mathcal{P} \prod_k \mathrm{d}^3 \mathbf{v}_k \mathrm{d}^3 \mathbf{B}_k = 1$. There are also joint distributions for different times, e.g. $\mathcal{P}(1, 2, \ldots, t_1; 1, 2, \ldots, t_2)$, and $\langle \mathbf{v}(t_1) \times \mathbf{B}(t_2) \rangle$ at location i would in terms of this $\mathcal{P}$ be equal to $\int \mathbf{v}_{i1} \times \mathbf{B}_{i2} \mathcal{P} \prod_k \mathrm{d}^3 \mathbf{v}_{k1} \mathrm{d}^3 \mathbf{B}_{k1} \prod_\ell \mathrm{d}^3 \mathbf{v}_{\ell 2} \mathrm{d}^3 \mathbf{B}_{\ell 2}$ and so on.

The time evolution of the distribution functions $\mathcal{P}$ is governed by a kind of Fokker–Planck equation in which the constraints posed by the MHD equations are built in, Section 1.8. This is a complicated matter, among other things because one should take the limit of zero cell size. My intention is merely to give the reader a flavour of the ensemble-average concept. In linear (kinematic) dynamo theory this distribution $\mathcal{P}$ is never needed because we may directly derive the equation for the average; $\mathcal{P}$ is just looming in the background and is only important for a correct understanding of certain aspects of the theory, such as turbulent diffusion, Section 1.6.2.

An important property is that when t_1 and t_2 are separated by more than a correlation time τ_c, quantities at these two times become statistically independent and the distribution function breaks in two. For example, $\mathcal{P}(i, j, t_1; i, j, t_2) = \mathcal{P}(i, j, t_1) \mathcal{P}(i, j, t_2)$, etc., thereby guaranteeing that (1.27) is obeyed. Relation (1.25) is trivially satisfied and (1.26) follows once it is realised that ∇ is now defined as a linear difference operation between cells.

The name ensemble derives from a popular visualisation of the distribution function. A series of independent draws from the distribution function $\mathcal{P}(1, 2, 3, \ldots, t)$ generates a representative set of copy systems at time t, all having the same mean flow and mean field but different fluctuating velocities and fields. Averages pertaining to position i may then be defined as averages over these copy systems labeled by k:

$$\langle Q \rangle_i = \lim_{N \to \infty} \frac{1}{N} \sum_{k=1}^N Q_k(i), \tag{1.30}$$

where $Q_k(i)$ is the value of the physical quantity Q at position i in copy system k (everything at time t). This definition highlights the problem of the interpretation of the ensemble average. Any ensemble member k may be picked as our physical system. But it is not clear in what sense the value of $\langle Q \rangle$ is representative for Q_k. In particular, for vectorial quantities such as the magnetic field it is not at all clear what $\langle \mathbf{B} \rangle$ (ensemble average) has to say about $\mathbf{B}$. Ergodic theory indicates that $\langle Q \rangle$ is equal to the time average of Q_k. However, the average of the field in the solar or terrestrial dynamo over a very long time is zero. This fits in with (1.29): if $\mathcal{P}$ is a stationary distribution where all initial condition effects have died out, then $\langle \mathbf{B} \rangle = 0$ since at any location i the value $\mathbf{B}$ must be equally probable as its opposite $-\mathbf{B}$.[4] This is nicely consistent, but at the same time rather useless. The issue will be resolved in Section 1.6. On the positive side, we note that the ensemble average does not require a separation of spatial scales.

1.3.2. Spatial averages

In this category we have the azimuthal average and volume averages:

$$\langle f \rangle = \frac{1}{2\pi} \int_0^{2\pi} f(\varphi) \, d\varphi; \quad \langle f \rangle = \frac{1}{V} \int_V f(\mathbf{r}) \, d^3\mathbf{r}. \tag{1.31}$$

The definition of the azimuthal average of a vector requires a little care, and for brevity I refer to Braginskii (1965a,b) who was the first to consider this type of average. The azimuthal average is the axisymmetric ($\ell = 0$) component of the scalar or vector field. The volume average is taken over a finite two- or three-dimensional computational box with periodic boundary conditions. In contrast to ensemble averages, these averages have a clear physical meaning, and they satisfy (1.25), (1.26) and (1.28), but (1.27) is not fulfilled.

It may help to regard these averages as averages over an incomplete ensemble, containing only the system itself and all azimuthally rotated or translated states of that same system (the periodic boundary conditions permit us to regard the system as infinite and periodic). The number of ensemble members N is finite: after each rotation or translation over a correlation length λ_c we have a 'new' system, so that $N \sim L/\lambda_c$ (azimuthal average), or $N \sim V/\lambda_c^3$ (volume average). It is straightforward to show that when f and g have uncorrelated fluctuating parts f_1 and g_1:

$$\langle fg \rangle = \langle f \rangle \langle g \rangle + \mathrm{O}(N^{-1/2}). \tag{1.32}$$

Validity of the equation for the mean derived in Section 1.2.2 requires therefore that $N \to \infty$. In other words, there should be a separation of spatial scales; the small scale λ_c should be much smaller than the size L of the system. It is qualitatively clear that in that case an azimuthal or volume average becomes approximately equal to an average over a complete ensemble, and we refer to Núñez and Galindo (1999) for a detailed analysis. We have no reason to believe that a separation of scales exists in dynamos, on the contrary, and the errors that we incur by using (1.14) or (1.16) are not known. Nevertheless, we shall see in Section 1.4.1 that the situation is not at all hopeless.

Specifically for magnetic fields Braginskii's (1965a,b) derivation of the dynamo equation for nearly axisymmetric dynamos must be mentioned here. The advantage of this method is the frontal attack: right from the beginning an azimuthal average of the induction equation is made. The disadvantage is that a rather special scaling of $\mathbf{u}$, $\mathbf{v}$ and $\mathbf{B}$ with the magnetic Reynolds number R_m is required (see further Moffatt, 1978; Roberts, 1994).

1.3.3. Traditional derivation

The purpose of this section is to make contact with a derivation of (1.14) that the reader may be more familiar with, using the first-order smoothing approximation (FOSA), and to show that property (1.27) is indispensable. We follow Krause and Rädler (1980, ch. 3), and split all quantities in a mean and a fluctuating component:

$$Q = Q_0 + Q_1, \tag{1.33}$$

$$Q_0 \equiv \langle Q \rangle \quad \text{and} \quad \langle Q_1 \rangle = 0, \tag{1.34}$$

$$\langle P_0 Q_1 \rangle = 0. \tag{1.35}$$

Relation (1.35) is equivalent to (1.28). Substitute $f = f_0 + f_1$ in (1.1) and infer in the usual way the evolution equations for f_0 and f_1, assuming that $A_0 = R$ and $A_1 = C$:

$$(\partial_t - R) f_0 = \langle C f_1 \rangle, \tag{1.36}$$

$$(\partial_t - R) f_1 = C f_0 + C f_1 - \langle C f_1 \rangle. \tag{1.37}$$

The argument then proceeds as follows. The term $g_1(t) \equiv C f_1 - \langle C f_1 \rangle$ in (1.37) is of order $|C| f_1$, and can be ignored in two cases:

1   When $|C| \tau_c \ll 1$ we have $g_1 \sim |C| f_1 \ll f_1 / \tau_c \sim \partial_t f_1$. In that case $\partial_t f_1 \simeq R f_1 + C f_0$ will be a good approximation.
2   Specifically for magnetic fields, when the dissipative term $\eta \nabla^2$ in R is large: $\eta \nabla^2 f_1 \gg |C| f_1 \sim g_1$. In other words, the diffusive eddy decay time λ_c^2 / η is much smaller than $|C|^{-1} \sim \lambda_c / v =$ eddy turnover time.[5]

The inhomogeneous equation $(\partial_t - R) f_1 = C f_0$ may be readily solved:

$$f_1 \simeq \exp(Rt) \left\{ f_1(0) + \int_0^t d\sigma \, \exp(-R\sigma) C(\sigma) f_0(\sigma) \right\}. \tag{1.38}$$

We insert this in (1.36), after having changed the integration variable to $\tau = t - \sigma$:

$$\left(\frac{\partial}{\partial t} - R \right) f_0 \simeq \langle C(t) \exp(Rt) f_1(0) \rangle$$

$$+ \int_0^t d\tau \, \langle C(t) \exp(R\tau) C(t - \tau) \rangle f_0(t - \tau). \tag{1.39}$$

So far we have only used the Reynolds conditions (1.25), (1.26) and (1.28) and either $|C| \tau_c \ll 1$ or $v \lambda_c / \eta \ll 1$. Relation (1.39) is almost the same as (1.16) since $f_0 = \langle f \rangle$. To proceed we must use (1.27), which holds if $\langle \cdot \rangle$ is the ensemble average. In that case the initial condition term vanishes, since $C(t)$ and $f_1(0)$ are uncorrelated. And for $t > \tau_c$ the integration limit t may be replaced by ∞ since the correlation function under the integral vanishes anyway. To get from there to (1.14) requires transformation to the interaction representation, which produces (1.11), and then we are on track to (1.14).

However, when $\langle \cdot \rangle$ is a spatial average the correlation functions will keep fluctuating as a function of time. Note that $h(t) \equiv \exp(Rt) f_1(0)$ is the solution of $\partial_t h = R h$ with $h(0) = f_1(0)$. When R is given by (1.2) the mean flow will amplify the initial field $f_1(0)$, but if it is purely toroidal a well-known anti-dynamo theorem says that $\exp(Rt) f_1(0) \to 0$ on a time scale L^2 / η ($L =$ size of the dynamo). In the geodynamo the initial condition would have died out after $\sim 10^5$ yr, which seems tolerable, but in stellar dynamos it may not yet have decayed. And in the presence of other types of mean flow there may be no decay at all. Likewise, $\langle C(t) \exp(R\tau) C(t - \tau) \rangle$ approaches zero also only after a time $\tau \gtrsim L^2 / \eta$. Hence we may not replace the upper integration limit t by ∞, and there is the prospect of very large, possibly diverging transport coefficients.

A way out could be to include an extra averaging over the initial condition $f_1(0)$ in the definition of the average. This operation may be performed directly in (1.39) since it is linear, and that will make the initial condition effectively zero, and maybe also $\langle C(t) \exp(R\tau) C(t - \tau) \rangle$ for $\tau \gtrsim \tau_d$. The physical motivation is that in practice the initial condition of the system is not known. This comes down to the statement that all solutions of (1.39) will 'look the same' after some time, regardless the initial condition, which is plausible, but unproven. Fortunately, numerical experiments show that the situation is much better than one might hope, see Section 1.4.1.

1.4. Turbulent transport of scalars and magnetic fields

In Section 1.2 a general transport equation for the mean has been derived which we shall now apply to a few specific problems. First, we summarise the elementary results for scalar and magnetic field transport for the simplest possible case: isotropic incompressible turbulence with a very short correlation time, so that $R\tau_c \ll 1$ and (1.15) may be applied.

In case of scalar transport f represents for example a concentration c or the temperature T, and substitution of (1.3) in (1.15) produces a well-known result:

$$\frac{\partial}{\partial t}\langle c\rangle = -\nabla \cdot \mathbf{u}\langle c\rangle + \eta\nabla^2\langle c\rangle + \nabla \cdot \int_0^\infty d\tau \langle \mathbf{v}^t \mathbf{v}^{t-\tau}\rangle \cdot \nabla\langle c\rangle$$

$$= -\nabla \cdot \mathbf{u}\langle c\rangle + \nabla \cdot (\eta + \beta)\nabla\langle c\rangle. \tag{1.40}$$

Time arguments appear as upper indices for brevity. Incompressibility permitted us to use the operator identity $\nabla \cdot \mathbf{v}^{t-\tau} = \mathbf{v}^{t-\tau} \cdot \nabla$, and the turbulent diffusion coefficient β is found to be equal to

$$\beta = \frac{1}{3}\int_0^\infty d\tau\langle \mathbf{v}^t \cdot \mathbf{v}^{t-\tau}\rangle \simeq \frac{1}{3}\langle v^2\rangle\tau_c. \tag{1.41}$$

Next consider vector transport, $f = \mathbf{B}$. Substitution of (1.2) in (1.15) gives

$$\frac{\partial}{\partial t}\langle \mathbf{B}\rangle = \nabla\times\left\{\mathbf{u}\times\langle \mathbf{B}\rangle + \int_0^\infty d\tau\langle \mathbf{v}^t\times\nabla\times\mathbf{v}^{t-\tau}\rangle\times\langle \mathbf{B}\rangle\right\} + \eta\nabla^2\langle \mathbf{B}\rangle. \tag{1.42}$$

The term with the integral is often referred to as the mean electromotive force $\langle \mathbf{E}\rangle$. It may be simplified by working out $\nabla\times\mathbf{v}^{t-\tau}\times\langle \mathbf{B}\rangle$ using that $\mathbf{v}$ and $\langle \mathbf{B}\rangle$ have zero divergence:

$$\langle \mathbf{E}\rangle \equiv \int_0^\infty d\tau\langle \mathbf{v}^t\times\nabla\times\mathbf{v}^{t-\tau}\rangle\times\langle \mathbf{B}\rangle \tag{1.43}$$

$$= \int_0^\infty d\tau\left\langle \mathbf{v}^t\times\left\{((\langle \mathbf{B}\rangle\cdot\nabla)\mathbf{v}^{t-\tau} - (\mathbf{v}^{t-\tau}\cdot\nabla)\langle \mathbf{B}\rangle)\right\}\right\rangle \tag{1.44}$$

$$= \alpha\langle \mathbf{B}\rangle - \beta\,\nabla\times\langle \mathbf{B}\rangle. \tag{1.45}$$

It is clear that $\langle \mathbf{E}\rangle$ will have two contributions, one $\propto \langle \mathbf{B}\rangle$ from the first term in (1.44), and one $\propto \nabla_i\langle B_j\rangle$ from the second term in (1.44). Remember that $\langle \mathbf{B}\rangle$ is outside the integral because its arguments are $\mathbf{r}$, t and that averaging and integration over time are two commuting operations. The final expression follows by using that the 'isotropic' second and third rank tensors $\langle a_i b_j\rangle$ and $\langle a_i\nabla_j b_k\rangle$ are equal to

$$\langle a_i b_j\rangle = \tfrac{1}{3}\langle \mathbf{a}\cdot\mathbf{b}\rangle\delta_{ij}; \quad \langle a_i\nabla_j b_k\rangle = \tfrac{1}{6}\langle \mathbf{a}\cdot\nabla\times\mathbf{b}\rangle\epsilon_{ijk}. \tag{1.46}$$

The coefficient α is proportional to the helicity of the turbulence:

$$\alpha = -\frac{1}{3}\int_0^\infty d\tau\langle \mathbf{v}^t\cdot\nabla\times\mathbf{v}^{t-\tau}\rangle \simeq -\frac{1}{3}\langle \mathbf{v}\cdot\nabla\times\mathbf{v}\rangle\tau_c. \tag{1.47}$$

Since for divergence-free vectors $\eta \nabla^2 = -\eta \nabla \times \nabla \times$, the transport equation for $\langle \mathbf{B} \rangle$ may now be written as

$$\frac{\partial}{\partial t} \langle \mathbf{B} \rangle = \nabla \times \{\mathbf{u} \times + \alpha - (\eta + \beta) \nabla \times\} \langle \mathbf{B} \rangle. \tag{1.48}$$

This is commonly referred to as the dynamo equation. The resistivity η is usually ignored because it is often much smaller than β.

1.4.1. Validity of the transport equations

Many mean field models of various dynamos have been constructed with (1.48). In this respect the equation is a real workhorse, that is still very much in demand today. It is almost certainly not valid in the simple form (1.48) because in real dynamos the turbulence is anisotropic and compressible, and the equation is nonlinear as the Lorentz force makes $\mathbf{u}$, α and β functions of the magnetic field. But so much of this is still unknown that the simplest form (1.48) of the dynamo equation is often preferred instead of a more complicated version.

An important condition for validity of both (1.48) and (1.40) is $|C|\tau_c \simeq \upsilon\tau_c/\lambda_c \ll 1$, often referred to as the quasilinear approximation or FOSA. This condition is also not satisfied as in real dynamos with fully developed turbulence $\upsilon\tau_c/\lambda_c$ is of order 1. In fact there are so many reasons why (1.48) and (1.40) should not be valid that it borders on mystery why they so often seem to catch the basic physics.

The reaction of researchers to this situation has been diverse. There are those who feel that the equations just model what we like to see, and not necessarily what really happens. This attitude of suspicion is particularly pronounced towards the dynamo equation. The problem is equally serious for scalar transport, but not usually perceived to be so. Consider, for example, the distribution of solar surface fields. Because the fields are practically vertical, the dynamo equation reduces to the scalar transport equation (1.40). Studies of surface diffusion of magnetic fields reproduce the observed surface distribution of solar magnetic fields rather well (e.g. Sheeley and Wang, 1994; Wang and Sheeley, 1994), in spite of the fact that all the usual objections such as a long correlation time and the presence of complicated nonlinear interactions between the surface fields apply here as well.

There may be many reasons why the equations are better than we think. For example, the relative error in (1.14) and (1.40) due to the omission of the $O(|C|\tau_c)$ term in (1.13), might just be a small rather than a large number times $|C|\tau_c$, so that (1.14) is still a reasonable approximation even for $\upsilon\tau_c/\lambda_c \sim 1$. This has to my knowledge never been investigated in the context of the dynamo equation. Strong supportive evidence is accumulating from several theoretical and numerical studies. Dittrich *et al.* (1984) have theoretically computed the mean field from the induction equation, by ensemble-averaging over randomly renovating flows, for zero mean flow. They show that the dynamo equation is basically correct if the correlation time is short, $\upsilon\tau_c/\lambda_c \ll 1$, as we already know. And if $\upsilon\tau_c/\lambda_c \sim 1$ it remains correct for spatial scales $\ell \gg \lambda_c$. This case has also been studied numerically by Drummond and Horgan (1986), with basically the same outcome. However they used almost maximally helical turbulence [$\sigma = 0.7$ in (1.50)]. Very promising are the numerical experiments of Gilbert *et al.* (1997). Their average is effectively an azimuthal average and a mean shear flow is included. Even though the turbulence model is not yet a random flow in the sense of Section 1.2, the results suggest that the basic picture of mean field theory is sound for the large spatial scales, even if the correlation time is long.

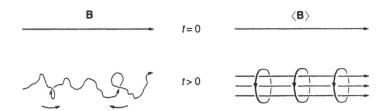

Figure 1.2 The physics of turbulent diffusion and the α-effect, as explained in the text. The vertical direction is up and $\alpha > 0$.

1.4.2. Physical mechanisms

We briefly discuss the physical significance of the various terms in (1.40) and (1.48).

Advection   The terms $-\nabla \cdot \mathbf{u}\langle c \rangle$ and $\nabla \times \mathbf{u} \times \langle \mathbf{B} \rangle$ say that the mean concentration $\langle c \rangle$ and the mean field $\langle \mathbf{B} \rangle$ are advected by the mean flow, just as the concentration c and field $\mathbf{B}$ are advected by the actual flow $\mathbf{u} + \mathbf{v}$.

Turbulent diffusion   Assuming β constant, the terms $\beta \nabla^2 \langle c \rangle$ and $\beta \nabla^2 \langle \mathbf{B} \rangle$ in (1.40) and (1.48) indicate that the mean concentration and mean field diffuse much more rapidly than c and $\mathbf{B}$ themselves, because usually $\beta \gg \eta$. The physical mechanism is turbulent mixing. In the case of magnetic fields the turbulence makes that the field lines get entangled, see Fig. 1.2. After averaging, the net result is that the field has spread over some distance. Therefore, even if the conductivity σ is infinite ($\eta = c^2/4\pi\sigma = 0$, no slip of field lines), the mean field behaves as if it is subject to a finite effective conductivity $c^2/4\pi\beta \propto \langle v^2 \rangle^{-1}$. Apparently, this effective conductivity and the associated decay time L^2/β decrease as the turbulence becomes more vigorous, which makes sense.

The α-effect   The field line displacements are not entirely random. Rising (sinking) cells expand (shrink) laterally as they adapt themselves to the ambient density. The Coriolis force makes these cells rotate around the vertical in opposite direction, so that the tube gets a helical structure on average, and this explains the α-term in (1.48): dropping all other terms we get $\partial_t \langle \mathbf{B} \rangle = \alpha \nabla \times \langle \mathbf{B} \rangle$, for α constant. This says that new $\langle \mathbf{B} \rangle$ will grow along $\nabla \times \langle \mathbf{B} \rangle$, i.e. in circles on the mantle of the tube, and that explains also why $\alpha \propto \langle \mathbf{v} \cdot \nabla \times \mathbf{v} \rangle$, the helicity of the turbulence. For small rotation angles around the vertical Krause (1968) obtained

$$\alpha \simeq \frac{\langle v^2 \rangle \tau_c^2}{H} \Omega \cos\theta, \tag{1.49}$$

(see also Stix, 1983); H = density scale height. Contrary to β, the coefficient α is zero in the absense of rotation. Apparently, α is positive in the Northern hemisphere and changes sign across the equator. The effect relies on a coupling of convection and rotation, which induces a preferred sense of rotation in vertically moving fluid parcels (Parker, 1955; Moffatt, 1978; Krause and Rädler, 1980). The fundamental point is that rotation breaks the reflectional symmetry of the turbulence which thereby acquires a nonzero helicity.

Things may be different when the correlation time is long because the rotation angle around the vertical may then be large. Likewise, in the absense of an appreciable vertical density gradient the lateral expansion may be induced by a boundary forcing the flow lines to diverge. In those cases the sign of α may differ from (1.49).

A useful order-of-magnitude estimate is

$$\alpha \sim \sigma |\mathbf{v}| \cdot | \nabla \times \mathbf{v}|\tau_c \sim \sigma v^2 \tau_c/\lambda_c \sim \sigma v. \tag{1.50}$$

Figure 1.3 A magnetic field may be thought of as an array of tiny advected arrows, and a scalar field as an array of advected dots. A dot senses the local flow, but a vector also feels the local shear of the flow. A vector is therefore sensitive to the correlated lifting and twisting process causing the α-effect, but a scalar is not because it has no intrinsic direction. The equation for scalar transport has therefore no α-like term.

Here σ is the correlation coefficient between $\mathbf{v}$ and $\nabla \times \mathbf{v}$, and $|\sigma| \leq 1$. At the last $\sim$ sign we used that in real dynamos $v\tau_c \sim \lambda_c$. The case $\sigma = \pm 1$ refers to maximally helical turbulence, for which $\mathbf{v}$ and $\nabla \times \mathbf{v}$ are always (anti) parallel.

1.4.3. Elementary properties

The co-operation of the three dynamo effects may be illustrated with a simple example. Consider an infinite space with a cartesian co-ordinate frame, $\mathbf{u} = u\mathbf{e}_y$, u is a linear function of x, z so that $\mathbf{a} \equiv \nabla u$ is a constant vector $\perp \mathbf{e}_y$. Assume translation symmetry, $\partial/\partial y \equiv 0$, and α, β constant ($\alpha > 0$). For $\langle \mathbf{B} \rangle$ we take the customary gauge

$$\langle \mathbf{B} \rangle = \nabla \times P\mathbf{e}_y + T\mathbf{e}_y, \tag{1.51}$$

with P, T functions of x, z, t; P determines the poloidal field $\perp \mathbf{e}_y$ and T the toroidal field $\| \mathbf{e}_y$. Substitution in (1.48) leads to

$$\frac{\partial P}{\partial t} = \alpha T + \beta \nabla^2 P. \tag{1.52}$$

$$\frac{\partial T}{\partial t} = (\mathbf{e}_y \times \mathbf{a}) \cdot \nabla P - \alpha \nabla^2 P + \beta \nabla^2 T. \tag{1.53}$$

These equations contain the essence of mean field dynamo action, see Fig. 1.4. The crucial term is αT in (1.52). If it were absent, then (1.52) predicts that $P \downarrow 0$ by turbulent diffusion, and then according to (1.53) also $T \downarrow 0$. The mean shear flow term $(\mathbf{e}_y \times \mathbf{a}) \cdot \nabla P$ (also called the Ω-term) and the term $\alpha \nabla^2 P$ may have widely-different relative magnitudes, and this leads to qualitatively different dynamos (Moffatt, 1978; Krause and Rädler, 1980):

1   $\alpha\Omega$-*dynamos*. The α-term in (1.53) is much smaller than the Ω-term. These dynamos tend to be periodic. The solar dynamo is thought to be an $\alpha\Omega$ dynamo.
2   α^2-*dynamos*. The Ω-term is much smaller than the α-term, so that the two α-terms are left over. These dynamos often have stationary solutions.
3   $\alpha^2\Omega$-*dynamos*. The Ω- and α-term in (1.53) are of comparable magnitude.

These properties are borne out by the plane wave solutions which we get by substitution of

$$(P, T) = (P_0, T_0) \exp\{i(\mathbf{k} \cdot \mathbf{r} - \omega t)\} \tag{1.54}$$

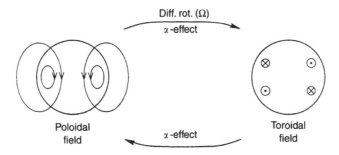

Figure 1.4 Mean field dynamo mechanism. New toroidal field is created from poloidal field by differential rotation and/or the α-effect. But new poloidal field can only be created from toroidal fields by the α-effect. A quasi-stationary state is possible because toroidal and poloidal fields are both subject to turbulent diffusion.

Table 1.1 Plane wave frequencies Ω and growth rates Γ

Case	$\Omega = \Re\omega$	$\Gamma = \Im\omega$
$\alpha\Omega^*$	$\pm(\alpha ks/2)^{1/2}$	$-\beta k^2 \pm (\alpha ks/2)^{1/2}$
α^2	0	$-\beta k^2 \pm \alpha k$

* $s = |\mathbf{a}|\sin\delta$; δ = angle between $\mathbf{k}$ and $\mathbf{a}$.

in (1.52) and (1.53), see Table 1.1. The waves with the lower signs are always damped, and play only a role during a transient state.

For real dynamos (1.48) must be solved numerically in a sphere or spherical shell, together with the appropriate boundary conditions. The fundamental mode with the largest growth rate Γ is the one that survives. Its growth rate is made effectively zero by fine-tuning the parameters in (1.48) by hand, or automatically, by incorporating some kind of nonlinear feedback in the dynamo equation. Overtones have increasingly smaller length scales L (larger wave vectors k) and the diffusion term $\beta\nabla^2 \sim -\beta/L^2$ makes their growth rates progressively smaller than that of the fundamental.

From a mathematical point of view the possibility of a periodic dynamo arises because the eigenvalue problem of associated with (1.48) is not self-adjoint. Various physical factors determine whether the fundamental mode of an $\alpha\Omega$ dynamo is periodic or not. A dynamo in a thin layer such as the solar convection zone tends to have a periodic fundamental that takes the form of a traveling wave in the direction of $\alpha(\nabla u) \times \mathbf{u}$, see Fig. 1.5. A spatial separation of the α and Ω effects favours a steady fundamental (Deinzer *et al.*, 1974), as does a meridional flow (Roberts, 1972). The role of the geometry of the dynamo has been investigated by Covas *et al.* (1999). The period of an $\alpha\Omega$ dynamo has a simple physical interpretation. Since the growth rate must be approximately zero, and $\Gamma = -\beta k^2 + \Omega$ according to Table 1.1, it follows that $\Omega \simeq \beta k^2 \simeq \beta/L^2$. Hence the period is of the order of the turbulent diffusion time scale across a typical dimension L of the dynamo, and overtones, if periodic, must have shorter periods.

In the same spirit we may estimate the parameters of the solar dynamo. From Table 1.1: $\Omega_\odot = (\alpha ks/2)^{1/2} \sim (\alpha\Delta\Omega_r/R_\odot)^{1/2}$, as $s \sim a \sim \Delta u/R_\odot \sim \Delta\Omega_r$ = magnitude of the differential rotation (about $6 \times 10^{-7}\,\text{s}^{-1}$), and $k \sim 1/R_\odot$. With $\Omega_\odot = 2\pi/22\,\text{yr}$, it follows

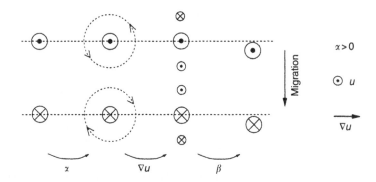

Figure 1.5 Mechanism of a traveling dynamo wave, after Stix (1976, 1978). The α-effect creates loops around a toroidal tube of mean field, and these are subsequently tilted by the mean shear flow. For presentation purposes each figure has been offset laterally, but all action is at the same horizontal position. Merging and annihilation by turbulent diffusion causes the whole wave pattern, which repeats itself along the vertical direction, to migrate along the plane $u = $ constant (Yoshimura, 1975).

that $\alpha \sim 10\,\mathrm{cm\,s^{-1}}$, much smaller than predicted by the simple formula (1.49). It is also much smaller than v, so that the correlation coefficient σ in (1.50) is small, of order 10^{-2}–10^{-3}. Moreover $\beta k^2 = \Omega_\odot$ (zero growth), from which $\beta \sim 10^{13}\,\mathrm{cm^2\,s^{-1}}$. These estimates are approximately correct even though we know now that the solar dynamo is not located in the convection zone, but rather in the narrow overshoot layer between the solar interior and the convection zone.

1.4.4. Paradoxes

Some of the properties of the mean field seem, on close scrutiny, rather strange. Three are mentioned here, as a warm-up for Section 1.6, to illustrate that it is not at all obvious what the mean field tells us about the real field.

The first is a well-known trivial consequence of the averaging. Equation (1.48) has axisymmetric solutions, which appears to be in conflict with Cowling's theorem. However, Cowling's theorem applies to the actual field, not to the mean field. And if the latter is axisymmetric, the former is in general not. So there is no conflict. This illustrates the strength of the mean field concept in that it leads to an enormous simplification. The price is that the information contained in the mean field is much less than that in the real field.

The second point is less trivial. The scalar transport equation (1.40) with the appropriate boundary conditions (e.g. zero flux at the boundaries) is self-adjoint and has therefore no periodic solutions.[6] The initial distribution decays away and the final result is a spatially uniform concentration. There is a continous loss of memory. But a periodic solution of the dynamo equation at zero growth rate is strictly periodic and has apparently an infinitely long phase memory, in spite of the fact that the realisation of the turbulence in adjacent periods is different, so that also the periods must be different. It follows that the physical reality of strictly periodic solutions of the dynamo equation is questionable.

A third strange fact is that for sufficiently large k all dynamo modes are damped by turbulent diffusion (the term $-\beta k^2$ in Table 1.1), although we know that the small spatial scales are generated copiously by the turbulence, in particular when η is small. This is also not a paradox. It is correct for ensemble-averaged fields, where it comes about through the averaging process. The issue is again to what extent the ensemble-averaged field has anything to say about the field of a single system. It will be argued in Section 1.6. that the large-scale component of the field is well modelled by the ensemble average for a limited period of time, but that small-scale fields are not at all well represented. The clue is that $1/\Gamma$ in Table 1.1 is a mode coherence time which *appears* as a damping due to the averaging. For spatially averaged field the results in Table 1.1 are simply not exact because of (1.32). The study of Gilbert *et al.* (1997) suggests that in this case also only the behaviour of the large-scale components is well captured.

1.5. Transport coefficients

Scalars and magnetic fields have the same turbulent diffusion coefficient β in the special case of incompressible isotropic turbulence with a short correlation time. But this is no longer true if the turbulence is anisotropic, because the simple relations (1.46) break down. On general grounds it is expected $\langle E \rangle$ may be expanded in a series of the form (Moffatt 1978):

$$\langle E_i \rangle = \alpha_{ij}\langle B_j \rangle + \beta_{ijk}\nabla_j\langle B_k \rangle + \gamma_{ijk\ell}\nabla_j\nabla_k\langle B_\ell \rangle + \cdots \tag{1.55}$$

and for scalar transport the mean diffusive flux $\langle \mathbf{F} \rangle = -\beta\nabla\langle c \rangle$ in (1.40) is to be replaced by

$$\langle F_i \rangle = -\beta_{ij}\nabla_j\langle c \rangle + \cdots . \tag{1.56}$$

Hence α is a second rank tensor, while β is of second rank for scalar transport but of third rank for transport of a vector. We shall see below that the terms with second and higher derivatives are often zero.

Anisotropic turbulence is the rule rather than the exception due to the vertical stratification and rotation of real dynamos. The tensors α_{ij} and β_{ijk} consist of combinations of the invariant tensors $\delta_{\ell m}$, $\epsilon_{\ell mn}$, of v_ℓ, $\nabla_\ell v_m$, and the two symmetry breaking vectors Ω_ℓ and $\nabla_\ell\rho$. This is an extensive topic for which the reader is referred to Krause and Rädler (1980, ch. 15) and Petrovay (1994). One is then faced with the problem that the the various tensor components are unknown, although they can be computed for specific models (see e.g. Rüdiger and Kitchatinov, 1993; Brandenburg, 1994; Ferriz-Mas *et al.*, 1994; Kitchatinov *et al.*, 1994; Ferrière, 1996; Brandenburg and Schmitt, 1998; Rüdiger and Arlt, Chapter 6, this volume; Schmitt, Chapter 4, this volume). Most authors include also the influence of the magnetic field on the transport coefficients. Here I shall restrict myself to the effect of (1) mean shear flows and (2) resistivity on the transport coefficients.

1.5.1. Shear flows and Galilean invariance

In the presence of a mean flow, expressions (1.41) and (1.47) lead to a paradox: α and β depend on $\mathbf{v}^t \equiv \mathbf{v}(\mathbf{r}, t)$ and $\mathbf{v}^{t-\tau} \equiv \mathbf{v}(\mathbf{r}, t - \tau)$, two vectors at the same Eulerian position $\mathbf{r}$ in the observer's frame, separated τ seconds in time. But in τ seconds the mean flow will move a different material point of the fluid to the position $\mathbf{r}$. The transport coefficients are therefore determined by the conditions in two material points that may be far removed from each other if the mean flow is fast, which is unphysical. Moreover, since the mean flow $\mathbf{u}$ depends on the velocity of the observer, we conclude that (1.41) and (1.47) violate Galilean invariance as their values depend on the choice of the observer's reference frame.

The remedy is to consider the general expression for the transport coefficients under first-order smoothing, and to start from (1.14) instead of (1.15). This means that $\langle E \rangle$ is no longer

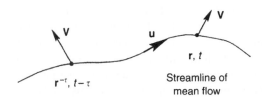

Figure 1.6 The transport coefficients at a given position depend on the correlation between the turbulent
velocity at that point, and at an earlier position of the same material point, corrected by the
displacement gradient matrix $D^{-\tau}$.

given by (1.43) but is equal to

$$\langle \mathbf{E} \rangle = \int_0^\infty \mathrm{d}\tau \, \langle \mathbf{v}^t \times \mathrm{e}^{R\tau} \, \mathbf{\nabla} \times \mathbf{v}^{t-\tau} \rangle \times \mathrm{e}^{-R\tau} \langle \mathbf{B} \rangle. \tag{1.57}$$

The exponential operators allow for the effect of a mean shear flow and resistivity on the
transport coefficients *and* they restore Galilean invariance. For $R = \mathbf{\nabla} \times \mathbf{u} \times$ (zero resistivity)
and $\mathbf{\nabla} \cdot \mathbf{u} = 0$ it has been shown that (Hoyng, 1985)

$$\mathrm{e}^{R\tau} \mathbf{\nabla} \times \mathbf{v}(\mathbf{r}, t - \tau) \times \mathrm{e}^{-R\tau} = \mathbf{\nabla} \times \{D^{-\tau} \mathbf{v}(\mathbf{r}^{-\tau}, t - \tau)\} \times, \tag{1.58}$$

where $\mathbf{r}^{-\tau}$ is a Lagrangian co-ordinate, viz. the position of a material point τ seconds before its
position was $\mathbf{r}$, under the action of the mean flow only, and $D^{-\tau}$ is the displacement gradient
matrix $\partial \mathbf{r} / \partial \mathbf{r}^{-\tau}$. The transport coefficients that follow from (1.57) and (1.58) depend only on
the physical conditions in one material point, see Fig. 1.6. A similar analysis can be made for
scalar transport, see Hoyng (1985). The condition of an incompressible mean flow is hardly
a restriction because the correlation function $\langle \cdot \rangle$ in (1.57) is zero for $\tau \gtrsim \tau_\mathrm{c}$. Since deviations
from incompressibility (e.g. a mean flow with a meridional component) become noticeable
only on time scales much longer than a correlation time, we may apply (1.58) also when
$\mathbf{\nabla} \cdot \mathbf{u} \neq 0$.

Resistivity can be ignored in $\exp(\pm R\tau)$ when (1) $|\mathbf{\nabla} \times \mathbf{u} \times| \gg |\eta \nabla^2|$ or $u/\lambda_\mathrm{c} \gg \eta/\lambda_\mathrm{c}^2$, i.e.
$u\lambda_\mathrm{c}/\eta \gg 1$,[7] and (2) $|\eta \nabla^2|\tau_\mathrm{c} \sim \tau_\mathrm{c}/\tau_\mathrm{r} \ll 1$, otherwise $\exp(\eta\tau \nabla^2)$ cannot be ignored. These
conditions are fulfilled in the solar dynamo, and in the geodynamo if we accept that the outer
core rotates differentially (Song and Richards, 1996; Su *et al.*, 1996). In these two important
cases it seems correct to take $\eta = 0$. Inserting (1.58) in (1.57) yields, for incompressible
turbulence,

$$\langle \mathbf{E} \rangle = \int_0^\infty \mathrm{d}\tau \, \langle \mathbf{v}^t \times \{ (\langle \mathbf{B} \rangle \cdot \mathbf{\nabla}) \tilde{\mathbf{v}}^{t-\tau} - (\tilde{\mathbf{v}}^{t-\tau} \cdot \mathbf{\nabla}) \langle \mathbf{B} \rangle \} \rangle, \tag{1.59}$$

with $\tilde{\mathbf{v}}^{t-\tau} = D^{-\tau} \mathbf{v}(\mathbf{r}^{-\tau}, t - \tau)$ ($\tilde{\mathbf{v}}^{t-\tau}$ has zero divergence when $\mathbf{v}^t$ does). Relation (1.59)
is easily elaborated in actual situations with the help of Lagrangian mean flow co-ordinates
(Hoyng, 1985). In this way we recover only the 'kinematic' influence of the shear flow $\mathbf{u}$
on the transport coefficients: they become anisotropic tensors even for isotropic turbulence.
On top of this come of course dynamic effects, since the shear flow alters the directional
properties of the turbulence (Urpin, 1999; Urpin and Brandenburg, 1999).

Note that relation (1.59) has the same structure as (1.44) so that $\langle \mathbf{E} \rangle$ contains no second
and higher order derivatives of the mean field, although the exponential operators contain
arbitrarily high derivatives. Of course $\exp(-R\tau)$ is the inverse of $\exp(R\tau)$ so that some
compensation is to be expected, but it is perhaps remarkable the compensation is complete.

1.5.2. Small shear and small resistivity

The generalisation of (1.58) to nonzero resistivity is not known, but we may consider the situation that $|\nabla \times \mathbf{u} \times |\tau_c \sim u\tau_c/\lambda_c$ and $|\eta\nabla^2|\tau_c \sim \tau_c/\tau_r$ are of comparable magnitude, and both small compared to unity. We may then expand $\exp(\pm R\tau) \simeq 1 \pm R\tau$. To keep the notation simple we start directly from (1.14) and write time arguments as upper indices as before:

$$
\left(\frac{\partial}{\partial t} - R\right)\langle f \rangle
$$

$$
\simeq \left\{\int_0^\infty d\tau \langle C^t C^{t-\tau}\rangle\right\}\langle f \rangle + \left\{\int_0^\infty \tau \, d\tau \langle C^t(RC^{t-\tau} - C^{t-\tau}R)\rangle\right\}\langle f \rangle, \qquad (1.60)
$$

with $R = \nabla \times \mathbf{u} \times + \eta\nabla^2$ and $C = \nabla \times \mathbf{v} \times$. To single out the new effects we take a highly simplified case: homogeneous, isotropic and incompressible turbulence $\mathbf{v}$ with zero helicity, $\langle v_i \nabla_j v_k \rangle = 0$, and a cartesian co-ordinate frame with y along $\mathbf{u}$, x along ∇u and $\partial^2 u/\partial x^2 = \partial u/\partial y = \partial u/\partial z = 0$. It is then a matter of working out the commutator $RC - CR$, and multiply again with another C. Many terms cancel or vanish as a result of the averaging. We give only the final result:[8]

$$
\frac{\partial}{\partial t}\langle \mathbf{B}\rangle = \nabla \times \mathbf{u} \times \langle \mathbf{B}\rangle + (\eta + \beta)\nabla^2\langle \mathbf{B}\rangle + \gamma\frac{\partial^2}{\partial x \partial y}\langle \mathbf{B}\rangle, \qquad (1.61)
$$

with

$$
\gamma = \frac{1}{3}\frac{\partial u}{\partial x}\int_0^\infty \tau \, d\tau \langle \mathbf{v}^t \cdot \mathbf{v}^{t-\tau}\rangle \simeq \left(\frac{\partial u}{\partial x}\tau_c\right)\beta. \qquad (1.62)
$$

The last term in (1.61) originates from the last term in (1.60), and is new. There is also a small resistive correction $\beta \to \beta(1 + \eta\tau_c/\lambda_c^2)$, which we have dropped. This result has been obtained by Urpin (1999) by a different technique. He shows that (1.61) may have growing solutions and argues that a mean field may be generated in the absence of helicity because the shear flow is effectively able to break the reflectional symmetry of the turbulence. In his analysis Urpin (1999) allows for dynamical effects, such as the perturbation of the turbulence by the shear flow, which are not taken into account above. The results nevertheless coincide exactly. Again, $\langle \mathbf{E}\rangle$ does not contain higher than first-order derivatives of the mean field, and this remains so if the helicity is nonzero (without proof).

1.5.3. Resistive effects

Galilean invariance is a strong argument in favour of (1.14) and (1.57) being correct. It was suggested above that we may handle some important cases either by ignoring resistivity or by expanding the exponential operators $\exp(\pm R\tau)$ to first order. However, that is not the end of the story, because the operator $\exp(-R\tau)$ is unbounded. It contains $\exp(-\tau\eta\nabla^2)$, and $\exp(-\tau\eta\nabla^2)\mathbf{b}$ is the solution of the diffusion equation $\partial_t \mathbf{B} = \eta\nabla^2\mathbf{B}$ advanced τ seconds backward in time from initial condition $\mathbf{b}$, which is known to be ill-posed. This is a problem for small spatial scales, as we shall now illustrate for transport of a scalar such as the temperature. We take zero mean flow:

$$
R = \eta\nabla^2; \quad C = -\nabla \cdot \mathbf{v}. \qquad (1.63)
$$

According to (1.14) we must evaluate the operator

$$\int_0^\infty d\tau \langle \nabla \cdot \mathbf{v}^t e^{\tau \eta \nabla^2} \nabla \cdot \mathbf{v}^{t-\tau} \rangle e^{-\tau \eta \nabla^2} = \nabla \cdot \int_0^\infty d\tau \langle \mathbf{v}^t e^{\tau \eta \nabla^2} \mathbf{v}^{t-\tau} e^{-\tau \eta \nabla^2} \rangle \cdot \nabla, \quad (1.64)$$

since $\mathbf{v}$ is supposed to have zero divergence and ∇_i commutes with $\exp(-\tau \eta \nabla^2)$. We represent $\mathbf{v}^t$ and $\mathbf{v}^{t-\tau}$ by their spatial Fourier transforms

$$\mathbf{v}^t = \int d^3\mathbf{q} \, \hat{\mathbf{v}} \exp(i\mathbf{q} \cdot \mathbf{r}); \quad \mathbf{v}^{t-\tau} = \int d^3\mathbf{q}' \, \hat{\mathbf{v}}' \exp(i\mathbf{q}' \cdot \mathbf{r}), \quad (1.65)$$

where $\hat{\mathbf{v}} = \hat{\mathbf{v}}(\mathbf{q}, t)$ and $\hat{\mathbf{v}}' = \hat{\mathbf{v}}(\mathbf{q}', t - \tau)$. Since $\nabla \exp(i\mathbf{q}' \cdot \mathbf{r})* = \exp(i\mathbf{q}' \cdot \mathbf{r})(\nabla * + i\mathbf{q}'*)$ we have the operator identity

$$f(\nabla) \exp(i\mathbf{q}' \cdot \mathbf{r}) = \exp(i\mathbf{q}' \cdot \mathbf{r}) f(\nabla + i\mathbf{q}'), \quad (1.66)$$

which we apply to

$$e^{\tau \eta \nabla^2} \mathbf{v}^{t-\tau} = \int d^3\mathbf{q}' \, \hat{\mathbf{v}}' e^{\tau \eta \nabla^2} \exp(i\mathbf{q}' \cdot \mathbf{r})$$

$$= \int d^3\mathbf{q}' \, \hat{\mathbf{v}}' \exp(i\mathbf{q}' \cdot \mathbf{r}) \exp\{\tau \eta (\nabla + i\mathbf{q}')^2\}. \quad (1.67)$$

The trick is to move the explicit $\mathbf{r}$-dependence to the left of the differential operators. It follows that

$$e^{\tau \eta \nabla^2} \mathbf{v}^{t-\tau} e^{-\tau \eta \nabla^2} = \int d^3\mathbf{q}' \, \hat{\mathbf{v}}' \exp(i\mathbf{q}' \cdot \mathbf{r}) \exp[\tau \eta \{(\nabla + i\mathbf{q}')^2 - \nabla^2\}. \quad (1.68)$$

We use (1.65) once more to infer that

$$\langle \mathbf{v}^t e^{-\tau \eta \nabla^2} \mathbf{v}^{t-\tau} e^{-\tau \eta \nabla^2} \rangle$$

$$= \iint d^3\mathbf{q} d^3\mathbf{q}' \langle \hat{\mathbf{v}} \hat{\mathbf{v}}' \rangle \exp\{i(\mathbf{q} + \mathbf{q}') \cdot \mathbf{r}\} \exp\{-\tau \eta (q'^2 - 2i\mathbf{q}' \cdot \nabla)\}. \quad (1.69)$$

For the correlation function we take the usual expression for incompressible, isotropic and reflectionally asymmetric turbulence:

$$\langle \hat{v}_i \hat{v}_j' \rangle = \{e(q)(\delta_{ij} - \hat{q}_i \hat{q}_j) - ih(q)\epsilon_{ijk} \hat{q}_k\} e^{-|\tau|/\tau_c} \delta(\mathbf{q} + \mathbf{q}'). \quad (1.70)$$

Here $\hat{\mathbf{q}} = \mathbf{q}/q$ is the unit vector along $\mathbf{q}$. The exponential time dependence is exemplary and may be replaced by some other function. To handle such a more complicated time dependence another Fourier transformation over time would be required, but that makes the notation more complicated without adding anything really new. With the help of (1.65) it may be verified that

$$\langle \mathbf{v}^t \cdot \mathbf{v}^{t-\tau} \rangle = e^{-|\tau|/\tau_c} \int_0^\infty E(q) \, dq; \quad E(q) = 8\pi q^2 e(q). \quad (1.71)$$

$$\langle \mathbf{v}^t \cdot \nabla \times \mathbf{v}^{t-\tau} \rangle = e^{-|\tau|/\tau_c} \int_0^\infty H(q) \, dq; \quad H(q) = 8\pi q^3 h(q). \quad (1.72)$$

$E(q)$ and $H(q)$ are the energy and helicity spectrum of the turbulence. The next step is to insert (1.70) in (1.69) and to compute, according to (1.64), the diffusion operator $D \equiv \nabla \cdot \int_0^\infty d\tau (1.69) \cdot \nabla$:

$$D = \int d^3\mathbf{q} \, e(q) \{\nabla^2 - (\hat{\mathbf{q}} \cdot \nabla)^2\} \int_0^\infty d\tau \exp\left[-\tau \left\{\eta(q^2 + 2i\mathbf{q} \cdot \nabla) + \frac{1}{\tau_c}\right\}\right]. \quad (1.73)$$

The helicity term does not contribute in the case of scalar transport. Relation (1.73) is a complicated differential operator, which may be reduced further by expanding the exponent for small ∇, i.e. large spatial scales (next section). More insight is gained by making a spatial Fourier transformation. The equation for the mean temperature is $\partial_t \langle T \rangle = (\eta \nabla^2 + D)\langle T \rangle$, or after transformation:

$$\frac{\partial}{\partial t} \langle T \rangle_k = (-\eta k^2 + D)\langle T \rangle_k. \quad (1.74)$$

where D is obtained from (1.73) by the substitution $\nabla \to i\mathbf{k}$. We introduce spherical co-ordinates in $\mathbf{q}$-space, taking the q_z-axis along $\mathbf{k}$:

$$D = -2\pi k^2 \int_0^\infty q^2 \, dq \, e(q) \int_0^\pi d\theta \sin^3 \theta$$
$$\times \int_0^\infty d\tau \exp\left[-\tau \left\{\eta(q^2 - 2qk\cos\theta) + \frac{1}{\tau_c}\right\}\right]$$
$$= -2\pi k^2 \int_0^\infty q^2 \, dq \, e(q) \int_0^\pi \frac{\sin^3 \theta \, d\theta}{\eta(q^2 - 2qk\cos\theta) + (1/\tau_c)} \quad (1.75)$$
$$\simeq -\frac{8\pi}{3} k^2 \int_0^\infty \frac{q^2 e(q) \, dq}{\eta q^2 + (1/\tau_c)}. \quad (1.76)$$

Relation (1.75) holds only if $q^2 - 2kq\cos\theta + 1/(\eta\tau_c) > 0$ for all q, which is true as long as $k < (\eta\tau_c)^{-1/2}$. The problem is therefore ill-posed for spatial scales smaller than $(\eta\tau_c)^{1/2} = $ diffusive skin depth in one correlation time. For these spatial scales the diffusion operator diverges. The integration over θ in (1.75) can be done analytically but there is no need for that. A reasonable assumption is that $k \ll (\eta\tau_c)^{-1/2}$. In that case $2qk\cos\theta$ may be ignored with respect to the other terms in the denominator of (1.75), and we obtain (1.76).

The only k-dependence in (1.76) is $-k^2$ in front, and hence the transformation back to configuration space yields the equation $\partial_t \langle T \rangle = (\eta + \beta)\nabla^2 \langle T \rangle$ with

$$\beta \simeq \frac{1}{3} \int_0^\infty \frac{E(q) \, dq}{\eta q^2 + (1/\tau_c)}. \quad (1.77)$$

The computations for magnetic fields are just lengthier with lots of outer products but the principle is exactly the same. For zero mean flow we obtain (1.45) with β given by (1.77) and

$$\alpha \simeq -\frac{1}{3} \int_0^\infty \frac{H(q) \, dq}{\eta q^2 + (1/\tau_c)}. \quad (1.78)$$

We thus recover the well-known result that α and β are weighted integrals of the helicity and energy spectrum of the turbulent flow.

A further simplification is possible by assuming that the spectra are sharply peaked near $q_0 \sim 1/\lambda_c$, permitting an approximate evaluation of the integrals, as follows:

$$\left.\begin{array}{l} \alpha \simeq -\frac{1}{3}\langle \mathbf{v} \cdot \mathbf{\nabla} \times \mathbf{v}\rangle \tau \\[2mm] \beta \simeq \frac{1}{3}\langle v^2\rangle \tau \end{array}\right\} \quad \text{with} \quad \frac{1}{\tau} = \frac{1}{\tau_r} + \frac{1}{\tau_c}, \tag{1.79}$$

where $\tau_r = \lambda_c^2/\eta$ is the eddy diffusive time scale. Diffusive effects reduce α and β because $\tau < \tau_c$. This is intuitively clear as the turbulence has less grip on the field lines. In the solar and geodynamo $\tau_r \gg \tau_c$ so that resistive effects on the transport coefficients appear negligible altogether, regardless of the mean flow.

1.5.4. The role of higher spatial derivatives

From a mathematical point of view the above treatment is unsatisfactory. The transport coefficients were found to diverge for spatial scales ℓ smaller than $(\eta\tau_c)^{1/2}$, and these – although usually much smaller than the correlation length λ_c – are always present at some level. A physicist would argue that these scales are 'irrelevant' as we are not interested in the behaviour of the mean on very small scales.

There is still the question for what spatial scales ℓ the approximate expression (1.76) is sufficient. We consider once more the scalar transport problem and compute the next order in the equation for the mean temperature:

$$\frac{\partial}{\partial t}\langle T\rangle = (\eta + \beta)\nabla^2\langle T\rangle + \zeta\nabla^4\langle T\rangle. \tag{1.80}$$

The coefficient ζ may be evaluated by expanding the denominator in (1.75) to second order in k (the first order term yields zero). The fourth order term in (1.80) may be ignored for spatial scales $\ell \gg (\zeta/\beta)^{1/2}$. This can be shown to amount to $\ell \gg \lambda_c$, and for magnetic fields the result is essentially the same. The bottom line is that we may ignore higher-than-second-order spatial derivatives in the equation for the mean if we restrict ourselves to spatial scales much larger than λ_c. And this is not too bad, because if one is interested in structure on length scales of order λ_c, there seems to be little point in employing averages.

Higher-order spatial derivatives do come into play, at least in principle, when higher-order correlation approximations are considered, i.e. when terms of higher order in $|C|\tau_c$ in (1.11) are included (Van Kampen, 1974; for a recent study see Dolginov and Silant'ev, 1992). Nicklaus and Stix (1988) have included all terms up to the fourth-order correlation, and found corrections to α and β, but no higher than second-order spatial derivatives in the dynamo equation (1.48), possibly as a result of the relatively simple case they studied (no mean shear flow and homogeneous, isotropic and gaussian turbulence). I expect, however, that also in this case higher order derivatives can be ignored for spatial scales $\ell \gg \lambda_c$, i.e. for the modelling of the large-scale structure. Otherwise, there would be a serious problem with the boundary conditions. Consider for example (1.80). Extra boundary conditions must be provided to ensure a unique solution. The physical meaning of these extra conditions and their very existence is not at all clear, and the same holds for a dynamo equation with higher-than-second-order spatial derivatives.

If this expectation is correct, there would be no room for a negative diffusion coefficient β because it renders the transport equation ill-posed. Several authors have pointed out that the diffusion coefficient β may be negative under certain circumstances, but this is certainly not generally accepted because it is unclear to what extent it is the result of cutting off a divergent higher-order-correlation expansion.

1.6. The meaning of the ensemble-averaged field

We pick up the discussion where we left it in Section 1.4.4. A periodic solution of (1.48) at zero growth rate depicts the dynamo as strictly periodic, with an infinite phase memory. But this is in conflict with the fact that the realisation of the turbulence underlying the dynamo action is different in each next period. The phase memory must therefore be finite. One might think that cause lies in the linearity of (1.48). Inclusion of a nonlinearity renders the solution indeed often irregular or chaotic (see e.g. Schmitt and Schüssler, 1989; Weiss and Tobias, 1997). However, that cannot be the real cause, as we may conceive of a linear dynamo experiment by replacing the magnetic field by a collection of needles as in Fig. 1.3. Such a system behaves conform (1.48) and hence the conflict remains.

1.6.1. The mean energy density

Not only is the existence of strictly periodic averages problematic, there is also the question of what happens to the magnetic energy. These two issues are intimately related. When mean fields of opposite polarity annihilate by turbulent diffusion, the mean field $\langle \mathbf{B} \rangle$ disappears, but $\mathbf{B}$ will not disappear immediately, the field lines just mix and fold up. And if $\eta = 0$ the field will stay forever and the magnetic energy density will grow.

To see what happens we consider the mean energy $\langle B^2 \rangle$, as was first done by Knobloch (1978a). More generally, we should consider $\langle \mathbf{BB} \rangle$, since no closed equation for $\langle B^2 \rangle$ exists in the presence of a mean flow. We set $\partial_t B_i B_j = (\partial_t B_i) B_j + B_i (\partial_t B_j)$ and use the induction equation. Assuming incompressibility and zero resistivity, the result is

$$\frac{\partial}{\partial t} B_i B_j = L_{ijk\ell} B_k B_\ell \tag{1.81}$$

with

$$L_{ijk\ell} = L^0_{ijk\ell} + L^1_{ijk\ell}. \tag{1.82}$$

and

$$L^0_{ijk\ell} = \delta_{ik}(\nabla_\ell u_j) + \delta_{j\ell}(\nabla_k u_i) - \delta_{ik}\delta_{j\ell} \, \mathbf{u} \cdot \nabla, \tag{1.83}$$

where $(\nabla_\ell u_j)$ and $(\nabla_k u_i)$ are matrices, not operators. And $L^1_{ijk\ell}$ is obtained from (1.83) by replacing $\mathbf{u}$ everywhere by $\mathbf{v}$. Equations (1.81) and (1.82) are of the form (1.1) with $R = L^0$ and $C = L^1$. We take a very short correlation time so that (1.15) can be used, and apply the general result (Hoyng, 1987b) to a simple dynamo: no mean flow, homogeneous and isotropic turbulence. In this case a closed equation for $\langle B^2 \rangle$ does exist:

$$\frac{\partial \epsilon}{\partial t} = 2\gamma \epsilon + \beta \nabla^2 \epsilon; \quad \epsilon \equiv \langle B^2 \rangle, \tag{1.84}$$

where γ is a new dynamo coefficient proportional to the mean vorticity

$$\gamma = \frac{1}{3} \int_0^\infty d\tau \langle (\nabla \times \mathbf{v}') \cdot (\nabla \times \mathbf{v}'^{-\tau}) \rangle \simeq \frac{1}{3} \langle | \nabla \times \mathbf{v} |^2 \rangle \tau_c. \tag{1.85}$$

The dynamo equation (1.48) becomes

$$\frac{\partial}{\partial t}\langle \mathbf{B}\rangle = \alpha\,\mathbf{\nabla}\times\langle\mathbf{B}\rangle + \beta\nabla^2\langle\mathbf{B}\rangle. \tag{1.86}$$

Equations (1.84) and (1.86) together admit some interesting conclusions. In a more comprehensive treatment resistive effects must be included in (1.84). This can be done (see e.g. Kim, 1999) but requires more advanced methods than the theory of Section 1.2. Moreover, the simple argument presented below would become very involved, while the conclusions are the same or even stronger, as resistivity is an extra sink of energy. We integrate (1.84) over the volume V of the dynamo:

$$\left(\frac{\partial}{\partial t} - 2\gamma\right)\int_V \epsilon\,\mathrm{d}^3\mathbf{r} = \beta\oint_{\partial V}\mathbf{\nabla}\epsilon\cdot\mathrm{d}^2\mathbf{r}. \tag{1.87}$$

This energy balance equation says that magnetic energy is generated by random field line stretching (term $\propto\gamma$) and escapes at the boundary into space. There is no volume term $\propto\beta$ which means that magnetic energy is not dissipated by turbulent diffusion, but merely relocated. There is of course no resistive dissipation because we took (were forced to take) $\eta = 0$.

And yet, the turbulent diffusion term $\beta\nabla^2\langle\mathbf{B}\rangle$ in the dynamo equation is equivalent to a strong resistive dissipation, apparently destroying mean field. Even more remarkable, when (1.84) has a stationary solution, all solutions of (1.86) decay exponentially! For a spherical dynamo volume of radius R, the argument goes as follows: $\partial_t\epsilon\simeq 0$ in (1.84) implies $\gamma\sim\beta/R^2$. The dynamo coefficients obey Schwarz's inequality $\alpha^2\leq\beta\gamma$ so that $\alpha^2\leq\beta(\beta/R^2)$ or $|\alpha R/\beta|\leq 1$. However, it is known that (1.86) has only decaying solutions when $|\alpha R/\beta|\leq 4.493$ (Krause and Rädler, 1980, ch. 14). To make the argument exact (1.84) is augmented with a boundary condition at $r = R$,

$$\frac{\partial\epsilon}{\partial r} = -a\epsilon, \tag{1.88}$$

which regulates the rate at which energy escapes from the dynamo [see Van Geffen and Hoyng, (1990) for details]. Both (1.84) and (1.86) can be solved analytically. The outcome is (1): $|\alpha R/\beta|\leq\pi/\sqrt{2}\simeq 2.22$, and (2): the damping time of the fundamental mode of (1.86) is $\lesssim 0.15 R^2/\beta$ (less than a diffusion time!), and depends only weakly on α and a (Hoyng, 1987b).[9]

The interpretation is straightforward. The dynamo is spherically symmetric, so that when the initial condition is forgotten, any possible field is an equally likely outcome as arbitrarily rotated states of that same field. The only conceivable outcome is $\langle\mathbf{B}\rangle = 0$, as we already concluded in Section 1.3.1. The dynamo does have a magnetic field since $\langle B^2\rangle$ is in principle nonzero if $\partial_t\epsilon = 0$, but mean field theory cannot predict its orientation. Van Geffen (1993a,b) has analysed a more realistic example of an axisymmetric dynamo in a spherical shell and shown that stationarity of $\langle\mathbf{BB}\rangle$ implies $\langle\mathbf{B}\rangle = 0$. Of course, $\langle\mathbf{B}\rangle = 0$ does not imply $\mathbf{B} = 0$.

1.6.2. Directional diffusion

The origin of this effect is directional diffusion, as the following example illustrates. Consider a vector $\mathbf{e}$ of constant length that is free to rotate in the x, y-plane about the z-axis with an angular velocity ω having a random element:

$$\frac{d\mathbf{e}}{dt} = \omega \times \mathbf{e}; \quad \omega = \omega_0 + \delta\omega(t). \tag{1.89}$$

This is of the type (1.1) with $R = \omega_0 \times$ and $C = \delta\omega(t) \times$. We assume a short correlation time, apply (1.15), using that $\delta\omega$ is perpendicular to $\mathbf{e}$ and $\langle\mathbf{e}\rangle$:

$$\frac{d}{dt}\langle\mathbf{e}\rangle = \left\{ \omega_0 \times + \int_0^\infty d\tau \langle \delta\omega(t) \times \delta\omega(t-\tau) \times \rangle \right\} \langle\mathbf{e}\rangle$$

$$= \omega_0 \times \langle\mathbf{e}\rangle - D\langle\mathbf{e}\rangle; \quad D \simeq \delta\omega_{\text{r.m.s.}}^2 \tau_c. \tag{1.90}$$

This says that $\langle\mathbf{e}\rangle$ rotates with angular velocity ω_0, but in addition the length of $\langle\mathbf{e}\rangle$ decreases exponentially, in spite of the fact that the length of $\mathbf{e}$ is constant. The reason is that all directions of $\mathbf{e}$ become eventually equally likely, whence $\langle\mathbf{e}\rangle \to 0$. The diffusive character underlying (1.90) can be made explicit by considering a set of copy systems all with the same ω_0 but with different realisations $\delta\omega(t)$. Let ϕ be the angle between $\mathbf{e}$ and the x-axis, and $f(\phi, t)\,d\phi$ the number of arrows with angle ϕ in the interval $d\phi$ around ϕ. The evolution of f is governed by the continuity equation in ϕ-space:

$$\frac{\partial f}{\partial t} = -\frac{\partial}{\partial \phi}\dot{\phi}f = -\omega_0 \frac{\partial f}{\partial \phi} - \delta\omega(t)\frac{\partial f}{\partial \phi}, \tag{1.91}$$

since $\dot{\phi} = \omega_0 + \delta\omega(t)$. The mean $\langle f \rangle$ is equal to the probability distribution $p(\phi, t)$ of ϕ (Van Kampen, 1992, section 16.5). Equation (1.91) is of the type (1.1) with $R = -\omega_0 \partial/\partial\phi$ and $C = -\delta\omega(t)\partial/\partial\phi$ and we may find $p = \langle f \rangle$ by using (1.15) once more:

$$\frac{\partial p}{\partial t} = -\omega_0 \frac{\partial p}{\partial \phi} + \left\{ \int_0^\infty d\tau \left\langle \delta\omega(t)\frac{\partial}{\partial \phi}\delta\omega(t-\tau)\frac{\partial}{\partial \phi} \right\rangle \right\} p$$

$$= -\omega_0 \frac{\partial p}{\partial \phi} + D\frac{\partial^2 p}{\partial \phi^2}. \tag{1.92}$$

This shows that the probability distribution of ϕ broadens diffusively on a time scale $1/D$. Starting with a given ϕ_0, we can no longer predict the direction of $\mathbf{e}$ after a time $1/D$. This is the essence of turbulent diffusion, as pointed out by Stix (1981): $\mathbf{B}$ is not dissipated, but the probability for a certain direction of $\mathbf{B}$ broadens diffusively. It is possible to prove this statement by introducing an ensemble probability distribution for $\mathbf{B}$. But in everyday practice this step is bypassed because it is unnecessary: if the starting equation is linear, like (1.89), one may directly infer the equation for the average from (1.14) or (1.15).

This line of reasoning is specific for a passively advected vector field, and does not apply to scalars such as the concentration because the notion of an intrinsic direction does not exist. There is also no energy concept associated with a scalar; the only conservation law is $\int \langle c \rangle d^3\mathbf{r} = $ constant (for zero flux $\mathbf{F}$ at the boundary). An exception should be made for scalars that are not positive definite, such as the angle ϕ of a pendulum. Here the same phenomenon occurs, viz. a constant value of $\langle \phi^2 \rangle$ implies that $\langle \phi \rangle$ is exponentially decaying (Van Kampen, 1976, section 16).

1.6.3. The information content of the ensemble-averaged field

These results suggest that the damping time of the mean field $\langle \mathbf{B} \rangle$ is to be interpreted as the coherence time of the actual field, and we substantiate that by deriving the equivalent of (1.90) for magnetic fields. The dynamo equation is written as $\partial_t \langle \mathbf{B} \rangle = \mathcal{D} \langle \mathbf{B} \rangle$ where $\mathcal{D}$ is not a simple model like the right-hand side of (1.48), but the real thing applicable to the system under study with all relevant anisotropies, radial dependencies etc. included. The eigenmodes of $\mathcal{D}$ obey

$$\mathcal{D} \mathbf{b}_i = \lambda^i \mathbf{b}_i; \quad i = 0, 1, \ldots, \tag{1.93}$$

where $i = 0$ indicates the fundamental mode. We may suppose that $\{\mathbf{b}_i\}$ is a complete function set and expand the field $\mathbf{B}$ (not $\langle \mathbf{B} \rangle$) as

$$\mathbf{B} = \sum_k a_k \mathbf{b}_k. \tag{1.94}$$

Taking the mean and the time derivative of (1.94) gives

$$\partial_t \langle \mathbf{B} \rangle = \sum_k \left(\frac{\mathrm{d}}{\mathrm{d}t} \langle a_k \rangle \right) \mathbf{b}_k$$

$$= \mathcal{D} \langle \mathbf{B} \rangle = \mathcal{D} \sum_k \langle a_k \rangle \mathbf{b}_k = \sum_k \lambda^k \langle a_k \rangle \mathbf{b}_k. \tag{1.95}$$

Since completeness means that $\sum_k d_k \mathbf{b}_k = 0$ implies $d_k \equiv 0$, it follows that

$$\frac{\mathrm{d}}{\mathrm{d}t} \langle a_k \rangle = \lambda^k \langle a_k \rangle, \tag{1.96}$$

(Hoyng and Schutgens, 1995). This relation gives a deep insight into the physical meaning of the dynamo equation: the mean of the i-th expansion coefficient of the field $\mathbf{B}$ evolves according to the i-th eigenvalue λ^i of the dynamo equation. These may be complex, but all $\Re \lambda^i$ are negative when ϵ, or more generally $\langle \mathbf{BB} \rangle$ is stationary. This statement has not been generally proven, but is very likely correct on the grounds expounded above. We have now a complete analogy between (1.96) and (1.90), the eigenvalues of which are $\pm i\omega_0 - D$. Summing all this up, the following interpretation/prescription suggests itself:

1   The dynamo equation has only exponentially damped solutions, and the eigenvalue of a mode of the dynamo equation determines the mean frequency and coherence time of that particular mode (multipole component) of the field: the mean frequency of mode i is $\Im \lambda^i$, and the frequency uncertainty (or inverse coherence time) is $-\Re \lambda^i$. For example, in a solar mean field model the dynamo frequency is $\Omega_\odot = \Im \lambda^0$ and the relative frequency variability is $\Delta \Omega_\odot / \Omega_\odot = -\Re \lambda^0 / \Im \lambda^0$. In a nonperiodic geodynamo model $-\Re \lambda^0$ would be equal to the mean reversal rate.

2   That all modes are damped would come out automatically if we knew the dynamo equation exactly ($\mathbf{u}$ and the tensors α and β as a function of position). Since this is not the case, one must adopt a model, and the 'working point' of the model may be defined by the requirement that $\langle \mathbf{BB} \rangle$ is stationary.

3   The dynamo equation is only able to predict the behaviour of model i for the duration of the coherence time $-1/\Re \lambda^i$. Since this becomes rapidly very small for overtones, the dynamo equation has virtually no predictive power for high overtones. These modes behave in practice as aperiodic noise because their coherence time is much shorter than the period.

Table 1.2 Main parameters of the interface-wave
dynamo of Ossendrijver and Hoyng (1997)

$B_{r.m.s.}$ (solar surface)	100 G
$B_{r.m.s.}$ (overshoot layer)	2×10^4–10^5 G
Energy flux density	
into corona	5×10^6 erg cm^{-2} s^{-1}
$2\pi/\Omega_\odot$	22 yr
$\Delta\Omega_\odot/\Omega_\odot$	0.1

The bottom line is that the dynamo equation for the mean field can only make probability statements about the large-scale field of the dynamo (fundamental mode and first few overtones), and that remains the case if an azimuthal average is used, Section 1.7. Admittedly, the method does not determine the relative r.m.s. magnitudes of the a_k, so that the distribution of magnetic energy over spatial scales (given by the mode number) remains unknown. A method that is in principle capable to handle this question is suggested in Section 1.8.

1.6.4. Applications

To see what this means in practice we discuss a plane two-layer dynamo model of Ossendrijver and Hoyng (1997) representing the overshoot layer and the overlying convection zone, after Parker's (1993) interface-wave dynamo. In the overshoot layer α is zero and β much smaller than in the convection zone. The most important boundary conditions and parameters are summarised in Table 1.2. The r.m.s. surface field is defined as $\langle B^2 \rangle^{1/2} \simeq (f B^2)^{1/2}$, using a filling factor $f = 0.01$ and a field of 1000 G. The value of the magnetic energy flux required for the heating of the solar corona sets the value of $(\beta/8\pi)\partial\langle B^2 \rangle/\partial r$ at the surface. The relative frequency uncertainty of 0.1 means that the solar cycle has a phase memory of 10 periods (220 yr). The eigenvalue of the fundamental mode of $\langle \mathbf{B} \rangle$ is known: $\lambda^0 = 2\pi(i - 0.1)/(22\,\text{yr})$.[10] Assuming translational invariance along one co-ordinate parallel to the layer, the equations for $\langle \mathbf{B} \rangle$ and the six independent components of $\langle \mathbf{BB} \rangle$ are solved.

It turns out to be impossible to find a solution compatible with the parameters in Table 1.2, because marginal stability of $\langle \mathbf{BB} \rangle$ requires $\beta \gtrsim 3 \times 10^{14}$ in the convection zone. Such a high value is needed because turbulent transport has to work very hard to get the magnetic energy (generated by the mean shear flow and random field line stretching) to the boundary where it can escape. But a correct cycle period is then impossible and the ratio of $B_{r.m.s.}$ at the bottom and the surface is too small. Ossendrijver and Hoyng (1997) conclude that an extra sink is needed for $\langle \mathbf{BB} \rangle$, the obvious candidate being resistive dissipation. Equations for $\langle B^2 \rangle$ allowing for resistive effects are now becoming available (e.g. Kim, 1999), and hopefully also for $\langle \mathbf{BB} \rangle$ in the presence of a mean flow. The idea of solving the dynamo equation for $\langle \mathbf{B} \rangle$ jointly with the equation for $\langle \mathbf{BB} \rangle$ remains a challenge (hopefully also to the reader), because it providers the possibility to link several hitherto unrelated quantities, like those in Table 1.2.

1.7. Fluctuating dynamo parameters

We now move on to the azimuthal average. The interpretation of $\langle \mathbf{B} \rangle$ is clear in this case: it is the large scale axisymmetric component of $\mathbf{B}$. The finite memory comes about in a straightforward fashion: the mean flow and the dynamo coefficients α (1.47) and β (1.41) are likewise defined as azimuthal averages. These must fluctuate in time, because there is only a finite number of convection cells on the circle over which the average is taken, and the cells are almost randomly renewed after a correlation time τ_c. Alternatively, the averaging may be seen as an average over an incomplete ensemble, cf. Section 1.3.2. The parameters in (1.48) therefore fluctuate on the short time scale τ_c, and that destroys the infinite phase memory. Fluctuations in α are probably the most important, because $\mathbf{v} \cdot \nabla \times \mathbf{v}$ can be positive and negative, contrary to v^2 (Otmianowska-Mazur et al., 1997). These fluctuations have recently been invoked to explain certain features of the solar cycle and the geomagnetic field. Note that the fluctuations appear as multiplicative noise. They do not determine the amplitude of $\langle \mathbf{B} \rangle$ since the dynamo equation remains linear and homogeneous. Nonlinear effects stay in control of the amplitude of the field.

Consider fluctuations $\alpha = \alpha_0 + \delta\alpha(t)$, where α_0 is the mean value of α over a long time. Their effect on the solar dynamo may be illustrated with the plane dynamo wave solution of Table 1.1 (Hoyng, 1993, 1996). The frequency $\Omega = (\alpha k s / 2)^{1/2}$ and growth rate $\Gamma = \Omega - \beta k^2$ of the wave will also vary, and if only α fluctuates we have

$$\delta\Gamma = \delta\Omega. \tag{1.97}$$

The wave will have zero average growth rate, and $\Gamma = \Gamma_0 + \delta\Gamma$ with $\Gamma_0 = 0$. An increase in wave amplitude ($\delta\Gamma > 0$) implies therefore $\delta\Omega > 0$, that is, a smaller wave period. This is really the whole story, except that it is simpler to use the phase ψ. We set $\psi = \psi_0 - \delta\psi$ and $\psi_0 = \Omega_\odot t + \text{const.}$[11] With $\Omega \equiv \Omega_\odot + \delta\Omega \equiv \dot{\psi} = \Omega_\odot - \delta\dot{\psi}$ it follows that $\delta\Omega = -\delta\dot{\psi}$. The wave amplitude A obeys $\dot{A} = \Gamma A = (\delta\Gamma)A$ or $\delta\Gamma = (\log A)^{\cdot}$. Substitution in (1.97) and integration yields

$$\delta\psi + \log A = \text{constant}. \tag{1.98}$$

This may be related to observations by supposing that A scales with the mean sunspot number R as $R \propto A^{1/\mu}$:

$$\delta\psi + \mu\log R = \text{constant}. \tag{1.99}$$

This says that if the solar cycle runs ahead of its reference phase $\Omega_\odot t$ (i.e. $\delta\psi < 0$) the cycle should be stronger (more sunspots). The solar cycle data obey this relation rather well, for $\mu \simeq 1.1$ (Hoyng, 1996). Operationally, the reference frequency $\Omega_\odot$ is determined first, by fitting all data to a single-frequency wave. The fact that stronger cycles usually last shorter, and vice versa, is evident in the sunspot record and was already known to Wolf (1861).[12] The magnitude of the phase fluctuations has the characteristic $t^{1/2}$ random walk behaviour:

$$\delta\psi_{\text{r.m.s.}} = \frac{\delta\alpha_{\text{r.m.s.}}}{2\alpha_0} \Omega_\odot \sqrt{2\tau_c t}. \tag{1.100}$$

The r.m.s. phase variation of the solar cycle over the last 300 yr is ~ 1.5 rad, and with $\tau_c \sim 15$ days it follows that $\delta\alpha_{\text{r.m.s.}}/\alpha_0 \sim 2$. This is probably a lower limit since (1.100) does not allow for nonlinear effects. The relative magnitude of the fluctuations is large, although their net effect in one wave period is reduced by a factor $(\Omega_\odot \tau_c)^{1/2} \sim 0.1$ due to cancellation. Dynamo waves seem rather robust – you have to shake pretty hard to get a modest perturbation of the period.

A plane wave analysis is only applicable for slow variations, with a typical time scale $\gg \Omega_\odot^{-1}$, while the fluctuations evolve on the short time scale $\tau_c \ll \Omega_\odot^{-1}$. Ossendrijver et al. (1996) have undertaken a more realistic study of a spherical dynamo model with constant β and $\delta\alpha$ a random function of time and position, and they basically confirm the results of the simple analysis above, see Fig. 1.7. The strict correlation between phase and amplitude (1.99) ceases to exist, but the value of $\delta\psi + \mu\log R$ changes only very slowly, on a time scale of 300 yrs or more. Linear dynamo wave physics seems therefore able to endow the solar dynamo with a long memory, and to mimic an internal clock mechanism as proposed by

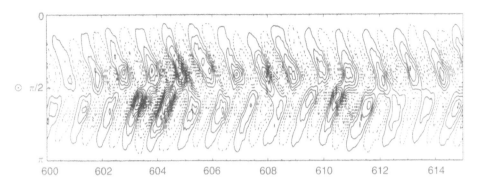

Figure 1.7 The toroidal mean field as a function of latitude of the overshoot layer dynamo model of Ossendrijver *et al.* (1996). Time is in units of the period of the fundamental dipole mode. The magnitude of the fluctuations is $\delta\alpha_{r.m.s.}/\alpha_0 \sim 3$. The phase-amplitude correlation is not well visible in this figure. Transiently excited overtones generate slow and fast modulations, North-South asymmetries, etc.

Dicke (1978). Whether a purely nonlinear dynamo model without fluctuations can produce such a phase-amplitude correlation is still an open question.

The second example refers to the geodynamo, which is believed to be an $\alpha\Omega$ dynamo with a non-periodic fundamental mode (Soward, 1991). The idea that reversals may be caused by some rapid variability in α had already been advanced by Parker (1969). Schmitt *et al.* (2001) and Hoyng *et al.* (2001) have analysed the behaviour of a nonlinear $\alpha\Omega$ mean field model, subject to a continuous level of fluctuations $\delta\alpha$. As might be expected, the field becomes variable, performs occasional fast reversals at random intervals, and overtones are excited.

A statistical analysis leads to a Fokker–Planck equation for the theoretical amplitude distribution $p(a, t)$ of the fundamental dipole mode. The physical picture is that the amplitude a behaves as a particle in a bistable potential, see Fig. 1.8. The $(1 - a^2)^2$ shape of the potential is a consequence of the assumed form of the α-quenching, $1 - \text{const.}|\langle \mathbf{B}\rangle|^2 \simeq 1 - a^2$ (the fundamental mode is usually dominant). The fluctuations also excite overtones, and reversals are solely due to the interaction of the dipole mode with these overtones.

The reversal rate is an exponentially increasing function of the amplitude of the random motion near the bottom of the well. The reason is that a number of favourable fluctuations $\delta\alpha$ is required, and on general statistical grounds the probability for that to occur (i.e. the waiting time) depends exponentially on the parameters. In geophysical parlance, a larger geomagnetic variation implies a larger reversal rate. Taking a current reversal rate of 2×10^5 yr the model predicts that the relative variability of the magnetic field between reversals has a variance of about 0.15, in agreement with the observations (Hoyng *et al.*, 2002).

1.7.1. Ensemble or azimuthal average?

It is perhaps necessary to point out that there is no conflict between the two averaging techniques. One may adopt either the ensemble view or the azimuthal average, according to need and intention, and we outline the following conceptual relation between the two. The azimuthal average is an average over an incomplete ensemble. A complete ensemble may be set up by considering a set of copy systems each with the same mean parameters but with different realisation of the turbulence, cf. Section 1.3.1. To make things more visual, arrange these systems in a horizontal row next to one another. Below each system we place the rotated

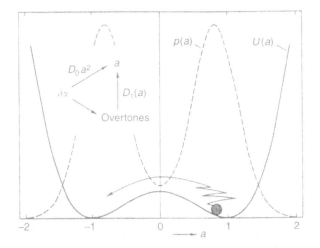

Figure 1.8 The geodynamo as a bistable oscillator. The amplitude a of the dipole mode behaves as a heavily damped particle in a bistable potential $U(a) \propto (1 - a^2)^2$. The central hill is due to supercritical dynamo action which forces the field away from zero, and the walls result from nonlinear α-quenching at large field amplitude. The fluctuations make the dipole amplitude perform a random motion near the bottom of one of the wells with amplitude distribution $p(a)$. Occasionally, the particle jumps to the other side: a geomagnetic reversal.

states of that system, in a vertical column. A column average is an azimuthal average, and produces mean field dynamos with a finite memory as in the previous section. Each column average evolves in time in a different way. One could say that each next column average produces a dynamo as in Fig. 1.7, but with a different initialisation of the random series. The ensemble average is the average over all columns and all rows, and as a result of phase mixing the ensemble mean field $\langle \mathbf{B} \rangle$ must approach zero.

1.8. Nonlinear mean field theory

Since this monograph is about nonlinear aspects of dynamo theory, I cannot get away without saying a few words about nonlinear effects. It is fair to say that a comprehensive theory of nonlinear effects does not yet exist. What we have is a great number of different and sometimes contradictory models. An important current issue beyond the scope of this review is the nonlinear quenching of α and β. Opinions differ particularly widely on α. Gruzinov and Diamond (1994) and Cattaneo and Hughes (1996) have argued that α-quenching sets in at mean field strengths a factor $\sqrt{R_m}$ smaller than equipartition. On the other hand, given the right kind of global anisotropy, Brandenburg and Schmitt (1998), Schmitt (2000) and Ferriz-Mas (1994) argue for a negative quenching, i.e. a magnetically enhanced α-effect.

The first remark is that it is questionable that nonlinear quenching should make α and β depend on $\langle \mathbf{B} \rangle$, because as I have argued there are good reasons to expect that in the ensemble approach $\langle \mathbf{B} \rangle = 0$. A dependence of α and β on $\langle \mathbf{B}\mathbf{B} \rangle$ seems more natural. Therefore, while it is clear that nonlinearities cannot be dispensed with, it seems that we do not yet fully understand how nonlinear effects are to be incorporated in our equations. Moreover, fluctuations are themselves a manifestation of nonlinearities (for example turbulent convection) and so, on a fundamental level, there is no difference between fluctuations and nonlinear effects.

In many physical situations *experience* has shown that on fast time scales the effect of the nonlinearities is practically indistinguishable from that of noise. Depending on the specific problem, this noise may be additive or multiplicative. For the longer time scales this is no longer true and proper inclusion of nonlinearities remains essential. This approach is widely followed in many areas of physics. It has the added advantage that several properties of the nonlinear system may now be evaluated theoretically, by making contact with the theory of stochastic processes.

The second remark is that the theory of Section 1.2 may be applied to nonlinear equations, by casting these into an equivalent linear form (Van Kampen, 1976). We illustrate the idea with the rotating vector model of Section 1.6.2. Assume that $\mathbf{e}$ is located at the origin, immersed in a linear shear flow $\mathbf{u} = (-\lambda y, 0, 0)$. Advection causes $\mathbf{e}$ to rotate nonuniformly. In addition there is a random component $\delta\omega(t)$. Instead of $\dot{\phi} = \omega_0 + \delta\omega(t)$ we now have

$$\dot{\phi} = \lambda \sin^2 \phi + \delta\omega(t). \tag{1.101}$$

The fluctuations dominate near $\phi = 0, \pi$. The evolution equation of $\mathbf{e} = (\cos\phi, \sin\phi)$ is nonlinear, but we may still write down the equation for the number density $f(\phi, t)$. Instead of (1.91) we get

$$\frac{\partial f}{\partial t} = -\frac{\partial}{\partial \phi}(\lambda \sin^2 \phi \, f) - \delta\omega(t)\frac{\partial f}{\partial \phi}. \tag{1.102}$$

Here comes the trick: (1.102) is a linear stochastic equation for f with $R = -(\partial/\partial\phi)\lambda \sin^2 \phi$ and $C = -\delta\omega(t)\partial/\partial\phi$, so that we may find the probability distribution $p = \langle f \rangle$ with (1.15):

$$\frac{\partial p}{\partial t} = -\frac{\partial}{\partial \phi}(\lambda \sin^2 \phi \, p) + D\frac{\partial^2 p}{\partial \phi^2}. \tag{1.103}$$

Once this (linear) Fokker-Planck equation is solved, one may find $\langle \phi \rangle = \int \phi p(\phi)\mathrm{d}\phi$, and all higher moments $\langle \phi^n \rangle$, the mean rotation period of $\mathbf{e}$, etc. (Hoyng, 1987a). The nonlinearity is hidden in the fact that a closed equation for $\langle \phi \rangle$ or $\langle \mathbf{e} \rangle$ no longer exists.

This idea may be generalised to the ensemble distribution function $\mathcal{P}(1, 2, 3, \ldots, t)$ from Section 1.3.1. The phase space density $\mathcal{F}$ evolves according to

$$\frac{\partial \mathcal{F}}{\partial t} = -\frac{\partial}{\partial \mathbf{B}_i}\dot{\mathbf{B}}_i \mathcal{F} - \frac{\partial}{\partial \mathbf{v}_i}\dot{\mathbf{v}}_i \mathcal{F}, \tag{1.104}$$

where $\dot{\mathbf{B}}_i$, and $\dot{\mathbf{v}}_i$ are given by the induction equation and the momentum equation, respectively. The latter is often assumed to have an (additive) random forcing term and averaging over these fluctuations yields the equation for the probability distribution $\mathcal{P} = \langle \mathcal{F} \rangle$, essentially by applying (1.15). In this description $\mathbf{v}_i$ and $\mathbf{B}_i$ are no longer dynamical variables but co-ordinates in phase space. The linear equation for $\mathcal{P}$ may be useful to obtain a (nonlinear) equation for $\langle \mathbf{B} \rangle$. The rotating vector example shows that such an equation will in general not exist in closed form, but perhaps it does within the context of a suitable approximation.

The last remark is a suggestion for an alternative method for modelling nonlinear dynamos with which I have played for some time (unsuccessfully), and maybe the reader gets the right idea. It turned out in Section 1.6.3. that the eigenfuctions $\mathbf{b}_k$ of the dynamo equation possess a preferred status among all function sets that one may choose to expand the field $\mathbf{B}$ of the dynamo with. Inserting (1.94) into the induction equation leads to the following system

$$\dot{a}_k = \sum_{\ell}\{R_{k\ell} + C_{k\ell}(t)\}a_\ell, \tag{1.105}$$

with

$$R_{k\ell} = \int_V \mathbf{b}_k^* \cdot R\mathbf{b}_\ell \, \mathrm{d}^3\mathbf{r} = \frac{4\pi}{c} \int_V \left\{ \mathbf{u} \cdot \mathbf{j}_k^* \times \mathbf{b}_\ell + \frac{4\pi\eta}{c}\mathbf{j}_k^* \cdot \mathbf{j}_\ell \right\} \mathrm{d}^3\mathbf{r}. \tag{1.106}$$

The eigenfunctions $\mathbf{b}_k$ and their adjoints $\mathbf{b}_k^*$ form a biorthogonal set on $V + E$, supposed to be normalised to unity, $\int_{V+E} \mathbf{b}_k^* \cdot \mathbf{b}_\ell \mathrm{d}^3\mathbf{r} = \delta_{k\ell}$, where V is the volume of the dynamo and E the exterior vacuum, and $\mathbf{j}_k = (c/4\pi)\nabla \times \mathbf{b}_k$ is the current generating $\mathbf{b}_k$. The second expression in (1.106) follows after a few elementary operations. The expression for $C_{k\ell}$ is similar except that it contains $\mathbf{v}(\mathbf{r}, t)$ instead of the mean flow $\mathbf{u}$.

Equation (1.105) describes the evolution of the mode coefficients a_k for a given velocity field. The first few coefficients $a_0, a_1, \ldots$ specify the evolution of the large scale (dynamo) field. This method of representing the field goes back to Elsasser (1946) and Bullard and Gellman (1954). The new element is that we know from (1.96) that higher mode coefficients evolve on progressively shorter time scales. Perhaps this renders it possible to extract a closed equation for these first few coefficients, by a multiple timescale analysis or by elimination of fast variables. The velocity field in the dynamo may be decomposed in a similar spirit, leading to a system of coupled large scale magnetic and velocity modes. The system is nonlinear with a random element, and such a formulation may be very useful for building nonlinear dynamo models.

The method may be regarded as an improvement to spectral theories in that it employs a natural function set to represent the field and the flow rather than the Fourier modes. The advantage is that the information on spatial structure of the field and flows remains intact. Convergence was an issue for the scheme of Bullard and Gellman (1954) because the existence of dynamo action was at stake, but is of less concern here. What matters is that the large scales are well represented. Note, finally, that Farrell and Ioannou (1999a,b) have argued that there are more efficient methods than eigenmode expansions, and their ideas may well be applicable in the present context.

1.9. Synopsis

The thread running through my story is that mean field dynamo theory is a statistical theory. The inexorable consequence of making an average is that the resulting equation for the mean has lost its predictive character. The idea that a solution of the dynamo equation represents the behaviour of the field for all times is misleading, for the same reason that the position of a Brownian particle cannot be predicted by its mean position for all times. This has nothing to do with the neglect of nonlinear effects, but is caused by the fact that the theory carries the seed of unpredictable variability from the very beginning, through the way it is formulated. In the end, only probability statements are possible, and against this background I have tried to identify the meaning and the information content of the mean field concept. Failure to recognize the probabilistic aspects of the theory leads to paradoxical results, and that may be partly responsible for its controversial status. Finally, I mention five problem areas where new insights are badly needed:

1   The relative error due to the omitted terms in the dynamo equation is $F v \tau_c / \lambda_c$. Maybe it is possible to prove that F is considerably smaller than unity also in the presence of a mean shear flow. That would eliminate the long-standing problem of a long correlation time, Section 1.4.1.

2   Extension of the analysis presented by Gilbert *et al.* (1997) to random flows, to show that hopefully also in this case the dynamo equation catches the physics of the large scale modes correctly.

3   Joint solution of the equations for $\langle \mathbf{B} \rangle$ and $\langle \mathbf{BB} \rangle$ including resistive effects. This would, for example, permit a determination of the phase memory of the solar dynamo and the mean reversal time of the geodynamo, Sections 1.6.3. and 1.6.4.

4   Ab initio calculations of the variability of the dynamo coefficients in the angular average approach, so that they need no longer be treated as a free parameter, Section 1.7.

5   Last but not least, a consistent theory of nonlinear effects Section 1.8.

Acknowledgments

I would like to thank Manuel Núñez and Eun-jin Kim for their helpful comments.

Notes

1   Farrell and Ioannou (1999a,b) have studied the magnetic structures that are preferentially excited by *additive* noise of fixed spatial structure in the induction equation.

2   Notation: $|C|$ is the r.m.s. size of C. For scalar and vectorial transport $|C| \sim v/\lambda_c =$ (eddy turnover time)$^{-1}$; $\lambda_c =$ correlation length or eddy size of the turbulence.

3   The meaning of averaged operators like $\langle C(t)C(t-\tau) \rangle$ becomes less obscure by noting that the averaging is always postponed. All vector operations are carried out first, and eventually the average condenses into the definition of α and β, see Section 1.4. The actual act of averaging takes place when a value for these transport coefficients is computed.

4   Unless an asymmetry is built into the ensemble, e.g. to allow for the effect of a boundary condition. An example would be a fossil dipole field in the core of the sun, which would generate a permanent asymmetry in the solar dynamo.

5   Due to the use of the interaction representation the existence of this second regime of validity is not immediately evident in Section 1.2. It applies when the Reynolds number $R_m = v\lambda_c/\eta$ for eddies is much smaller that unity (large resistivity). In the solar dynamo and the geodynamo $v\lambda_c/\eta \gg 1$.

6   Barring exceptional cases such as periodic boundary conditions (in time), or a periodic source such as in the solar surface flux distribution problem.

7   One might think the proper condition is $u/L \gg \eta/\lambda_c^2$, but the operators $\eta\nabla^2$ and $\nabla \times u \times$ do not commute and get mixed in the series expansion of $\exp(\pm R\tau)$.

8   If the helicity is nonzero, the shear flow has also a large effect of α (Hoyng, 1985).

9   The fact that the dynamo is 'invisible' (Rädler, 1982; Rädler and Geppert, 1999) does not prevent radiation of magnetic energy into space (a 'corona'), see also Hollerbach *et al.* (1998, section 5). Nevertheless it may be objected that our dynamo is highly artificial. The following *gedanken* experiment shows that neither inhomogeneity nor nonlinear and resistive effects will lead to a different view. Consider a very large number of numerical geodynamo experiments à la Glatzmaier and Roberts (1995) running in parallel, with the same boundary conditions but different initial conditions (do not worry about computer time). After a long time the field averaged over all experiments will be zero, as half of them will be in one polarity state, the other half in the reversed state. Another example is the dynamo of Gilbert *et al.* (1988) which has a field, but no mean field since $\alpha = 0$.

10   One may try to constrain the parameters further by determining the next eigenvalue λ^1. One way to do this is to suppose that modulation effects such as the Gleissberg cycle and North-South asymmetries are due to interference of the fundamental mode with the first overtone.

11   A plus sign, $\psi = \psi_0 + \delta\psi$, would be consequent, but the minus sign agrees with the cited literature.

12   In the instantaneous cycle amplitude and frequency profiles of Paluš and Novotná (1999) the effect shows up as a correlation between the two profiles, in that amplitude appears to run ahead of frequency.

References

Bourret, R.C., "Propagation of randomly perturbed fields," *Can. J. Phys.* **40**, 782–790 (1962).

Braginskii, S.I., "Self-excitation of a magnetic field during the motion of a highly conducting fluid," *Soviet Phys. JETP* **20**, 726–735 (1965a).

Braginskii, S.I., "Theory of the hydromagnetic dynamo," *Soviet Phys. JETP* **20**, 1462–1471 (1965b).

Brandenburg, A., "Solar dynamos: computational background," in *Lectures on Solar and Planetary Dynamos* (Eds. M.R.E. Proctor and A.D. Gilbert), Cambridge UP, pp. 117–159 (1994).

Brandenburg, A., Jennings, R.L., Nordlund, Å., Rieutord, M., Stein, R.F. and Tuominen, I., "Magnetic structures in a dynamo simulation," *J. Fluid Mech.* **306**, 325–352 (1996).

Brandenburg, A. and Schmitt, D., "Simulations of an alpha-effect due to magnetic buoyancy," *Astron. Astrophys.* **338**, L55–L58 (1998).

Bullard, E.C. and Gellman, H., "Homogeneous dynamos and terrestrial magnetism," *Phil. Trans. R. Soc. Lond.* A **247**, 213–278 (1954).

Cattaneo, F. and Hughes, D.W., "Nonlinear saturation of the turbulent α effect," *Phys. Rev. E* **54**, 4532–4535 (1996).

Childress, S. and Gilbert, A.D., *Stretch, Twist, Fold: The Fast Dynamo*, Springer, Berlin (1995).

Christensen, U., Olson, P. and Glatzmaier, G.A., "A dynamo model interpretation of geomagnetic field structures," *Geophys. Res. Lett.* **25**, 1565–1568 (1998).

Covas, E., Tavakol, R., Tworkowski, A., Brandenburg, A., Brooke, J. and Moss, D., "The influence of geometry and topology on axisymmetric mean-field dynamos," *Astron. Astrophys.* **345**, 669–679 (1999).

Deinzer, W., Von Kusserow, H.-U. and Stix, M., "Steady and oscillatory $\alpha\omega$-dynamos," *Astron. Astrophys.* **36**, 69–78 (1974).

Dicke, R.H., "Is there a chronometer hidden deep in the sun?," *Nature* **276**, 676–680 (1978).

Dittrich, P., Molchanov, S.A., Sokoloff, D.D. and Ruzmaikin, A.A., "Mean magnetic field in renovating random flow," *Astron. Nachr.* **305**, 119–125 (1984).

Dolginov, A.Z. and Silant'ev, N.A., "Diffusion of scalar and vector fields in compressible turbulent media," *Geophys. Astrophys. Fluid Dynam.* **63**, 139–178 (1992).

Drummond, I.T. and Horgan, R.R., "Numerical simulation of the α-effect and turbulent magnetic diffusion with molecular diffusivity," *J. Fluid Mech.* **163**, 425–438 (1986).

Elsasser, W.M., "Induction effects in terrestrial magnetism. Part I. Theory," *Phys. Rev.* **69**, 106–116 (1946).

Farrell, B.F. and Ioannou, P.J., "Optimal excitation of magnetic fields," *Astrophys. J.* **522**, 1079–1087 (1999a).

Farrell, B.F. and Ioannou, P.J., "Stochastic dynamics of field generation in conducting fluids," *Astrophys. J.* **522**, 1088–1099 (1999b).

Ferrière, K., "Alpha-tensor and diffusivity tensor due to supernovae and superbubbles in the galactic disk near the sun," *Astron. Astrophys.* **310**, 438–455 (1996).

Ferriz-Mas, A., Schmitt, D. and Schüssler, M., "A dynamo effect due to instability of magnetic flux tubes," *Astron. Astrophys.* **289**, 949–956 (1994).

Glatzmaier, G.A. and Roberts, P.H., "A three-dimensional self-consistent computer simulation of a geomagnetic field reversal," *Nature* **377**, 203–209 (1995).

Gilbert, A.D., Frisch, U. and Pouquet, A., "Helicity is unnecessary for alpha effect dynamos, but it helps," *Geophys. Astrophys. Fluid Dynam.* **42**, 151–161 (1988).

Gilbert, A.D., Soward, A.M. and Childress, S., "A fast dynamo of $\alpha\omega$-type," *Geophys. Astrophys. Fluid Dynam.* **85**, 279–314 (1997).

Gruzinov, A.V. and Diamond, P.H., "Self-consistent theory of mean-field electrodynamics," *Phys. Rev. Lett.* **72**, 1651–1653 (1994).

Hollerbach, R., Galloway, D.J. and Proctor, M.R.E., "On the adjustment to the Bondi-Gold theorem in a spherical-shell fast dynamo," *Geophys. Astrophys. Fluid Dynam.* **87**, 111–132 (1998).

Hoyng, P., "Exact evaluation of the effect of an arbitrary mean flow in kinematic dynamo theory," *J. Fluid Mech.* **151**, 295–309 (1985).

Hoyng, P., "Turbulent transport of magnetic fields I. A simple mechanical model," *Astron. Astrophys.* **171**, 348–356 (1987a).

Hoyng, P., "Turbulent transport of magnetic fields II. The role of fluctuations in kinematic theory," *Astron. Astrophys.* **171**, 357–367 (1987b).

Hoyng, P., "Mean field dynamo theory," in *The Sun, A Laboratory for Astrophysics* (Eds. J.T. Schmelz and J.C. Brown), Kluwer, Dordrecht, pp. 99–138 (1992).

Hoyng, P., "Helicity fluctuations in mean field theory: an explanation for the variability of the solar cycle?," *Astron. Astrophys.* **272**, 321–339 (1993).

Hoyng, P., "Is the solar cycle timed by a clock?," *Solar Phys.* **169**, 253–264 (1996).

Hoyng, P. and Schutgens, N.A.J., "Dynamo spectroscopy," *Astron. Astrophys.* **293**, 777–782 (1995).

Hoyng, P., Ossendrijver, M.A.J.H. and Schmitt, D., "The geodynamo as a bistable oscillator," *Geophys. Astrophys. Fluid Dynam.* **94**, 263–314 (2001).

Hoyng, P., Schmitt, D. and Ossendrijver, M.A.J.H., "A theoretical analysis of the observed variability of the geomagnetic dipole field," *Phys. Earth Planet. Inter.* **130**, 143–157 (2002).

Kageyama, A., and Sato, T., "Generation mechanism of a dipole field by a magnetohydrodynamic dynamo," *Phys. Rev.* E **55**, 4617–4626 (1997).

Kim, E., "Analytic results on a simple nonlinear dynamo model," in *Stellar Dynamos: Nonlinearity and Chaotic Flows* (Eds. M. Núñez and A. Ferriz-Mas), *A.S.P. Conf. Ser.* **178**, pp. 69–78 (1999).

Kitchatinov, L.L., Pipin, V.V. and Rüdiger, G., "Turbulent viscosity, magnetic diffusivity, and heat conductivity under the influence of rotation and magnetic field," *Astron. Nachr.* **315**, 157–170 (1994).

Knobloch, E., "The root-mean-square magnetic field in turbulent diffusion," *Astrophys. J.* **220**, 330–334 (1978a).

Knobloch, E., "Turbulent diffusion of magnetic fields," *Astrophys. J.* **225**, 1050–1057 (1978b).

Krause, F., *Habilitationsschrift*, Univ. Jena (1968).

Krause, F. and Rädler, K.-H., *Mean-Field Magnetohydrodynamics and Dynamo Theory*, Pergamon Press, Oxford (1980).

Kuang, W. and Bloxham, J., "An earth-like numerical dynamo model," *Nature* **389**, 371–374 (1997).

Moffatt, H.K., *Magnetic Field Generation in Electrically Conducting Fluids*, Cambridge UP (1978).

Monin, A.S. and Yaglom, A.M., *Statistical Fluid Mechanics*, Vol 1, MIT Press, Cambridge (1973).

Nicklaus, B. and Stix, M., "Corrections to first order smoothing in mean-field electrodynamics," *Geophys. Astrophys. Fluid Dynam.* **43**, 149–166 (1988).

Nordlund, Å., Brandenburg, A., Jennings, R.L., Rieutord, M., Ruokolainen, J., Stein, R.F. and Tuominen, I., "Dynamo action in stratified convection with overshoot," *Astrophys. J.* **392**, 647–652 (1992).

Nordlund, Å., Galsgaard, K. and Stein, R.F., "Magnetoconvection and magnetoturbulence," in *Solar Surface Magnetism* (Eds. R.J. Rutten and C.J. Schrijver), Kluwer, Dordrecht, pp. 471–498 (1994).

Núñez, M. and Galindo, F., "Averages in mean-field magnetohydrodynamics," *Phys. Lett.* A **260**, 377–380 (1999).

Ossendrijver, A.J.H., Hoyng, P. and Schmitt, D., "Stochastic excitation and memory of the solar dynamo," *Astron. Astrophys.* **313**, 938–948 (1996).

Ossendrijver, A.J.H. and Hoyng, P., "Mean magnetic field and energy balance of Parker's surface-wave dynamo," *Astron. Astrophys.* **324**, 329–343 (1997).

Otmianowska-Mazur, K., Rüdiger, G., Elstner, D. and Arlt, R., "The turbulent EMF as a time series and the 'quality' of dynamo cycles," *Geophys. Astrophys. Fluid Dynam.* **86**, 229–247 (1997).

Paluš, M. and Novotná, D., "Sunspot cycle: a driven nonlinear oscillator?," *Phys. Rev. Lett.* **83**, 3406–3409 (1999).

Parker, E.N., "Hydromagnetic dynamo models," *Astrophys. J.* **122**, 293–314 (1955).

Parker, E.N., "The occasional reversal of the geomagnetic field," *Astrophys. J.* **158**, 815–827 (1969).

Parker, E.N., "A solar dynamo surface wave at the interface between convection and nonuniform rotation," *Astrophys. J.* **408**, 707–719 (1993).

Petrovay, K., "Theory of passive magnetic field transport," in *Solar Surface Magnetism* (Eds. R.J. Rutten and C.J. Schrijver), Kluwer, Dordrecht, pp. 415–440 (1994).

Pouquet, A., Frisch, U. and Léorat, J., "Strong MHD helical turbulence and the nonlinear dynamo effect," *J. Fluid Mech.* **77**, 321–354 (1976).

Rädler, K.-H., "On dynamo action in the high-conductivity limit," *Geophys. Astrophys. Fluid Dynam.* **20**, 191–211 (1982).

Rädler, K.-H. and Geppert, U., "Turbulent dynamo action in the high-conductivity limit: a hidden dynamo," in *Stellar Dynamos: Nonlinearity and Chaotic Flows* (Eds. M. Núñez and A. Ferriz-Mas), *A.S.P. Conf. Ser.* **178**, pp. 151–163 (1999).

Roberts, P.H., "Kinematic dynamo models," *Phil. Trans. R. Soc. Lond. A* **271**, 663–698 (1972).

Roberts, P.H., "Fundamentals of dynamo theory," in *Lectures on Solar and Planetary Dynamos* (Eds. M.R.E. Proctor and A.D. Gilbert), Cambridge UP, pp. 1–58 (1994).

Rogachevskii, I. and Kleeorin, N., "Intermittency and anomalous scaling for magnetic fluctuations," *Phys. Rev. E* **56**, 417–426 (1997).

Rüdiger, G. and Kitchatinov, L.L., "Alpha-effect and alpha-quenching," *Astron. Astrophys.* **269**, 581–588 (1993).

Rüdiger, G. and Arlt, R., "Physics of the solar cycle," this volume, Ch. 6, 147–193 (2003).

Schmitt, D., "Dynamo action of magnetostrophic waves," this volume, Ch. 4, 83–121 (2003).

Schmitt, D. and Schüssler, M., "Non-linear dynamos I. One-dimensional model of a thin layer dynamo," *Astron. Astrophys.* **223**, 343–351 (1989).

Schmitt, D., Ossendrijver, M.A.J.H. and Hoyng, P., "Magnetic field reversals and secular variation in a bistable geodynamo model," *Phys. Earth Planet. Int.* **125**, 119–124 (2001).

Sheeley, N.R. and Wang, Y.-M., "Returning to the random walk," in *Solar Surface Magnetism* (Eds. R.J. Rutten and C.J. Schrijver), Kluwer, Dordrecht, pp. 379–383 (1994).

Song, X. and Richards, P.G., "Seismological evidence for differential rotation of the earth's inner core," *Nature* **382**, 221–224 (1996).

Soward, A.M., "The earth's dynamo," *Geophys. Astrophys. Fluid Dynam.* **62**, 191–209 (1991).

Stix, M., "Dynamo theory and the solar cycle," in *Basic Mechanisms of Solar Activity* (Eds. V. Bumba and J. Kleczek), Reidel, Dordrecht, pp. 367–388 (1976).

Stix, M., "The solar dynamo," in *Pleins Feux sur la Physique Solaire* (Eds. S. Dumont and J. Rösch), Éditions du CNRS, Paris, pp. 37–61 (1978).

Stix, M., "Theory of the solar cycle," *Solar Phys.* **74**, 79–101 (1981).

Stix, M., "Helicity and α-effect of simple convection cells," *Astron. Astrophys.* **118**, 363–364 (1983).

Su, W., Dziewonski, A.M. and Jeanloz, R., "Planet within a planet: rotation of the inner core of earth," *Science* **274**, 1883–1887 (1996).

Urpin, V., "Mean electromotive force and dynamo action in a turbulent flow," *Astron. Astrophys.* **347**, L47–L50 (1999).

Urpin, V. and Brandenburg, A., "Magnetic drift processes in differentially rotating turbulence," *Astron. Astrophys.* **345**, 1054–1058 (1999).

Van Geffen, J.H.G.M., "Distribution of magnetic energy in $\alpha\Omega$-dynamos, II: a solar convection zone dynamo," *Geophys. Astrophys. Fluid Dynam.* **71**, 223–241 (1993a).

Van Geffen, J.H.G.M., "Distribution of magnetic energy in $\alpha\Omega$-dynamos III. A localized dynamo," *Astron. Astrophys.* **274**, 534–542 (1993b).

Van Geffen, J.H.G.M. and Hoyng, P., "Turbulent transport of magnetic fields V. Distribution of magnetic energy in a simple α^2-dynamo," *Geophys. Astrophys. Fluid Dynam.* **53**, 109–123 (1990).

Van Kampen, N.G., "A cumulant expansion for stochastic linear differential equations. II," *Physica* **74**, 239–247 (1974).

Van Kampen, N.G., "Stochastic differential equations," *Phys. Reports* **24C**, 171–228 (1976).

Van Kampen, N.G., *Stochastic Processes in Physics and Chemistry*, North-Holland, Amsterdam (1992).

Wang Y.-M. and Sheeley, N.R., "The rotation of photospheric magnetic fields: a random walk transport model," *Astrophys. J.* **430**, 399–412 (1994).

Weiss, N.O. and Tobias, S.M., "Modulation of solar and stellar activity cycles," in *Solar and Heliospheric Plasma Physics*, (Eds. G.M. Simnett, C.E. Alissandrakis, L. Vlahos), Springer, Berlin, pp. 25–47 (1997).

Wolf, R., *Mittheilungen über die Sonnenflecken* **12**, 41–82 (1861). See p. 76.

Yoshimura, H., "Solar-cycle dynamo wave propagation," *Astrophys. J.* **201**, 740–748 (1975).

2 Fast dynamos

David Galloway

School of Mathematics and Statistics, University of Sydney, NSW 2006 Sydney, Australia, E-mail: dave@maths.usyd.edu.au

A dynamo is termed fast if it grows on the turnover timescale of the flow rather than on the ohmic diffusion timescale. In astrophysics the latter is often much too long to allow significant build-up of magnetic fields during the lifetime of the Universe, and for this reason fast dynamos are sought to explain the generation of fields in stars and larger-scale objects.

This chapter reviews the topic of fast dynamos, mainly from a numerical point of view (this emphasis reflects the author's own background, and also the existence of an excellent book by Childress and Gilbert dealing mainly with analytic aspects). In the Introduction, the history and motivation of the search for fast dynamos are established, and this leads into Section 2.2 which explains the necessity for the flow to be chaotic (in the sense of dynamical systems theory) before a fast dynamo is possible. The third section reviews the various numerical experiments which have been mounted to provide evidence that fast dynamos actually exist, and describes the physical processes whereby the field is generated. The fourth, short section mentions some of the fast dynamos for which analytic solutions have been found: these flows cheat the chaos by localising it in artificial features outside which piecewise integration is possible, often using asymptotic methods.

Section 2.5 addresses what happens when fast dynamos enter the nonlinear regime and the Lorentz force is allowed to limit the growth of the field strength by reacting back on the motion. A number of calculations are described and for some of these it is possible to advance scaling laws to determine the ultimate strength of the fields. So far these examples all suggest that the effectiveness of the dynamo goes down as the (kinetic) Reynolds number of the flow goes up. The difficulties this poses for astrophysical dynamos are confronted but remain largely unsolved, even at high magnetic Reynolds number.

There is a short conclusion which summarises the achievements of fast dynamo theory and describes likely areas where further progress can be expected.

2.1. Introduction

The distinction between fast and slow dynamos was introduced by Russian dynamo theorists, starting with the paper by Vainshtein and Zeldovich (1972). In that and most subsequent papers the discussion has been mainly kinematic, though the role of the Lorentz force has begun to attract attention in recent years, and will be addressed in the penultimate section of the current chapter.

The idea is that the induction equation has two naturally occurring timescales, the turnover time of the flow being used for the dynamo, and the electromagnetic diffusion time. If the equation is scaled with length scale L and timescale L/U, where L is the size of the astrophysical object and U is (say) the RMS velocity, we arrive at

$$\frac{\partial \mathbf{B}}{\partial t} = \nabla \times (\mathbf{U} \times \mathbf{B}) + \frac{1}{R_m} \nabla^2 \mathbf{B}, \qquad (2.1)$$

where $R_m \equiv UL/\eta$ is the magnetic Reynolds number. Here η, the magnetic diffusivity, is $1/\mu_0\sigma$, and σ is the fluid conductivity, assumed uniform. In these units the turnover time is of order 1 and the diffusion time is of order R_m. A dynamo is termed fast if it grows on the turnover timescale and slow if it grows on the diffusion timescale. In the kinematic case, we take $\mathbf{U}$ to be a given velocity field, and (1) is then linear. This means solutions can be decomposed into eigenmodes and the left-hand side of (1) can be replaced by $s\mathbf{B}$, where s, the growthrate, is an appropriate eigenvalue. If s has a positive real part, the eigenmode grows and the flow acts as a dynamo. If s has a non-zero imaginary part, the mode is oscillatory.

In the limit $R_m \rightarrow \infty$ the diffusion timescale becomes huge and disappears from the picture: any dynamo that grew with such a timescale would have growthrate tending to zero in this limit. So an alternative, and more practically manageable, definition of a fast dynamo is as one whose growthrate tends to a finite positive value as $R_m \rightarrow \infty$. In fact this limit is a wonderfully subtle affair that is at the heart of this whole topic. When '$R_m = \infty$', field lines are frozen in. But the limit is singular in the mathematical sense: if the diffusion term is omitted altogether, the spatial order of the induction equation changes from second to first. When R_m is large but finite, it might be anticipated that boundary layer effects could be important, and this turns out to be very much the case.

Zeldovich and colleagues (see their book 'Magnetic Fields in Astrophysics', 1983) argued that slow dynamos would be no use at all in astrophysics, where the magnetic Reynolds number is so huge that any associated field would have grown by a only small factor during the lifetime of the Universe. Additionally, the 22-year period of the solar cycle is more suggestive of the turnover timescale than the diffusion timescale, which is of the same order as the Sun's age. These remarks were enough to send theoreticians off on a scramble to look for flows that would act as fast dynamos. A number of working dynamos were known by this time, but oddly enough on close inspection they all turned out to be slow. We now know there was a good reason for this – it turns out that to act as a fast dynamo a flow has to be chaotic, and such flows are not the ones that spring to mind when looking for tractable models.

For an excellent recent treatment of fast dynamo theory, see the book by Childress and Gilbert (1995). The point of view there is mainly analytic: in an attempt to complement this, and to reflect the current author's own background, the present chapter will mostly concentrate on a description of numerical work – the experimental evidence that needs to be incorporated into a general theoretical framework.

2.2. Fast dynamos and chaos

The earliest fast dynamo that was publicised as such is the Rope Dynamo, from Vainshtein and Zeldovich (1972). (Essentially the same idea had been introduced earlier by Alfvén.) Take a loop of magnetic field embedded in a highly conducting fluid and perform the sequence of operations shown in Fig. 2.1. In the first phase, the loop is stretched till it has twice its original length: assuming flux freezing, this doubles the field strength. Then twist the loop half a turn to get a figure-of-eight. Finally fold one half back on the other. The end result is a doubled loop with the same cross section as the original and twice the field strength. In principle these operations could take place as part of a more general fluid flow whose turnover time is that of

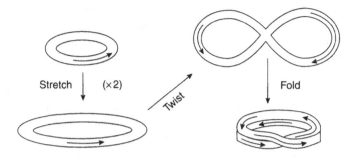

Figure 2.1 Vainshtein and Zeldovich's rope dynamo (see the text for details).

this stretch-twist-fold (STF) cycle. Repeating over and over again, the field strength grows by two every turnover, so that $B(t = n) = 2^n B(t = 0) = e^{n \log 2} B(t = 0)$. Thus the growth rate is log 2, provided the conductivity is high enough for flux freezing to hold, and then the dynamo is fast.

If this simple an example can do the trick, the reader may wonder what all the fuss is about. Certainly its co-inventor, Zeldovich, would become impatient with anyone who questioned what to him seemed so straightforward and obvious. But there is a possible snag to all this, and it centres on what happens at the knot where the crossover occurs. Near it, there will be strong gradients, and diffusion will eventually bite no matter how high the conductivity. There is a conflict between the $t \to \infty$ and the $R_m \to \infty$ limits: we should really be taking the first limit before taking the second when testing for self-exciting dynamo action. There is an intriguing paper by Moffatt and Proctor (1985), who exhibit a piecewise continuous flow mimicking this STF cycle. After just two or three iterations the details of the knot and the loop's cross-section become so contorted that any confident assertion about what will happen as $t \to \infty$ followed by $R_m \to \infty$ seems unwise to say the least. Some reconnection of the field at the knot has to happen in order to reestablish a simple field topology: Zeldovich was happy to wave his hands about that, whereas others have worried. Such reconnection cannot occur without a finite conductivity. The situation is similar to that in the Sun's corona, where resistive effects must occur in very small regions in order that the magnetic field topology can reconnect and relax. In fact, fast dynamos and fast reconnection are just two sides of the same coin.

The stretching part of STF is crucial: it is the actual part of the cycle where the amplification occurs, and suggests that flows with a lot of it are likely to be good fast dynamo candidates. Flows where stretching takes place exponentially fast are termed chaotic by nonlinear dynamical systems theorists. This idea can be quantified by objects known as Lyapunov exponents. Take two neighbouring fluid elements embedded in the flow, with relative vector displacement defining a so-called tangent vector. If the modulus of the tangent vector grows like $e^{\lambda t}$ as we follow the flow evolution, λ will in general fluctuate depending on location in the flow domain, but over long times it converges to an average value. This is a Lyapunov exponent, more formally defined as

$$\lambda = \lim_{t \to \infty} \frac{1}{t} \log_e \frac{\ell(t)}{\ell(0)},$$

where $\ell(t)$ is the length of the tangent vector, assumed infinitesimal at all finite times. Because in a three-dimensional flow there are three independent choices for the initial tangent vector, there are actually three Lyapunov exponents, and it can be shown that when the flow is incompressible they sum to zero. If one or more exponent is positive, the flow is chaotic. For more on this and other aspects of relevant dynamical systems theory, see Ott (1993).

Now imagine that this tangent vector is a piece of magnetic field line, and that there is perfect conductivity so that the field lines move with the fluid. Clearly exponential growth of the tangent vector is equivalent to exponential growth of magnetic field, explaining in loose terms why chaos is a necessary requirement for a flow to exhibit fast dynamo action (modulo the sorts of worries discussed in the last paragraph but one). Another way of seeing this is to view the fluid flow as a mapping taking any material point which is at $\mathbf{a}$ at time $t = 0$ to $\mathbf{x}(\mathbf{a}, t)$ at time t. A nearby point $d\mathbf{a}$ away will go to $\mathbf{x}(\mathbf{a} + d\mathbf{a}, t)$. Taylor expanding the latter expression to first order and writing $d\mathbf{x} = \mathbf{x}(\mathbf{a} + d\mathbf{a}, t) - \mathbf{x}(\mathbf{a}, t)$, we obtain (in suffix notation)

$$dx_i = \frac{\partial x_i}{\partial a_j} \, da_j.$$

By comparison, the well-known Cauchy solution to the diffusionless induction equation reads

$$B_i(\mathbf{x}, t) = \frac{\partial x_i}{\partial a_j} B_j(\mathbf{a}, 0).$$

These two equations are the same, showing again that exponential stretching of line elements is the same as exponential stretching of magnetic fields, for perfect electrical conductivity.

This deals with the stretching. But it is the folding that gives rise to the nagging worries about what an arbitrarily small diffusion might do, and if the flow takes place in a compact domain, folding is an inevitable part of the story. Magnetic fields are unlike line elements in that they have polarity. Field lines that started out far from each other will inevitably be brought close together, in fact arbitrarily close, if the flow is ergodic and one waits long enough. In general they will be misaligned and diffusion is inevitable at some level. Does this destroy the picture just outlined? What matters is the relative extent to which fields reinforce rather than cancel. Du and Ott (1993) give a heuristic description of this in terms of something they call a cancellation exponent. This is too technical to describe further here, but is summarised on pp190ff of Childress and Gilbert (1995). More fundamentally, there are two theorems, due to Vishik (1989) and Klapper and Young (1995). The first bounds the growth rate of any fast dynamo from above by the largest Lyapunov exponent of the flow, when the diffusion is arbitrarily small but finite. However, included in the spectrum of possible Lyapunov exponents are those associated with isolated stagnation points. One can have positive values for these in flows which would not otherwise be regarded as chaotic: steady axisymmetric stagnation point flow is a case in point. Klapper and Young proved a stronger result: the growth rate is bounded above by a quantity called the topological entropy. The exact definition of the latter is complicated, but loosely speaking it can be viewed as a Lyapunov exponent weighted by the relative fraction of time a typical particle trajectory spends in each part of the flow domain, so it is a better quantitative measure of chaos for these purposes. It effectively gives weight zero to the contributions from any hyperbolic stagnation points, since the time to attain or leave such points is infinite and typical fluid particles never get there. The end conclusion of all this is that chaos has been shown to be a necessary though not sufficient condition for fast dynamo action.

This has dire consequences for those trying to model fast dynamos. Chaos is equivalent to non-integrability of the flow, meaning particle trajectories cannot be evaluated in terms of simple functions or integrals. So any straightforward attempt to produce an analytic fast dynamo is doomed to failure. This is why the early dynamos were all slow: their authors used integrable flows, not wanting to shoot themselves in the foot before they even started. To find fast dynamos it is necessary either to proceed numerically, or to do something analytic but extremely clever.

2.3. Numerical calculations

Chaotic flows tend by nature to be unmanageable: turbulent flows, for instance, admit no simple rules whereby either the velocity field or the particle trajectories can be written down. However, there are chaotic flows known where at least the velocity can be written down simply, even though the particle trajectories have to be found numerically. Most popular for our purpose are the so-called ABC flows, whose velocities are given by the innocuous-looking formula

$$\mathbf{U} = (A \sin z + C \cos y, \ B \sin x + A \cos z, \ C \sin y + B \cos x).$$

This is defined in an infinite domain with 2π periodicity in all three directions, and if none of A, B or C vanish there are regions with chaotic particle trajectories (if one or more are zero, trajectories can be integrated in terms of elliptic functions).

How do we know when the flow is chaotic? For this flow family, the answer is: only from numerical experiments. These flows were introduced by Arnold (1965), determined to be chaotic by Hénon (1966), and analysed in detail by Dombre *et al.* (1986). One straightforward way to detect chaos is to calculate the Lyapunov exponents: there are well-established numerical algorithms to do this, which typically yield values ranging from zero for $ABC = 0$ to 0.2 or 0.3 in the best cases (the flows are normalised so that $A^2 + B^2 + C^2 = 3$). The flow incompressibility means that at least one exponent is zero, and the other two are equal and opposite: a positive value means chaos is present. A more revealing method is to make Poincaré sections, following particle trajectories repeatedly from one face of the cube to an opposite face, identifying planes modulo 2π. Every time the particle crosses the plane, a dot is drawn. Regions where the dots form closed curves are integrable, and the curves are called KAM-tori after Kolmogorov, Arnold and Moser who gave the theorem accounting for their existence. The regions where the dots look ergodic are the chaotic parts of the domain. Dombre *et al.* (1986) contains numerous examples of such plots for a variety of A, B, C values, and in the current chapter examples are given for other flows in Figs. 2.4 and 2.7.

Given that the ABC flows are chaotic, they are good candidates for fast dynamos. Arnold and Korkina (1983) were the first to investigate this, followed by Galloway and Frisch (1984, 1986). The idea is to solve the induction equation (1) prescribing the ABC velocity field for **U**, which is assumed inexorable (the kinematic approximation). One tries to make R_m as big as possible and hopes to obtain evidence that the growthrate tends to a finite positive value. The problem is linear and can either be solved as an eigenvalue problem or as a timestepping problem. Arnold and Korkina did the former, using an inverse iteration method. Galloway and Frisch did the latter, exploiting the fact that the flows contain only components with wavenumber 1, so that $\nabla \times (\mathbf{U} \times \mathbf{B})$ can be cheaply evaluated when a spectral method

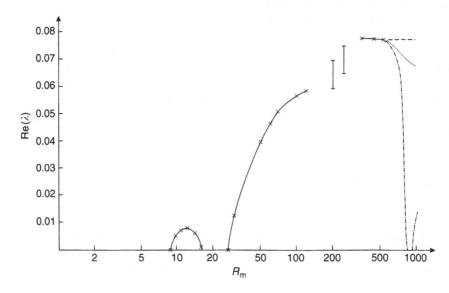

Figure 2.2 Growthrates for the $1:1:1$ ABC dynamo, plotted as a function of R_m. Note the two windows of dynamo action, where the generated fields have different symmetries (from Galloway and Frisch, 1986).

is used. At that time, it was possible to reach R_m values up to 500, using $(64)^3$ resolution. In a timestepping method, a log/linear plot of the total magnetic energy against time yields an overall slope equal to twice the real part of the growthrate, and if there are oscillations their frequency gives twice the imaginary part. The results of this procedure are shown in Fig. 2.2, which gives the observed real part of the growthrate as a function of R_m.

Regarding this plot, we see that it is something of a dog's dinner as far as giving evidence for fast dynamo action goes – asymptopia for high R_m has clearly not been reached. Since then, Lau and Finn (1993) have gone up to $R_m = 1000$ and claim to have somewhat more compelling evidence, although this is rather an act of faith; see also Dorch (2000) and Archontis (2000). The point is that the $1:1:1$ flow turns out to be non-optimal for this purpose: it happens to have chaotic regions and Lyapunov exponents that are very small. The set $A:B:C = 5:2:2$ has much more chaos and does provide better evidence (Galloway and O'Brian, 1993), with minimal change in the growthrate between $R_m = 200$ and 800.

Nonetheless, the $1:1:1$ case is much the most studied, and is interesting for a number of reasons. Fig. 2.2 shows that there are two windows of dynamo action, the first for R_m between 8 and 18 and the second for $R_m > 27$. Arnold originally selected $1:1:1$ for study because as a mathematician he was interested in the symmetries of the flow and its associated magnetic fields: the case of equal A, B, C has a bigger symmetry group and appealed to him for that reason. In the first window all the symmetries are preserved, and the frequency is relatively high: in the second, symmetries are broken and the frequency is lower, possibly becoming zero around $R_m = 300$, according to Lau and Finn. It may well be that excessive symmetry is a disincentive to dynamo action (cf. Cowling's Theorem), so perhaps it is not surprising that the mode in the second window has less of it.

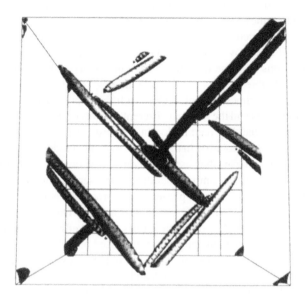

Figure 2.3 Isosurface plot of the surface $|\mathbf{B}^2| = 0.17 B_{\mathrm{max}}^2$, for the mode in the second dynamo window with $R_{\mathrm{m}} = 100$ and $A : B : C = 1 : 1 : 1$. A moment has been chosen where double cigars are clearly visible. The periodicity is apparent, and the two cigars associated with the stagnation point at $(7\pi/4, 7\pi/4, 7\pi/4)$ appear at top right (from Galloway and O'Brian, 1993).

Passing on to the nature of the associated eigenfunction, one comes up against visualisation problems. For these chaotic problems the field lines are often a mess, and one useful thing is to plot isosurfaces of magnetic intensity B^2 as in Fig. 2.3. This is a surface on which B^2 reaches some specified fraction of its maximum over the whole cube (here the fraction is 0.17), at that particular instant in time. The obvious things in Fig. 2.3 are the extended cigar-like features, and the question is why are they there? For $A : B : C = 1 : 1 : 1$ the answer is that they are related to stagnation points in the flow pattern. The existence of stagnation points is dependent on whether A^2, B^2 and C^2 can form an acute-angled triangle, with $C^2 < A^2 + B^2$, etc. If they can, there are eight stagnation points; if not, there are none. When there are eight stagnation points, these divide into two classes. To look at these, one can linearise the flow in the neighbourhood of each of these points, and determine the nature of the fixed points, which are all saddle-points with three real eigenvalues. These eigenvalues always sum to zero because the flow is incompressible. If two are negative and one positive, two eigendirections suck fluid in along a plane, and the third eigendirection pushes fluid out along an axis. In the jargon of the subject, which is worth learning on this point, there is a two-dimensional stable manifold and a one-dimensional unstable manifold. If two eigenvalues are positive and one negative all these statements are reversed, i.e. there is a two-dimensional unstable manifold and a one-dimensional stable one. We will refer to these different sorts of stagnation point as Type 1 and Type 2, respectively. The importance of the distinction between them was emphasised by Childress and Soward (1985).

The Type 1 stagnation points catch any magnetic fields in their vicinity, sweeping them in toward the axis and stretching them along it. This accounts for the cigar-like concentrations that are observed. The characteristic size of the flux concentrations scales as $R_{\mathrm{m}}^{-1/2}$, a first indication of the part played by boundary layers in this story. As $R_{\mathrm{m}} \to \infty$ these features get smaller and smaller, but they are always there! The Type 2 stagnation points have the opposite tendency: they tend to remove fields from their vicinity. These basic processes were studied for non-axisymmetric magnetic fields embedded in a steady axisymmetric stagnation point flow by Galloway and Zheligovsky (1994), and solutions were found in terms of confluent hypergeometric functions. In some ways the behaviour is that of a fast dynamo, although the examples there are undoubtedly cheating because the flow is unbounded for large r, z.

Why are the cigars sometimes multiple? This is a little complicated but is connected with the various flow symmetries and the fact that different pockets of the flow domain can feed the same stagnation point – what they feed in does not necessarily have the same sign, and one gets aligned cigars of opposite polarity. In this second dynamo window, the eigenfunctions are symmetry-breaking, which means they are degenerate – several exist with the same complex growthrate. Starting from generic initial conditions, one sees a complicated superposition with different amplitudes and different phases. Such processes have recently been greatly clarified using better computer visualisation techniques, by Dorch (2000) and Archontis (2000), who manage to exploit plots of the field lines to good effect. These authors also shed light on the scale-separated ABC dynamos first discussed in Galanti *et al.* (1992). In these, the magnetic field is periodic not over one ABC cell but over several of them. Features from one cell can be then fed into neighbouring cells, and in certain circumstances the growth rates can be larger than in the non-scale-separated case. The reader is referred to the above papers for the current state of the art for fully three-dimensional ABC dynamos.

How can more convincing examples than this $1 : 1 : 1$ ABC flow be constructed? As mentioned, $A : B : C = 5 : 2 : 2$ does better, with a growthrate varying by around 1.5% between $R_m = 200$ and 800. For these values, there are no stagnation points and the eigenfunctions consist of more extensive flux sheets, also with characteristic thickness scaling as $R_m^{-1/2}$. Another three-dimensional flow that looks a good candidate is the 'sines' flow, the $1 : 1 : 1$ ABC flow with the cosines left out. This has big chaotic regions and cigar-like eigenfunctions concentrated around stagnation points. With half as many terms it is twice as cheap to compute. (Fig. 2.5 contains a plot of the growthrate up to $R_m = 800$ as a sideline to its main point.) Recently, Archontis (2000) has reexamined this flow with more computing power, and has also found intriguing results for a nonlinear version of the problem. All these studies suffer from a need for large computational resources to deal with the combination of the three-dimensional flow and the need to resolve its associated **B**. Time and storage requirements scale very unfavourably in three dimensions and quite substantial increases in computer power allow only modest increases in R_m. Can we somehow make do with a two-dimensional calculation?

It is known that two-dimensional steady flows cannot have chaotic regions, but two-dimensional unsteady ones can. This suggests we might be able to get fast dynamos using the latter. Take for instance the CP (for circularly polarised) flow

$$\mathbf{U} = \left[A \sin(z + \sin t) + C \cos(y + \cos t), \; A \cos(z + \sin t), \; C \sin(y + \cos t) \right],$$

which is an ABC flow first made integrable by setting $B = 0$ and then made non-integrable again by introducing the time-dependent phasing (Galloway and Proctor, 1992). This has three components but depends on only two space coordinates, a situation sometimes referred to loosely as 2.5-dimensional. Fig. 2.4 shows a Poincaré section for this flow: note the KAM curves as well as the extensive chaotic regions.

There is a two-dimensional analogue of Cowling's theorem (see Zeldovich *et al.*, 1983), telling us that any dynamo this velocity generates must depend on all three space coordinates.

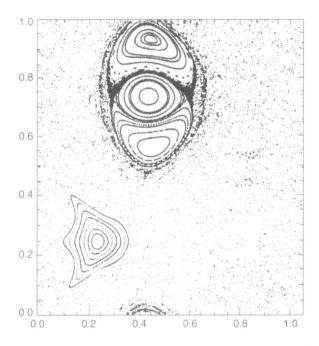

Figure 2.4 Poincaré section for the CP flow in the (y, z) plane: the x-component of **U** has been ignored. A dot is drawn every time a particle trajectory has advanced another 2π in time, with (y, z) values taken modulo 2π in space. Several initial conditions have been followed forward in order to show both the chaotic regions and the KAM tori.

But the beauty of this flow and others like it is that we are free to specify the x-dependence: if we take modes of the form $\mathbf{B}(x, y, z, t) = e^{ikx}\mathbf{B}_k(y, z, t)$, then each $\mathbf{B}_k$ evolves independently of all the others. This is easily verified by substituting this form into the induction equation for **U** as given above. So we can fix k and solve the induction equation with a two-dimensional code. Some experimentation is necessary, but the k with the best growthrate is $k = 0.57$. Following the dynamo with this k up to higher and higher R_m yields Fig. 2.5, showing that the CP flow has an almost constant growthrate between $R_m = 100$ and $R_m = 20,000$ (since then, we have confirmed the same value for $R_m = 50,000$). This and a very similar dynamo studied by Otani (1993) provide the best empirical evidence to date that fast dynamos exist. In relation to the theorems announced earlier, it should be noted that for the CP flow given, the largest Lyapunov exponent is 0.23 (note there is an error in the value given in Galloway and Proctor (1992), which is out by a factor 2π). The growthrate real part is 0.297, and the topological entropy is around 0.33. So this flow (as well as several others) manages to defeat Vishik's (1989) result because of the stagnation points, but bears out Klapper and Young (1995).

Looking at the eigenfunction, Fig. 2.6 shows contours of B_x in the plane $x = 0$. Again the flow has stagnation points, but now they are not fixed in space but instead precess around in circles. A fluid particle leaving the neighbourhood of one stagnation point along its unstable direction (unstable manifold) approaches another stagnation point along its stable manifold. But everything is now unsteady, and the fluid element makes increasingly wild excursions

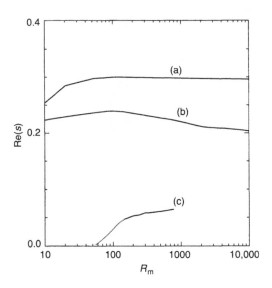

Figure 2.5 Real part of the growthrate as a function of R_m for the CP flow with $k = 0.57$ is shown in (a). (The growthrate of a related 2.5-dimensional flow is shown in (b), and (c) gives the growthrate for the 'sines' flow $\mathbf{U} = (\sin z, \sin x, \sin y)$.) (From Galloway and Proctor, 1992.)

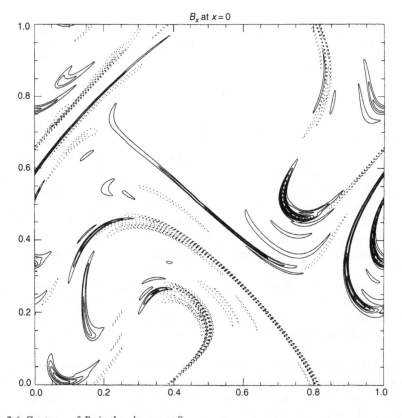

Figure 2.6 Contours of B_x in the plane $x = 0$.

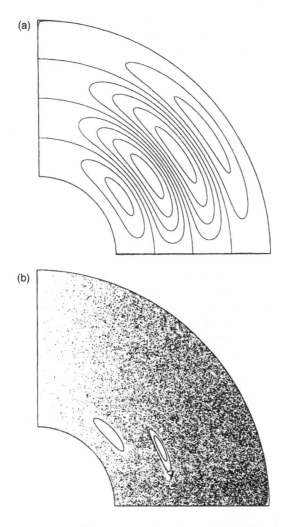

Figure 2.7 (a) Instantaneous streamlines, at $t = 0 \bmod 8$, for the spherical shell flow given in the text (meridional part only). (b) Poincaré section with points plotted at $t = 0 \bmod 8$ (from Hollerbach *et al.*, 1995).

to one side or another as it approaches, like a mad version of the zigzag stitch on a sewing machine. The magnetic field is like the thread, and is pulled backwards and forwards in increasingly violent folds whose adjacent sides are positive and negative. In addition to stretching and folding, there is strong shearing; thus such dynamos are termed 'stretch-fold-shear' (SFS). At finite R_m the folds of this concertina have characteristic thickness scaling as $R_m^{-1/2}$, and as $R_m \to \infty$ the result is a fractal. The marvellous thing is that all this can happen while the growthrate tends to a finite positive limiting value. The associated process is known in nonlinear dynamics as heteroclinic tangling, because heteroclinic orbits are orbits that join different stagnation points. The discussion here is necessarily intuitive, but more details can be found in Childress and Gilbert (1995, chapter 8).

The last calculation can be adapted to yield a candidate fast dynamo in a more astrophysically relevant geometry, the spherical shell. It turns out that all the same 2.5-dimensional tricks work if we replace the dependence of **B** on e^{ikx} in the Cartesian case by a dependence on $e^{im\phi}$ in the spherical case, where (r, θ, ϕ) are spherical polar coordinates. The same idea of having an axisymmetric but time-dependent velocity field leads us to take

$$\mathbf{U} = \nabla \times [r^{-1} f(r, t) \sin \theta \cos \theta \mathbf{e}_\phi] + r\omega(r, t) \sin \theta \mathbf{e}_\phi,$$

with $f(r, t) = 0.5(r-0.5)(r-1.5) \sin[4\pi r + \sin(0.25\pi t)]$ describing the meridional motion and $\omega(r, t) = 2\sin[4\pi r + \sin(0.25\pi t)]$ the rotation. This motion is confined to the spherical shell $0.5 \leq r \leq 1.5$, and has time period 8: flow contours and Poincaré sections are shown in Fig. 2.7. Again, there is plenty of chaos, and the interior stagnation points are shaken around by the time dependence.

The magnetic field is determined by using an expansion in Associated Legendre Polynomials $P_n^m(\theta)$ with $m = 1$ as the prescribed azimuthal dependence, and finite differences in r. The boundary conditions are that the interior **B** matches to a potential field at $r = 0.5$ and $r = 1.5$. The resulting growth rate as a function of R_m is shown in Fig. 2.8, along with a formula that seems to fit it extremely well. On this basis, the dynamo looks fast, with an asymptotic growth rate of 0.21 – but even at $R_m = 500,000$ the value has not settled down! It is an interesting puzzle as to why this dynamo is so much more grudging than its Cartesian cousin. The eigenfunctions for the three components of the field are shown for two R_m values in Fig. 2.9; again, there is evidence for the heteroclinic tangling process, though it is less obvious than in the Cartesian version (Hollerbach et al., 1995, with the help of a large Cray T-3D).

It remains inconceivable that a flow such as the last one would actually arise in reality, and more recently, there have been attempts to use flows that could arise out of natural physical

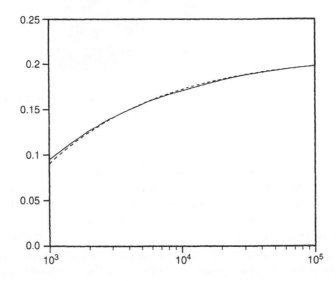

Figure 2.8 Solid line: observed growthrate for the dipolar $m = 1$ mode, plotted against R_m on a log scale, for the candidate spherical fast dynamo. The dotted line shows the curve $0.21 - 0.12(1000/R_m)^{1/2}$ (from Hollerbach et al., 1995).

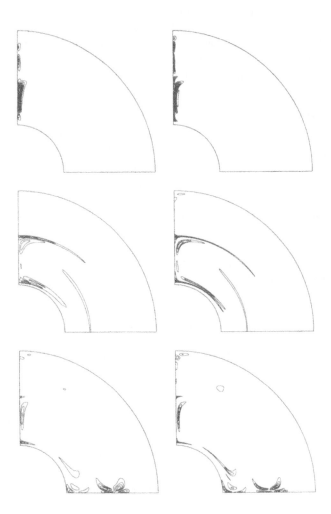

Figure 2.9 Contours of field components B_r (top), B_θ (centre) and B_ϕ (bottom), for the candidate $m = 1$ spherical fast dynamo. Left: $R_m = 10,000$; right: $R_m = 40,000$ (from Hollerbach *et al.*, 1995).

processes. Thus Kim *et al.* (1999) have analysed dynamo action for flows that are asymptotic solutions of the problem of convection in a rotating cylindrical annulus (cf. Busse, 1970). At the magnetic Reynolds numbers for which computations are possible, the flows are not in fact dynamos, but arguments are given as to why they might be expected to yield fast dynamos for much higher R_m values. Ponty *et al.* (2000) consider a flow intended to model the solar tachocline, a layer at the base of the Sun's convection zone where helioseismology tells us there are sharp changes in the rotation rate. A key role is played by the Ekman layer that arises above the bottom boundary. This is subject to hydrodynamic and/or thermal instabilities, and these are assumed to equilibrate and retain the form of the linear eigenfunctions in the nonlinear regime. They are then used as the basis for the calculation of kinematic dynamos.

Depending on the type of instability and the latitude (the calculations are done in the so-called f-plane approximation), both slow and fast dynamos appear to be possible. Again, it is hard to reach the appropriate asymptotic regime numerically, and the identification of the type of dynamo is based as much on the nature of the solutions, which in some cases show heteroclinic tangling, as on detailed measurements of the growthrates.

2.4. Some analytic results

What kinds of tricks have been used to circumvent the need for non-integrability and make analytical progress? In nearly all cases, the models treated have been almost integrable, and the non-integrability has been sneaked in via some singular feature. This means the flows used are often open to criticism because they have artificial aspects, but after seeing those used by the numerical people, the reader may feel this is the pot calling the kettle black. Childress and Gilbert devote most of their 400 pages to analytic calculations: here we single out two models as being indicative. The first is a helical dynamo due originally to Ponomarenko (1973); the second is the Roberts (1972) cell, actually an integrable ABC flow in its simplest form. There is also a class of models which use mappings rather than actual flows. In all of these the object is to break the problem up into piecewise-integrable sections or outer flows and boundary layers, and then try to match the bits together. Often, great ingenuity is involved.

Ponomarenko's dynamo, which was shown to be fast by Gilbert (1988), uses a cylindrical geometry, and in its most basic form has the wonderfully simple velocity field

$$\mathbf{U} = (0, r\omega, W), \quad (r < a),$$
$$= 0, \qquad\qquad (r > a),$$

where we use cylindrical polar coordinates (r, θ, z) and ω, W are constants. Inside $r = a$ the particle paths are helices and are easily integrable; outside integration is even easier! At any finite R_m the induction equation using this $\mathbf{U}$ can be solved in each subdomain and the solutions can be patched at $r = a$, using the continuity of the field components and appropriate derivatives. For $r > a$ one takes $\mathbf{B} \to 0$ as $r \to \infty$ as a boundary condition, and at $r = 0$ one demands the solution is regular. Because $\mathbf{U}$ is independent of θ and z, one can use separable solutions of the form

$$\mathbf{B} = \mathbf{b}(r)e^{im\theta + ikz + st},$$

where each pair of (m, k) values will evolve independently of any others and have its own growthrate s. The cylindrical geometry means that the solutions involve (modified) Bessel Functions, and the eigenvalues s are determined by the properties of their zeroes. At high R_m the latter's asymptotic forms can be exploited. The case where (m, k) satisfy $m\omega + kW = 0$ is slightly easier algebraically and not fundamentally different from the general case. Note that because m, k contribute terms of order m^2/R_m, k^2/R_m in the diffusion term, the natural scaling so that these keep on mattering for higher and higher R_m is $m, k \sim R_m^{1/2}$. This fact shows up in the asymptotic formula for s, which is

$$s(m, k, R_m) \sim \pm R_m^{-1/3} \left(\frac{im\omega}{2a} \right)^{2/3} - R_m^{-1} \left(k^2 + \frac{m^2}{a^2} \right).$$

Any fixed (k, m) is a dynamo with a growth rate tending to zero as $R_m \to \infty$, because for large enough R_m the first term on the right-hand side dominates and the plus sign is a dynamo.

However, as R_m increases a new larger (m, k) pair can pick up the baton. The formula shows that as long as (m, k) are selected so they scale like $R_m^{1/2}$, the growthrate asymptotes to a finite value – in other words, the dynamo is fast.

The fastness of this dynamo relies on the fact that there is a tangential discontinuity in velocity at $r = a$. This gives diffusion an arbitrarily small scale to exploit. If the discontinuity is smoothed over a finite distance ϵ, for instance by using $1 - \tanh[(r - a)/\epsilon]$ as a velocity profile, the dynamo looks fast until R_m is of order $\epsilon^{-1/2}$, but for higher R_m still, it becomes slow. The fastness is thus completely due to the discontinuity. Note again the subtlety of the double limits $\epsilon \to 0$ and $R_m \to \infty$, and the importance of the order in which they are taken. For recent work on generalisations of slow Ponomarenko dynamos, see Gilbert and Ponty (2000).

The Roberts Cell exhibits similar pathologies, though with a twist in its tail. In its simplest form, this velocity field is just the integrable ABC flow

$$\mathbf{U} = (\sin z + \cos y, \cos z, \sin y),$$

although it is often expressed slightly differently in coordinates rotated by $\pi/4$. This flow was shown to be a dynamo by Roberts (1972), and was the subject of an ingenious analysis by Soward (1987). Like the CP flow referred to in Section 2.3, the problem is 2.5-dimensional, with modes expressible as

$$\mathbf{B}(x, y, z, t) = e^{ikx}\mathbf{B}_k(y, z, t).$$

Soward showed that the field was confined in boundary layers at the edges of each cell, the whole array of cells again being treated as a 2π-periodic network. By a virtuoso performance with matched asymptotic expansions, he was able to establish that the most unstable mode has a k scaling as $R_m^{1/2}$ and that the associated growthrate scales (incredibly) as

$$s = \frac{\log(\log R_m)}{\log R_m}.$$

This is not fast, but is certainly as close as makes no difference. For example, if $R_m = 10^{20}$, the growthrate is around 0.1, in turnover time units! Clearly, a kinematic dynamo based on this flow would be astrophysically viable, even though technically it is slow. (Of course, that such a flow could arise naturally is inconceivable.) Soward was also able to show that the addition of a logarithmic singularity in vorticity at the stagnation points is sufficient to make this dynamo fast, although again any fixed finite k is eventually slow, and again the fastness is explicitly due to the flow singularity.

The last two examples are the most interesting analytic fast dynamos that work with what could be considered as real flows. There are also a fair number of more artificial dynamos based on mappings rather than actual flows. Some of these are very ingenious: an extensive discussion is given in Childress and Gilbert (1995), and there is also an excellent review by Bayly (1994).

2.5. Including the Lorentz force

At the large values of R_m for which fast dynamos are relevant, the Lorentz force rapidly gets out of hand. We have seen that features with characteristic size $R_m^{-1/2}$ are typical in both numerical and analytical examples. If a fluxrope or sheet has length scale $R_m^{-1/2}$ and

field strength B, the current goes like $R_m^{1/2}B$, and unless the associated vectors are perfectly aligned, the Lorentz force scales like $R_m^{1/2}B^2$. Objections similar to this have caused
Vainshtein and Cattaneo (1992) and Kulsrud and Anderson (1993) to criticize dynamo theory
in general, and mean field theory in particular, as giving out at this level far before enough
flux has been produced to match observations. A key quantity is the ratio $q = \langle B^2 \rangle / \langle B \rangle^2$, the
issue being that at high R_m this adopts a huge value, so that the mean field is minuscule. The
problem is particularly extreme for the mean field dynamos commonly used to explain the
fields observed in spiral galaxies (though see Brandenburg, 1994 for the case for the defence).

The calculations underlying these criticisms are mostly based on models of magnetoconvection with an imposed mean field. It is useful to look at non-mean-field dynamos to
see whether a flow which generates its own field can get round these objections. The basic
hope (Galloway *et al.*, 1995) is that once the stage is reached where the Lorentz force begins
to halt the flow locally in $R_m^{-1/2}$-thickness structures, the chaotic flow elsewhere will keep
shovelling more field in so that the structures fatten into stagnant plugs with $R_m^{-1/2}$-thickness
edges. This process continues until saturation occurs when the field is globally and not just
locally hindered. We have started to compute such dynamos (Podvigina and Galloway, in
preparation). The models used are the three-dimensional ABC dynamos referred to earlier,
modified to include the Lorentz force. Similar work was performed by Galanti *et al.* (1992).
The only difference is that we are considering the case of high magnetic and low kinetic
Reynolds numbers, in an attempt to have a flow which is stable (i.e. an attractor) if left to its
own devices, but which nonetheless wants to be a dynamo. The equations being solved are
the induction equation

$$\frac{\partial \mathbf{B}}{\partial t} = \nabla \times (\mathbf{U} \times \mathbf{B}) + \frac{1}{R_m}\nabla^2 \mathbf{B}$$

and the momentum equation

$$\frac{\partial \mathbf{U}}{\partial t} + \mathbf{U} \cdot \nabla \mathbf{U} = -\nabla P + \mathbf{J} \times \mathbf{B} + \mathbf{F} + \frac{1}{R_e}\nabla^2 \mathbf{U}.$$

Everything is 2π-periodic and the force is necessary to stop the ABC flow from running
down due to viscosity: thus

$$\mathbf{F} = (A\sin z + C\cos y,\ B\sin x + A\cos z,\ C\sin y + B\cos x)/R_e.$$

In the absence of magnetic fields the ABC flow is a solution to the Navier–Stokes equation
with this forcing field. In the case $A:B:C = 1:1:1$, this solution is hydrodynamically
unstable when the kinetic Reynolds number R_e exceeds 13.09 (Galloway and Frisch, 1987).

Some results are shown in Fig. 2.10 (evolution of total kinetic and magnetic energies, in
common units) and Figs. 2.11 and 2.12 (structure of the solution at two different times or isosurface levels). The kinetic and magnetic Reynolds numbers based on the non-magnetic
velocity are 5 and 400, respectively. The ABC parameters are 1, 1, 1 though we have
computed other cases with and without stagnation points.

The main comments about these results are:

1   At these R_m values, the magnetic field generated is substantial and its total energy is
    comparable with, but at most times greater than, that of the flow.
2   There is no steady state. At the beginning of the calculation, the seed field grows at the
    rate predicted by linear theory. Flux ropes develop around stagnation points. These grow

in strength until the Lorentz force bites. Thereafter, structures come and go around the various stagnation points (there are four of the appropriate kind), building up in strength until they are destroyed by instabilities and/or fast reconnection. This process keeps repeating, giving a dynamo which is erratically cyclic.

3   Once the calculation has settled down, there are times when the magnetic field is growing on the turnover timescale, and times when it is being destroyed. During the latter the field is undergoing fast reconnection. There has been some debate over when and how this process is possible, with a key role being played by the boundary conditions (see Priest and Forbes, 2000). The current calculations show that fast reconnection does occur in natural circumstances, where the system finds its own boundary conditions through the assumed periodicity.

4   The ratio q is still large in these calculations – in fact it is infinite because there is no mean field! But this is irrelevant for this kind of dynamo, in which the mean field is a conserved quantity of the calculation. Indeed, any astrophysical object has zero mean field when averaged over a sufficiently large volume of space – see the contribution by Hoyng in this volume for a discussion of what is involved in defining a mean field. Although the question remains open as to the effectiveness of mean field dynamos, there certainly appear to be other dynamos that can fill substantial fractions of space with strong fields, at least at R_m's of 400 and probably at much higher values too.

However, in astrophysics R_e is large as well as R_m, and the natural question is whether the above behaviour persists in this case too. Here the picture is less rosy. It is possible to derive scaling laws for the ratio of total magnetic to total kinetic energies over one fundamental ABC cube. For the moment, we ignore the time dependence and assume that the following

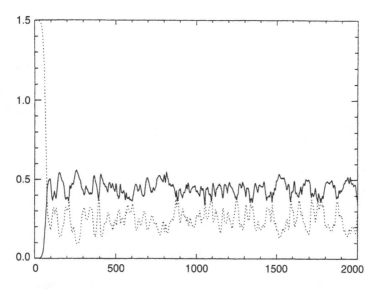

Figure 2.10 Evolution of total magnetic (solid) and kinetic (dashed) energies as a function of time, for the dynamo discussed in this section. The values are normalised so that with no magnetic field the kinetic energy is 1.5, and a seed magnetic field is added at $t = 0$. Time is measured in units of the flow turnover time; in these units, the magnetic diffusion time is of order R_m, here equal to 400.

Figure 2.11 Plots of a magnetic energy isosurface at time 1963 for the solution described in the text. Illustrated is the surface on which the magnetic energy level reaches 10% of its maximum value at that particular time.

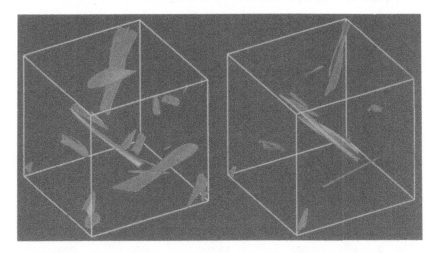

Figure 2.12 Plots of magnetic energy isosurfaces at times 1963 (left) and 2000 (right) for the solution described in the text. Illustrated are the surfaces on which the magnetic energy level reaches 25% of its maximum value at that particular time.

arguments hold in some average sense. Observe that the rate of working by the forcing must exceed the ohmic dissipation or the viscous dissipation when taken separately (cf. the arguments for field strength limits in magnetoconvection given by Galloway *et al.*, 1977). This provides an upper bound for the ohmic dissipation, and hence for the ratio of energies that we are after. The work rate is the integral of $\mathbf{F} \cdot \mathbf{u}$ over the cube, and we will get an overestimate if we assume that $\mathbf{u}$ is the undisturbed ABC flow ignoring the back reaction. Working in dimensional units with a length scale L for the cube, this quantity is

$$W = \frac{3F_0^2 L^5 \rho}{4\pi^2 \nu},$$

where ρ is the density, F_0 is the magnitude of the forcing, and ν is the kinematic viscosity. It is easy to verify that with no magnetic field this is equal to the viscous dissipation. The ohmic dissipation depends on whether the flow is just coming out of the kinematic regime, or whether a plug of field with a stagnant interior and a peripheral fall-off has formed. In the former case the flux cigars have characteristic radius $R_{\mathrm{m}}^{-1/2} L$ and length L, with associated total ohmic dissipation of order

$$\frac{B^2}{\mu_0^2 \sigma \epsilon^2 L^2} 4\pi \epsilon^2 L^3,$$

assuming 4 cigars. Equating this to the power available from the force gives an upper bound for the field strength in the cigars, and hence for the total magnetic energy. The total kinetic energy will be less than, but of the same order as, that in the absence of magnetic field, and this can also be calculated in terms of F_0. The result is that the ratio of total magnetic energy to total kinetic energy scales like $1/R_{\mathrm{e}}$, independent of the magnetic Reynolds number. This is bad news, particularly as all the input hypotheses are likely to overestimate the true ratio.

Almost as bad news results if instead field is allowed to build up in fat cigars of order ΔL in radius, with outer sheaths of thickness order $R_{\mathrm{m}}^{-1/2} L$ in which all the ohmic dissipation occurs. This idea is based on the analogous process which occurs in axisymmetric magnetoconvection (Galloway and Moore, 1979); the scenario only makes sense if $\Delta \gg \epsilon$. The ohmic dissipation is now of order

$$\frac{B^2}{\mu_0^2 \epsilon^2 \sigma} 2\pi \Delta \epsilon L,$$

per cigar. Equating this to the available power gives a maximum field strength which is reduced by a factor of $(\epsilon/\Delta)^{1/2}$ compared to the earlier case. However, the increased volume occupied by the field more than outweighs this when the magnetic energy is calculated. The best possible case is if Δ is of order 1, and then it is found that the ratio of total magnetic to kinetic energies scales as $(\nu/\eta)^{1/2} R_{\mathrm{e}}^{-1/2}$. This is better than the earlier case, but unless one can think of reasons why $\nu \gg \eta$ – not normally the case in most of astrophysics – there is still a problem.

Whilst no quantitative comparison has been attempted, these scaling laws reflect the general trend in the numerical results published by Galanti *et al.* (1992). They also agree with the results of alternative arguments presented independently by Brummell *et al.* (2000). These authors describe the results of a series of calculations where a three-dimensional time-dependent ABC forcing is used, and the back reaction of the Lorentz force is again included.

Boundary layer arguments are used to derive a scaling law which is the same as that for the fat-cigar case above. The most surprising result is that there are cases where an initial phase of kinematic growth is followed by a nonlinear phase where the flow changes to quench the dynamo, which then dies away completely. The flow starts as one thing, and evolves via a long but transient dynamo phase into something completely different which is a non-dynamo! The non-magnetic problem apparently has two stable solutions, and a seed magnetic field triggers the transition from one to the other.

What is the reason for the disappointing scaling laws, and can anything be done about it? The cause is the assumption that there is a nice laminar flow with a huge kinetic Reynolds number. In the non-magnetic case, there is a corresponding build-up of a vast amount of kinetic energy; alternatively stated, it requires only a tiny force to sustain the flow. This would not happen in reality; even at moderate Reynolds numbers, the ABC flow is unstable and changes into something else, and any real astrophysical flow is likely to be turbulent and to have orders of magnitude more viscous dissipation than a simple laminar flow at a Reynolds number of 10^{10} or more. It is the kinetic energy that is overestimated, rather than the magnetic field being underestimated. The only long-term way of improving this is to develop a satisfactory theory of MHD turbulence, which may be akin to waiting for the ark to come home. In the meantime, some kind of modelling recipe is necessary. This has been highlighted by Archontis (2000), who recognised the nature of the problem and gave a recipe for adjusting the forcing level. The procedure is quite ad hoc, but gives better results than doing nothing. An extreme position would be to take a Schatzmannian point of view and say that most astrophysical phenomena have an effective Reynolds number of around 100, because that is the order of magnitude estimate of the value which leads to instabilities on the next scale down. Then the fat-cigar scaling law gives an energy ratio of order a tenth, which is not a big problem. This is a Draconian point of view, but it makes more sense than using the laminar estimates.

2.6. Conclusion

At this stage one can say that the kinematic aspects of the fast dynamo story seem to be more or less established: fast dynamos exist, and most of the examples display $R_m^{-1/2}$ thickness features which can be intricately mixed together in the high R_m limit. These may be either cigars or sheets, depending on the geometry and the presence or absence of hyperbolic stagnation points. Some of the generation mechanisms, such as heteroclinic tangling, have been identified and qualitatively understood. The subject has achieved enough maturity that it now has a good textbook (Childress and Gilbert, 1995).

In astrophysics the sort of contrived flows so far used cannot be expected to arise naturally. The real flows out there are genuinely turbulent. This is good, because turbulent flows are chaotic, both in the everyday and mathematical senses of the word – any dynamos they produce are likely to be fast. But calculating them numerically is a formidable task. So far there are tentative beginnings at using more realistic flows, and simulations of localised parts of a star. As computers evolve, the models will become progressively more realistic.

Still under debate are the effects of the Lorentz force, and the basic input that fast dynamo theory needs to provide to parametrised models such as α–ω dynamos. At the moment, the latter are the only means we have to estimate the global behaviour of stellar magnetic fields, yet we know that the basic assumptions made in deriving values for the regeneration coefficient α and the turbulent diffusion coefficient β are not satisfied in typical astrophysical objects. The fastness of the actual dynamo process enters in the values given to these quantities. So understanding how small-scale motions conspire to yield these mean values is an important and largely uncharted problem for fast dynamo theory.

Concerning the effects of the Lorentz force on both large- and small-scale field generation, the sorts of model described in the last section face the difficulty that they give too little magnetic flux to be astrophysically viable in the high kinetic Reynolds number limit. This is true even when the magnetic Reynolds number is very high. The probable reason is that the estimates for the actual viscous dissipation are much too small, due to the assumption of laminar viscosities. The development of better theories of MHD turbulence is ultimately the only way to solve this difficulty; meanwhile, all we can do is use reasonable recipes to bump up the diffusion coefficients so that they give better agreement with the dissipative processes that seem to be going on in real astrophysical objects.

Acknowledgements

This chapter started as a set of notes I prepared for two extended lectures given as part of a course 'Topics in Solar Physics' organised at JILA, University of Colorado, Boulder, by Tom Bogdan and Paul Charbonneau. I thank both for their agreement in allowing me to use the notes as the basis for this chapter.

References

Archontis, V., "*Linear, Non-Linear and Turbulent Dynamos*," PhD Thesis, University of Copenhagen (see www.astro.ku.dk/ bill) (2000).

Arnold, V.I., "Sur la topologie des écoulements stationnaires des fluides parfaits," *C. R. Acad. Sci. Paris* **261**, 17–20 (1965).

Arnold, V.I. and Korkina, E.I., "The growth of a magnetic field in a steady incompressible flow," *Vest. Mosk. Un. Ta. Ser. 1, Matem. Mekh.* **3**, 43–46 (1983).

Bayly, B.J., "Maps and dynamos," in: *Lectures on Solar and Planetary Dynamos* (Eds. M.R.E. Proctor and A.D. Gilbert), Cambridge University Press, pp. 305–329 (1994).

Brandenburg, A., "Solar dynamos: computational background," in: *Lectures on Solar and Planetary Dynamos* (Eds. M.R.E. Proctor and A.D. Gilbert), Cambridge University Press, pp. 117–159 (1994).

Brummell, N.H., Cattaneo, F. and Tobias, S.M., "Linear and nonlinear dynamo properties of time-dependent ABC flows," *Fluid Dynam. Res.* **28**, 237–265 (2001).

Busse, F.H., "Thermal instabilities in rapidly rotating systems," *J. Fluid Mech.* **44**, 441–460 (1970).

Childress, S. and Gilbert, A.D., *Stretch, Twist, Fold: The Fast Dynamo*, Springer (1995).

Childress, S. and Soward, A.M., "On the rapid generation of magnetic fields," in: *Chaos in Astrophysics* (Eds. J.R. Buchler, J.M Perdang and E.A. Spiegel), Reidel, pp. 223–244 (1985).

Dombre, T., Frisch, U., Greene, J.M., Hénon, M., Mehr, A. and Soward, A.M., "Chaotic streamlines in the ABC flows," *J. Fluid Mech.* **167**, 353–391 (1986).

Dorch, S.B.F., "On the structure of the magnetic field in a kinematic ABC flow dynamo," *Physica Scripta* **61**, 717–722 (2000).

Du, Y. and Ott, E., "Growth rates for fast kinematic dynamo instabilities of chaotic fluid flows," *J. Fluid Mech.* **257**, 265–288 (1993).

Galanti, B., Pouquet, A. and Sulem, P.-L., "Linear and nonlinear dynamos associated with ABC flows," *Geophys. Astrophys. Fluid Dynam.* **66**, 183–208 (1992).

Galloway, D.J. and O'Brian, N.R., "Numerical calculations of dynamos for ABC and related flows," in: *Solar and Planetary Dynamos* (Eds. M.R.E. Proctor, P.C. Matthews and A.M. Rucklidge), Cambridge University Press, pp. 105–113 (1993).

Galloway, D.J. and Frisch, U., "A numerical investigation of magnetic field generation in a flow with chaotic streamlines," *Geophys. Astrophys. Fluid Dynam.* **29**, 13–19 (1984).

Galloway, D.J. and Frisch, U., "Dynamo action in a family of flows with chaotic streamlines," *Geophys. Astrophys. Fluid Dynam.* **36**, 53–83 (1986).

Galloway, D.J. and Frisch, U., "A note on the stability of a family of space-periodic Beltrami flows," *J. Fluid Mech.* **180**, 557–564 (1987).

Galloway, D. J. and Moore, D. R., "Axisymmetric convection in the presence of a magnetic field," *Geophys. Astrophys. Fluid Dynam.* **12**, 73–106 (1979).

Galloway, D.J. and Proctor, M.R.E., "Numerical calculations of fast dynamos in smooth velocity fields with realistic diffusion," *Nature* **356**, 691–693 (1992).

Galloway, D.J. and Zheligovsky, V.A., "On a class of non-axisymmetric flux rope solutions to the electromagnetic induction equation," *Geophys. Astrophys. Fluid Dynam.* **76**, 253–264 (1994).

Galloway, D.J., Hollerbach, M.R.E. and Proctor, M.R.E., "Fine structure in fast dynamo computations," in: *Small-Scale Structures in Three Dimensional Hydro- and Magnetohydrodynamic Turbulence* (Eds. M. Meneguzzi, A. Pouquet and P.-L. Sulem), Springer Lecture Notes in Physics **462**, pp. 341–346 (1995).

Galloway, D. J., Proctor, M. R. E. and Weiss, N. O., "Formation of intense magnetic fields near the surface of the Sun," *Nature* **266**, 686–689 (1977).

Gilbert, A.D., "Fast dynamo action in the Ponomarenko dynamo," *Geophys. Astrophys. Fluid Dynam.* **44**, 214–258 (1988).

Gilbert, A.D. and Ponty, Y., "Dynamos on stream surfaces of a highly conducting fluid," *Geophys. Astrophys. Fluid Dynam.* **93**, 55–95 (2000).

Hénon, M., "Sur la topologie des lignes de courant dans un cas particulier," *C. R. Acad. Sci. Paris* **262**, 312–314 (1966).

Hollerbach, R., Galloway, D.J. and Proctor, M.R.E. "Numerical evidence of fast dynamo action in a spherical shell," *Phys. Rev. Lett.* **74**, 3145–3148 (1995).

Kim, E.-J., Hughes, D.W. and Soward, A.M., "An investigation into high conductivity dynamo action driven by rotating convection," *Geophys. Astrophys. Fluid Dynam.* **91**, 303–332 (1999).

Klapper, I. and Young, L.S., "Bounds on the fast dynamo growth rate involving topological entropy," *Comm. Math. Phys.* **173**, 623–646 (1995).

Kulsrud, R.M. and Anderson, S.W., "Magnetic fluctuations in fast dynamos," in: *Solar and Planetary Dynamos* (Eds. M.R.E. Proctor, P.C. Matthews and A.M. Rucklidge), Cambridge University Press, pp. 195–202 (1993).

Lau, Y.-T. and Finn, J.M., "Fast dynamos with finite resistivity in steady flows with stagnation points," *Phys. Fluids* B **5**, 365–375 (1993).

Moffatt, H.K. and Proctor, M.R.E., "Topological constraints associated with fast dynamo action," *J. Fluid Mech.* **154**, 493–507 (1985).

Otani, N.F., "A fast kinematic dynamo in two-dimensional time-dependent flows," *J. Fluid Mech.* **253**, 327–340 (see also the 1989 abstract referred to therein) (1993).

Ott, E., *Chaos in Dynamical Systems*, Cambridge University Press (1993).

Ponomarenko, Y.B., "On the theory of hydromagnetic dynamos," *Zh. Prikl. Mekh. & Tekh. Fiz.* (USSR) **6**, 47–51 (1973).

Ponty, Y., Gilbert, A.D. and Soward, A.M., "Kinematic dynamo action in large magnetic Reynolds number flows driven by shear and convection," *J. Fluid Mech.* **435**, 261–287 (2001).

Priest, E.R. and Forbes, T.G., *Magnetic Reconnection*, Cambridge University Press (2000).

Roberts, G.O., "Dynamo action of fluid motions with two-dimensional periodicity," *Phil. Trans. R. Soc. Lond.* A **271**, 411–454 (1972).

Soward, A.M., "Fast dynamos in a steady flow," *J. Fluid Mech.* **180**, 267–295 (1987).

Vainshtein, S.I and Zeldovich, Ya.B., "Origin of magnetic fields in astrophysics," *Sov. Phys. Usp.* **15**, 159–172 (1972).

Vainshtein, S.I. and Cattaneo, F., "Nonlinear restrictions on dynamo action," *Ap. J.* **393**, 199–203 (1992).

Vishik, M.M., "Magnetic field generation by the motion of a highly conducting fluid," *Geophys. Astrophys. Fluid Dynam.* **48**, 151–167 (1989).

Zeldovich, Ya.B, Ruzmaikin, A.A. and Sokoloff, D.D., *Magnetic Fields in Astrophysics*, Gordon and Breach (1983).

3 On the theory of convection in the Earth's core

Stanislav Braginsky[1] and Paul H. Roberts[2]

[1] *Institute of Geophysics and Planetary Physics, University of California, Los Angeles, CA 90095, USA, E-mail: sbragins@igpp.ucla.edu*
[2] *Institute of Geophysics and Planetary Physics, University of California, Los Angeles, CA 90095, USA, E-mail: roberts@math.ucla.edu*

A general strategy is presented for the study of convection in a turbulent-fluid system, such as the Earth's core, in which the adiabatic density differences across the system are much larger than the density differences that drive the convection. This situation is drastically different from the laboratory, where the density differences due to convection are the greater, and where the Boussinesq approximation is valid. In the case considered here, the anelastic approximation deals satisfactorily with the large basic density differences across the system. Turbulent transport of large scale fields such as entropy is evaluated through the application of a local description of turbulence. The resulting theory is applied to the Earth's core, and the system of equations obtained by Braginsky and Roberts (Geophys. Astrophys. Fluid Dynam. **79**, 1, 1995) is recovered; insights that have emerged since that paper was written are added.

3.0. Introduction

This work is mainly a distillation of a paper with a similar title by Braginsky and Roberts (1995), which will be referred to as 'BR'. Glatzmaier and Roberts (1996a, 1997) based simulations of the geodynamo on BR. We shall call these 'simulations A', where the 'A' stands for 'anelastic'. Most models of core MHD assume that the physical properties of the Earth's core are uniform, and we call these 'simulations B', where 'B' stands for 'Boussinesq'; see, e.g. Glatzmaier and Roberts (1995a,b, 1996b, 1998), Kuang and Bloxham (1997) and Sakuraba and Kono (1999). They also incorporate in an essential way the secular cooling of the Earth over geological time. They do not take into account the dynamical effects of the Earth's variable rotation, and in particular the so-called Poincaré force. Appendix A lists some of the parameters describing the core, and also summarizes notation that is used here. Fig. 3.1 is a rough sketch of the Earth's interior.

We believe that convection in the core is driven by the gravitational force $\mathbf{g}\rho$, where $\mathbf{g}$ is the gravitational acceleration (including the centrifugal force arising from the Earth's rotation $\mathbf{\Omega}$) and ρ is the density. In fact, convection in the core is so vigorous that all advecting quantities, such as the specific entropy, S, are well mixed ($\nabla S \approx \mathbf{0}$). We add a suffix a to variables when they refer to an adiabatic state in hydrostatic equilibrium, which will be our 'reference state'. The variation of core structure with depth is assessed by a parameter $\epsilon_a \equiv \delta\rho_a/\rho_a$, which is essentially the difference $\delta\rho_a$ in density ρ_a at the inner-core boundary (ICB) and the core-mantle boundary (CMB) divided by the mean density, and is of order 0.2. The smallness of this parameter makes it reasonable to represent the core by a Boussinesq model. This is a

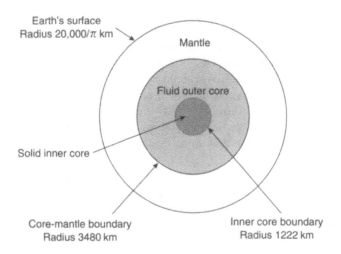

Figure 3.1 A sketch of the Earth's interior.

mathematically convenient approximation, but one that nevertheless does not properly allow for the inhomogeneity of the system. We shall not consider the Boussinesq approximation further in this chapter; for a detailed derivation and discussion, see Section 8 of BR.

To estimate how thoroughly the core is mixed, we introduce convective quantities as additions to the adiabatic reference values; for example $\rho_c = \rho - \rho_a$. The main part, $\mathbf{g}_a$, of the gravitational acceleration $\mathbf{g}$ is exactly balanced by the pressure term $\rho_a^{-1}\nabla P_a$ and therefore does not cause convection. We may estimate the importance of the remainder by equating $\mathbf{g}_a\rho_c/\rho_a$ in order of magnitude to the Coriolis acceleration $2\mathbf{\Omega} \times \mathbf{V}$, where $\mathbf{V}$ is the fluid velocity. Adopting the common estimate $V \sim 5 \times 10^{-4}\,\mathrm{m\,s^{-1}}$ for $\mathbf{V}$, we obtain $\rho_c/\rho_a \sim \epsilon_c$, where $\epsilon_c \equiv 2\Omega V/g \sim 10^{-8}$. Thus $\epsilon_c = O(10^{-7}\epsilon_a)$, a situation in stark contrast to the laboratory, in which usually $\epsilon_a \ll \epsilon_c$. This means that the Boussinesq model for the core has no direct physical justification; it is at best an approximation, for $\epsilon_a \ll 1$, to the inhomogeneous anelastic model. As is usual in convection theory, we describe convection in two steps: first we select a convenient reference state (as we have done above), and second we study departures from that reference state associated with convection. It is clear that as the core cools our reference state changes secularly on a geological timescale, t_a, so that $S_a = S_a(t_a)$.

The extreme smallness of ϵ_c makes possible a tremendous simplification of the theory: the thermodynamics can be 'linearized' about the reference state. For example, whenever ρ appears in a factor that involves a convective quantity, we may replace it by ρ_a; thus $\rho\mathbf{V}$ in the mass continuity equation is (see below) replaced by $\rho_a\mathbf{V}_c$. Since even the convective timescale is long compared with the time taken by seismic waves to cross the core, we may also neglect $\partial\rho_c/\partial t$ in comparison with $\nabla \cdot (\rho_a\mathbf{V}_c)$, and this leads to the so-called 'anelastic' continuity equation (see Section 3.1). The gravitational and pressure forces (per unit mass) appear in the combination $\mathbf{f} = \mathbf{g} - \rho^{-1}\nabla P$ in the equation of motion, and hydrostatic balance requires that $\mathbf{f}_a = \mathbf{0}$, so that all that remains is the convective part $\mathbf{f}_c$. This consists of two parts, $\mathbf{g}_c$ and $\mathbf{g}_a - \rho^{-1}\nabla P = \rho_a^{-1}\nabla P_a - \rho^{-1}\nabla P$, the first of which recognizes that the Earth is a self-gravitating body and that the density ρ_c therefore produces a change U_c in the gravitational potential U. Braginsky and Roberts (1995) noticed the two contributions can be conveniently dealt with together by introducing the 'reduced pressure' $\Pi_c = P_c/\rho_a + U_c$; see Section 3.3. The reduced pressure plays a useful role in core convection theory: ∇U_c and the part of the convective buoyancy force that arises from the perturbation ∇P_c in pressure do not create convective motions, and they are neatly and automatically absorbed into a gradient term $\nabla \Pi_c$, leaving behind the Archimedean buoyancy force $C\mathbf{g}_a$ which, by creating convective circulations, plays a central role in convection theory. BR called C the 'co-density', an acronym for convection originating density. It is actually a fractional change in density, and in the geophysical context it involves a further complication: buoyancy is produced in the core not only thermally but also chemically, by differences in composition created by the light admixture released at the ICB as the inner core grows, as was first pointed out by Braginsky (1964).

Though the core is 'an uncertain mixture of all the elements', there seems to be little doubt that it is abundant in iron. There is general agreement that the fluid outer core (FOC) is significantly less dense than iron would be at core pressures, and that alloying elements must be present. There is no consensus on what the alloying elements are, or even what the predominant light constituent is. From a theoretical point of view, the basic physics is satisfied by assuming the simplest possible case, in which there is only one alloying element its mass fraction being $\xi = \xi_a + \xi_c$, where ξ_a is the uniform composition of the reference state and ξ_c is created by convection; ξ_a, like S_a, depends on t_a only. Seismological models of the Earth's interior show that the density of the solid inner core (SIC) is closer to that of pure iron than is that of the FOC. There is a density jump $\Delta\rho_{ICB}$ across the ICB of about 0.6 gm cm^{-3}, which is presumed to arise mainly because the value ξ_N of ξ at the top of the SIC is less than ξ_a. This is naturally the case if, as is now believed, the ICB is a freezing interface. A simple phase diagram that would lead to this conclusion is shown on the left-hand side of Fig. 3.2; the right-hand side indicates its relationship to the core. A phase diagram usually shows the solidus and liquidus of a material as curves in ξT-space. This is because they are generally

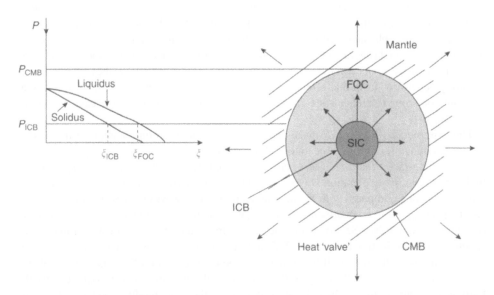

Figure 3.2 Sketch of a phase diagram of core material (left), and its relationship to the core itself (right).

discussed in laboratory contexts where variations in pressure, P, are unimportant. In reality the liquidus and solidus depend on the thermodynamic state of the material. They should therefore appear as surfaces in three dimensions. The traditional diagrams showing them as curves are merely intersections of these surfaces with the appropriate constant-P plane. In the context of the core it is appropriate to plot the solidus and liquidus in ξPS-space and to project them onto the plane $S = S_a$. This is the way they are shown in Fig. 3.2, where we focus attention on the left-hand solidus and liquidus only. On descending through the FOC from the CMB, we eventually encounter the ICB, where the pressure is such that mixed phases can co-exist. The fluid alloy freezes onto the SIC, releasing latent heat and light constituent as it does so. These sources of buoyancy establish core convection that stirs the fluid, making S_a and ξ_a uniform, and generating the Earth's magnetic field by dynamo action.

As in many other applications of convection theory, motions in the Earth's core are on many length and time scales. That this must be the case is clear when we consider quantities such as the Péclet number, $Pe = VL/\kappa^T$, and the compositional Péclet number, $Pc = VL/\kappa^\xi$, of the macroscale defined by the characteristic macroscale velocity V and length L and by typical molecular diffusivities κ^T and κ^ξ of heat and composition. Taking $L = 10^6$ m and $V = 5 \times 10^{-4}$ m s^{-1} as before, we find that $Pe \sim 10^8$ and $Pc \sim 10^{12}$. This shows that molecular transport of heat and composition is almost totally ineffective on the macroscale. It is also true of the momentum transport as measured by the hydrodynamic Reynolds number, $Re = VL/\nu \sim 10^9$, where ν is the molecular kinematic viscosity. It is clear that, to transport heat, composition and momentum on the macroscale, the core has to develop small-scale, rapidly varying motions (turbulence) superimposed on the comparatively slow macroscale convection. It should be stressed however that, since the magnetic diffusivity η is large ($\eta \sim 2$ m^2 s^{-1}), the magnetic Reynolds number, $R_m = UL/\eta$, is only moderately large: $R_m \sim 200$.

Core turbulence presents the theory of core convection with its greatest challenge. On the one hand, we require the theory to be sufficiently simple so that it be integrated by computer. On the other hand it must describe the evolution of the large-scale fields by incorporating the effects of the sub-grid scale fields and motions in a physically consistent way. Can this be done, even though the theory of turbulence has not yet reached a stage in which a deductive approach is available? The answer is, 'Yes, in the same sense as we can incorporate the effects of molecular motions in the equations governing the average hydrodynamic motion'. From general principles of thermodynamics and linearity, we can establish expressions for the fluxes of heat, chemical composition and momentum together with the expression for entropy production. In the case of molecular transport, the fluxes of heat, composition and momentum contain unknown kinetic coefficients: the thermal conductivity, the compositional diffusivity and the viscosity. The theoretical calculation of these transport coefficients can be extremely difficult, but this does not diminish the truth and utility of the expressions for the fluxes and entropy production. The same is true when we separate the average convective state from the local turbulence and treat the latter in analogy with the molecular motion. While the determination of the transport coefficients is even more difficult, the general expressions for the fluxes and entropy production are true and useful. We shall show later how these may be derived.

Although a consistent theory of turbulence does not exist, the following order of magnitude estimates may be useful. Let the turbulence be on a length scale of L^t and a velocity scale of $V^t \sim V$, and suppose that the turbulent mixing coefficient, $\kappa^t \sim \frac{1}{3}V^t L^t$, is of the same order as the greatest molecular coefficient $\eta \sim 2$ m^2 s^{-1}. Then $L^t \sim 3\eta/V^t \sim 10^4$ m $\sim 10^{-2}L$. This estimate demonstrates that small scale turbulence is capable of performing the necessary diffusive mixing of S_c and ξ_c in the core on the diffusional time scale, $\tau_\eta = L^2/\eta$, of the geodynamo. Braginsky and Meytlis (1990) model the small scale turbulence mechanism by generalizing the ideas and estimates provided by Braginsky (1964). They demonstrate that the turbulence is highly anisotropic so its diffusive effect on S_c and ξ_c should be represented by a tensor $\overleftrightarrow{\kappa}$ and not a scalar κ^1; see also Appendix C of BR. It can be shown from these estimates (see below) that the relevant microscale magnetic Reynolds number is very small, so that the inductive effects of the microscale on the macroscale magnetic field is negligible; there is no enhancement of η and no α-effect from the small scale turbulence.

We may call the model we have just outlined *local* turbulence theory. We divide a field such as S_c into a macroscale part, $\overline{S}_c$, and a microscale part, S', e.g. $S_c = \overline{S}_c + S'$. We then develop expressions for the macroscale fluxes of entropy, composition and momentum

Table 3.0 Strategy: Development of governing convection equations

1: Basic equations
for **V**, **B**, S, ξ, ...

2: Reference state – well mixed
for $S_a(t_a)$, $\xi_a(t_a)$, $\mathbf{V}_a(t_a)$, ...
t_a = slow (geological) timescale

3: Full (anelastic) convection equations for
$\mathbf{V}_c \approx \mathbf{V}$, **B**, $S_c = S - S_a$, $\xi_c = \xi - \xi_a$, ... which
depend on t_a and the convective timescale, t_c

4: Scale separation into mean convection and turbulence:
$\mathbf{V} = \overline{\mathbf{V}} + \mathbf{V}'$, $\mathbf{B} = \overline{\mathbf{B}} + \mathbf{B}'$, $S_c = \overline{S}_c + S'$, $\xi_c = \overline{\xi}_c + \xi'$, ...

5: Local turbulence theory
(For including effects of $\mathbf{V}'$, $\mathbf{B}'$, ...
into theory of mean convection)

6A: Full anelastic theory
governing mean quantities
$\overline{\mathbf{V}}, \overline{\mathbf{B}}, \overline{S}_c, \overline{\xi}_c$, ...

6B: 'Boussinesq' approximation
governing mean quantities
$\overline{\mathbf{V}}, \overline{\mathbf{B}}, \overline{S}_c, \overline{\xi}_c$, ...

7A: Simulations A
of the geodynamo

7AB: Thermodynamic
efficiency (BR)

7B: Simulations B
of the geodynamo

at position **r** and time t created by the turbulence. In a local theory, these depend on the macroscale properties of the system, but only at the same **r** and t.

The steps required to develop a workable theory of core convection and the geodynamo are now apparent; they are summarized in Table 3.0. Once the final anelastic equations have been derived (see step 6A), a reduction to a Boussinesq-type theory is possible (step 6B). The steps leading from 6A to 6B may be found in BR and are not repeated here. The anelastic and Boussinesq-like theories have interesting thermodynamic implications concerning the thermodynamic efficiency of convective systems, regarded as heat engines (step 7AB). These are discussed in Section 7 of BR, but will not be described here.

3.1. Basic equations

The arguments that BR employed to complete the strategy set out in Table 3.0 are fully described in BR and will therefore only be summarized here. The starting point is the set of basic equations, sometimes also called 'the primitive equations'. These are well known and are given in Table 3.1, which also contains some notation.

The expressions for σ^S and $\mathbf{I}^q$ given in Table 3.1 are very general. They are required from considerations of energy conservation and thermodynamics; they do not assume any particular forms for the constitutive relations; see Landau and Lifshitz (1987), or BR, or any book on the thermodynamics of irreversible processes. It is just this generality that makes it possible to employ the same expressions for σ^S and $\mathbf{I}^q$ that describe molecular transport and that model small scale turbulent transport, by exploiting in each case the appropriate constitutive relations. The equations in this table do not rely on any specific constitutive relations connecting the fluxes $\mathbf{I}^S$, $\mathbf{I}^\xi$ and the momentum flux to the relevant inhomogeneities of the fluid state such as ∇S, $\nabla \xi$ and $\nabla \mathbf{V}$. Such relations are needed in order to obtain a closed system of equations, solved subject to appropriate boundary conditions.

Table 3.1 Basic equations

$$d_t \mathbf{V} = -\rho^{-1} \nabla P + \mathbf{g} - 2\mathbf{\Omega} \times \mathbf{V} + \mathbf{F}^B + \mathbf{F}^\nu, \quad \partial_t \rho = -\nabla \cdot (\rho \mathbf{V})$$
$$\partial_t \mathbf{B} = \nabla \times (\mathbf{V} \times \mathbf{B}) - \nabla \times (\eta \nabla \times \mathbf{B}), \quad \nabla \cdot \mathbf{B} = 0$$
$$\rho d_t \xi = -\nabla \cdot \mathbf{I}^\xi, \quad \rho d_t S = -\nabla \cdot \mathbf{I}^S + \sigma^S$$
$$T \sigma^S = Q^R + Q^V + Q^J - \mathbf{I}^S \cdot \nabla T - \mathbf{I}^\xi \cdot \nabla \mu, \quad \mathbf{I}^q = T \mathbf{I}^S + \mu \mathbf{I}^\xi$$
$$\nabla^2 U_N = 4\pi k_N \rho$$

Notation

$\partial_t = \partial / \partial t$ = Eulerian time derivative;

$d_t = \partial_t + \mathbf{V} \cdot \nabla$ = Lagrangian (motional) derivative; $\quad \mathbf{V}$ = fluid velocity;

$\mathbf{g}$ = effective gravitational field = $\mathbf{g}_N - \mathbf{\Omega} \times (\mathbf{\Omega} \times \mathbf{r})$; $\quad k_N$ = gravitational constant;

$\mathbf{g}_N = -\nabla U_N$ = true gravitational field; $\quad U_N$ = gravitational potential;

$\mathbf{\Omega}$ = angular velocity of reference frame (mantle);

$\mathbf{B}$ = magnetic field; $\quad \mathbf{J}$ = electric current density = $\nabla \times \mathbf{B}/\mu_0$;

$\mathbf{F}^\nu$ = viscous force/unit mass; $\quad \mathbf{F}^B$ = Lorentz force/unit mass = $\mathbf{J} \times \mathbf{B}/\rho$;

$\mathbf{I}^\xi$ = flux of composition; $\quad \mathbf{I}^S$ = entropy flux; $\quad \mathbf{I}^q$ = heat flux;

T = temperature; $\quad \mu$ = chemical potential; $\quad \sigma^S$ = volumetric entropy source;

Q = volumetric heat sources from radioactivity (R), viscosity (V), Joule losses (J);

μ_0 = permeability of free space.

Table 3.2 Reference state

$$\nabla \xi_a = 0, \quad \nabla S_a = 0$$
$$\rho_a^{-1} \nabla P_a = -\nabla U_a \equiv \mathbf{g}_a, \quad \nabla^2 U_a = 4\pi k_N \rho_a - 2\Omega^2$$
$$\nabla \cdot (\rho_a \mathbf{V}_a) = -\dot{\rho}_a$$

Elementary consequence:

$$\rho_a^{-1} \nabla \rho_a = \mathbf{g}_a / v_P^2, \quad \nabla T_a = \alpha^S \mathbf{g}_a, \quad \nabla \mu_a = \alpha^\xi \mathbf{g}_a.$$

3.2. The reference state

The reference state (step 2 of Table 3.0) is described in Table 3.2, in which all variables are functions of t_a, the slow geological time scale; $\mathbf{g}_a$, P_a, ρ_a and μ_a depend on $\mathbf{r}$ but ξ_a and S_a do not. The overdot, here and later, is reserved for differentiation with respect to t_a, an operation also denoted by ∂_t^a. We note that $\partial_t^a = O(V_a/L)$ is extremely small – the inverse of billions of years – so that $\mathbf{V}_a$ (which represents a tiny deviation from hydrostatic balance, required by mass conservation) is extremely small and will be neglected in Section 3.3.

It is possible to express hydrostatic equilibrium in the following interesting way (Glatzmaier and Roberts, 1996a). Since from thermodynamics

$$d\varepsilon^H = \rho^{-1} \, dp + T \, dS + \mu \, d\xi,$$

where ε^H is the specific enthalpy, it follows that in the reference state, in which ∇S_a and $\nabla \mu_a$ are zero, $\nabla \varepsilon_a^H = \rho_a^{-1} \nabla P_a$. Hence, from the hydrostatic equation, Π_a, like S_a and ξ_a, depends only on t_a and not on $\mathbf{r}$:

$$\nabla \Pi_a = 0, \quad \text{where} \quad \Pi_a = \varepsilon^H(P_a, S_a, \xi_a) + U_a.$$

To derive the elementary consequences shown in Table 3.2, we observe first that, since

$$\rho_a = \rho(P_a, S_a, \xi_a), \quad T_a = T(P_a, S_a, \xi_a), \quad \mu_a = \mu(P_a, S_a, \xi_a)$$

and since S_a and μ_a are constant, the surfaces of constant density, temperature and chemical potential coincide in the reference state with those of constant P_a and U_a. Now make use of the thermodynamic relation

$$\frac{d\rho}{\rho} = \frac{dP}{\rho v_P^2} - \alpha^\xi \, dS - \alpha^S \, d\xi,$$

where

$$\alpha^\xi = -\frac{1}{\rho} \left(\frac{\partial \rho}{\partial \xi} \right)_{P,S}, \quad \alpha^S = -\frac{1}{\rho} \left(\frac{\partial \rho}{\partial S} \right)_{P,\xi} = \frac{\alpha T}{c_p},$$

and v_P is the speed of sound in the fluid core $\left(v_P^2 = (\partial P/\partial \rho)_{S,\xi} \right)$, α is the coefficient of thermal expansion $(\alpha = -\rho^{-1}(\partial \rho/\partial T)_{P,\xi})$ and c_p is the specific heat at constant pressure $(c_p = T(\partial S/\partial T)_{P,\xi})$. The first of the elementary consequences displayed in the table now follows from the hydrostatic equation. The other two can be obtained similarly from

$$dT = (\alpha^S/\rho) \, dP + (T/c_p) \, dS - (h^\xi/c_p) \, d\xi,$$

$$d\mu = (\alpha^\xi/\rho) \, dP + (h^\xi/c_p) \, dS + \mu^\xi \, d\xi.$$

where $h^\xi = T(\partial S/\partial \xi)_{PT} = -T(\partial \mu/\partial T)_{P\xi}$ and $\mu^\xi = (\partial \mu/\partial \xi)_{PS}$. A geophysically more familiar form of ∇T_a is

$$T_a^{-1}\nabla T_a = \gamma \mathbf{g}_a/v_P^2, \quad \gamma = \alpha v_P^2/c_p = \text{the Grüneisen parameter.}$$

Because of the rotation of the Earth, the equipotential surfaces are oblate. The degree of flattening is measured by a further dimensionless parameter, $\epsilon_\Omega = \Omega^2 L/g \sim 10^{-3}$. Veronis (1973) analyzed the magnitude of the errors ($\sim\epsilon_\Omega$) that result from ignoring this effect. It is clear that they are not large, and to include rotational flattening would add severe complications to the theory without any compensating enlightenment. We shall therefore later set $\epsilon_\Omega \equiv 0$, and so neglect the centrifugal force even though we retain the much smaller Coriolis force. We shall therefore be dealing with a spherical Earth in which $\mathbf{g} = \mathbf{g}_N$ instead of the theoretically far more complicated spheroidal Earth. The spatial dependence of reference state variables is then on distance r from the geocenter alone. Both BR and Simulations A used numerical values obtained from the PREM model of Dziewonski and Anderson (1981) which also depends on r alone; see Appendix A.

3.3. Full convection equations

The full convection equations (step 3 of Table 3.0) are the subject of Table 3.3. We write

$$S = S_a + S_c, \quad \xi = \xi_a + \xi_c, \quad \mathbf{g} = \mathbf{g}_a + \mathbf{g}_c, \quad U = U_a + U_c, \dots,$$

where $S_c = O(\epsilon_c S_a) \dots$. The magnetic field $\mathbf{B}$ is created by the convection and therefore does not appear in the reference state; a suffix c on $\mathbf{B}$ would therefore be superfluous. Since $\mathbf{V}_a = O(\epsilon_c \mathbf{V}_c)$, we may neglect $\mathbf{V}_a$ and understand $\mathbf{V}$ to be the same as $\mathbf{V}_c$. We shall employ two timescale methods, writing

$$\partial_t = \partial_t^a + \partial_t^c.$$

Because $\partial_t^a = O(\epsilon_c \partial_t^c)$, we shall, except where confusion might arise, write ∂_t instead of ∂_t^c and d_t instead of d_t^c for $\partial_t^c + \mathbf{V} \cdot \nabla$. We substitute into the basic equations of Section 3.1, use the reference state variables of Section 3.2, and retain only the largest terms in ϵ_c. We then obtain the equations set out in Table 3.3, in which mass continuity is approximated by the anelastic equation. This is justified because the term $\partial_t \rho_c = O(\rho_c V_c/L)$ is negligibly small on the convective time scale compared with $\rho_a V_c/L$.

Braginsky and Roberts (1995) pointed out that the anelastic equation of motion could be written in a simpler form; see Table 3.4.

To derive this result, we again use the thermodynamic relation set out in Section 3.2:

$$\frac{\rho_c}{\rho_a} = \frac{P_c}{\rho_a v_P^2} - \alpha^S S_c - \alpha^\xi \xi_c = \frac{P_c}{\rho_a v_P^2} + C.$$

Table 3.3 Full convection equations

$d_t \mathbf{V} = -\rho_a^{-1}\nabla P_c + \mathbf{g}_c + \mathbf{g}_a \rho_c/\rho_a - 2\mathbf{\Omega} \times \mathbf{V} + \mathbf{F}^B + \mathbf{F}^\nu$

$\rho_a d_t \xi_c = -\nabla \cdot \mathbf{I}^\xi - \rho_a \dot{\xi}_a, \quad \rho_a d_t S_c = -\nabla \cdot \mathbf{I}^S - \rho_a \dot{S}_a + \sigma^S$

$\partial_t \mathbf{B} = \nabla \times (\mathbf{V} \times \mathbf{B}) - \nabla \times (\eta \nabla \times \mathbf{B}), \quad \nabla \cdot \mathbf{B} = 0$

$\mathbf{g}_c = -\nabla U_c, \quad \nabla^2 U_c = 4\pi k_N \rho_c, \quad \nabla \cdot (\rho_a \mathbf{V}) = 0$

Table 3.4 Simplified equation of motion

$d_t \mathbf{V} = -\nabla \Pi_c + C\mathbf{g}_a - 2\mathbf{\Omega} \times \mathbf{V} + \mathbf{F}^B + \mathbf{F}^v$

where

$\Pi_c = P_c/\rho_a + U_c$ = reduced pressure perturbation

$C = -\alpha^S S_c - \alpha^\xi \xi_c$ = co-density, or equivalently

$\rho_a C = \rho_c - P_c/v_P^2$

It then follows that the perturbation of the sum $\mathbf{f} = \mathbf{g} - \rho^{-1}\nabla P$ of the pressure and gravitational forces can be written in a simple form:

$$\mathbf{f}_c = -\frac{1}{\rho_a}\nabla P_c - \nabla U_c + \frac{\rho_c}{\rho_a}\mathbf{g}_a = -\nabla \Pi_c - \frac{P_c}{\rho_a^2}\nabla \rho_a + \left(\frac{P_c}{\rho_a v_P^2} + C\right)\mathbf{g}_a$$

$$= -\nabla \Pi_c + \frac{P_c}{\rho_a^2}(-\nabla \rho_a + \rho_a \mathbf{g}_a) + C\mathbf{g}_a = -\nabla \Pi_c + C\mathbf{g}_a.$$

We note that $\Pi_a + \Pi_c$ is the linearized form of $\varepsilon^H(P, S_a, \xi_a) + U$, where only variations in P cause ε^H to change.

Our form of the equation of motion enjoys two great advantages over the primitive form of the momentum equation. First, it obviates the need to calculate the perturbation, $\mathbf{g}_c$, in $\mathbf{g}$ created by the perturbation, ρ_c, in density. If $\mathbf{g}_c$ is deemed to be interesting, it can be evaluated after the main calculation has been completed. Second, it establishes that density variations produced by pressure variations, P_c, will not drive convection. This simplification allows P_c/v_P^2 to be split off from ρ_c and absorbed into the gradient term, so leaving the remaining variations in ρ_c, those due to changes in entropy and composition from which convection originates, to be included in the co-density $C = \rho_a^{-1}(\rho_c - P_c/v_P^2)$. Since the co-density is strictly speaking a relative density (relative to ρ_a), it is dimensionless. It is remarkable that the co-density is produced by the inhomogeneities in entropy and composition alone, but not by the pressure inhomogeneity. Thus the co-density is generated by thermal and compositional sources and removed mostly by the turbulent mixing processes. From now onwards we shall refer to $\mathbf{F}^\alpha = C\mathbf{g}_a$ as the 'Archimedean buoyancy force'. This is the driving force of core convection and the geodynamo.

The role of the co-density in the theory of convection of a stratified fluid may be clarified through the well-known parcel argument; see, e.g. section 4 of Landau and Lifshitz (1987). Suppose that a small parcel of fluid lying in the gravitational field $g_r = -g < 0$ is displaced vertically from $r = r'$ to $r = r''$. When the diffusivities are very small (as they are in Earth's core), we can move the parcel so slowly that it is continuously in mechanical equilibrium with its environment but so rapidly that all forms of diffusion are negligible. Then S and ξ are conserved, and retain their initial values S' and ξ'. When it reaches its final level r'', its pressure has adapted to its environment, so that its density is $\rho_P = \rho(P'', S', \xi')$. The difference between the density $\rho'' = \rho(P'', S'', \xi'')$ of the environment and that of the parcel, $\delta\rho = \rho'' - \rho_P$, is approximately $-\rho(\alpha^S \partial_r S + \alpha^\xi \partial_r \xi)\delta r$, for small $\delta r = r'' - r'$. This can also be written as the difference $\rho'' - \rho' \approx \partial_r \rho \delta r$ in the densities of the environment at r'' and r' minus the change in density, $\rho_P - \rho' = v_P^{-2}\delta P = v_P^{-2}\partial_r P\delta r = \rho(g_r/c_P^2)\delta r$, due to the different pressures $\delta P = P'' - P'$ at those points. We have here used the hydrostatic equation $\partial_r P = g_r \rho$. If $\delta r > 0$ and the parcel is heavier than its environment ($\delta\rho/\delta r < 0$), the

Archimedean force $g_r \delta\rho$ tends to return the parcel back to its original level; the fluid is then convectionally stable; if $\delta\rho/\delta r > 0$, the displacement of the parcel will increase, and the fluid is unstable.

Equating the force $-g_r \delta\rho$ and the inertia, we obtain $\rho d_t^2 \delta r = -N^2 \rho \delta r$, where $N^2 = g_r(\rho^{-1}\partial_r\rho - g_r/v_p^2) = -g_r(\alpha^S \partial_r S + \alpha^\xi \partial_r \xi)$ defines the Brunt–Väisälä frequency N. If we assume that α^S and α^ξ are constants, $N^2 = g_r \partial_r C$ involves only the gradient of the co-density. The parcel argument makes it obvious why the gradients of S and ξ alone determine the stability of a stratified fluid: the pressure of the parcel can adjust to the pressure of the environment, but S and ξ cannot. The parcel argument also shows how simple the mechanism of convectional instability is; it relies directly and solely on the gradients ∇S and $\nabla \xi$. This instability is the cause of small scale (local) turbulence in the core, which tends to mix entropy and composition, thus smoothing ∇S and $\nabla \xi$. In the next section, we incorporate this effect into the system of governing equations.

3.4. Scale separation

Scale separation (step 4 of Table 3.0) is the subject of Table 3.5. In this table and in the remainder of the paper we shall, to simplify the notation, dispense with the suffix c except where ambiguity might arise; where the suffix c is absent it is generally implied. We shall write

$$\mathbf{V} = \overline{\mathbf{V}} + \mathbf{V}', \quad \mathbf{B} = \overline{\mathbf{B}} + \mathbf{B}', \quad \Pi = \overline{\Pi} + \Pi', \quad C = \overline{C} + C', \quad S_c = \overline{S}_c + S', \ldots$$

where $\overline{\mathbf{V}}, \overline{\mathbf{B}}, \ldots$ are ensemble averages of $\mathbf{V}, \mathbf{B} \ldots$ over the turbulence, while $\mathbf{V}', \mathbf{B}', \ldots$ are the parts of $\mathbf{V}, \mathbf{B}, \ldots$ associated with the turbulence.

We first derive the equation governing the mean flow, $\overline{\mathbf{V}}$, by averaging the momentum equation over the turbulent ensemble. It is easy to average linear terms such as the viscous force, $\rho_a \mathbf{F}^\nu = \nabla \cdot \overset{\leftrightarrow}{\pi}$, where $\overset{\leftrightarrow}{\pi}$ is the viscous stress tensor and is linear in $\mathbf{V}$. If, as in our case, $\mathbf{F}^\nu$ is sufficiently well approximated by $\mathbf{F}^\nu = \nu^m \nabla^2 \mathbf{V}$, where ν^m is the molecular kinematic viscosity, it is clear that the mean of $\mathbf{F}^\nu$ is simply $\mathbf{F}^{\nu m} = \nu^m \nabla^2 \overline{\mathbf{V}}$.

Table 3.5 Scale separation

$\overline{d}_t \overline{\mathbf{V}} = -\nabla\overline{\Pi} + \overline{C}\mathbf{g}_a - 2\,\Omega \times \overline{\mathbf{V}} + \overline{\mathbf{F}}^B + \overline{\mathbf{F}}^\nu, \quad \nabla \cdot (\rho_a \overline{\mathbf{V}}) = 0$

$\partial_t \overline{\mathbf{B}} = \nabla \times (\overline{\mathbf{V}} \times \overline{\mathbf{B}}) - \nabla \times (\eta \nabla \times \overline{\mathbf{B}}), \quad \nabla \cdot \overline{\mathbf{B}} = 0$

$\rho_a \overline{d}_t \overline{\xi} = -\nabla \cdot \mathbf{I}^{\xi t} - \rho_a \dot{\xi}_a$

$\rho_a \overline{d}_t \overline{S} = -\nabla \cdot \mathbf{I}^{St} + \nabla \cdot [(K^T/T_a)\nabla T_a] - \rho_a \dot{S}_a + \overline{\sigma}^S$

$\overline{C} = -\alpha^S \overline{S} - \alpha^\xi \overline{\xi}$

Definition of average quantities

$\overline{d}_t = \partial_t + \overline{\mathbf{V}} \cdot \nabla = $ Lagrangian derivative following only the large scale motion;

$\rho_a \overline{\mathbf{F}}^B = \overline{\mathbf{J}} \times \overline{\mathbf{B}}, \quad \overline{\mathbf{F}}^\nu = \overline{\mathbf{F}}^{\nu m} + \overline{\mathbf{F}}^{\nu t}, \quad \overline{\mathbf{F}}^{\nu t} = \overline{\mathbf{F}}^{Vt} + \overline{\mathbf{F}}^{Bt};$

$\rho_a \mathbf{F}^{Vt} = \nabla \cdot \overset{\leftrightarrow}{\pi}{}^{Vt}, \quad \overset{\leftrightarrow}{\pi}{}^{Vt} = -\rho_a \overline{\mathbf{V}'\mathbf{V}'} = $ Hydrodynamic Reynolds stress;

$\rho_a \mathbf{F}^{Bt} = \nabla \cdot \overset{\leftrightarrow}{\pi}{}^{Bt}, \quad \overset{\leftrightarrow}{\pi}{}^{Bt} = \mu_0^{-1} \overline{\mathbf{B}'\mathbf{B}'} = $ Magnetic Reynolds Stress;

$\mathbf{I}^{\xi t} = \rho_a \overline{\xi'\mathbf{V}'}, \quad \mathbf{I}^{St} = \rho_a \overline{S'\mathbf{V}'};$

$\overline{\sigma}^S = \sigma^R + \sigma^T + \sigma^J + \sigma^V + \sigma';$

$\sigma^R = Q^R/T_a, \sigma^T = K^T(\nabla T_a/T_a)^2, \quad \sigma^J = \mu_0 \eta \overline{\mathbf{J}}^2/T_a, \quad \sigma' = -(\mathbf{I}^{St} \cdot \nabla T_a + \mathbf{I}^{\xi t} \cdot \nabla \mu_a)/T_a.$

Some care is necessary in averaging quadratic terms. The mean of the inertial force, $\rho_a \mathbf{F}^V = -\rho_a \mathbf{V} \cdot \nabla \mathbf{V}$, is

$$\rho_a \overline{\mathbf{F}}^V = \rho_a \overline{\mathbf{F}}^{\overline{V}} + \rho_a \overline{\mathbf{F}}^{V\prime},$$

where

$$\rho_a \overline{\mathbf{F}}^{\overline{V}} = -\rho_a \overline{\mathbf{V}} \cdot \nabla \overline{\mathbf{V}}. \quad \rho_a \overline{\mathbf{F}}^{V\prime} = -\rho_a \overline{\mathbf{V}' \cdot \nabla \mathbf{V}'} = -\nabla \cdot (\rho_a \overline{\mathbf{V}' \mathbf{V}'}).$$

The last term expresses $\rho_a \overline{\mathbf{F}}^{V\prime}$ as the divergence of the (hydrodynamic) *Reynolds stress* tensor, $\rho_a \overline{\mathbf{V}' \mathbf{V}'}$. In a similar way, the mean of the Lorentz force, $\rho_a \mathbf{F}^B = \mathbf{J} \times \mathbf{B}$, is

$$\rho_a \overline{\mathbf{F}}^B = \rho_a \overline{\mathbf{F}}^{\overline{B}} + \rho_a \overline{\mathbf{F}}^{B\prime},$$

where

$$\rho_a \overline{\mathbf{F}}^{\overline{B}} = \overline{\mathbf{J}} \times \overline{\mathbf{B}}, \quad \rho_a \overline{\mathbf{F}}^{B\prime} = \overline{\mathbf{J}' \times \mathbf{B}'} = \nabla \cdot \left(\frac{\overline{\mathbf{B}' \mathbf{B}'}}{\mu_0}\right) - \nabla \left(\frac{\overline{\mathbf{B}'^2}}{2\mu_0}\right).$$

The term $-\nabla(\overline{\mathbf{B}'^2}/2\mu_0)$ in the final expression for $\rho_a \overline{\mathbf{F}}^{B\prime}$ can be absorbed into $\nabla \overline{\Pi}$ and is therefore insignificant; it is omitted in Table 3.5. This leaves $\rho_a \overline{\mathbf{F}}^{B\prime}$ as the divergence of the magnetic Reynolds stress tensor, $\overline{\mathbf{B}' \mathbf{B}'}/\mu_0$.

Turbulent eddies transfer momentum in a similar way to molecules and, in local turbulence theory (see Section 3.6), the most significant effect of the kinetic Reynolds stress is to strongly augment the molecular viscous stress on the mean flow. The magnetic Reynolds stress plays a similar role. It is convenient to recognize this by collapsing the two kinds of Reynolds stress into one, and to use ν, suggestive of viscosity, to denote the result:

$$\rho_a \overline{\mathbf{F}}^{\nu\prime} = \rho_a \overline{\mathbf{F}}^{V\prime} + \rho_a \overline{\mathbf{F}}^{B\prime} = \nabla \cdot (-\rho_a \overline{\mathbf{V}' \mathbf{V}'} + \overline{\mathbf{B}' \mathbf{B}'}/\mu_0).$$

The net viscous force per unit mass on the mean flow is then $\rho_a \overline{\mathbf{F}}^{\nu m} + \rho_a \overline{\mathbf{F}}^{\nu\prime} = \rho_a \overline{\mathbf{F}}^{\nu}$ (say).

The viscous energy dissipation by the mean flow, $T_a \sigma^V$ in Table 3.4, is (except in thin Ekman layers) small compared with the ohmic dissipation, $T_a \sigma^J$, from the mean field. The mean fluxes, $\overline{\mathbf{I}}^S = \mathbf{I}^{Sm} + \mathbf{I}^{S\prime}$ and $\overline{\mathbf{I}}^\xi = \mathbf{I}^{\xi m} + \mathbf{I}^{\xi\prime}$, of entropy and composition are created both by molecular and turbulent mechanisms. The molecular thermal diffusivity is small ($\kappa^T \sim 10^{-6}\eta$) but the adiabatic gradient ∇T_a is about six orders of magnitude greater than ∇T_c and the flux $\mathbf{I}^{Sm} = -(K^T/T_a)\nabla T_a$, where $K^T = \rho c_p \kappa^T$ is the thermal conductivity, is therefore significant. This conduction of heat 'down the adiabat' of the reference state and the associated entropy production rate, $\sigma^T = -T_a^{-1} \mathbf{I}^{Sm} \cdot \nabla T_a$, must therefore both be retained. The molecular diffusion of ξ and the associated thermal and baro-diffusion are proportional to $\kappa^\xi \sim 10^{-9}\eta$, which is very small. We may therefore discard $\mathbf{I}^{\xi m}$ and the associated entropy production. Terms involving the viscosity, both molecular and turbulent, are small and can be neglected except possibly in boundary layers.

With the help of the expressions for the gradients in the reference state given in Table 3.2, we see that the turbulent dissipation $T_a \sigma^\prime$, given in Table 3.5, is

$$T_a \sigma^\prime = -\mathbf{I}^{S\prime} \cdot \nabla T_a - \mathbf{I}^{\xi\prime} \cdot \nabla \mu_a = -\mathbf{I}^{S\prime} \cdot (\alpha^S \mathbf{g}_a) - \mathbf{I}^{\xi\prime} \cdot (\alpha^\xi \mathbf{g}_a) = \mathbf{g}_a \cdot \mathbf{I}^{C\prime}.$$

where

$$\mathbf{I}^{Ct} = -\alpha^S \mathbf{I}^{St} - \alpha^\xi \mathbf{I}^{\xi t} = -\rho_a \overline{(\alpha^S S' + \alpha^\xi \xi')\mathbf{V}'} = \rho_a \overline{C'\mathbf{V}'}$$

is the co-density flux. This expression for $T_a\sigma^t$ shows that the turbulent dissipation is precisely equal to the rate of working of the buoyancy force on the microscale, $\mathbf{g}_a \cdot \mathbf{I}^{Ct}$. And clearly since σ^t must be positive, so must $\mathbf{g}_a \cdot \mathbf{I}^{Ct}$; see Section 3.5. The turbulent dissipation $Q^t = \mathbf{g}_a \cdot \mathbf{I}^{Ct}$ creates entropy at the rate $\sigma^t = Q^t/T_a$. The energy Q^t released by buoyancy supplies the viscous and Joule heat requirements of the microscale but, because $\nu_m \ll \eta$, the latter dominates: $Q^t \approx \mu_0\eta\overline{\mathbf{J}'^2}$. This makes a significant contribution to the global energy balance, though it is frequently ignored when estimates are made of the net Joule losses of the core, where it is often supposed that only $\overline{\sigma^J}$ matters.

Because η is so large, the associated turbulent magnetic Reynolds number R^t_m is tiny, and $\mathbf{B}' \sim R^t_m\overline{\mathbf{B}}$ is extremely small. The mean emf, $\overline{\mathbf{V}' \times \mathbf{B}'}$, created by the microscale does not significantly affect the macroscale magnetic field; there is no turbulent magnetic diffusivity (or turbulent α-effect): see below. Therefore $\overline{\mathbf{B}}$ is governed by equations of the same simple forms as $\mathbf{B}$ is.

With this preamble it should now be apparent that when we average the convection equations of Section 3.3, we obtain the equations shown in Table 3.5.

3.5. Local turbulence theory

All forms of turbulence originate from the instability of laminar states. Many papers seek the origin of instabilities in the core, and several mechanisms have been identified and studied. One of these is especially simple: buoyancy instability, which arises when heavy fluid lies above light. This arises naturally through the cooling of the Earth. The advance of the ICB releases material that is lighter both because it is hotter and because it is less rich in iron. Thus an unstable top heavy density stratification is created that is inherently unstable, giving rise to motions in which light material moves upwards to be replaced by denser descending fluid. Buoyancy instability is so simple that weak dissipative processes in the core cannot prevent it; they can only retard its growth. The molecular diffusivities ν, κ^T and κ^ξ of momentum, heat and composition are so tiny in the core that the instability can develop even on very short length scales. It is then natural to suppose that the resulting turbulence is also on such a short length scale that it transports mean momentum, entropy and composition in much the same way (though much more effectively) as molecular diffusion does, so that the fluxes of these quantities are then related to the gradients of the mean fields also.[1] This is sometimes called the 'Reynolds analogy' and the resulting theory is called 'local turbulence theory'.

The analysis of BR, like those of Braginsky (1964) and Braginsky and Meytlis (1990), is qualitative in nature. It rests on a simplified linear stability analysis of convection, and on some heuristic assumptions on how that linear theory would be modified if it were generalized to finite amplitudes. The simplified analysis concerns the stability of a plane layer of conducting fluid rotating with angular velocity $\mathbf{\Omega} = \Omega\mathbf{1}_z$ about the upward vertical Oz in the presence of a uniform horizontal magnetic field $\mathbf{B} = B_y\mathbf{1}_y$, which plays the role of a zonal field $\mathbf{B}_\phi = B_\phi\mathbf{1}_\phi$ in spherical geometry; a simple buoyancy instability develops because of an applied downward temperature gradient. (We use $\mathbf{1}_q$ for the unit vector in the direction of increasing coordinate q.) For the branch of solutions we examine[2] the Coriolis force suppresses motions perpendicular to $\mathbf{\Omega}$ and the Lorentz force impedes flows perpendicular to **B**. Thus the fastest growing perturbations are those that are elongated in both the z- and y-directions, and are smallest in the x-direction.

We assume therefore that the predominant turbulent cells have a 'plate-like' form with dimensions $\ell_y \sim \ell_z \equiv \ell_\parallel$ and $\ell_x \equiv \ell_\perp \ll \ell_\parallel$; the corresponding characteristic turbulent velocities are $V'_y \sim V'_z \equiv V'_\parallel$ and $V'_x \equiv V'_\perp \sim (\ell_\perp/\ell_\parallel)V'_\parallel \ll V'_\parallel$. We visualize the turbulence in the following way. On the background of locally unstable stratification ($\mathbf{g}_a \cdot \nabla \overline{C} < 0$; see below), perturbations in the form of packets of plate-like cells grow exponentially. When the amplitudes of the convective cells become sufficiently large for strongly nonlinear effects to come into play, the cells are destroyed and smoothed out. In their place new packets of cells are born, grow and die in their turn, in a never ending sequence. The entire fluid, or more precisely all the fluid in which $\mathbf{g}_a \cdot \nabla \overline{C} < 0$, is filled by such turbulent cells, in various stages of development.

While our qualitative estimates (Appendix C of BR),

$$\ell_\parallel \sim 50\,\text{km}, \quad \ell_\perp \sim 2\,\text{km}, \quad V'_\parallel \sim 2 \times 10^{-5}\,\text{m s}^{-1},$$

are highly uncertain, the basic premise of the local theory looks very plausible because the joint action of buoyancy, Coriolis forces and Lorentz forces in generating plate-like cells is readily understood. The numerical simulations of St Pierre (1996) provide further corroboration.

This picture of local turbulence, if correct, has far-reaching consequences. First, the relevant magnetic Reynolds number, R'_m, of the turbulence is very small, and the perturbations of the magnetic field are tiny. They can be estimated from equating $\overline{\mathbf{B}} \cdot \nabla \mathbf{V}'$ and $\eta \nabla^2 \mathbf{B}'$ in order of magnitude. This gives $\overline{B}V'_\parallel/\ell_\parallel \sim \eta B'_\parallel/\ell_\perp^2$, or $B'_\parallel \sim R'_m \overline{B}$ where $R'_m \sim V'_\parallel \ell_\perp^2/\ell_\parallel \eta \sim \ell_\perp V'_\perp/\eta$. If we assume that the turbulent diffusivity is $\kappa'_\perp \sim \ell_\parallel V'_\parallel \sim \eta$, then $R'_m \sim \ell_\perp^2/\ell_\parallel^2$, which is very small for the plate-like cells. This implies that the turbulent resistivity is negligibly small, and the same is true of the turbulent α-effect. The averaged equations for the magnetic field have the same simple forms as they had before averaging.

The main effect, and a very strong effect too, of small-scale turbulence is that of enhancing the transport of heat (entropy) and composition. And because both the entropy and composition are mixed by the turbulent motions in the same way, the tensor diffusivity is the same for both: $\overset{\leftrightarrow St}{\kappa} = \overset{\leftrightarrow \xi t}{\kappa} \equiv \overset{\leftrightarrow t}{\kappa}$ (say); see Table 3.6. Numerically, this tensor is rather uncertain, but rough estimates, such as $\kappa_{zz} \sim \kappa_{\phi\phi} \equiv \kappa_\parallel \sim \eta$ and $\kappa_{ss} \equiv \kappa_\perp \sim 10^{-2}\eta$, look reasonable. We should also recognize that $\overset{\leftrightarrow t}{\kappa}$ is not a constant; it depends on $\overline{\mathbf{B}}$ and $\nabla \overline{C}$. According to Braginsky and Meytlis (1990), $\kappa_\parallel$ is proportional to $\partial_z \overline{C}$ and increases rapidly with $\overline{B}$. The details need further investigation of course, but it is clear that $\overset{\leftrightarrow t}{\kappa}$ is inhomogeneous and strongly anisotropic.

An additional restriction implied by our local picture of small-scale turbulence was noted in Section 3.4; σ' cannot be negative. Using the expression given there for $\mathbf{I}^{Ct}$ and the forms for $\mathbf{I}^{St}$ and $\mathbf{I}^{\xi t}$ shown in Table 3.6, we see that

$$\mathbf{I}^{Ct} = \alpha^S \rho_a \overset{\leftrightarrow t}{\kappa} \cdot \nabla \overline{S}_c + \alpha^\xi \rho_a \overset{\leftrightarrow t}{\kappa} \cdot \nabla \overline{\xi}_c = \rho_a \overset{\leftrightarrow t}{\kappa} \cdot (\alpha^S \nabla \overline{S}_c + \alpha^\xi \nabla \overline{\xi}_c),$$

which, if α^S and α^ξ can be assumed constant, gives

$$\mathbf{I}^C = -\rho_a \overset{\leftrightarrow t}{\kappa} \cdot \nabla \overline{C}, \quad \text{and} \quad T_a \sigma' = -\rho_a \mathbf{g}_a \cdot \overset{\leftrightarrow t}{\kappa} \cdot \nabla \overline{C}.$$

Table 3.6 Local turbulence theory

$$\mathbf{I}^{St} \equiv \rho_a \overline{S'\mathbf{V}'} = -\rho_a \overset{\leftrightarrow^t}{\boldsymbol{\kappa}} \cdot \nabla \overline{S}_c, \quad \mathbf{I}^{\xi t} \equiv \rho_a \overline{\xi'\mathbf{V}'} = -\rho_a \overset{\leftrightarrow^t}{\boldsymbol{\kappa}} \cdot \nabla \overline{\xi}_c$$

$$\overline{\sigma}^S = \sigma^R + \sigma^T + \sigma^J + \sigma^t, \quad T_a\sigma^t = \rho_a \mathbf{g}_a \cdot \overline{C'\mathbf{V}'} = \mathbf{g}_a \cdot \mathbf{I}^{Ct}$$

Thus, in the simple case of isotropic $\overset{\leftrightarrow^t}{\boldsymbol{\kappa}}$, we may divide the core into regions of stable stratification where $\mathbf{g} \cdot \nabla \overline{C} > 0$ and regions of unstable stratification where $\mathbf{g} \cdot \nabla \overline{C} < 0$. Local turbulence can occur only in the latter because, by the second law of thermodynamics, σ^t cannot be negative anywhere. Similarly, in the anisotropic case, turbulence occurs only where $\mathbf{g}_a \cdot \overset{\leftrightarrow^t}{\boldsymbol{\kappa}} \cdot \nabla \overline{C} < 0$, and elsewhere we must take $\overset{\leftrightarrow^t}{\boldsymbol{\kappa}} \equiv \mathbf{0}$. This restriction can be rather inconvenient for numerical work, but is dictated by the demand that entropy production should be nonnegative everywhere. Of course, in reality turbulence can arise even in stably stratified regions, either by (nonlocal) penetration from adjacent turbulent regions, or through other physical causes. This cannot be dealt with by the present local theory, and we could only return to the general expressions for $\mathbf{I}^S$, $\mathbf{I}^{\xi}$ and the associated entropy production, as given by molecular theory (e.g. Landau and Lifshitz, 1987).

As we saw in Section 3.4, Reynolds stresses of both kinetic and magnetic type contribute to the turbulent viscosity which, though very difficult to estimate, must be highly anisotropic. We may assume however that the viscosity (turbulent and molecular alike) is significant only in thin layers. On areas of the CMB and ICB where $\mathbf{\Omega}$ has a nonzero normal component, thin Ekman layers develop of thickness $\delta_{\Omega} \sim \sqrt{(\nu^m/\Omega)}$, where ν^m is the molecular viscosity. The turbulent viscosity is ineffective in these layers because the mixing length in the layer is less than δ_{Ω}. The turbulent viscosity, $\nu_{\perp}^t \sim \ell_{\perp} V_{\perp}^t \sim 10^{-2}\eta$, is however much more significant than the molecular viscosity, $\nu^m \sim 10^{-6}\eta$, in internal shear layers surrounding the tangent cylinder, i.e. the imaginary cylinder parallel to $\mathbf{\Omega}$ and touching the inner core on its equator.

It should be stressed that the local turbulence considered here differs strongly from the more commonly encountered form of hydrodynamic turbulence, in which kinetic energy is injected on the largest scales and cascades 'down the spectrum' to small length scales where it is transformed into heat by viscosity. Local turbulence in the core acquires its energy from the work done by the Archimedean force on the growing cells. This energy is dissipated ohmically by the turbulent electric currents in the same cells, i.e. on the same length scale; viscous dissipation is of secondary significance. This is why the turbulent dissipation is given by $Q^t \approx \mu_0 \eta \overline{J'^2}$; see Appendix C of BR. This is why we may assume that the spectrum of the local turbulence has a maximum for the dominating cells, i.e. those for which the instability has the largest growth rate.

The energy dissipation Q^t can be easily estimated: $Q^t \sim \rho_a |\mathbf{g}_a \cdot \overset{\leftrightarrow^t}{\boldsymbol{\kappa}} \cdot \nabla \overline{C}| \sim \rho_a g_a \overline{C} \kappa_{\parallel}/L$. This may be compared with the rate of working of the Archimedean force driving the convection $A \sim \rho_a g_a \overline{C} \cdot \mathbf{V}_M \sim \rho g \overline{C} V_M$, where V_M is a typical meridional velocity $\mathbf{V}_M$. It follows that $Q^t/A \sim \kappa_{\parallel}/V_M L$. If we assume that $\kappa_{\parallel} \sim \eta$ and $V_M L \sim \eta$, we obtain $Q^t \sim A$. A more accurate estimate may involve a rather large numerical constant.

The mathematical consequences of the local turbulence theory described here are summarized in Table 3.6.

3.6. Governing convection equations

We have at last completed the anelastic theory that determines the mean MHD of the core, i.e. the equations that govern $\overline{\mathbf{V}}, \overline{\mathbf{B}}, \overline{S}_c, \overline{\xi}_c, \ldots$ (step 6A of Table 3.1), but we shall soon dispense with the overbars; see Table 3.7.

Table 3.7 Governing convection equations

$$d_t \mathbf{V} = -\nabla \Pi + C\mathbf{g} - 2\,\mathbf{\Omega} \times \mathbf{V} + \mathbf{F}^B + \mathbf{F}^\nu, \quad \nabla \cdot (\rho_a \mathbf{V}) = 0$$
$$\partial_t \mathbf{B} = \nabla \times (\mathbf{V} \times \mathbf{B}) - \nabla \times (\eta \nabla \times \mathbf{B}), \quad \nabla \cdot \mathbf{B} = 0$$

$$\rho_a d_t \xi_c = -\nabla \cdot \mathbf{I}^\xi - \rho_a \dot{\xi}_a. \qquad\qquad \mathbf{I}^\xi = -\rho_a \overset{\leftrightarrow}{\kappa}{}' \cdot \nabla \xi_c$$

$$\rho_a d_t S_c = -\nabla \cdot \mathbf{I}^S - \rho_a \dot{S}_a + \sigma^S. \qquad \mathbf{I}^S = -\rho_a \overset{\leftrightarrow}{\kappa}{}' \cdot \nabla S_c$$
$$C = -\alpha^S S_c - \alpha^\xi \xi_c$$

Simplified Notation
$$\overline{d}_t \to d_t, \overline{\mathbf{V}} \to \mathbf{V}, \overline{\mathbf{B}} \to \mathbf{B}, \mathbf{g}_a \to \mathbf{g}, \overline{S}_c \to S_c, \overline{\xi}_c \to \xi_c, \overline{\Pi} \to \Pi,$$
$$\mathbf{I}^{St} \to \mathbf{I}^S, \mathbf{I}^{\xi t} \to \mathbf{I}^\xi, \mathbf{F}^{\overline{B}} \to \mathbf{F}^B, \ldots$$

Definition of entropy production rate
$$\sigma^S = \sigma^R + \sigma^T_- + \sigma^J + \sigma^\nu + \sigma^t, \quad \sigma^t = \rho_a \mathbf{g} \cdot \overset{\leftrightarrow}{\kappa}{}' \cdot (\alpha^S \nabla S_c + \alpha^\xi \nabla \xi_c) / T_a$$
$$\sigma^T_- = -T_a^{-1} \nabla \cdot \mathbf{I}^q_a = \sigma^T - \nabla \cdot [T_a^{-1} \mathbf{I}^q_a], \quad \mathbf{I}^q_a = -K^T \nabla T_a$$

Note

Entropy production by molecular conduction down the adiabatic gradient is $\sigma^T = K^T (T_a^{-1} \nabla T_a)^2$; it is positive, but σ^T_- need not be, and is in fact negative.

Solutions are required to satisfy a number of boundary conditions, the most challenging of which arise at the ICB. These require knowledge of the properties of the liquidus sketched in Fig. 3.2, and govern the release of latent heat and light constituent; they therefore determine the fluxes $\mathbf{I}^{St}$ and $\mathbf{I}^{\xi t}$ at the ICB. They raise complicated issues, and the numerical values of the pertinent coefficients are very uncertain. The topic was analyzed both in Section 3.6 and Appendix E of BR but will not be discussed here.

The magnetic field and the tangential components of the electric field are required to be continuous across the ICB and the FOC. There are several choices for the remaining boundary conditions. Glatzmaier and Roberts (1996a, 1997) adopted the no-slip conditions on the CMB and ICB. The SIC was allowed to turn about the geographical axis, which is parallel to $\mathbf{\Omega}$, in response to the viscous and electromagnetic torques to which the FOC subjects it. They assumed the SIC to be electrically conducting, with the same magnetic diffusivity as the FOC.

The mass flux of each constituent across the CMB, $r = R_1$, was supposed to be zero, so that $I_1^\xi \equiv I_r^{\xi t}(R_1) = 0$. Using the expression $\mathbf{I}^q = T_a \mathbf{I}^S + \mu_a \mathbf{I}^\xi$ from Table 3.1 and $I_1^\xi = 0$, we see that the turbulent entropy flux, I_1^{St}, on the CMB is given by $T_a I_1^{St} = I_1^q + (K^T \partial_r T_a)_1$, where the last term is evaluated on $r = R_1$. We should recall here that S_c and ξ_c stand for the small and smoothly varying quantities $\overline{S}_c$ and $\overline{\xi}_c$, while $\mathbf{I}^S$ and $\mathbf{I}^\xi$ stand for the turbulent fluxes, $\mathbf{I}^{St}$ and $\mathbf{I}^{\xi t}$. Thus $\mathbf{I}^{St}$ does not include the molecular flux of entropy down the adiabat, $-T_a^{-1} K^T \nabla T_a$. Glatzmaier and Roberts (1996a) specified the heat flow I_1^q from core to mantle to be 7.2TW, of which 5.2TW was the flux down the adiabat. They examined alternatives in a later paper (Glatzmaier and Roberts, 1998).

Glatzmaier and Roberts again adopted no-slip conditions in their B-simulations, but Kuang and Bloxham (1997) supposed instead that the ICB and CMB are stress-free.

It is interesting to compare the equations displayed in Table 3.6 with the corresponding equations used in atmospheric and ocean physics, and in discussions of the Earth's global oscillations. Our choice ($\nabla S_a = \nabla \xi_a = 0$) of the adiabatic state as reference state, which led to the simple expression shown in Table 3.4 for the combined gravity-pressure force, $\mathbf{f} = \mathbf{g} - \rho^{-1} \nabla P$, is motivated by the assumption that the core is well mixed. The Earth's atmosphere, oceans and mantle are not well mixed; see, e.g. figs. (6.4.2) and (6.4.3) in Pedlosky's (1979) book, which display the variation of N^2 in the atmosphere and ocean, and also fig. 2.1 of Monin's (1990) book which depicts various temperature and salinity distributions in the oceans. Under these circumstances, it is natural to choose a reference state that is not adiabatic.

Let us denote quantities in such a state by the suffix 0 and convective perturbations away from that state by 1. In the equilibrium state $\mathbf{f}_0 = \mathbf{g}_0 - \rho_0 \nabla P_0$ is zero, and the Brunt Väisälä frequency N_0 is given by $N_0^2 = \mathbf{g}_0 \cdot (\rho_0^{-1} \nabla \rho_0 - v_P^{-2} \mathbf{g}_0) = -\mathbf{g}_0 \cdot (\alpha^S \nabla S_0 + \alpha^\xi \nabla \xi_0)$, where the coefficients v_P, α^S and α^ξ, here and below, are evaluated in the new reference state. The gravity-pressure force arises from the perturbation and is

$$\mathbf{f}_1 = -\nabla \left(\frac{P_1}{\rho_0} + U_1 \right) + C_1 \mathbf{g}_0 - N_0^2 \left(\frac{P_1}{\rho_0 g_0^2} \right) \mathbf{g}_0 = \mathbf{f}_1^{(1)} + \mathbf{f}_1^{(2)} + \mathbf{f}_1^{(3)} \text{ (say)},$$

where $C_1 = -\alpha^S S_1 - \alpha^\xi \xi_1$. The first two terms are reminiscent of the expression $\mathbf{f}_c = -\nabla \Pi_c + C \mathbf{g}_a$ obtained earlier for the adiabatic reference state. The third term, $\mathbf{f}_1^{(3)} = -N_0^2 (P_1/\rho_0 g_0^2) \mathbf{g}_0$ is zero when the 0-state is adiabatic ($N_0 = 0$). Otherwise the ratio $|\mathbf{f}_1^{(3)}|/|\mathbf{f}_1^{(1)}|$ may be estimated as $N_0^2 H/g_0 \sim \beta H/L_0$, where $L_0 \sim v_P^2/g_0$ is a scale height for the density ρ_0, H is the thickness of the convecting layer (so that $\nabla (P_1/\rho_0) \sim (P_1/\rho_0) H^{-1}$), and $\beta = (N_0 v_P/g_0)^2 = 1 - \partial_r \rho_0/\partial_r \rho_a$ is a coefficient that is sometimes called the 'stability factor'; $\beta < 0$ means stability. For a thin layer in which $H \ll L_0$, the density changes only slightly across the layer, and $\mathbf{f}_1^{(3)}$ is small compared with $\mathbf{f}_1^{(1)}$. The Boussinesq approximation is applicable in this case: ρ_0 is assumed to be constant, and the approximation $\mathbf{f}_1 = -\rho_0^{-1} \nabla P_1 - \nabla U_1 + C_1 \mathbf{g}_0$ is used, which resembles the expression $\mathbf{f}_c = -\nabla \Pi_c + C \mathbf{g}_a$, though with ρ_a replaced by the constant ρ_0.

The validity of the Boussinesq approximation is obvious for small systems where the pressure has a negligible effect on the density, so that the Archimedean force, $C_1 \mathbf{g}_0$, is proportional to the quantity we have called the co-density. In this context, αT_1 usually replaces $\alpha^S S_1$ in the expression for C_1; e.g. see section 56 of Landau and Lifshitz (1987). A systematic derivation of the Boussinesq equations for geophysical systems has been carried out by Spiegel and Veronis (1960), Veronis (1962, 1973) and others. Veronis (1973) assumed that the convecting layer is thin but supposed that the reference state is adiabatic, and he was led to the Boussinesq equations with constant ρ_0. These equations provide an adequate description of the Earth's oceans, in which the maximum observed variations in ρ_0 are about 5%. The pressure term in the Boussinesq equation of motion is written as $-\rho_0^{-1} \nabla P_1$, where ρ_0 is constant. It is remarkable that the generalization to the anelastic approximation is achieved for the adiabatic reference state simply by moving the density ρ_a, which depends on r, inside the gradient term as $-\nabla (P_1/\rho_a)$! The density ρ_0 varies significantly in the Earth's atmosphere. In this case the thinness of the convecting layer is commonly used to justify simplifications of the governing equations. Various approximations have been devised, and it is sometimes possible to find the special combination $-\nabla (P_1/\rho_0)$; e.g. see Pedlosky (1979), where the vertical component of $-\nabla (P_1/\rho_0)$ may be seen in the rather elaborate equation (6.5.32).

Monin's (1990) book considers the fluid mechanics of atmosphere, oceans and core. In section 3 of chapter 2, he points out that 'If ρ_0 varies greatly with height within the layer being considered (for instance, in the convective layer of the Sun, by a factor of 10^6), then it may be convenient to introduce ρ_{00} from the relation $\rho_{00}^{-1} \mathbf{g}_0 \cdot \nabla \rho_{00} = N_0^2$ (where ρ_{00} is analogous to the potential density), as well as $\mathbf{f}_1 = -\rho_{00}^{-1} \nabla (\rho_{00} P_1/\rho_0) - \alpha^S S_1 \mathbf{g}_0$. For small $N_0^2 H/g_0$ (where H is the thickness of the layer) we can assume that $\rho_{00} \approx$ const, and the first term on the right-hand side will be approximately potential, while the second term is proportional to the linearized entropy'. (We have slightly modified his text, mainly to bring it into conformity with our notation.)

The gradient $\nabla(P_1/\rho_0 + U_1) = -\mathbf{f}_1^{(1)}$ of the reduced pressure, has often been employed in papers that deal with the subseismic (i.e. anelastic) oscillations of the Earth's core relative to a basic density distribution, ρ_0, that is not necessarily adiabatic; see, e.g. Smylie and Rochester (1981) and Crossley (1984). These authors retained the $\mathbf{f}_1^{(1)}$ and $\mathbf{f}_1^{(2)} \equiv C_1 \mathbf{g}_0$ in the expression for $\mathbf{f}_1$ displayed above, but they neglected $\mathbf{f}_1^{(3)}$. This is a good approximation if the stability factor β is sufficiently small. For the motions they had under consideration, they could neglect all forms of diffusion, so that $C_1 = -\mathbf{u} \cdot \nabla C_0$, leading them to $\mathbf{f}_1^{(2)} = -\mathbf{1}_r u_r N_0^2$. This approach facilitates the analysis of core oscillations for specified N_0, but it is inadequate for the study of core convection.

It may be seen from this short review of the literature that all the ideas that led BR to the simplification of the equation of motion shown in Tables 3.4 and 3.7, could be found in earlier, well-known publications. Surprisingly, however, we have not seen any previous publication in which all the ideas were combined to give our simple equation (for the case where there is a significant change in the equilibrium density). Its simplicity is a result of the choice of the adiabatic state as the reference state, and in the compressible case this choice is reasonable if and only if the fluid is well mixed. (In the Boussinesq case $\rho_0 \approx$ constant, the choice of reference state is less crucial; almost any choice is acceptable.) We have two reasons for believing that the core is well mixed. First, geomagnetic observations suggest that $V \sim 5 \times 10^{-4}\, \mathrm{m\, s^{-1}}$ is typical of core flow speeds, and the simple estimate $Cg \sim 2\Omega V$ of the co-density then gives $C \sim 10^{-8}$, which is very small indeed. It is beyond belief that the individual contributions $\alpha^S S_c$ and $\alpha^\xi \xi_c$ to $C = -\alpha^S S_c - \alpha^\xi \xi_c$ could cancel each other out to this precision, and it is much more plausible that $\alpha^S S_c$ and $\alpha^\xi \xi_c$ are individually also of order 10^{-8}. If we take $\alpha^S \sim 10^{-4}\, \mathrm{kg\, J^{-1}\, K}$ and $\alpha^\xi \sim 0.5$ (see BR), we obtain $S_c \sim 10^{-4}\, \mathrm{J\, kg^{-1}\, K^{-1}}$ and $\xi_c \sim 10^{-8}$, which represent very small departures from the well mixed state. Second, local turbulence provides an effective mechanism for homogenizing S and ξ. We have argued in Section 3.5 that, on the macroscale, the turbulent diffusivity for both S_c and ξ_c is of order $\eta \sim 2\, \mathrm{m^2\, s^{-1}}$ (except in the direction perpendicular to $\mathbf{\Omega}$ and $\mathbf{B}$). The secular cooling of the SIC creates an outward flux of light constituent that BR estimates to be $I_r^\xi \sim 10^{-9}\, \mathrm{kg\, m^{-2}\, s^{-1}}$ on the ICB. Supposing this to be typical of the entire FOC, and taking $I_r^\xi \sim -\rho_a \kappa_\| \partial \xi_c/\partial r$, we see that $|\partial \xi_c/\partial r| \sim 10^{-13}\, \mathrm{m^{-1}}$, which suggests that $\xi_c \sim 10^{-7}$. The A-simulations of Glatzmaier and Roberts, where $\kappa' = \eta$ was assumed, demonstrated quantitatively both that C is small and that $\alpha^S S_c$ and $\alpha^\xi \xi_c$ have similar magnitudes.

It is of interest to estimate $\mathbf{g} \cdot \nabla C$ in the core; this is similar to N^2 but is predominantly negative because in the main part of the core $\partial C/\partial r > 0$, corresponding to convective instability. Taking $L \sim 10^3\, \mathrm{km}$ as a characteristic length and $C \sim 10^{-8}$, we have $(gC/L)^{1/2} \sim 3 \times 10^{-7}\, \mathrm{s^{-1}}$. This contrasts strongly with the situations in the oceans and atmosphere, where N is of the order of, or greater than, the Coriolis frequency $2\Omega = 1.4 \times 10^{-4}\, \mathrm{s^{-1}}$.

It should be noted that neither of our reasons for supposing the core to be well mixed applies to a thin layer at the top of the core, just below the solid CMB. If some small fraction of the light admixture released at the ICB is incompletely mixed as it rises through the main body of the FOC, its C (though small) may greatly exceed 10^{-8} when it reaches the CMB, and it may be captured there by its own buoyancy, to form part of a thin but very stably stratified layer. A similar accumulation of light material could result from percolation from the mantle to the core. Braginsky (1993) estimated both the thickness H of this layer, and its Brunt–Väisälä frequency N. He did this by making use of the observed 65-year period in the variations in the geomagnetic dipole and in the length of the day. He demonstrated that both these phenomena can be explained in terms of MAC wave oscillations of the stably stratified layer, and he obtained $H \sim 80\,\mathrm{km}$ and $N \sim 2\Omega$. These values make the dynamical properties of this layer rather similar to those of the oceans on the surface of the Earth; see Braginsky (1999). If the existence of this layer is confirmed, it will be a 5th ocean of the Earth, but one that is upside-down and situated beneath the mantle rather than on top of it. Its possible influence on convection in the core and on the geodynamo is beyond the scope of the present chapter.

3.7. The future

When the final equations of Section 3.6 are solved, they are usually simplified by ignoring radioactive sources ($\sigma^R = 0$), by using isotropic turbulent diffusivities ($\kappa^t_{ij} = \kappa^t \delta_{ij}$), and by neglecting the centrifugal force ($\epsilon_\Omega = 0$). In addition, the large-scale inertial forces are often neglected ($d_t \mathbf{V} = \mathbf{0}$). For numerical reasons, the viscous force has to be retained, and even enhanced beyond reasonable turbulent values; hyperdiffusion is also commonly required.

The theory can also be simplified by adopting the Boussinesq approximation; see entry 6B of Table 3.0. This consists of taking $\epsilon_a = 0$, so that the reference state and all its associated functions depend only on t_a. The way that the theory of Section 3.5 can be reduced to Boussinesq form is discussed in detail in section 3.8 of BR. To date most simulations of the geodynamo have used the Boussinesq approximation, and have even specified continuous buoyancy sources, so that the system is completely steady, i.e. independent of t_a see, e.g. Simulations B. Usually compositional buoyancy is also omitted on the grounds that its effect can be qualitatively similar to that of thermal buoyancy. Nevertheless, simulations A did not make these simplifications and integrated inhomogeneous systems closely based on the ideas presented here and in BR.

Though modern supercomputers are very powerful, the apparently simple road to a geo-dynamo model (equations → choice of numerical method → construction of a computer program → number crunching → final result) contains significant pitfalls. These are of two main types. First, some of the parameters needed for a reliable simulation are very poorly, or completely, unknown. These include the radioactive heating, Q^R, the heat leaving the core, I_1^q, and how it is distributed over the CMB, and several physical constants related to freezing on the ICB that influence the boundary conditions determining the co-density flux, $I_2^C \equiv I_r^C(R_2, \theta, \phi)$, where $r = R_2$ is the ICB. Second, and possibly even more serious, is the lack of an accurate description of turbulence in the core. Our approach, a local description, may not be sufficient. Other types of turbulence are conceivable, based on other instability mechanisms that operate on longer length scales. Whether these are significant or not is an open question that we have not addressed here. It can be investigated using the equations that we have developed in this paper, and it is clear that the small scale turbulence we have analyzed will have a stabilizing effect on all larger scale motions including the turbulence driven by the alternative instability mechanisms we have just mentioned. This matter can, perhaps, be investigated by modeling these mechanisms numerically.

Local turbulence is generally on too short a length scale and on too short a time scale to be resolvable by global numerical simulations, but enlightenment may be possible through direct numerical simulations (DNS) of small sub-regions in the core, across which the large-scale fields are almost uniform and are supposed known, in much the same way as the simple planar model described in Section 3.5. A preliminary DNS has recently been reported by Matsushima *et al.* (1999). Perhaps experimental investigations will also be conducted in the future that will help to elucidate this central issue. When the parameterization of turbulence by such theoretical and experimental means is in a more highly developed state, its results will be incorporated into global simulations.

No matter how great our efforts to derive correctly the equations governing core MHD and the geodynamo, our final theory will contain some unknown parameters (UP). These include the radioactive heat source, Q^R, the turbulent diffusivities, $\overset{\leftrightarrow}{\kappa}^I$, and the fluxes I_1^q and I_2^C, that determine the 'feeding' of the geodynamo from the top and the bottom of the FOC. These parameters can only be found by fitting the geodynamo solution to the set of available observational parameters (OP); see the discussion by Braginsky (1997). If the number of OP exceeds the number of UP, we will be able to estimate and cross check the UP. In this way we will confirm that our model is indeed realistic. We will also be able to derive significant information about the physical state and properties of the core, and will be able to measure the turbulent transport coefficients.

The success of geodynamo simulations can only be judged by very careful comparisons of their findings with the known observational facts about the geomagnetic field, past and present. This will provide the acid test of whether the journey along the road to a realistic geodynamo model has been successfully completed.

Acknowledgements

One of us (PHR) wishes to thank NSF for support under Grant NSF EAR97-25627, during the tenure of which this chapter was written.

Appendix A: some notation and numerical values

We summarize here some of the values and notation used in the main body of this chapter. This notation is in several respects slightly simpler than that employed in BR, a paper that introduced several forms of average. BR used $\langle Q \rangle^I$ for the mean of a field Q over the turbulent ensemble and Q^+ for the fluctuating part of Q. In the present chapter, we have not used many averages, and could employ $\overline{Q}$ instead of $\langle Q \rangle^I$ and Q' in place of Q^+; we have also adopted P instead of p, Π in place of P, and (to conform to the usual way that the longitudinal sound wave is denoted in seismology) v_P instead of u_S.

Table 3.A.1 presents some notation and some values of pertinent geophysical quantities. The well-determined parameters are from the Preliminary Earth Reference Model (PREM) of Dziewonski and Anderson (1981); the thermodynamic parameters and less well-determined parameters are from appendix E of BR.

Appendix B: corrections to BR

General remarks:
Our paper, submitted to *Geophysical and Astrophysical Fluid Dynamics* in April 1994, appeared in its final revised form first as a widely distributed preprint,

Braginsky, S.I. and Roberts, P.H., "Equations governing convection in Earth's core and the Geodynamo," IGPP Report, UCLA, November 17, 1994, and then as the published paper referenced earlier. Its main results were presented by one of us (SIB) at the SEDI Meeting held in Whistler, British Columbia, Canada 11–14 August 1994. Buffett and Lister also presented a talk at the Whistler meeting entitled 'The relative importance of thermal and compositional convection in the dynamo problem'. This agreed with one of our principal conclusions, that thermal buoyancy is as important as compositional

Table 3.A.1 Notation and geophysical magnitudes

Well-determined parameters

$R_E = 6.371 \times 10^6$ m	Average radius of the Earth
$R_1 = 3.480 \times 10^6$ m	Radius of the fluid outer core (FOC)
$R_2 = 1.2215 \times 10^6$ m	Radius of the solid inner core (SIC)
$\rho_0 = 10.9 \times 10^3$ kg m^{-3}	Mean density of the FOC
$\rho_1 = 9.9 \times 10^3$ kg m^{-3}	Density of the FOC at the CMB
$\rho_2 = 12.166 \times 10^3$ kg m^{-3}	Density of the FOC at the ICB
$\rho_N = 12.764 \times 10^3$ kg m^{-3}	Density of the SIC at the ICB
$\rho(0) = 13.088 \times 10^3$ kg m^{-3}	Density at the geocenter
$\Delta\rho = \rho_N - \rho_2 = 0.6 \times 10^3$ kg m^{-3}	Density jump at the ICB (relatively poorly known)
$g_1 = 10.68$ m s^{-2}	Acceleration due to gravity at the CMB
$g_2 = 4.40$ m s^{-2}	Acceleration due to gravity at the ICB
$P_1 = 135.75$ GPa	Pressure at the CMB
$P_2 = 328.85$ GPa	Pressure at the ICB
$P(0) = 363.85$ GPa	Pressure at the geocenter
$v_{P1} = 8.065$ km s^{-1}	Seismic velocity in the FOC at the CMB
$v_{P2} = 10.356$ km s^{-1}	Seismic velocity in the FOC at the ICB

Thermodynamic parameters

$T_1 = 4000\,^\circ$K	Temperature of the CMB
$T_2 = 5300\,^\circ$K	Temperature of the ICB
$T_0 = 4590\,^\circ$K	Average temperature of the FOC
$\Delta T_{12} \equiv T_2 - T_1 = 1300\,^\circ$K	Temperature contrast across the FOC
$\gamma_1 = 1.35$	Grüneisen parameter at the CMB
$\gamma_2 = 1.27$	Grüneisen parameter at the ICB
$\alpha_1 = 1.8 \times 10^{-5}\,^\circ$K^{-1}	Thermal coefficient of volume expansion at the CMB
$\alpha_2 = 1.0 \times 10^{-5}\,^\circ$K^{-1}	Thermal coefficient of volume expansion at the ICB
$c_{p1} = 866$ J kg$^{-1}\,^\circ$K^{-1}	Specific heat at constant pressure at the CMB
$c_{p2} = 842$ J kg$^{-1}\,^\circ$K^{-1}	Specific heat at constant pressure at the ICB
$\alpha_1^S = 8.47 \times 10^{-5}$ kg J$^{-1}\,^\circ$K	Entropy coefficient of volume expansion at CMB
$\alpha_2^S = 6.28 \times 10^{-5}$ kg J$^{-1}\,^\circ$K	Entropy coefficient of volume expansion at ICB

Less well-determined parameters

$h^\xi = -10^7$ J kg^{-1}	Heat of reaction
$h_L = 10^6$ J kg^{-1}	Latent heat of crystallization
$\eta = 2$ m^2 s^{-1}	Magnetic diffusivity of FOC
$\eta_N = 1.5$ m^2 s^{-1}	Magnetic diffusivity of SIC
$\kappa^T = 4.8 \times 10^{-6}$ m^2 s^{-1}	Thermal diffusivity of FOC
$\kappa^\xi = 10^{-9}$ m^2 s^{-1}	Compositional diffusivity of FOC
$\nu = 10^{-6}$ m^2 s^{-1}	Kinematic viscosity of FOC
$\xi_a = 0.15$	Mass fraction of light constituent in FOC
$\xi_N = 0.1$	Mass fraction of light constituent in SIC
$\Delta T_m = 700\,^\circ$K	Depression of melting point through alloying

buoyancy in powering the geodynamo, a conclusion also drawn in the first of the following related works:

Buffett, B.A., Huppert, H.E., Lister, J.R. and Woods, A.W., "Analytical model for solidification of the Earth's core," *Nature*, **356**, 329–331 (1992).
Lister, J.R. and Buffett, B.A., "The strength and efficiency of thermal and compositional convection in the geodynamo," *Phys. Earth Planet. Inter.* **91**, 17–30 (1995).
Buffett, B.A., Huppert, H.E., Lister, J.R. and Woods, A.W., "On the thermal evolution of the Earth's core," *J. Geophys. Res.* **101**, 7989–8006 (1996).

The significant difference between these works and ours is that they are principally concerned with general balances within an evolving Earth and do not derive the set of equations and boundary conditions governing the geodynamo. It may be particularly noticed that the rate of working of the Archimedean force is the sum of a 'macroscopic' part, which can be used to power the geodynamo, and a 'microscopic' part that is squandered uselessly as heat through the Joule dissipation of the small-scale electric currents associated with the local turbulence; see Section 3.5. The microscopic part is explicitly contained in our formulation but is not isolated in the works cited above.

Major corrections

In what follows we use the notation of BR and not the slightly different notation employed in the present chapter.

1. Page 25. *The Reynolds analogy.* Equation (4.34) would have been better written as

$$\langle \sigma^S \rangle^t = \sigma^{\langle v \rangle^t} + \sigma^{\langle J \rangle^t} + \sigma^R + \sigma^T - (\mathbf{I}^{St} \cdot \nabla T_a + \mathbf{I}^{\xi t} \cdot \nabla \mu_a)/T_a. \tag{4.34}$$

The use in our paper of the simpler notation $\langle \sigma^v \rangle^t$ instead of $\sigma^{\langle v \rangle^t}$, and of $\langle \sigma^J \rangle^t$ instead of $\sigma^{\langle J \rangle^t}$, gave the erroneous impression that both the macroscale and microscale parts of **V** and **J** were used to evaluate these viscous and ohmic sources of entropy, whereas only the macroscale fields (i.e. the fields averaged over the turbulent ensemble) are relevant in the application of the Reynolds analogy. The microscale contributions (i.e. the contributions from the local turbulence) are given by the last term in (4.34). This term has the same form as the corresponding term that arises from molecular diffusion, but (true to the Reynolds analogue) involves the turbulent fluxes $\mathbf{I}^{St}$ and $\mathbf{I}^{\xi t}$, instead of the molecular fluxes. According to Appendix C of BR,

$$-(\mathbf{I}^{St} \cdot \nabla T_a + \mathbf{I}^{\xi t} \cdot \nabla \mu_a)/T_a = \langle \sigma^j \rangle^t, \tag{a}$$

which is the ohmic entropy source from the turbulence, The expression (a) confirms the validity of the Reynolds analogy in our case. The viscous source from the microscale, $\langle \sigma^v \rangle^t$, is absent in (a) only because the magnetic Prandtl number, v/η, is very small. Otherwise this source of entropy would be added into the right-hand side of (a); see the discussion above (C22), an equation that should itself be replaced by

$$Q^v \equiv 2\rho_0 v \langle (e_{ij} e_{ij}) \rangle^t \sim (v/\eta)(\langle v^2 \rangle^t / \langle b^2 \rangle^t) Q^j \sim (v/\eta) Q^j. \tag{C22}$$

where e_{ij} is the rate of strain tensor defined by the turbulent velocity **v**.

2. Page 41. *The freezing condition on the SIC.* Equation (6.38) should be replaced by

$$\partial_t R_{2c}/R_2 = r_{2p} \partial_t p_c - r_{2S} \partial_t S_c - r_{2\xi} \partial_t \xi_c, \tag{6.38}$$

or (with appropriately defined zero levels for R_{2c}, p_c, S_c and ξ_c) by

$$R_{2c}/R_2 = r_{2p}p_c - r_{2S}S_c - r_{2\xi}\xi_c. \tag{6.38}$$

Minor corrections

1. Line below (2.32): (2.23) → (2.32).
2. '$= \mu_H + U_a$' should be deleted from footnote 5 on page 14.
3. Line below (6.22a,b), $I_2 \to I_2^\xi$.
4. 2 lines below (6.31b): (6.29b) → (6.29a,b).
5. 5 lines below (6.38): 'is emanating' → 'emanating'.
6. 7 lines below (6.42): (6.40c) → (6.32c).
7. Equation (8.23b) should be $A_2 I_2^X = -\mathcal{V}_{12}\sigma_2^X$.
8. In (8.25a), $\alpha_1^S \to \alpha_0^S$.
9. Delete t_{20}/t_2 from (8.40).
10. Although α^ξ/α_T^ξ and c_p/c_v are similar in magnitude (both being about 1.1), α^ξ/α_T^ξ in (D22) is incorrect and should be replaced by $\alpha^\xi/\alpha_T^\xi = 1 - \alpha h^\xi/\alpha_T^\xi c_p$.
11. Several small errors appear above (E1). For clarity, some lines above (E1) are given here in corrected form (see also Table E2):
    'Convenient tabulations have been provided by Stacey (1992), who gave for example $\gamma_1 \equiv \gamma(R_1) = 1.44$ and $\gamma_2 \equiv \gamma(R_2) = 1.27$. In later work (Stacey, 1994), he modified several of his estimates, and in particular took, as we shall, $\gamma_1 = 1.35$ and $\gamma_2 = 1.27\ldots$. Given the temperature $T_2 = 5300\,°K$ of the ICB...'.
12. Page 96. '$\kappa_i^T = 5.7 \times 10^{-6}\,\mathrm{m}^2\,\mathrm{s}^{-1}$' → '$\kappa_i^T = 4.76 \times 10^{-6}\,\mathrm{m}^2\,\mathrm{s}^{-1}$'.
13. 6 lines before the end of Appendix E: $\xi_2 h^\xi/c + pT_a \to \xi_{2N} h^\xi/c + pT_a$.

Notes

1 For an alternative approach, see Moffatt and Loper (1994).
2 We are not concerned here with motions and fields of large scale, such as MAC waves, where the Proudman–Taylor constraint exerted by the Coriolis forces is relaxed by the Lorentz forces, and the corresponding two-dimensional constraint exerted by the field $\overline{\mathbf{B}}$ is relaxed by the Coriolis force, so that three-dimensionality is largely restored. We are considering here small-scale motions, on which the magnetic field adds, in effect, only a further magnetic friction to oppose the flow; it does not release the Proudman–Taylor constraint.

References

Braginsky, S.I., "Magnetohydrodynamics of Earth's core," *Geomag. Aeron.* **4**, 698–712 (1964).

Braginsky, S.I., "MAC-oscillations of the hidden ocean of the core," *J. Geomagn. Geoelectr.* **45**, 1517–1538 (1993).

Braginsky, S.I., "On a realistic geodynamo model," *J. Geomagn. Geoelectr.* **49**, 1035–1048 (1997).

Braginsky, S.I., "Dynamics of the stably stratified ocean at the top of the core," *Phys. Earth Planet. Inter.* **111**, 21–34 (1999).

Braginsky, S.I. and Meytlis, V.P., "Local turbulence in the Earth's core," *Geophys. Astrophys. Fluid Dynam.* **55**, 71–87 (1990).

Braginsky, S.I. and Roberts, P.H., "Equations governing convection in Earth's core and the Geodynamo," *Geophys. Astrophys. Fluid Dynam.* **79**, 1–97 (1995). (This is referred to as 'BR' in the text.)

Crossley, D.J., "Oscillatory flow in the liquid core," *Phys. Earth Planet. Inter.* **36**, 1–16 (1984).

Dziewonski, A.M. and Anderson, D.L., "Preliminary reference Earth model," *Phys. Earth Planet. Inter.* **25**, 297–356 (1981).

Glatzmaier, G.A. and Roberts, P.H., "A three-dimensional convective dynamo solution with rotating and finitely conducting inner core and mantle," *Phys. Earth Planet. Inter.* **91**, 63–75 (1995a).

Glatzmaier, G.A. and Roberts, P.H., "A simulated geomagnetic reversal," *Nature*, **377**, 203–208 (1995b).

Glatzmaier, G.A. and Roberts, P.H., "An anelastic evolutionary geodynamo simulation driven by composition and thermal convection," *Physica D***97**, 81–94 (1996a).

Glatzmaier, G.A. and Roberts, P.H., "Magnetic sounding of planetary interiors," *Phys. Earth Planet. Inter.* **98**, 207–220 (1996b).

Glatzmaier, G.A. and Roberts, P.H., "Simulating the geodynamo," *Contemp. Phys.* **38**, 269–288 (1997).

Glatzmaier, G.A. and Roberts, P.H., "Dynamo theory then and now," *Int. J. Engng. Sci.* **36**, 1325–1338 (1998).

Kuang, W. and Bloxham, J., "An earth-like numerical dynamo model," *Nature*, **389**, 371–374 (1997).

Landau, L.D. and Lifshitz, E.M., *Fluid Mechanics*, 2nd edn, Pergamon, Oxford (1987).

Matsushima, M., Nakajima, T. and Roberts, P.H., "The anisotropy of turbulence in the Earth's core," *Earth Planets Space* **51**, 277–286 (1999).

Moffatt, H.K. and Loper, D.E., "The magnetostrophic rise of a buoyant parcel in the Earth's core," *Geophys. J. Int.* **117**, 394–402 (1994).

Monin, A.S., *Theoretical Geophysical Fluid Dynamics*, Kluwer, Dordrecht (1990).

Pedlosky, J., *Geophysical Fluid Dynamics*, Springer, Berlin (1979).

Sakuraba, A. and Kono, M., "Effect of the inner core on the numerical solution of the magnetohydrodynamic dynamo," *Phys. Earth Planet. Inter.* **111**, 105–121 (1999).

Smylie, D.E. and Rochester, M.G., "Compressibility, core dynamics and the subseismic wave equation," *Phys. Earth Planet. Inter.* **24**, 308–319 (1981).

Spiegel, E.A. and Veronis, G., "On the Boussinesq approximation for a compressible fluid," *Astrophys. J.* **131**, 442–447 (1960).

St Pierre, M.G., "On the local nature of turbulence in the Earth's outer core," *Geophys. Astrophys. Fluid Dynam.* **83**, 293–306 (1996).

Veronis, G., "The magnitude of the dissipation term in the Boussinesq approximation," *Astrophys. J.* **135**, 655–656 (1962).

Veronis, G., "Large scale ocean circulation," *Adv. Appl. Mech.* **13**, 2–92 (1973).

4   Dynamo action of magnetostrophic waves

Dieter Schmitt

Max-Planck-Institut für Aeronomie, Max-Planck-Str. 2, D-37191 Katlenburg-Lindau, Germany, E-mail: schmitt@linmpi.mpg.de

The stability of magnetostrophic waves is considered in a horizontal thin plane layer of a perfectly conducting fluid, which is stratified according to gravitation and permeated by a variable toroidal magnetic field. The layer rotates rigidly around an axis inclined to the horizontal plane. It is shown that unstable magnetostrophic waves, driven by magnetic buoyancy, are capable of inducing an electromotive force parallel to the toroidal magnetic field. This dynamic effect is an alternative to the kinematic α-effect and of importance for the dynamo theory of strong magnetic fields. For parameters of the lower convection zone of the sun the process leads to an effective α of a few cm/s. The dynamo action for various angles of inclination is discussed.

4.1. Introduction

Magnetic fields are responsible for many phenomena observed on the sun. The origin of the field is generally ascribed to inductive processes in its interior. Dynamo theory describes how a magnetic field is generated by electric currents which are induced by motions in an electrically conducting fluid. Two processes are most important, differential rotation and helical flows. The first winds up a poloidal magnetic field and generates a toroidal component. The latter is realized in rotating turbulent matter and regenerates the poloidal field components. It is most crucial for a dynamo.

The field generation mechanism was first described by Parker (1955). Rising eddies in the stratified convection zone expand and, in order to conserve angular momentum and due to the action of the Coriolis force, rotate. This cyclonic motion bends magnetic field lines to loops which are twisted and form field components perpendicular to the original field. The effect of these small-scale motions on the large-scale magnetic field has been systematically investigated within the framework of mean-field electrodynamics which was established by Steenbeck *et al.* (1966). They formally showed that helical motions drive a mean electric current parallel or antiparallel to the mean magnetic field. This current is represented by an additional term in the induction equation for the mean magnetic field, called the α-effect. Furthermore the mean field is subject to enhanced turbulent diffusion. Mean-field theory is presented in detail in the textbooks by Moffatt (1978), Parker (1979), and Krause and Rädler (1980). The statistical aspects of the theory are discussed by Hoyng (Chapter 1).

Combining the α-effect and differential rotation in an $\alpha\Omega$-dynamo and making suitable assumptions about these effects in the convection zone, the global properties of the solar magnetic field can be represented (e.g. Steenbeck and Krause, 1969; and many others, see reviews by Rädler, 1990 and Rüdiger and Arlt, Chapter 6) like the 22-year activity cycle, Maunder's butterfly diagram and Hale's polarity rules. The general agreement of the calculated fields with the observed patterns provided confidence that the basic ideas are correct. This can however be questioned in the light of recent developments.

Simulations of magnetoconvection (reviewed by Galloway and Weiss, 1981; Proctor and Weiss, 1982; Hughes and Proctor, 1988; Proctor, 1992; Cattaneo, 1994) suggest that the majority of the solar magnetic flux in the convection zone is concentrated in small-scale intermittent features such as observed on the solar surface (Stenflo, 1989; Solanki, 1993). These flux concentrations are difficult to store in the convection zone for times comparable to the solar cycle. Several processes, most notably magnetic buoyancy, transport magnetic flux from the bottom to the top of the convection zone in times of the order of one month, much too short for the dynamo to generate the field (Parker, 1975; Schüssler, 1977, 1979).

Another problem of locating the dynamo in the convection zone is the nearly strict obeyance of the polarity rules for bipolar active regions. Their appearance at the surface is difficult to explain with a field originating in the turbulent convection zone but suggests a well-ordered strong toroidal magnetic field.

Helioseismology shows that differential rotation does not at all dominate over convective motions in the convection zone proper. The oscillation data imply that the main convection zone rotates like the solar surface with no significant radial gradient, and that the deep interior rotates almost rigid at a rate between the equatorial and polar rates on the surface (Brown and Morrow, 1987; Libbrecht, 1988; Schou *et al.*, 1992; Tomczyk *et al.*, 1995). A strong radial gradient of angular velocity occurs in a transition region between the base of the convection zone and the top of the interior. Dynamo models in the convection zone with only a latitudinal gradient of angular velocity do not show migration towards the equator (Köhler, 1973; Prautzsch, 1993), but this is demanded by the observed butterfly diagram.

For these and other reasons it has been suggested that the bulk of the solar magnetic flux is stored in the overshoot layer below the convection zone proper (Spiegel and Weiss, 1980; van Ballegooijen, 1982; Schüssler, 1984). The differential rotation there is able to wind up a strong toroidal field (Schüssler, 1987; Fisher *et al.*, 1991). Magnetic buoyancy is reduced because of the subadiabatic stratification (Spruit and van Ballegooijen, 1982), thus enabling storage of strong magnetic fields (Ferriz-Mas, 1996). Finally turbulent diffusivity is supposed to be reduced in the overshoot region.

The structure of the (toroidal) field in the overshoot layer is not clear. It can be diffusively distributed (as it is assumed in this contribution) or in form of flux tubes. In the latter case fields with strengths up to 10^5 G are stably supported (Moreno-Insertis, 1992; Moreno-Insertis *et al.*, 1992; Ferriz-Mas and Schüssler, 1993, 1994, 1995). For even stronger fields a kink instability sets in which leads to flux loss from the overshoot region (Moreno-Insertis, 1986). Parts of the flux tube enter the convection zone, are floated upward by buoyancy in about a month, and finally emerge at the surface, while other parts are still rooted down in the overshoot region. At the base of the convection zone tubes as strong as 10^5 G are needed to avoid poleward slip (Choudhuri and Gilman, 1987; Choudhuri, 1989; Chou and Fisher, 1989). With such field strengths the tubes emerge at low latitudes as in the case of sunspots (Schüssler *et al.*, 1994). Also the observed small tilt with respect to the east-west direction and the asymmetry of bipolar active regions is then obtained (D'Silva and Choudhuri, 1993; Fan *et al.*, 1993, 1994; Moreno-Insertis, 1994; Moreno-Insertis *et al.*, 1994; Caligari *et al.*, 1995).

In the case of a more homogeneously distributed field we expect equipartition field strength which is of the order of 10^4 G for the lower convection zone. Magnetic buoyancy instabilities may be a means of breaking up the diffusive field into flux tubes (Hughes, 1992), resulting again in loss of flux from the layer. Thus the generation of toroidal magnetic field by the

Ω-effect may be limited. The tube may subsequently undergo further field amplification in the convection zone (Moreno-Insertis *et al.*, 1995).

The regeneration of the poloidal field is now an unsolved problem because the strong fields involved in the overshoot region resist the turbulent flow, and the kinematic α-effect is not longer applicable. Since the global features of the magnetic cycle can be understood in terms of an $\alpha\Omega$-dynamo, the structure of the underlying equation seems to be a good representation, but the α-effect needs to be reconsidered.

As we will see, the magnetic buoyancy instability is not only important for the escape of magnetic flux but also provides a dynamic explanation of the α-effect. A localized horizontal magnetic field in a gravitationally stratified fluid is potentially unstable due to magnetic buoyancy instability if the field strength decreases rapidly enough with height, i.e. naturally in the upper parts of the layer. In this Rayleigh–Taylor like instability potential energy of the extra mass supported against gravity is released (Newcomb, 1961; Gilman, 1970; Taylor, 1973; Acheson, 1979).

Of special interest with respect to the toroidal field at the base of the solar convection zone is the magnetic buoyancy instability in a rotating system. For simplicity reasons we assume a constant angular velocity Ω. We further restrict to the magnetohydrodynamically fast rotating case where

$$V^2 \ll \Omega^2 H^2 \ll a^2,$$

which is the relevant one in the lower convection zone; $V = B/\sqrt{\rho}$ means the Alfvén speed, H the scale height and a the (isothermal) sound speed.

Modes that do not bend field lines are stabilized by fast rotation (Acheson and Gibbons, 1978). For distorted field lines instability occurs if the magnetic field strength falls off with height faster than the density does (Acheson and Gibbons, 1978), i.e. for

$$\frac{d}{dz} \log\left(\frac{B}{\rho}\right) < 0.$$

This condition (for isothermal disturbances in an inviscid and thermally and electrically ideal conducting fluid) is not much influenced by rotation. The fastest growing modes have small but nonzero wavenumbers in the direction of the field and large horizontal wavenumbers perpendicular to it. The growth rate is of the order of $V^2/\Omega H^2$, and is considerably reduced to that without rotation (of the order of V/H). In this respect fast rotation stabilizes the magnetic buoyancy instability. Influences of adiabatic disturbances, of Ohmic and thermal dissipation and of the geometry are discussed e.g. in Acheson (1978).

The instability takes the form of slow magnetostrophic waves. In order to understand the nature of these waves consider the simplest case of an infinite and incompressible medium of constant density ρ, which is permeated by a homogeneous magnetic field **B** and rotates with constant angular velocity Ω around a fixed axis. Neglecting dissipative effects, small perturbations spread as plane waves with angular frequency ω and wavevector **k** related by the dispersion relation (Acheson and Hide, 1973)

$$\omega^2 \pm \frac{2\Omega \cdot \mathbf{k}}{\kappa}\omega - (\mathbf{V} \cdot \mathbf{k})^2 = 0,$$

where $\mathbf{V} = \mathbf{B}/\sqrt{\rho}$ is the Alfvén velocity and $\kappa = |\mathbf{k}|$. Solutions are given by

$$\omega_\pm^2 = (\mathbf{V}\cdot\mathbf{k})^2 + \frac{1}{2}\left\{\frac{(2\mathbf{\Omega}\cdot\mathbf{k})^2}{\kappa^2} \pm \sqrt{\frac{(2\mathbf{\Omega}\cdot\mathbf{k})^4}{\kappa^4} + \frac{4(\mathbf{V}\cdot\mathbf{k})^2(2\mathbf{\Omega}\cdot\mathbf{k})^2}{\kappa^2}}\right\}.$$

It follows $\omega_-^2 \le (\mathbf{V}\cdot\mathbf{k})^2 \le \omega_+^2$ with $\omega_-^2 = \omega_+^2 = (\mathbf{V}\cdot\mathbf{k})^2$ for $\mathbf{\Omega}\cdot\mathbf{k} = 0$. For increasing Ω, ω_+^2 increases and ω_-^2 decreases monotonously, such that the product $\omega_+^2\omega_-^2 = (\mathbf{V}\cdot\mathbf{k})^4$ remains independent of Ω. In the case of slow rotation, i.e. $\kappa^2(\mathbf{V}\cdot\mathbf{k})^2/(2\mathbf{\Omega}\cdot\mathbf{k})^2 \gg 1$

$$\omega_\pm^2 \approx (\mathbf{V}\cdot\mathbf{k})^2\left(1 \pm \left|\frac{2\mathbf{\Omega}\cdot\mathbf{k}}{\kappa\mathbf{V}\cdot\mathbf{k}}\right|\right).$$

The Coriolis force causes a small split of the Alfvén frequency and the restoring force is mainly the Lorentz forces. In case of fast rotation, i.e. $\kappa^2(\mathbf{V}\cdot\mathbf{k})^2/(2\mathbf{\Omega}\cdot\mathbf{k})^2 \ll 1$, which is relevant for the lower convection zone of the sun, we find

$$\omega_+^2 \approx \frac{(2\mathbf{\Omega}\cdot\mathbf{k})^2}{\kappa^2} \gg (\mathbf{V}\cdot\mathbf{k})^2 \gg \frac{(\mathbf{V}\cdot\mathbf{k})^4\kappa^2}{(2\mathbf{\Omega}\cdot\mathbf{k})^2} \approx \omega_-^2$$

and the two frequencies are widely separated. One solution corresponds to inertial waves and the Coriolis force is the restoring agent. The other solution describes magnetostrophic waves which are characterized by an approximate balance of Coriolis and Lorentz force, leaving only a weak net restoring force. The frequency is much smaller than the Alfvén frequency which, by assumption, is much smaller than the inertial wave frequency. Formally, magnetostrophic waves are obtained by neglecting the first term in the dispersion relation. The energy of magnetostrophic waves is essentially of magnetic origin, and the waves are highly dispersive.

Magnetostrophic waves have been studied in the literature under various aspects, especially in connection with thermally driven magnetoconvection and the geodynamo (e.g. Taylor, 1963; Braginsky, 1967, 1980; Eltayeb, 1972; Acheson, 1972, 1973; Fearn, 1979; Soward, 1979; Fearn and Proctor, 1984). For recent reviews, see Fearn (1998) and Braginsky and Roberts (this monograph).

Here we are interested in unstable magnetostrophic waves driven by magnetic buoyancy instability due to an unstable vertical gradient of a horizontal magnetic field in a gravitationally stratified compressible fluid, which was first considered by Acheson and Gibbons (1978) and Acheson (1978, 1979). The induction action of these waves is based on the helical character of growing modes and was first proposed by Moffatt (1978, section 10.7). The aim of the present paper is a detailed study of the dynamo action of unstable magnetostrophic waves relevant for the solar dynamo at the base of the convection zone.

Of special interest is that the effect is of dynamical nature, applicable to strong magnetic fields which resist distorsion by convective flows. The velocity is not prescribed but follows from the present forces and the interaction of magnetic and velocity field is taken into account. Convection is not necessary, the driving mechanism being the magnetic buoyancy instability.

This aspect of a dynamic theory of dynamo action was originally put forward by Moffatt who studied random inertial waves (Moffatt, 1972, 1978). Wälder *et al.* (1980) considered sound and gravity waves in a rotating stratified fluid. There, however, the Lorentz force does not play any role.

Magnetically driven instabilites in connection with the dynamo mechanism have experienced a revival in recent years. The Balbus–Hawley instability (Balbus and Hawley, 1991, see also Velikhov, 1959; Chandrasekhar, 1960) provides a means to explain turbulence in accretion disks (which is needed for the angular momentum transport and the mass accretion rate necessary to account for the released energy). Brandenburg *et al.* (1995) have shown that large-scale magnetic fields can be generated from the fluid motions associated with this instability. Guided by numerical simulations and by observations of stellar activity Brandenburg (1998, 1999) proposes a magnetic α-effect in cool stars. In the present article the magnetic buoyancy instability and its dynamo action are presented in the magnetostrophic limit. A related investigation outside this limit has recently been carried out by Thelen (1997). Brandenburg and Schmitt (1998) confirmed the existence of an α-effect by magnetic buoyancy in a numerical simulation.

An induction effect has also been derived from an instability of thin magnetic flux tubes (Ferriz-Mas *et al.*, 1994, see also Hanasz and Lesch, 1997) which is important for a stellar dynamo based on flux tubes (Schüssler, 1980, 1993). An application to the solar dynamo can be found in Schmitt *et al.* (1996), who explain the occurrence of grand Maunder minima by the lower bound of magnetic field strength for the flux tube instability and its dynamo effect. For more details see Schüssler and Ferriz-Mas (Chapter 5).

The plan of this article is as follows. In Section 4.2 a Cartesian model of the toroidal magnetic field layer in the overshoot region is introduced. In Section 4.3 an eigenvalue equation for magnetostrophic waves in this layer is derived. The dynamo action of these waves is considered in Section 4.4. It follows that only unstable waves are capable of induction action. Section 4.5 deals with an analytical solution of the eigenvalue problem for special cases by means of local analysis and provides conditions for instability. In Section 4.6 the eigenvalue problem is solved numerically, the global properties of unstable magnetostrophic waves are derived and their dynamo action in a thin rotating magnetic layer is determined. Section 4.7 concludes with a summary of the results and discusses their relevance for the theory of the solar magnetic field.

Sections 4.2–4.6 are an abridged English version of an investigation by Schmitt (1985). The present paper is meant to make the results of this study accessible to a wider audience and is motivated by an increasing interest in the subject indicated through recent publications.

4.2. Cartesian model of a rotating toroidal magnetic layer

Consider a layer of isothermal, inviscid, compressible fluid, of density $\rho_0(z)$, which is a perfect conductor of electricity and heat and rotates with constant angular velocity $\mathbf{\Omega} = \Omega(-\sin\theta, 0, \cos\theta)$, where $\Omega > 0$ and Ox, Oy, Oz are interpreted as south, east and vertically upwards directions, respectively, of a rotating Cartesian frame of reference with origin O in the lower convection zone at colatitude θ (Fig. 4.1). The fluid is infinite in directions x and y but bounded by plane walls at $z = z_1, z_2$. A variable magnetic field $\mathbf{B} = (0, B(z), 0)$ permeats the fluid, which is therefore in equilibrium under gravity $\mathbf{g} = (0, 0, -g)$, provided

$$(a^2\rho_0 + B^2/2)' + \rho_0 g = 0. \qquad (4.1)$$

Here a is the constant isothermal sound speed and, throughout the paper, a prime denotes ordinary differentiation of a function with respect to its argument, in this case z. The pressure $p_0(z)$ has been already removed from (4.1) using the isothermal equation of state $p_0 = a^2\rho_0$. If we specify the profile $B(z)$, (4.1) determines the stratification $\rho_0(z)$. We can rewrite (4.1)

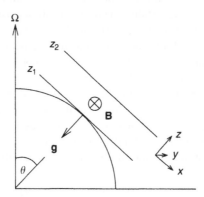

Figure 4.1 Schematic view of the equilibrium model.

in the form

$$\rho_0'/\rho_0 + (a^2 + V^2/2)^{-1}(g + VV') = 0, \tag{4.2}$$

where $V(z) = B(z)/[\rho_0(z)]^{1/2}$ is the Alfvén speed; (4.2) determines the stratification once the dependence of V on height has been prescribed. We shall assume that z_1 and z_2 are no more than a scale height apart, and take g to be constant.

The system we have just described is potentially unstable by the mechanism of magnetic buoyancy. The energy available for instability is evident. If the magnetic field decreases with height, extra mass is supported against gravity by the magnetic pressure gradient. In case of instability, the associated additional potential energy is released by downward transport of mass. At the same time, the energy stored in the magnetic field is released by upward transport of magnetic flux.

In the lower convection zone, the Alfvén speed is much smaller than the sound speed. Assuming a magnetic field strength of $B = 10^4$ G, a ratio of $V^2/a^2 = 10^{-7} \ll 1$ is obtained. Even for steep relative field gradients B'/B, (4.1) and (4.2) can then be approximated by

$$\rho_0'/\rho_0 + g/a^2 = 0. \tag{4.3}$$

4.3. Eigenvalue equation of magnetostrophic waves

Small perturbations of the equilibrium are described by the linearized MHD equations. We restrict ourselves and consider only magnetostrophic waves by applying a number of approximations.

Regarding dissipative effects we choose the most simple case by considering an inviscid and ideal electrically and thermally conducting fluid. We are thus restricted to perturbations whose time scales are much larger than the thermal diffusion time scale, but small enough to be able to neglect viscous and Ohmic diffusion effects. Through isothermal perturbations in an isothermal fluid the Brunt–Väisälä frequency vanishes and internal gravity waves are filtered out of consideration.

Magnetostrophic waves are slow waves with phase speeds much smaller than the Alfvén speed and occur for magnetohydrodynamically fast rotating fluids, i.e. in those cases where the Alfvén speed is much smaller than the rotational speed. In the lower convection zone the latter is again much smaller than the sound speed. The frequencies of the respective waves are thereby ordered according to

$$\omega_{ms} \ll \omega_A \ll \omega_\Omega \ll \omega_{ac}.$$

Introducing the anelastic approximation by neglecting $\partial\rho/\partial t$ in the continuity equation (magneto-)acoustic waves are filtered out without influencing the description of the other waves. Changes are of the order of V^2/a^2 relative to 1. Furthermore we apply the magnetostrophic approximation by neglecting $\partial\mathbf{u}/\partial t$ in the momentum equation. This filters out inertial and Alfvén waves and is only allowed for waves with much smaller frequencies.

Under these conditions the linearized MHD equations are the equations of motion, continuity, state, and induction

$$2\rho_0\mathbf{\Omega}\times\mathbf{u} = -\nabla p + (\nabla\times\mathbf{B})\times\mathbf{b} + (\nabla\times\mathbf{b})\times\mathbf{B} + \rho\mathbf{g},$$
$$\nabla\cdot(\rho_0\mathbf{u}) = 0,$$
$$p = a^2\rho, \tag{4.4}$$
$$\frac{\partial\mathbf{b}}{\partial t} = \nabla\times(\mathbf{u}\times\mathbf{B}),$$
$$\nabla\cdot\mathbf{b} = 0.$$

Here $\mathbf{u} = (u, v, w)$, $\mathbf{b} = (b_x, b_y, b_z)$, p and ρ are perturbations of the velocity, the magnetic field, the pressure and density with respect to a rotating frame of reference. The centrifugal force is negligible compared to the much larger gravitation.

All coefficients of the linear homogeneous system (4.4) are functionals of equilibrium quantities and therefore depend only on the height z. Thus a Fourier ansatz for the variables $\psi \in \{\mathbf{u}, \mathbf{b}, p, \rho\}$, namely

$$\psi(x, y, z, t) = \text{Re}[\hat{\psi}(z)\exp\{i(kx + my - \omega t)\}], \tag{4.5}$$

is possible with, in general complex, angular frequencies $\omega = \omega_R + i\omega_I$ and wavenumbers k and m in x- and y-directions, respectively. We thereby obtain a system of ODE's:

$$-2\rho_0\Omega\cos\theta\hat{v} = -ik\hat{p} - ikB\hat{b}_y + imB\hat{b}_x,$$
$$2\rho_0\Omega(\cos\theta\hat{u} + \sin\theta\hat{w}) = -im\hat{p} + B'\hat{b}_z,$$
$$-2\rho_0\Omega\sin\theta\hat{v} = -\hat{p}' - (B\hat{b}_y)' + imB\hat{b}_z - \hat{\rho}g,$$
$$ik\hat{u} + im\hat{v} + \hat{w}' + \frac{\rho_0'}{\rho_0}\hat{w} = 0, \tag{4.6}$$
$$\hat{p} = a^2\hat{\rho},$$
$$-i\omega\hat{b}_x = imB\hat{u},$$
$$-i\omega\hat{b}_y = -ikB\hat{u} - (B\hat{w})',$$
$$-i\omega\hat{b}_z = imB\hat{w}.$$

A prime again denotes the derivative with respect to z. The solenoidal condition for the magnetic field perturbations is automatically accomplished by the Fourier ansatz together with the induction equation.

By applying the above approximations we restrict ourselves to waves with

$$|\omega|^2 \ll \omega_A^2 = m^2 V^2. \tag{4.7}$$

This requires that both the wavenumber m and Alfvén speed V do not vanish. We will see in the next section that these requirements are also necessary for dynamo action and we thus assume them hereafter. It should be pointed out that the approximations do not guarantee solutions which satisfy (4.7) and its validity must be verified *a posteriori*.

For the further reduction of the system (4.6) the transformation to field variables (Eckart, 1960, p. 55; Moffatt, 1978, section 10.1; Wälder *et al.*, 1980)

$$(\tilde{u}, \tilde{v}, \tilde{w}) = \rho_0^{1/2}(\hat{u}, \hat{v}, \hat{w}), \quad \tilde{\rho} = \rho_0^{-1/2}\hat{\rho},$$
$$\tilde{p} = a^2\tilde{\rho}, \quad (\tilde{b}_x, \tilde{b}_y, \tilde{b}_z) = V(\hat{b}_x, \hat{b}_y, \hat{b}_z) \tag{4.8}$$

is very helpful. For constant Alfvén speed V the coefficients of the system of equations for field variables resulting from (4.6) would be constants. We do not however restrict ourselves to constant Alfvén speed.

By successive elimination of all variables

$$\tilde{b}_x = -\frac{m}{\omega} V^2 \tilde{u},$$

$$\tilde{b}_y = -\frac{m}{\omega} V^2 \tilde{v} - \frac{i}{\omega}\left(\frac{g}{a^2} + \frac{B'}{B}\right) V^2 \tilde{w},$$

$$\tilde{b}_z = -\frac{m}{\omega} V^2 \tilde{w},$$

$$\tilde{\chi} = \tilde{p} + \tilde{b}_y = a^2\tilde{\rho} + \tilde{b}_y = \frac{i}{m} 2\Omega \cos\theta \tilde{u} - \frac{mV^2}{\omega}\tilde{v} + i\left(\frac{2}{m}\Omega \sin\theta - \frac{g}{a^2}\frac{V^2}{\omega}\right)\tilde{w}, \tag{4.9}$$

$$\tilde{v} = -\frac{k}{m}\tilde{u} - \frac{i}{2m}\frac{g}{a^2}\tilde{w} + \frac{i}{m}\tilde{w}',$$

$$\tilde{u} = \frac{1}{k^2 + m^2}\left[\left\{\frac{ik}{2}\frac{g}{a^2} - \frac{\omega}{mV^2}2\Omega\left(ik\sin\theta + \frac{1}{2}\frac{g}{a^2}\cos\theta\right)\right\}\tilde{w}\right.$$

$$\left. + \left(ik + \frac{\omega}{mV^2}2\Omega\cos\theta\right)\tilde{w}'\right],$$

we obtain the homogeneous ODE

$$P\widetilde{w}'' + Q\widetilde{w}' + R\widetilde{w} = 0, \tag{4.10a}$$

for $\widetilde{w}$, where

$$P = -4\Omega^2 \cos^2\theta\omega^2 + m^4 V^4,$$

$$Q = 8\Omega^2 \left(ik \sin\theta \cos\theta + \cos^2\theta \frac{V'}{V} \right) \omega^2$$

$$- 4i\frac{g}{a^2}mkV^2\Omega \cos\theta\,\omega + 2\frac{V'}{V}m^4 V^4, \tag{4.10b}$$

$$R = 4\Omega^2 \left(k^2 \sin^2\theta + \frac{1}{4}\frac{g^2}{a^4}\cos^2\theta - \frac{g}{a^2}\frac{V'}{V}\cos^2\theta - 2ik\frac{V'}{V}\sin\theta\cos\theta \right) \omega^2$$

$$- 2\frac{g}{a^2}mV^2\Omega \sin\theta(m^2 + 2k^2)\omega$$

$$- m^4 V^4 \left\{ k^2 + m^2 - \frac{1}{4}\frac{g^2}{a^4} + \frac{k^2}{m^2}\frac{g}{a^2}\left(\frac{V'}{V} - \frac{1}{2}\frac{g}{a^2} \right) \right\},$$

subject to

$$\widetilde{w}(z_1) = \widetilde{w}(z_2) = 0. \tag{4.10c}$$

The coefficients P, Q, R are complex, depend on z, and we solve (4.10a) for no flow through the bottom and top boundaries (4.10c). Problem (4.10) forms an eigenvalue problem with eigenvalues ω and eigenvectors $\widetilde{w}$.

For real frequencies ω and variable Alfvén speed V there may exist so called critical levels z_c at which $P(z_c) = 0$. As seen later we are primarily interested in complex frequencies where these levels do not occur, so we do not need to consider possible singular behaviour any further.

The form of (4.10a) suggests another transformation

$$\bar{w} = \exp\left\{ \int_{z_1}^z \frac{Q(z')}{2P(z')}dz' \right\} \widetilde{w} \tag{4.11}$$

which, in general, is complex because of the corresponding properties of the quantities P and Q. Note that the wave numbers k and m and the angular frequencies ω enter into the transformation. Only at the equator, $\theta = \pi/2$, the transformation considerably simplifies to $\bar{w} = V\widetilde{w} = B\hat{w}$. Thus (4.10a) reduces to

$$S\bar{w}'' + T\bar{w} = 0, \tag{4.12a}$$

where

$$S = -P^2 = -\left(16\Omega^4 \cos^4\theta\omega^4 - 8\Omega^2 \cos^2\theta m^4 V^4\omega^2 + m^8 V^8\right),$$

$$T = \frac{1}{4}Q^2 - PR + \frac{1}{2}PQ' - \frac{1}{2}P'Q$$

$$= 16\Omega^4 \cos^4\theta \left(2\frac{V'^2}{V^2} - \frac{V''}{V} - \frac{g}{a^2}\frac{V'}{V} + \frac{1}{4}\frac{g^2}{a^4}\right)\omega^4$$

$$- 8\Omega^3 \sin\theta \cos^2\theta \frac{g}{a^2}m^3 V^2\omega^3 \qquad (4.12b)$$

$$- 4\Omega^2 m^4 V^4 \left[\cos^2\theta \left\{6\frac{V'^2}{V^2} + \frac{k^2}{m^2}\frac{g}{a^2}\left(\frac{V'}{V} + \frac{1}{2}\frac{g}{a^2}\right)\right.\right.$$

$$\left.\left. - \frac{g}{a^2}\frac{V'}{V} + k^2 + m^2\right\} + k^2 \sin^2\theta\right]\omega^2$$

$$+ 2\Omega \sin\theta \frac{g}{a^2}(m^2 + 2k^2)m^5 V^6\omega$$

$$+ m^8 V^8 \left\{\frac{V''}{V} + \frac{k^2}{m^2}\frac{g}{a^2}\left(\frac{V'}{V} - \frac{1}{2}\frac{g}{a^2}\right) + k^2 + m^2 - \frac{1}{4}\frac{g^2}{a^4}\right\}$$

with boundary conditions

$$\bar{w}(z_1) = \bar{w}(z_2) = 0. \qquad (4.12c)$$

The eigenvalue ω enters at fourth order in (4.12). At the equator, however, the angular frequency ω occurs only at second order. For given equilibrium and wave numbers k and m a whole set of fundamental mode and overtones solves the eigenvalue problem (4.12). To each overtone there exist four (at the equator two) eigenvalues ω. The eigenfunctions $\bar{w}$ to these eigenvalues differ from each other but usually display the same number of knots. This is however no longer the case for the velocity amplitudes $\tilde{w}$ and $\hat{w}$ because of the complex transformation (4.11), except at the equator.

The coefficients of ω in S and T are real. Thus the frequencies ω occur as real quantities or in conjugate complex pairs. The existence of complex solutions means instability. In the next section we see that unstable waves are of special interest for dynamo action. The conditions for instability are studied in Section 4.5.

4.4. α-effect of magnetostrophic waves

Before actually solving the eigenvalue problem derived in the previous section we first consider the conditions necessary for dynamo action. We therefore derive the mean electromotive force parallel to the toroidal magnetic field of equilibrium induced by the perturbations. By analogy to kinematic dynamo theory we define the α-effect by

$$\alpha\langle B\rangle = \langle \mathbf{u} \times \mathbf{b}\rangle_y, \qquad (4.13)$$

where $\mathbf{u}$ and $\mathbf{b}$ are the velocity and magnetic field of the magnetostrophic wave. The brackets denote an average over the fluid layer whose volume extends over one wave length in horizontal directions and is confined between z_1 and z_2. First we consider a single wave.

Following (4.5) a perturbation quantity is written

$$\psi = \mathrm{Re}\left[\hat{\psi}(z)e^{i(kx+my-\omega t)}\right],$$

$$= \frac{1}{2}\left[\hat{\psi}(z)e^{i(kx+my-\omega_R t)+\omega_I t} + \hat{\psi}^*(z)e^{-i(kx+my-\omega_R t)+\omega_I t}\right],$$

where * denotes the complex conjugate, and subscripts R and I the real and imaginary part, respectively. With transformation (4.8) and the abbreviation $() = (kx + my - \omega_R t)$ we find

$$(\mathbf{u} \times \mathbf{b})_y = wb_x - ub_z,$$

$$= \frac{1}{4B}\left[\tilde{w}\tilde{b}_x e^{2i()+2\omega_I t} + \tilde{w}\tilde{b}_x^* e^{2\omega_I t} + \tilde{w}^*\tilde{b}_x e^{2\omega_I t} + \tilde{w}^*\tilde{b}_x^* e^{-2i()+2\omega_I t}\right.$$

$$\left. - \tilde{u}\tilde{b}_z e^{2i()+2\omega_I t} - \tilde{u}\tilde{b}_z^* e^{2\omega_I t} - \tilde{u}^*\tilde{b}_z e^{2\omega_I t} - \tilde{u}^*\tilde{b}_z^* e^{-2i()+2\omega_I t}\right].$$

After elimination of the magnetic field perturbations, we obtain

$$(\mathbf{u} \times \mathbf{b})_y = \frac{1}{4}m\frac{B}{\rho_0}\left(\frac{1}{\omega} - \frac{1}{\omega^*}\right)\left(\tilde{u}^*\tilde{w} - \tilde{u}\tilde{w}^*\right)e^{2\omega_I t}. \tag{4.14}$$

Finally, expressing the x-component $\tilde{u}$ by the z-component $\tilde{w}$ of the velocity perturbation we end with

$$(\mathbf{u} \times \mathbf{b})_y = \frac{1}{2}\frac{m}{k^2+m^2}\frac{\omega_I}{|\omega|^2}\frac{B}{\rho_0}\left[\left(\frac{4k\Omega\sin\theta}{mV^2}\omega_R - k\frac{g}{a^2} + \frac{2g\Omega\cos\theta}{a^2mV^2}\omega_i\right)\tilde{w}\tilde{w}^*\right.$$

$$- \left(k + \frac{2\Omega\cos\theta}{mV^2}\omega_I\right)\left(\tilde{w}\tilde{w}^{*\prime} + \tilde{w}^\prime\tilde{w}^*\right)$$

$$\left. + \frac{2i\Omega\cos\theta}{mV^2}\omega_R\left(\tilde{w}^\prime\tilde{w}^* - \tilde{w}\tilde{w}^{*\prime}\right)\right]e^{2\omega_I t},$$

$$= \frac{1}{2}\frac{m}{k^2+m^2}\frac{\omega_I}{|\omega|^2}B\left[\frac{4\Omega}{mV^2}\left(k\sin\theta\omega_R + \frac{g}{a^2}\cos\theta\omega_i\right)\hat{w}\hat{w}^*\right.$$

$$- \left(k + \frac{2\Omega\cos\theta}{mV^2}\omega_I\right)\left(\hat{w}\hat{w}^{*\prime} + \hat{w}^\prime\hat{w}^*\right)$$

$$\left. + \frac{2i\Omega\cos\theta}{mV^2}\omega_R\left(\hat{w}^\prime\hat{w}^* - \hat{w}\hat{w}^{*\prime}\right)\right]e^{2\omega_I t}. \tag{4.15}$$

Note that these expressions are real. Given the equilibrium quantities and the horizontal wave numbers k and m, the angular frequencies ω are obtained as eigenvalues of problem (4.12) together with the eigenfunctions $\tilde{w}$. The velocity amplitudes $\tilde{w}$ and $\hat{w}$ are readily obtained with the help of the transformations (4.11) and (4.8).

Note that the requirements for magnetostrophic waves, $B \neq 0$ (and $V \neq 0$), $\Omega \neq 0$ and $m \neq 0$ are also necessary for dynamo action. At the equator $k \neq 0$ is needed for non-vanishing $\tilde{u}$ and $(\mathbf{u} \times \mathbf{b})_y$.

The most important condition for dynamo action however is $\omega_I \neq 0$. Only for complex angular frequencies a phase difference occurs between the perturbations $\mathbf{u}$ and $\mathbf{b}$ forming the magnetostrophic wave. In the studies of dynamo action of inertial waves by Moffatt (1972) and of sound and gravity waves by Wälder *et al.* (1980) the phase lag was due to Ohmic diffusion of waves with real frequencies. The Coriolis force causes a phase difference also between the velocity components $\tilde{u}$ and $\tilde{w}$ which is responsible for the non-vanishing of the second parenthesis in (4.14). This phase lag carries over to the magnetic field components $\tilde{b}_x$ and $\tilde{b}_z$. The velocity and magnetic field perturbation vectors thus screw in space and time, the wave is helical. As $\alpha \sim \exp(2\omega_I t)$, ω_I must be positive, implying unstable waves for inductive action.

We must further demand that a random superposition of magnetostrophic waves displays a non-zero α-effect, i.e. the contributions of the various waves do not cancel. At the equator $\theta = \pi/2$ this is not the case. Consider two unstable magnetostrophic waves that only differ in the sign of the wavenumber k but have same parameters otherwise. Since k only occurs quadratically in the coefficients S and T of (4.12) these waves have the same eigenvalues ω and eigenfunctions $\tilde{w}$. At the equator they further do not differ in the coefficients P and Q and thus in the velocity components $\tilde{w}$ and $\hat{w}$. According to (4.15) each of these waves induces an α of the same absolute value but of different sign. Thus the total α-effect of magnetostrophic waves is zero at the equator.

Outside the equator, $\theta \neq \pi/2$, we expect non-vanishing dynamo action. Two unstable waves that again only differ in the sign of the wavenumber k, thus travelling in the $+x$ and $-x$ directions, still have the same eigenfrequencies ω and eigenfunctions $\tilde{w}$, but differ in Q and thus in $\tilde{w}$ and $\hat{w}$. They possess different induction action that do not add to zero.

To calculate the dynamo action of two waves with different sign of k, the sign change does not need to be carried out. Complex frequencies ω always occur as complex conjugate pairs. Comparing an unstable wave after a change of sign of k with the corresponding stable wave without the sign change reveals that S, T and $\tilde{w}$, further P and Q and thus $\tilde{w}$ and $\hat{w}$ transform into their complex conjugate values and the time-independent part α_0 of α, defined by

$$\alpha = \alpha_0 e^{2\omega_I t}, \tag{4.16}$$

is the same for both waves.

By similar considerations one finds that wave pairs with wavenumbers $(+k, -m)$ and $(-k, +m)$ and with $(+k, +m)$ and $(-k, -m)$ induce the same α-effect, respectively. Thus, outside the equator, it is not possible by simple considerations to find a second wave that compensates the dynamo action of the other.

To obtain the net dynamo action of magnetostrophic waves for a given equilibrium magnetic field layer, all wavenumbers k and m have to be varied and the whole spectrum of unstable modes has to be considered. This is not an easy task. Growth rates vary from the fundamental mode to overtones and for the various wavenumbers k and m. Because of the exponential time factor in (4.16) the induction action of the most unstable modes will dominate in the linear theory considered here.

After this consideration, unstable modes with same growth rate ω_I always occur in pairs, those with different signs of wavenumbers k or m. Their total dynamo action is the sum of the individual contributions

$$\alpha^T = \alpha(+k, +m) + \alpha(-k, +m) = \alpha_0^T e^{2\omega_I t} \tag{4.17}$$

which, in general, does not vanish, except at the equator.

Finally, let us consider the symmetry property with respect to the equator. To this end compare a wave at colatitude $\pi - \theta$ and wavenumber $+k$ with the corresponding wave at colatitude θ and wavenumber $-k$. We find that besides S, T, ω and $\bar{w}$ also P and Q and thus $\tilde{w}$ and $\hat{w}$ are identical. After (4.15) both waves induce an α of same absolute value but different sign. Together with the discussion above we find that α^T is antisymmetric with respect to the equator:

$$\alpha^T(\theta) = -\alpha^T(\pi - \theta), \quad \text{especially} \quad \alpha^T(\pi/2) = 0. \tag{4.18}$$

Before calculating the dynamo action of magnetostrophic waves, the eigenvalue problem (4.12a) with its eigenfrequencies ω and eigenvelocities $\bar{w}$, and the velocity amplitudes $\tilde{w}$ and $\hat{w}$ must be computed. In the next section we first analytically consider the conditions for instability and the properties of unstable magnetostrophic waves. Later, the eigenvalue problem will be solved numerically and the dynamo action of magnetostrophic waves in a thin magnetic layer at the bottom of the solar convection zone is calculated.

4.5. Local stability analysis

In Section 4.3 we formulated an eigenvalue problem for the description of magnetostrophic waves in a rotating magnetic field layer. Here we solve the eigenvalue problem analytically by means of a local analysis.

The coefficients S and T of the eigenvalue problem (4.12) in general depend on height z. In a local analysis these are considered as local constants. Then the simple Fourier ansatz

$$\bar{w} = \bar{w}_0 \sin \frac{N\pi}{d}(z - z_1), \quad d = z_2 - z_1, \quad N = 1, 2, 3, \ldots \tag{4.19}$$

which already obeys the boundary conditions, solves the problem (4.12) and we obtain the dispersion relation

$$
16\Omega^4 \cos^4\theta \left(\frac{N^2\pi^2}{d^2} + \frac{1}{4}\frac{g^2}{a^4} + 2\frac{V'^2}{V^2} - \frac{V''}{V} - \frac{g}{a^2}\frac{V'}{V} \right) \omega^4
$$

$$
- 8\Omega^3 \sin\theta \cos^2\theta \frac{g}{a^2} m^3 V^2 \omega^3
$$

$$
- 4\Omega^2 m^4 V^4 \left\{ \cos^2\theta \left(\frac{2N^2\pi^2}{d^2} + k^2 + m^2 + \frac{1}{2}\frac{k^2}{m^2}\frac{g^2}{a^4} + 6\frac{V'^2}{V^2} \right. \right.
$$

$$
\left. \left. - \frac{g}{a^2}\frac{V'}{V} + \frac{k^2}{m^2}\frac{g}{a^2}\frac{V'}{V} \right) + k^2 \sin^2\theta \right\} \omega^2
$$

$$
+ 2\Omega \sin\theta \frac{g}{a^2}(m^2 + 2k^2)m^5 V^6 \omega
$$

$$
+ m^8 V^8 \left(\frac{N^2\pi^2}{d^2} + k^2 + m^2 - \frac{1}{4}\frac{g^2}{a^4} - \frac{1}{2}\frac{k^2}{m^2}\frac{g^2}{a^4} + \frac{V''}{V} + \frac{k^2}{m^2}\frac{g}{a^2}\frac{V'}{V} \right) = 0. \tag{4.20}
$$

It describes the local behaviour of a wave at a given location z_0 by determining the frequency and growth rate ω for given equilibrium quantities at z_0 and chosen wavenumbers k and m and order N of the overtone. The approximation of local analysis is only valid if the coefficients vary only slightly in the interval $[z_1, z_2]$ respectively for one wavelength along z. Although this is not the case for a thin magnetic field layer with steep field gradients, the local approach nevertheless yields valuable insights. Local stability criteria are usually also globally valid.

As already mentioned earlier, slow wave solutions only exist for $V \neq 0$, $\Omega \neq 0$ and $m \neq 0$. Instead of the pair of wavenumbers (k, m) it proves helpful to use the pair (λ, m) where λ is the ratio of both wavenumbers:

$$\lambda = k/m. \tag{4.21}$$

Furthermore we must keep in mind that only solutions with $|\omega|^2 \ll m^2 V^2$ are correctly determined by (4.20) and describe magnetostrophic waves.

The coefficients of ω in the dispersion relation (4.20) are real. Therefore, solutions ω are either real, the waves are neutrally stable, or conjugate complex, describing growing and decaying waves. Since only unstable waves are subject to dynamo action, the main emphasis is laid on criteria for instability and the properties of unstable magnetostrophic waves.

4.5.1. Equator

At the equator $\theta = \pi/2$ the dispersion relation (4.20) considerably reduces to

$$\omega^2 - \frac{mV^2}{2\Omega}\left(2 + \frac{1}{\lambda^2}\right)\frac{g}{a^2}\omega - \frac{m^2 V^4}{4\Omega^2}\left\{\left(\frac{V'}{V} - \frac{1}{2}\frac{g}{a^2}\right)\frac{g}{a^2}\right.$$
$$\left. + \left(1 + \frac{1}{\lambda^2}\right)m^2 + \frac{1}{\lambda^2}\left(\frac{N^2\pi^2}{d^2} - \frac{1}{4}\frac{g^2}{a^4} + \frac{V''}{V}\right)\right\} = 0. \tag{4.22}$$

Here $\lambda \neq 0$ is assumed; for $\lambda = 0$ there exist only real solutions which are not of main interest here.

For instability the roots of (4.22) must be complex. The condition for instability is

$$\left(\frac{V'}{V} + \frac{1}{2}\frac{g}{a^2}\right)\frac{g}{a^2} + \left(1 + \frac{1}{\lambda^2}\right)m^2 + \frac{1}{\lambda^2}\left\{\frac{N^2\pi^2}{d^2} + \frac{1}{4}\left(3 + \frac{1}{\lambda^2}\right)\frac{g^2}{a^4} + \frac{V''}{V}\right\} < 0. \tag{4.23}$$

A necessary condition for instability is the first term to be negative:

$$\frac{V'}{V} + \frac{1}{2}\frac{g}{a^2} = \frac{B'}{B} - \frac{1}{2}\frac{\rho_0'}{\rho_0} + \frac{1}{2}\frac{g}{a^2} = \frac{B'}{B} - \frac{\rho_0'}{\rho_0} = \frac{d}{dz}\left(\log\frac{B}{\rho_0}\right) < 0, \tag{4.24}$$

meaning that the magnetic field must fall off with height faster than the density does. The second term of (4.23) is positive and stabilizing. It represents the action of magnetic tension. The third term also stabilizes. The second derivative of V may be negative and reduces the stabilizing action of its other contributions. For large values of λ this terms becomes less important.

A steep gradient of the magnetic field strength after (4.24) drives the instability. It is a magnetic buoyancy instability under the action of rotation. Relation (4.23) accounts for the stabilizing and destabilizing effects and is necessary and sufficient for local instability. For a given unstable equilibrium stratification after (4.24) the following relations must be fulfilled

$$\lambda^2 > \lambda_c^2 = -\left(\frac{V'}{V} + \frac{1}{2}\frac{g}{a^2}\right)^{-1}\frac{a^2}{g}\left\{\frac{\pi^2}{d^2} + \frac{1}{4}\left(3 + \frac{1}{\lambda^2}\right)\frac{g^2}{a^4} + \frac{V''}{V}\right\},$$

$$1 \le N \le N_c = \mathcal{I}\left[\frac{\lambda d}{\pi}\left\{-\left(\frac{V'}{V} + \frac{1}{2}\frac{g}{a^2}\right)\frac{g}{a^2} - \frac{1}{\lambda^2}\left(\frac{1}{4}\left(3 + \frac{1}{\lambda^2}\right)\frac{g^2}{a^4} + \frac{V''}{V}\right)\right\}^{1/2}\right],$$

(4.25)

$$0 < m^2 < m_c^2 = \left(1 + \frac{1}{\lambda^2}\right)^{-1}\left[-\left(\frac{V'}{V} + \frac{1}{2}\frac{g}{a^2}\right)\frac{g}{a^2}\right.$$
$$\left. - \frac{1}{\lambda^2}\left\{\frac{N^2\pi^2}{d^2} + \frac{1}{4}\left(3 + \frac{1}{\lambda^2}\right)\frac{g^2}{a^4} + \frac{V''}{V}\right\}\right].$$

Here $\mathcal{I}$ means maximum integer less than. These relations demonstrate that, besides steep gradients of the magnetic field strength, instability is favoured for large values of λ, for the fundamental mode and low order rather than higher overtones, and for intermediate wavenumbers m.

The frequencies and growth rates of unstable modes are given by

$$\omega_R = \frac{mV^2}{2\Omega}\left(1 + \frac{1}{2\lambda^2}\right)\frac{g}{a^2},$$

$$\omega_I = \frac{mV^2}{2\Omega}\left[-\left(\frac{V'}{V} + \frac{1}{2}\frac{g}{a^2}\right)\frac{g}{a^2} - m^2\left(1 + \frac{1}{\lambda^2}\right)\right.$$

(4.26)

$$\left. - \frac{1}{\lambda^2}\left\{\frac{N^2\pi^2}{d^2} + \frac{1}{4}\left(3 + \frac{1}{\lambda^2}\right)\frac{g^2}{a^4} + \frac{V''}{V}\right\}\right]^{1/2}.$$

Note that $\omega_R m > 0$, i.e. unstable modes only propagate in one direction along the magnetic field. The growth rate is proportional to V^2/Ω. In this respect rotation stabilizes. Of special interest is the dependence of the growth rate on wavenumber m. Note that ω_I increases with increasing m until it reaches its maximum value ω_{I0} at $m_0 = m_c/\sqrt{2}$ and falls off thereafter, becoming zero at $m = m_c$. Weak bending of field lines brakes the stabilizing action of the Coriolis force for $m = 0$ and favours the instability, whereas for larger values of m the stabilizing action of magnetic tension wins. The maximum growth rate increases linearly with the negative magnetic field gradient. The condition of validity of the magnetostrophic approximation is fulfilled as long as

$$\frac{V^2}{4\Omega^2}\frac{g}{a^2}\left|\frac{d}{dz}\log B\right| \ll 1.$$

4.5.2. Pole

At the pole the dispersion relation (4.20) is biquadratic in ω and some analytical results can be derived. For $\theta = 0$ it follows that

$$
16\Omega^4 \left\{ \frac{N^2\pi^2}{d^2} + \left(\frac{V'}{V} - \frac{1}{2}\frac{g}{a^2} \right)^2 + \frac{V'^2}{V^2} - \frac{V''}{V} \right\} \omega^4
$$

$$
- 4\Omega^2 m^4 V^4 \left\{ \frac{2N^2\pi^2}{d^2} - \frac{1}{24}\frac{g^2}{a^4} + \left(\sqrt{6}\frac{V'}{V} - \frac{1}{2\sqrt{6}}\frac{g}{a^2} \right)^2 \right.
$$

$$
\left. + m^2 \left(1 + \lambda^2 \right) + \lambda^2 \frac{g}{a^2} \left(\frac{V'}{V} + \frac{1}{2}\frac{g}{a^2} \right) \right\} \omega^2
$$

$$
+ m^8 V^8 \left\{ \frac{N^2\pi^2}{d^2} + m^2(1 + \lambda^2) - \frac{1}{4}\frac{g^2}{a^4} + \frac{V''}{V} + \lambda^2 \frac{g}{a^2} \left(\frac{V'}{V} - \frac{1}{2}\frac{g}{a^2} \right) \right\} = 0, \quad (4.27)
$$

which is abbreviated as $a_1\omega^4 + a_2\omega^2 + a_3 = 0$ and yields solutions

$$
\omega^2 = \frac{1}{2a_1} \left(-a_2 \pm \sqrt{a_2^2 - 4a_1 a_3} \right). \quad (4.28)
$$

The discriminant $a_2^2 - 4a_1 a_3$ is positive, therefore ω^2 is real and monotonous instability occurs when $\omega^2 < 0$. Since $a_1 > 0$ except for unrealistic large positive values of V'', at least one of the solutions ω^2 of (4.28) is negative if $a_2 > 0$ or $a_3 < 0$. The first case is a subcase of the second, from which we derive the necessary condition for instability:

$$
\frac{V'}{V} - \frac{1}{2}\frac{g}{a^2} = \frac{V'}{V} + \frac{1}{2}\frac{\rho_0'}{\rho_0} = \frac{B'}{B} < 0. \quad (4.29)
$$

At the pole a less stringent condition is obtained than at the equator and a simple decrease of magnetic field strength with height may lead to instability. The wavenumbers of unstable modes are constrained by

$$
\infty > \lambda^2 > \lambda_c^2 = - \left(\frac{V'}{V} - \frac{1}{2}\frac{g}{a^2} \right)^{-1} \frac{a^2}{g} \left(\frac{\pi^2}{d^2} - \frac{1}{4}\frac{g^2}{a^4} + \frac{V''}{V} \right),
$$

$$
1 \le N \le N_c = \mathcal{I} \left[\frac{\lambda d}{\pi} \left\{ - \left(\frac{V'}{V} - \frac{1}{2}\frac{g}{a^2} \right) \frac{g}{a^2} - \frac{1}{\lambda^2} \left(\frac{V''}{V} - \frac{1}{4}\frac{g^2}{a^4} \right) \right\}^{1/2} \right], \quad (4.30)
$$

$$
0 < m^2 < m_c^2 = \frac{\lambda^2}{1 + \lambda^2} \left[- \left(\frac{V'}{V} - \frac{1}{2}\frac{g}{a^2} \right) \frac{g}{a^2} - \frac{1}{\lambda^2} \left\{ \frac{N^2\pi^2}{d^2} - \frac{1}{4}\frac{g^2}{a^4} + \frac{V''}{V} \right\} \right],
$$

and can be discussed in the same manner as we did for the equatorial case. The growth rate of instability is difficult to obtain analytically and we refer to the numerical solution of the disperison relation in Section 4.5.4.

4.5.3. Limit $\theta \to \pi/2$

At arbitrary colatitude θ the dispersion relation (4.20) is of fourth order in ω and difficult to discuss analytically. Only in the limit $\theta \to \pi/2$ a clarifying remark is possible. In this case the coefficients of ω^3 and ω^4 approach zero while the other coefficients stay finite. This means that two of the four solutions ω grow without limit to $\pm\infty$. They disregard the validity of the magnetostrophic approximation and are wrongly described by (4.20). Without the approximation two stable Alfvén waves would have been obtained which are however of no interest for dynamo action. The remaining two solutions approach those at the equator.

4.5.4. Numerical solution of the dispersion relation

We first introduce non-dimensional variables by scaling the length by of a^2/g, the Alfvén speed by V_0 and the frequency by $V_0^2 g^2/(2\Omega a^4)$. We then put, without loss of generality, the non-dimensional Alfvén speed and the thickness of the layer to 1.

With the values for the lower convection zone $g = 270 \, \mathrm{m\,s}^{-2}$, $a^2/g = 5 \times 10^7 \, \mathrm{m}$, $\Omega = 3 \times 10^{-6} \, \mathrm{s}^{-1}$ and $V_0^2/a^2 = 10^{-7}$ or $B_0 = 10^4 \, \mathrm{G}$, the time unit is $2\Omega a^4/(V_0^2 g^2) = 130 \, \mathrm{d}$. The requirement for application of the magnetostrophic approximation reads $|\omega|^2 \ll 4\Omega^2 a^4/(V_0^2 g^2)m^2 = 67 m^2$ in non-dimensional form.

Two of the four solutions ω are always real and represent stable waves which are not further considered. The properties of unstable waves are displayed in the following graphs. Figs. 4.2 and 4.3 show the already discussed stability diagrams, i.e. the curves of marginal stability

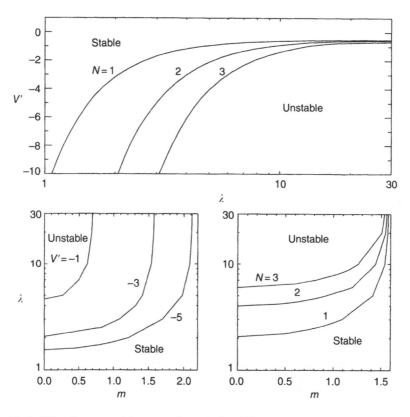

Figure 4.2 Stability diagrams with curves of marginal stability at the equator $\theta = 90°$.

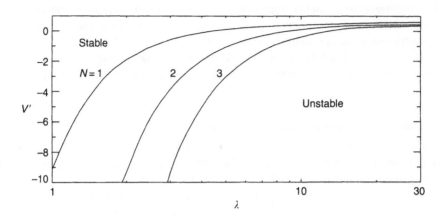

Figure 4.3 Stability diagram at the pole $\theta = 0°$.

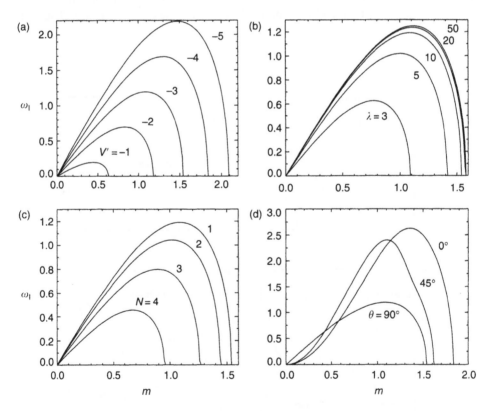

Figure 4.4 Growth rates ω_1 versus wavenumber m: (a) $\theta = 90°$, $V' = -1, -2, -3, -4, -5$, $\lambda = 10$, $N = 1$; (b) $\theta = 90°$, $V' = -3$, $\lambda = 3, 5, 10, 20, 50$, $N = 1$; (c) $\theta = 90°$, $V' = -3$, $\lambda = 10$, $N = 1, 2, 3, 4$; (d) $\theta = 90°, 45°, 0°$, $V' = -3$, $\lambda = 10$, $N = 1$.

for various parameters. The pole is more unstable than the equator, and the curves for other latitudes are in between.

The growth rate ω_I versus wavenumber m in direction of the magnetic field is displayed in Fig. 4.4. The general behaviour and the physical mechanisms are already discussed for the equator. Steep negative magnetic field gradients, large values of wavenumber ratio λ and low order overtones all favour instability. This is the case for all latitudes, and at the pole the growth rate is larger than at the equator. One important difference is that there is a finite growth rate for $\lambda \to \infty$ near the equator while growth is unlimited for $\lambda \to \infty$ at higher latitudes.

Most important is the existence of a certain wavenumber m_0 with maximum growth rate. These values are collected in Tables 4.1 and 4.2. Note m_0 is larger for steeper field gradients. This is true for all latitudes. Also m_0 usually increases from equator to pole. There is however a slight decrease at middle latitudes for large values of λ. The dependence on λ is weak and differs for the various cases. For overtones m_0 is smaller.

Table 4.1 Wavenumbers m_0 of the fundamental mode $N = 1$ with maximum growth rate ω_{10} for $\lambda = 10$ and various Alfvén speed gradients

θ	V'					
	0	-1	-2	-3	-4	-5
90°		0.44	0.83	1.09	1.30	1.48
80°	0.09	0.45	0.83	1.08	1.29	1.46
70°	0.18	0.40	0.82	1.05	1.24	1.41
60°	0.26	0.49	0.75	0.90	1.10	1.25
50°	0.33	0.60	0.84	1.03	1.19	1.35
40°	0.39	0.68	0.96	1.17	1.35	1.52
30°	0.44	0.76	1.04	1.27	1.46	1.63
20°	0.48	0.82	1.08	1.32	1.52	1.70
10°	0.50	0.86	1.11	1.35	1.55	1.74
0°	0.51	0.87	1.12	1.36	1.56	1.75

Table 4.2 Wavenumbers m_0 with maximum growth rate ω_{10} for $V' = -3$

θ	λ						
	3	5	10				50
	$N = 1$	$N = 1$	$N = 1$	$N = 2$	$N = 3$	$N = 4$	$N = 1$
90°	0.77	1.00	1.09	1.02	0.89	0.67	1.11
80°	0.84	1.01	1.08	1.02	0.90	0.69	1.10
70°	0.95	1.03	1.05	0.98	0.87	0.74	1.07
60°	1.05	1.08	0.90	0.90	0.90	0.85	0.90
50°	1.11	1.17	1.03	1.06	1.05	0.95	1.00
40°	1.15	1.25	1.17	1.19	1.14	1.02	1.12
30°	1.17	1.30	1.27	1.27	1.21	1.06	1.22
20°	1.19	1.34	1.32	1.32	1.25	1.09	1.27
10°	1.19	1.36	1.35	1.35	1.27	1.11	1.30
0°	1.20	1.37	1.36	1.36	1.28	1.11	1.31

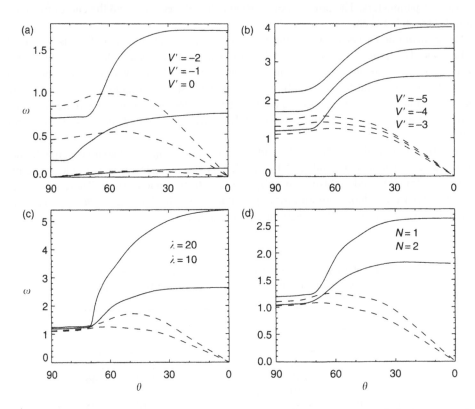

Figure 4.5 Maximum growth rate ω_{I0} (solid) and corresponding frequency ω_R (dashed) versus colatitudes θ: (a) $\lambda = 10$, $N = 1$, $V' = 0, -1, -2$; (b) $\lambda = 10$, $N = 1$, $V' = -3, -4, -5$; (c) $V' = -3$, $\lambda = 10, 20$, $N = 1$; (d) $V' = -3$, $\lambda = 10$, $N = 1, 2$.

The maximum growth rate ω_{I0} at m_0 increases from equator to pole. This increase is weak near the equator, steep at middle latitudes and weak again near the pole. This is displayed for various parameter values of V', λ and N in Fig. 4.5. Figs. 4.6 and 4.7 illustrate again the dependence of the growth rate on the Alfvén speed gradient V' and on the wavenumber ratio λ at four different colatitudes.

4.6. Unstable magnetostrophic waves and its dynamo action

After having discussed the local analysis we now solve the eigenvalue problem (4.12) numerically by matrix methods to study the global properties of unstable magnetostrophic waves and to determine their dynamo action in a thin rotating magnetic layer. This layer shall represent the toroidal magnetic field in the overshoot region at the bottom of the convection zone.

We first introduce non-dimensional variables in the same manner as in the last section. The complex velocity amplitude $\hat{w}$ is normalized by its maximum absolute value. To determine the magnitude of α we set $|\hat{w}|_m = 1 \, \mathrm{m \, s^{-1}}$ which is roughly a hundredth of the convective velocity. For the non-dimensional Alfvén speed we take the Gaussian profile

$$V(z) = \exp\left\{-(z/\epsilon)^2\right\}. \tag{4.31}$$

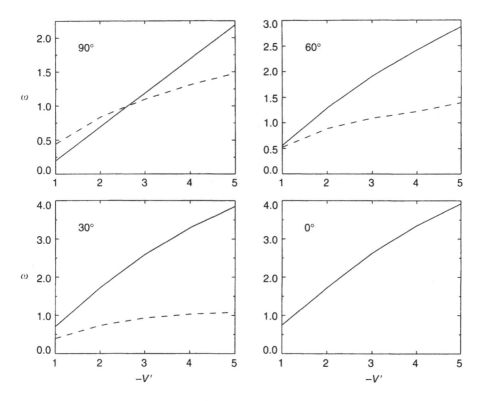

Figure 4.6 Maximum growth rate ω_{I0} (solid) and corresponding frequency ω_R (dashed) of the unstable fundamental mode $N = 1$ versus Alfvén speed gradient V' for ratio of horizontal wave numbers $\lambda = 10$ at four colatitudes $\theta = 90°, 60°, 30°$ and $0°$.

The half width ϵ is a free parameter which is varied. Most results refer to $\epsilon = 0.3$ which corresponds to a thickness of the magnetic layer of about 2×10^7 m. The boundaries are taken far off at $z_1 = -1$ and $z_2 = 1$, which is 10^8 m apart (see Fig. 4.8). They do not influence the description of the unstable magnetostrophic waves within the magnetic layer.

4.6.1. Numerics

For the numerical treatment another transformation

$$\bar{w}(z) = P(z)\breve{w}(z) \tag{4.32}$$

of the eigenfunction is convenient (Knölker and Stix, 1983) which transforms the non-dimensional form of (4.12) into

$$S\breve{w}'' + S'\breve{w}' + (T - PP'')\breve{w} = 0 \tag{4.33a}$$

with boundary conditions

$$\breve{w}(z_1) = \breve{w}(z_2) = 0. \tag{4.33b}$$

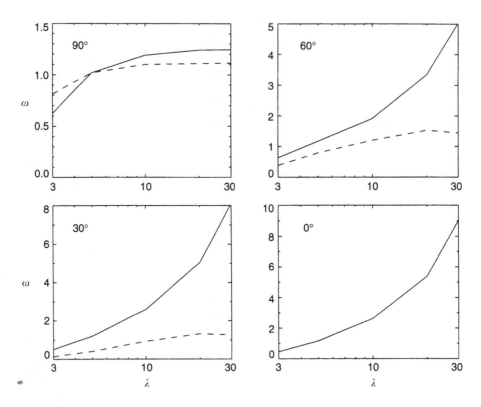

Figure 4.7 Maximum growth rate ω_{I0} (solid) and corresponding frequency ω_R (dashed) of the unstable fundamental mode $N = 1$ versus ratio λ of horizontal wave numbers for Alfvén speed gradient $V' = -3$ at four colatitudes $\theta = 90°, 60°, 30°, 0°$.

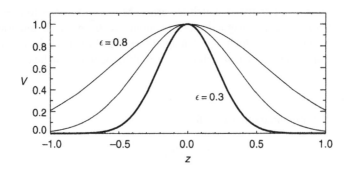

Figure 4.8 Vertical profile of the Alfvén speed $V(z)$ for $\epsilon = 0.3, 0.5, 0.8$.

Then $\breve{w}$ is expanded in a Fourier series

$$\breve{w}(z) = \sum_{v=1}^{\infty} c_v \sin \frac{v\pi}{d}(z - z_1), \qquad d = z_2 - z_1, \qquad (4.34)$$

which already takes care of the boundary conditions. By standard techniques this results in a matrix eigensystem for the expansion coefficients c_ν as eigenvector. This system is a fourth order polynomial in the eigenvalue ω. By incorporating the eigenvalue ω into the eigenvector c_ν, an eigensystem linear in ω can be obtained with a big matrix which contains the matrices of the polynomial eigensystem as submatrices (Garbow *et al.*, 1977). Although this large matrix is no longer symmetric, its submatrices are symmetric by virtue of the above transformation (4.32). The eigenvalues ω and eigenvectors c_ν are then determined with the help of the appropriate EISPACK routines (Smith *et al.*, 1976; Garbow *et al.*, 1977) which are based on numerical algorithms developed by Wilkinson (1965) and Wilkinson and Reinsch (1971). The eigenvectors c_ν are normalized and the eigenfunctions $\breve{w}$, $\bar{w}$, $\tilde{w}$ and $\hat{w}$ are readily obtained and used to calculate the dynamo coefficient α according to (4.15). Test calculations and convergence tests were performed. We further made sure that the boundary conditions were far off the region of instability and their location did not influence the results, neither the eigenvalues nor the eigenfunctions.

4.6.2. *Global properties of unstable magnetostrophic waves*

We solved the eigensystem (4.33) and derived the global properties of unstable magneto-strophic waves for the Gaussian profile (4.31) of the Alfvén speed $V(z)$. Most results refer to $\epsilon = 0.3$. As in Section 4.5 we introduce $\lambda = k/m$. As eigenfunction we describe the velocity amplitude $\hat{w}(z)$ in z-direction.

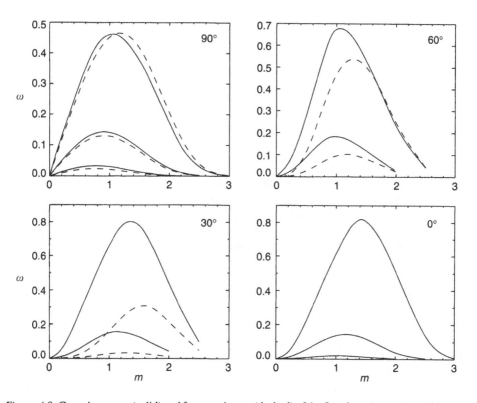

Figure 4.9 Growth rates ω_I (solid) and frequencies ω_R (dashed) of the first three (or two) unstable modes versus wavenumber m for $\epsilon = 0.3$, $\lambda = 10$ and four colatitudes $\theta = 90°$, $60°$, $30°$ and $0°$.

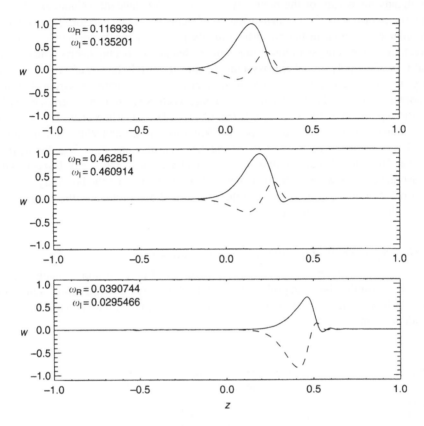

Figure 4.10 Velocity amplitude $\hat{w}(z) = \hat{w}_R + i\hat{w}_I$ ($\hat{w}_R$ solid, $\hat{w}_I$ dashed), norm $\max(|\hat{w}_R|, |\hat{w}_I|) = 1$ of the unstable fundamental mode for $\epsilon = 0.3$, $\lambda = 10$ and $m = 0.2$, $m_0(= 1.1)$, 2.5 (from top to bottom) at $\theta = 90°$, together with the eigenvalues $\omega = \omega_R + i\omega_I$.

One of the important results of the local analysis was the dependence of the growth rate ω_I as function of wavenumber m in the direction of the magnetic field. In Fig. 4.9 the growth rates ω_I and freqencies ω_R of the first three (or two) unstable modes at four colatitudes are displayed versus wavenumber m. The constant parameters are $\lambda = 10$ and $\epsilon = 0.3$. The local behaviour, the existence and location m_0 of a maximum growth rate as well as the variation of the maximum growth rate for various parameters was also globally recovered. Only the decrease of ω_I for $m > m_0$ is less steep. The reason lies in the fact that unstable waves with $m > m_0$ originate from slightly higher levels (see Fig. 4.10) where the Alfvén speed gradient is steeper and thus more unstable than for waves at $m = m_0$. In contrast to the local analysis of Section 4.5 the frequency ω_R also decreases at larger m. The reason is again the contribution height of the instability. At the equator, local analysis gives $\omega_R \sim mV^2$, but V decreases rapidly with height respectively $m > m_0$.

Frequencies and growth rates of unstable overtones are much smaller. Fig. 4.11 shows that overtones originate from greater heights too. The plots of the eigenfunctions clearly demonstrate that unstable magnetostrophic waves originate from and are confined to the upper regions of the magnetic layer with its unstable field gradient.

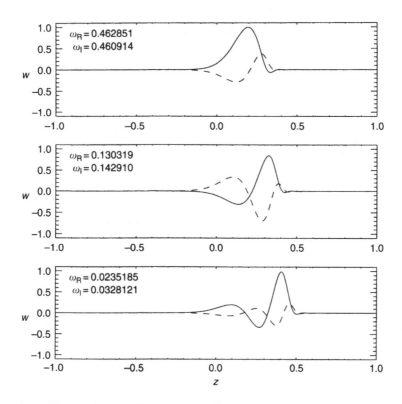

Figure 4.11 Velocity amplitudes $\hat{w}(z)$ of the unstable fundamental mode (top) and the first (middle) and second (bottom) overtone for $\epsilon = 0.3$, $\lambda = 10$ and $m = m_0$ at $\theta = 90°$.

Table 4.3 Wavenumbers m_0 with maximum growth rate ω_{10} for $\epsilon = 0.3$

θ	λ						
	5	10			20		50
Mode	F	F	1	2	F	1	F
90°	0.9	1.1	0.9	0.8	1.1	1.0	1.2
80°	0.9	1.0	0.9		1.1	1.0	1.2
70°	0.9	0.9	0.9		0.9	0.9	1.0
60°	1.0	1.0	1.0		1.0	1.0	0.9
50°	1.1	1.1	1.1		1.1	1.1	1.0
40°	1.1	1.2	1.1		1.2	1.2	1.2
30°	1.1	1.3	1.1		1.3	1.3	1.3
20°	1.2	1.4	1.2		1.4	1.4	1.4
10°	1.2	1.4	1.2		1.5	1.4	1.4
0°	1.2	1.4	1.2	1.0	1.5	1.4	1.5

Table 4.4 Wavenumbers m_0 of the fundamental mode with maximum growth rate ω_{I0} for $\lambda = 10$

θ	ϵ					
	0.3	0.4	0.5	0.6	0.7	0.8
90°	1.1	0.9	0.8	0.7	0.7	0.6
80°	1.0		0.8			0.6
70°	0.9		0.6			0.5
60°	1.0	0.8	0.7	0.6	0.6	0.5
50°	1.1		0.8			0.6
40°	1.2		1.0			0.7
30°	1.3	1.2	1.1	1.0	0.9	0.8
20°	1.4		1.1			0.9
10°	1.4		1.1			0.9
0°	1.4	1.3	1.1	1.1	1.0	0.9

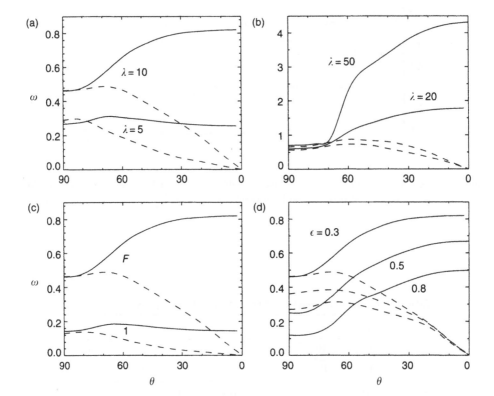

Figure 4.12 Maximum growth rate ω_{I0} (solid) and corresponding frequency ω_R (dashed) versus colatitude θ. (a) fundamental mode, $\epsilon = 0.3$, $\lambda = 5, 10$; (b) fundamental mode, $\epsilon = 0.3$, $\lambda = 20, 50$; (c) fundamental mode and first overtone, $\epsilon = 0.3$, $\lambda = 10$; (d) fundamental mode, $\lambda = 10$, $\epsilon = 0.3, 0.5, 0.8$.

For other values of λ and ϵ and for other latitudes these findings also apply. Tables 4.3 and 4.4 provide the wavenumbers m_0 with maximum growth rates ω_{I0} for various parameters of θ, λ and ϵ and for fundamental modes and overtones. The local behaviour is recovered.

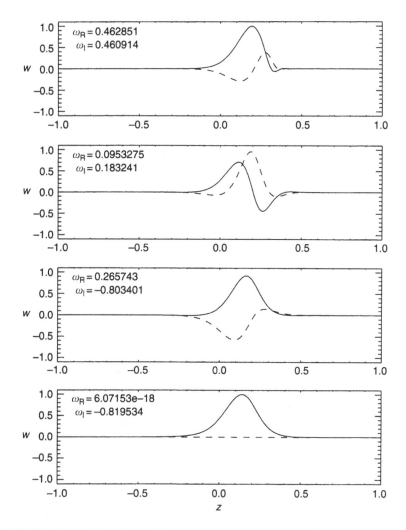

Figure 4.13 Velocity amplitude $\hat{w}(z)$ of the unstable fundamental mode for $\epsilon = 0.3$, $\lambda = 10$, $m = m_0$ and $\theta = 90°, 60°, 30°, 0°$.

Fig. 4.12 provides the maximum growth rate ω_{I0} at wavenumber m_0 versus colatitude θ for various values of λ and ϵ. As for the local analysis the growth rate increases from equator to pole. The rise is small near the equator, large at middle latitudes and small again near the pole. It is larger for larger values of λ and for the fundamental mode compared to overtones, but relatively independent of the magnetic field gradient. The frequency vanishes at the pole.

The variation of the unstable eigenfunction with colatitude is provided in Fig. 4.13. According to the instability condition the instability occurs at slightly deeper levels at the pole compared to the equator.

Fig. 4.14 illustrates the change of the maximum growth rate ω_{I0} with ratio λ for four colatitudes. Here the behaviour near the equator is different from middle and high latitudes.

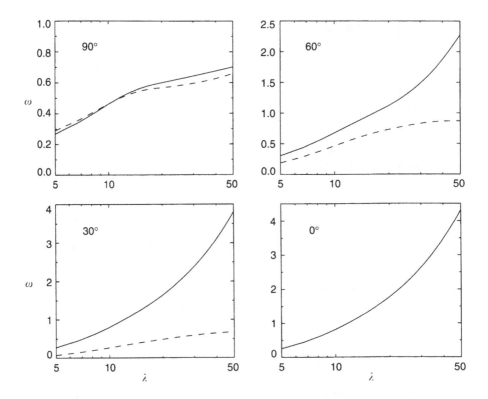

Figure 4.14 Maximum growth rate ω_{I0} (solid) and corresponding frequency ω_R (dashed) of the unstable fundamental mode versus ratio λ of horizontal wavenumbers for $\epsilon = 0.3$ and $\theta = 90°$, $60°$, $30°$, $0°$.

At the equator there exists a finite limit of ω_{I0}, at the pole it diverges for increasing λ. The velocity amplitude $\hat{w}(z)$ in Fig. 4.15 shows that for larger λ the unstable wave is more concentrated around a certain level which is shifted slightly to deeper layers.

For a thicker magnetic layer with a smaller field strength gradient the growth rate decreases (Fig. 4.16) and the instability extends to a wider layer (Fig. 4.17) which is, according to the most unstable levels, shifted slightly upwards. In Figs. 4.15 and 4.17 the equator is selected representatively for all latitudes.

Altogether the globally obtained growth rates ω_I and frequencies ω_R and their parameter variation are in good accordance to their local values if the local equilibrium quantities are taken at the location which contributes the most to the instability.

Finally in Figs. 4.18–4.20 the velocity amplitudes $\hat{w}$ are given for an unstable wave and its stable counterpart. As discussed in Section 4.4 the velocity of the stable counterpart can be used to calculate the dynamo action of the corresponding unstable wave with negative wavenumber $-k$. At the equator the velocity distribution of unstable wave and stable counterpart do not differ and the total α-effect vanishes. The differences at other latitudes lead to the non-vanishing dynamo action which is reported in the next subsection.

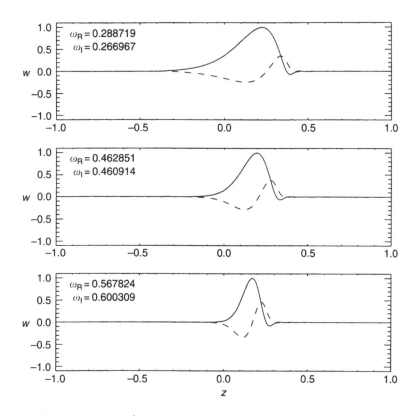

Figure 4.15 Velocity amplitude $\hat{w}(z)$ of the unstable fundamental mode for $\epsilon = 0.3$, $\lambda = 5, 10, 20$ (from top to bottom), $m = m_0$ and $\theta = 90°$.

4.6.3. Dynamo action of unstable magnetostrophic waves

Given the eigenvalues ω and eigenfunctions $\hat{w}(z)$ we derive the α-effect of unstable magnetostrophic waves after (4.15). According to the discussion in Section 4.4 there always exists a second wave to every unstable wave with the same growth rate but different propagation direction along the x-axis. Both waves contribute to the total α^T given by (4.17).

An absolute value of α^T cannot be given because the velocity amplitudes cannot be fixed in linear theory. The following results always refer to a maximum velocity amplitude of $|\tilde{w}|_m = 1 \, \mathrm{m \, s^{-1}}$, and the unit of α^T in the graphs is $1 \, \mathrm{cm \, s^{-1}}$. Because of the exponential growth in time the most unstable modes contribute the most. These are the fundamental modes with maximum growth rate ω_{I0} at wavenumber m_0. Only the time-independent part α_0^T is shown in the following.

In Fig. 4.21 the contributions of both members of an unstable wave pair are separately plotted versus colatitude θ. The other parameters are $\epsilon = 0.3$, $\lambda = \pm 10$ and $m = m_0$. The two contributions cancel at the equator and result in a negative α^T north of the equator, followed by a sign change around $\theta \approx 60°$ and a positive α^T thereafter towards the north pole. Note α^T is antisymmetric with respect to the equator and we thus only display the northern hemisphere.

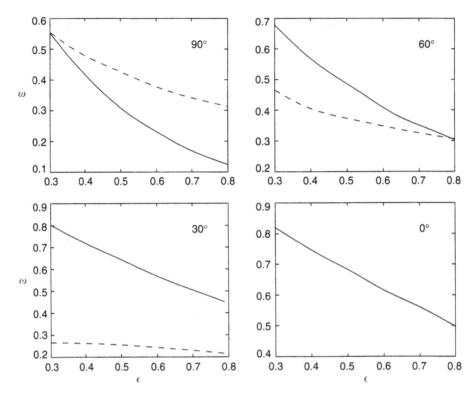

Figure 4.16 Maximum growth rate ω_{10} (solid) and corresponding frequency ω_R (dashed) of the unstable fundamental mode versus half-thickness ϵ for $\lambda = 10$ and $\theta = 90°$, $60°, 30°, 0°$.

This behaviour was qualitatively found for all free parameters (see Fig. 4.22). For smaller values of λ the positive amplitudes of α_0^T at high latitudes dominate the negative ones near the equator. This ratio changes for larger λ and for $\lambda > 15$ the negative values near the equator clearly dominate. This is also the case for larger values of ϵ for which the zero crossing is also shifted slightly towards the pole.

4.7. Discussion and conclusions

We considered magnetostrophic waves in a horizontal thin plane layer of a perfectly conducting fluid, which is stratified according to gravitation, permeated by a variable horizonal magnetic field and rotating rigidly around an inclined axis. The configuration intends to represent that of the toroidal magnetic field in the overshoot region at the bottom of the solar convection zone where most of the magnetic flux is expected. The main aim was to study unstable magnetostrophic waves which are driven by a vertical field gradient. Magnetostrophic waves require the Alfvén frequency smaller than the rotation frequency. They are slow waves characterized by an approximate balance of the Coriolis and the Lorentz force.

A sufficient decrease of the magnetic field with height leads to instability which is interpreted as a magnetic buoyancy instability in the presence of rotation. In this Rayleigh–Taylor

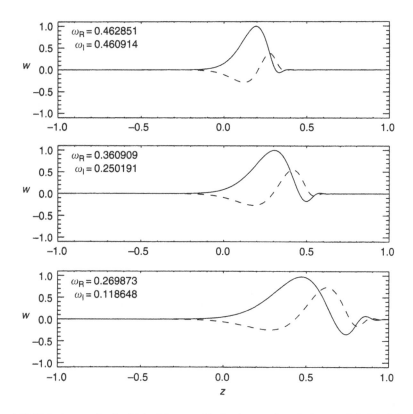

Figure 4.17 Velocity amplitude $\hat{w}(z)$ of the unstable fundamental mode for $\epsilon = 0.3, 0.5, 0.8$ (from top to bottom), $\lambda = 10$, $m = m_0$ and $\theta = 90°$.

like instability potential energy of extra mass supported against gravity is released by downward transport of mass and upward transport of magnetic flux. The instability naturally occurs in the upper part of the magnetic layer in the overshoot region.

The properties of the excited unstable magnetostrophic waves were discussed. The growth rate has a maximum at a certain value of the wavenumber in the direction of the magnetic field. For smaller values rotation acts stabilizing, for larger values this is achieved by magnetic tension. The optimal wavenumber depends mainly on the field gradient. Typical values are of the order of the inverse of one scale height. The other horizontal wavenumber perpendicular to the field has a lower limit and is typically much larger. An upper limit cannot be given because of neglection of dissipative processes. In the vertical direction, the wave is confined to the unstable parts of the layer. Growth rates and frequencies are proportional to V^2/Ω where V is the Alfvén speed and Ω the rotation frequency. The maximum growth rate depends on the relative magnetic field gradient, the ratio of horizontal wavenumbers, and the latitude. Overtones are less unstable than the fundamental mode. Unstable waves only travel in one direction along the magnetic field. Typical values of growth rate and frequency are of the order of the inverse of one year for the lower convection zone of the sun.

We especially showed that unstable magnetostrophic waves are capable of inducing an electromotive force parallel to the horizontal magnetic field. These waves are helical; their

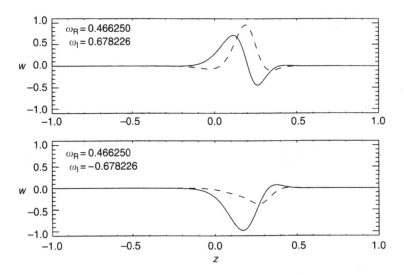

Figure 4.18 Velocity amplitude $\hat{w}(z)$ of the unstable fundamental mode (top) and the corresponding stable mode (bottom) for $\epsilon = 0.3$, $\lambda = 10$, $m = m_0$ and $\theta = 60°$.

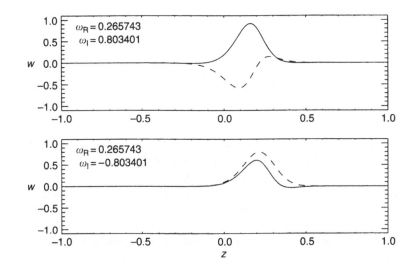

Figure 4.19 Same as Fig. 4.18 for $\theta = 30°$.

growth in amplitude causes a phase shift between the perturbations of magnetic field and velocity which leads to an electromotive force parallel or antiparallel to the toroidal field. This induction effect is of dynamic nature. The velocities are not prescribed but follow from the present forces in the momentum equation and the back reaction of the magnetic field on the velocity through the Lorentz force is taken into account, albeit in a linear description. This dynamic α-effect is able to generate a poloidal magnetic field out of a strong toroidal field. It

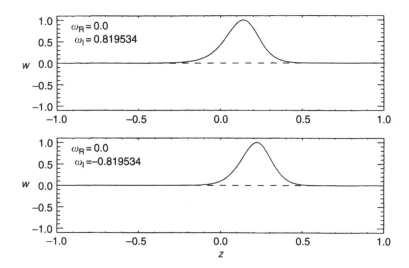

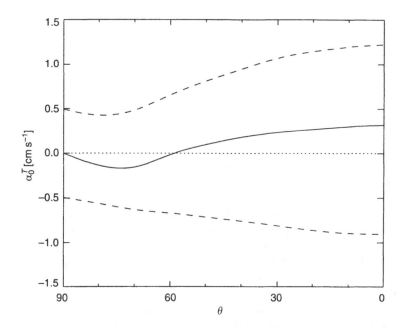

Figure 4.20 Same as Fig. 4.18 for $\theta = 0°$.

Figure 4.21 Individual contributions of the two most unstable waves (dashed) and the resulting total α-effect (solid) versus colatitude θ for $\epsilon = 0.3$, $\lambda = \pm 10$ and $m = m_0$.

is therefore an alternative to the kinematic α-effect and of importance for the dynamo theory of strong magnetic fields, especially for the solar dynamo processes in the overshoot region. It is emphasized that this effect does not need convection, it follows from a naturally occuring magnetic buoyancy instability.

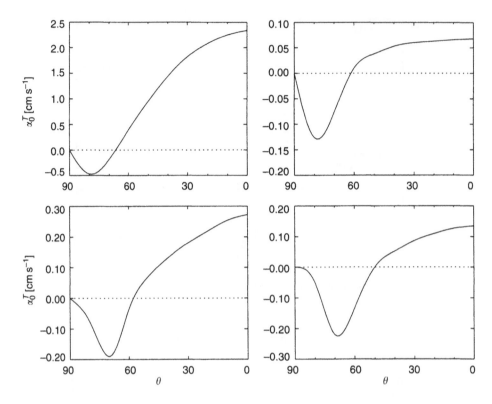

Figure 4.22 Total α-effect versus colatitude θ for $\epsilon = 0.3$ and $\lambda = \pm 5$ (upper left) and $\lambda = \pm 20$ (upper right), and for $\lambda = \pm 10$ and $\epsilon = 0.5$ (lower left) and $\epsilon = 0.8$ (lower right), each for $m = m_0$.

The magnitude of α cannot be given because of the linear treatment. If we assume the z-component of the perturbation velocity of the order of 1 m s^{-1}, a hundredth of the convective velocities in the lower convection zone, we find an α of the order of 1 cm s^{-1}. This is a perfect magnitude for the operation of the solar dynamo (Köhler, 1973).

Of special interest is the dependence of the induction constant α on colatitude θ. Conventionally, $\alpha = \alpha_0 \cos\theta$, $\alpha_0 > 0$, is assumed (e.g. Moffatt, 1978, sect. 9.12) because of the effective Coriolis force. Based on unstable magnetostrophic waves we found a more complicated behaviour. Again α is antisymmetric with respect to the equator where it vanishes. Here however α is found to be negative just north of the equator, it changes its sign around $\theta \approx 60°$ and is positive towards the north pole. This non-monotonous variation of $\alpha(\theta)$ is the consequence of a superposition of the two most unstable magnetostrophic waves which travel north- and southward with equal enthusiasm.

Such a latitude dependence of the α-effect has remarkable consequences for the solar dynamo. The dynamo wave of an $\alpha\Omega$-dynamo propagates along contours of constant angular velocity Ω (Yoshimura, 1975). For an equatorward migration, as is suggested by the butterfly diagram of sunspots, radial gradients of Ω are needed and the sign of $\alpha \cdot \partial\Omega/\partial r$ must be negative in the northern hemisphere.

If $\partial\Omega/\partial r$ would be positive throughout, the sign change of α leads to two branches of wave propagation in the butterfly diagram (Schmitt, 1987). One branch migrates from middle latitudes towards the equator and is identified with the ordinary sunspot branch. The other branch is much weaker and migrates from middle latitudes towards the poles. It can be identified with the behaviour of polar fields.

Helioseismology suggests that the main convection zone rotates like the solar surface with no significant radial gradient, and that the deep interior rotates almost rigidly at a rate intermediate between the equatorial and the polar rate on the surface. A radial gradient occurs in the transitional region between the bottom of the convection zone and the top of the radiative interior. This gradient is positive near the equator and negative near the poles. At the poles the gradient seems steeper by a factor of approximately 2.

With such a rotational law a conventional dynamo in the convection zone does not exhibit solar-like behaviour (Köhler, 1973). Also a dynamo in the overshoot region with the conventional α-profile does not succeed (Prautzsch, 1993). But combining the helioseismological rotation profile with the α-effect of magnetostrophic waves in an overshoot layer dynamo, results in sunspot-like butterfly diagrams. Here it is important to compensate the steep Ω-gradient near the poles by an α concentrated around the equator (Schmitt, 1993; Prautzsch, 1997), as is derived for magnetostrophic waves with a large horizontal wavenumber ratio. The magnetic field is then generated only in a small region around the equator at the bottom of the convection zone. A problem of overshoot layer dynamos with $\alpha < 0$ and $\partial\Omega/\partial r > 0$ near the equator may be the phase relation between the toroidal and radial field components (Schlichenmaier and Stix, 1995).

An alternative picture of the solar dynamo has been presented by Parker (1993) with the α-effect acting in the highly diffusive convection zone and the differential rotation in the less diffusive overshoot region at its bottom where most of the field is concentrated. The migration direction and the phase relation can be further influenced by meridional circulation (Durney, 1995, 1996; Choudhuri *et al.*, 1995).

Shortcomings of the present theory are the assumption of ideal electrical conductivity, the restriction to rigid rotation and the linear treatment. It is relatively easy to show that Ohmic dissipation exhibits a damping effect on the instability (Acheson and Gibbons, 1978; Acheson, 1978, 1979). With dissipation, steeper magnetic field gradients are needed for instability. Furthermore dissipation yields an upper bound to the horizontal wavenumber perpendicular to the magnetic field, beyond which instability is suppressed.

Differential rotation in the overshoot region generates the toroidal field on which the instability is based. For self-consistent dynamo models on the basis of the induction action of unstable magnetostrophic waves the knowledge of the influence of differential rotation is highly desirable. Depending on various circumstances it stabilizes or destabilizes magnetic buoyancy or is itself a source of instability, a complicated matter of ongoing research (van Ballegooijen, 1983; Fearn and Proctor, 1983; Fearn, 1989; Fearn and Weiglhofer, 1992; Foglizzo and Tagger, 1994, 1995; Terquem and Papaloizou, 1996; Fearn *et al.*, 1997; Brandenburg *et al.*, 2001).

The linear treatment presented here excludes the determination of the strength of the induction effect of the magnetic buoyancy instability. A non-linear numerical simulation (Brandenburg and Schmitt, 1998) confirms the existence of the α-effect but lacks the confirmation of the non-monotonous latitude dependence. This may be due to different excitation conditons of the instability outside the magnetostrophic limit, or to the missing of one of the two most unstable modes, points which need further clarification.

References

Acheson, D.J., "On the hydrodynamic stability of a rotating fluid annulus," *J. Fluid Mech.* **52**, 529–541 (1972).

Acheson, D.J., "Hydromagnetic wavelike instabilities in a rapidly rotating fluid," *J. Fluid Mech.* **61**, 609–624 (1973).

Acheson, D.J., "On the instability of toroidal magnetic fields and differential rotation in stars," *Phil. Trans. Roy. Soc.* A**289**, 459–500 (1978).

Acheson, D.J., "Instability by magnetic buoyancy," *Solar Phys.* **62**, 23–50 (1979).

Acheson, D.J. and Gibbons, M.P., "Magnetic instabilities of a rotating gas," *J. Fluid Mech.* **85**, 743–757 (1978).

Acheson, D.J. and Hide, R., "Hydromagnetics of rotating fluids," *Rep. Prog. Phys.* **36**, 159–221 (1973).

Balbus, S.A. and Hawley, J.F., "A powerful local shear instability in weakly magnetized disks. I. Linear analysis. II. Nonlinear evolution," *Astrophys. J.* **376**, 214–233 (1991).

Braginsky, S.I., "Magnetic waves in the Earth's core," *Geomag. Aeron.* **7**, 851–859 (1967).

Braginsky, S.I., "Magnetic waves in the core of the Earth II," *Geophys. Astrophys. Fluid Dynam.* **14**, 189–208 (1980).

Braginsky, S. I. and Roberts, P. H., "On the theory of convection in the Earth's core," this volume, Ch. 3, 60–82 (2003).

Brandenburg, A., "Theoretical basis of stellar activity cycles," in: *Tenth Cambridge Workshop on Cool Stars, Stellar Systems, and the Sun* (Eds. R. Donahue and J. Bookbinder), Astron. Soc. Pac. Conf. Ser., Vol. 154, pp. 173–191 (1998).

Brandenburg, A., "Simulations and observations of stellar dynamos: evidence for a magnetic alpha-effect," in: *Stellar Dynamos: Nonlinearity and Chaotic Flows* (Eds. M. Nunez and A. Ferriz-Mas), Astron. Soc. Pac. Conf. Ser., Vol. 178, pp. 13–21 (1999).

Brandenburg, A. and Schmitt, D., "Simulations of an alpha-effect due to magnetic buoyancy," *Astron. Astrophys.* **338**, L55–L58 (1998).

Brandenburg, A., Bigazzi, A. and Subramanian, K., "The helicity constraint in turbulent dynamos with shear," *Mon. Not. R. Astron. Soc.* **325**, 685–692 (2001).

Brandenburg, A., Nordlund, A. and Stein, R.F., Torkelsson, U., "Dynamo-generated turbulence and large-scale magnetic fields in a Keplerian shear flow," *Astrophys. J.* **446**, 741–754 (1995).

Brandenburg, A., Jennings, R.L., Nordlund, A., Rieutord, M., Stein, R.F. and Tuominen, I., "Magnetic structures in a dynamo simulation," *J. Fluid Mech.* **306**, 325–352 (1996).

Brown, T.M. and Morrow, C.A., "Depth and latitude dependence of solar rotation," *Astrophys. J.* **314**, L21–L26 (1987).

Caligari, P., Moreno-Insertis, F. and Schüssler, M., "Emerging flux tubes in the solar convection zone. I. Asymmetry, tilt, and emergence latitude," *Astrophys. J.* **441**, 886–902 (1995).

Cattaneo, F., "Magnetoconvection," in: *Solar Magnetic Fields* (Eds. M. Schüssler and W. Schmidt), Cambridge University Press, pp. 261–275 (1994).

Chandrasekhar, S., "The stability of non-dissipative Couette flow in hydromagnetics," *Proc. Natl. Acad. Sci.* **46**, 253–257 (1960).

Chou, D.-Y. and Fisher, G.H., "The influence of the Coriolis force on flux tubes rising through the solar convection zone," *Astrophys. J.* **341**, 533–548 (1989).

Choudhuri, A.R., "The evolution of loop structures in flux rings within the solar convection zone," *Solar Phys.* **123**, 217–239 (1989).

Choudhuri, A.R. and Gilman, P.A., "The influence of the Coriolis force on flux tubes rising through the solar convection zone," *Astrophys. J.* **316**, 788–800 (1987).

Choudhuri, A.R., Schüssler, M. and Dikpati, M., "The solar dynamo with meridional circulation," *Astron. Astrophys.* **303**, L29–L32 (1995).

Durney, B.R., "On a Babcock-Leighton dynamo model with a deap-seated generating layer for the toroidal magnetic field," *Solar Phys.* **160**, 213–235 (1995).

Durney, B.R., "On a Babcock-Leighton dynamo model with a deap-seated generating layer for the toroidal magnetic field II," *Solar Phys.* **166**, 231–260 (1996).

D'Silva, S.Z. and Choudhuri, A.R., "A theoretical model for tilts of bipolar magnetic regions," *Astron. Astrophys.* **272**, 621–633 (1993).

Eckart, C., *Hydrodynamics of Oceans and Atmosphers*, Pergamon Press, Oxford (1960).

Eltayeb, I.A., "Hydromagnetic convection in a rapidly rotating fluid layer," *Proc. Roy. Soc. Lond.* A**326**, 229 (1972).

Fan, Y., Fisher, G.H. and DeLuca, E.E., "The origin of morphological asymmetries in bipolar active regions," *Astrophys. J.* **405**, 390–401 (1993).

Fan, Y., Fisher, G.H. and McClymont, A.N., "Dynamics of emerging active region flux loops," *Astrophys. J.* **436**, 907–928 (1994).

Fearn, D.R., "Thermal and magnetic instabilities in a rapidly rotating fluid sphere," *Geophys. Astrophys. Fluid Dynam.* **14**, 103–126 (1979).

Fearn, D.R., "Differential rotation and thermal convection in a rapidly rotating magnetic system," *Geophys. Astrophys. Fluid Dynam.* **49**, 173–193 (1989).

Fearn, D.R., "Hydromagnetic flow in planetary cores," *Rep. Prog. Phys.* **61**, 175–235 (1998).

Fearn, D.R. and Proctor, M.R.E., "The stabilizing role of differential rotation on hydromagnetic waves," *J. Fluid Mech.* **128**, 21–36 (1983).

Fearn, D.R. and Proctor, M.R.E., "Self-consistent dynamo models driven by hydromagnetic instabilities," *Phys. Earth Planet. Inter.* **36**, 78–84 (1984).

Fearn, D.R. and Weiglhofer, W.S., "Magnetic instabilities in rapidly rotating spherical geometries. III. The effect of differential rotation," *Geophys. Astrophys. Fluid Dynam.* **67**, 163–184 (1992).

Fearn, D.R., Lamb, C.J., McLean, D.R. and Ogden, R.R., "The influence of differential rotation on magnetic instability, and nonlinear magnetic instability in the magnetostrophic limit," *Geophys. Astrophys. Fluid Dynam.* **86**, 173–200 (1997).

Ferriz-Mas, A., "On the storage of magnetic flux tubes at the base of the solar convection zone," *Astrophys. J.* **458**, 802–816 (1996).

Ferriz-Mas, A. and Schüssler, M., "Instabilities of magnetic flux tubes in a stellar convection zone. I. Equatorial flux rings in differentially rotating stars," *Geophys. Astrophys. Fluid Dynam.* **72**, 209–247 (1993).

Ferriz-Mas, A. and Schüssler, M., "Waves and instabilities of a toroidal magnetic flux tube in a rotating star," *Astrophys. J.* **433**, 852–866 (1994).

Ferriz-Mas, A. and Schüssler, M., "Instabilities of magnetic flux tubes in a stellar convection zone. II. Flux rings outside the equatorial plane," *Geophys. Astrophys. Fluid Dynam.* **81**, 233–265 (1995).

Ferriz-Mas, A., Schmitt, D. and Schüssler, M., "A dynamo effect due to instability of magnetic flux tubes," *Astron. Astrophys.* **289**, 949–956 (1994).

Fisher, G.H., McClymont, A.N. and Chou, D.-Y., "The stretching of magnetic flux tubes in the convective overshoot region," *Astrophys. J.* **374**, 766–772 (1991).

Foglizzo, T. and Tagger, M., "The Parker instability in disks with differential rotation," *Astron. Astrophys.* **287**, 297–319 (1994).

Foglizzo, T. and Tagger, M., "The Parker-shearing instability in azimuthally magnetized disks," *Astron. Astrophys.* **301**, 293–308 (1995).

Galloway, D.J. and Weiss, N.O., "Convection and magnetic fields in stars," *Astrophys. J.* **243**, 945–953 (1981).

Garbow B.S., Boyle J.M., Dongarra J.J. and Moler C.B., *Matrix Eigensystem Routines, EISPACK Guide Extension*, Springer (1977).

Gilman, P.A., "Instability of magnetohydrostatic stellar interiors from magnetic buoyancy I," *Astrophys. J.* **162**, 1019–1029 (1970).

Hanasz, M. and Lesch, H., "The galactic dynamo effect due to Parker-shearing instability of magnetic flux tubes. I. General formalism and the linear approximation," *Astron. Astrophys.* **321**, 1007–1020 (1997).

Hoyng, P., "The field, the mean, and the meaning," this volume, Ch. 1, pp. 1–36 (2003).

Hughes, D.W., "The formation of flux tubes at the base of the convection zone," in: *Sunspots, Theory and Observations* (Eds. J.H. Thomas and N.O. Weiss), Kluwer, Dordrecht, pp. 371–384 (1992).

Hughes, D.W. and Proctor, M.R.E., "Magnetic fields in the solar convection zone: Magnetoconvection and magnetic buoyancy," *Ann. Rev. Fluid Mech.* **20**, 187–223 (1988).

Knölker, M. and Stix, M., "A convenient method to obtain stellar eigenfrequencies," *Solar Phys.* **82**, 331–341 (1983).

Köhler, H., "The solar dynamo and estimates of the magnetic diffusivity and the α-effect," *Astron. Astrophys.* **25**, 467–476 (1973).

Krause, F. and Rädler, K.-H., *Mean-Field Magnetohydrodynamics and Dynamo Theory*, Pergamon Press, Oxford (1980).

Libbrecht, K.G., "Solar p-mode frequency splittings," in: *Seismology of the Sun and Sun-like Stars* (Ed. E.J. Rolfe), ESA-SP 286, pp.131–136 (1988).

Moffatt, H.K., "An approach to a dynamic theory of dynamo action in a rotating conducting fluid," *J. Fluid Mech.* **53**, 385–399 (1972).

Moffatt, H.K., *Magnetic Field Generation in Electrically Conducting Fluids*, Cambridge Univiserity Press (1978).

Moreno-Insertis, F., "Nonlinear time-evolution of kink-unstable magnetic flux tubes in the convection zone of the sun," *Astron. Astrophys.* **166**, 291–305 (1986).

Moreno-Insertis, F., "The motion of magnetic flux tubes in the convection zone and the subsurface origin of active regions," in: *Sunspots: Theory and observations* (Eds. J.H. Thomas and N.O. Weiss), Kluwer, Dordrecht, pp. 385–410 (1992).

Moreno-Insertis, F., "The magnetic field in the convection zone as a link between the active regions on the surface and the field in the solar interior," in: *Solar Magnetic Fields* (Eds. M. Schüssler and W. Schmidt, Cambridge University Press, pp. 117–135 (1994).

Moreno-Insertis, F., Caligari, P. and Schüssler. M., "'Explosion' and intensification of magnetic flux tubes," *Astrophys. J.* **452**, 894–900 (1995).

Moreno-Insertis, F., Schüssler, M. and Caligari, P., "Active region asymmetry as a result of the rise of magnetic flux tubes," *Solar Phys.* **153**, 449–452 (1994).

Moreno-Insertis, F., Schüssler, M. and Ferriz-Mas, A., "Storage of magnetic flux tubes in a convective overshoot region," *Astron. Astrophys.* **264**, 686–700 (1992).

Newcomb, W.A., "Convective instability induced by gravity in a plasma with a frozen-in magnetic field," *Phys. Fluids* **4**, 391–396 (1961).

Parker, E.N., "Hydromagnetic dynamo models," *Astrophys. J.* **122**, 293–314 (1955).

Parker, E.N., "The generation of magnetic fields in astrophysical bodies. X. Magnetic buoyancy and the solar dynamo," *Astrophys. J.* **198**, 205–209 (1975).

Parker, E.N., *Cosmical Magnetic Fields*, Clarendon Press, Oxford (1979).

Parker, E.N., "A solar dynamo surface-wave at the interface between convection and nonuniform rotation," *Astrophys. J.* **408**, 707–719 (1993).

Prautzsch, T., "The dynamo mechanism in the deep convection zone of the Sun," in: *Solar and Planetary Dynamos* (Eds. M.R.E. Proctor, P.C. Matthews and A.M. Rucklidge), Cambridge University Press, pp. 249–256 (1993).

Prautzsch, T., *Zum Entstehungsort solarer Magnetfelder*, PhD thesis, Universität Göttingen (1997).

Proctor, M. R. E., "Magnetoconvection," in: *Sunspots: Theory and Observation* (Eds. J. H. Thomas and N. O. Weiss), Kluwer, Dordrecht, pp. 221–241 (1992).

Proctor, M.R.E. and Weiss, N.O., "Magnetoconvection," *Rep. Prog. Phys.* **45**, 1317–1379 (1982).

Rädler, K.-H., "The solar dynamo," in: *Inside the Sun* (Eds. G. Berthomien and M. Cribier), Kluwer, Dordrecht, IAU-Coll. 121, pp. 385–402 (1990).

Rüdiger, G. and Arlt, R., "The physics of the solar cycle," this volume, Ch. 6, pp. 147–194 (2003).

Schlichenmaier, R. and Stix, M., "The phase of the radial mean field in the solar dynamo," *Astron. Astrophys.* **302**, 264–270 (1995).

Schmitt, D., "Dynamo action of magnetostrophic waves," in: *The Hydrodynamics of the Sun* (Eds. T.D. Guyenne and J.J. Hunt), ESA SP-220, pp. 223–224 (1984).

Schmitt, D., "*Dynamowirkung magnetostrophischer Wellen*," PhD thesis, Universität Göttingen (1985).

Schmitt, D., "An alpha-omega-dynamo with an alpha-effect due to magnetostrophic waves," *Astron. Astrophys.* **174**, 281–287 (1987).

Schmitt, D., "The solar dynamo," in: *The Cosmic Dynamo*, (Eds. F. Krause, K.-H. Rädler and G. Rüdiger), Kluwer, IAU-Symp. No. 157, pp. 1–12 (1993).

Schmitt, D., Schüssler, M. and Ferriz-Mas, A., "Intermittent solar activity by an on-off dynamo," *Astron. Astrophys.* **311**, L1–L4 (1996).

Schou, J., Christensen-Dalsgaard, J. and Thompson, M. J., "The resolving power of current helioseismic inversion for the sun's internal rotation," *Astrophys. J.* **385**, L59–L62 (1992).

Schüssler, M., "On buoyant magnetic flux tubes in the solar convection zone," *Astron. Astrophys.* **56**, 439–442 (1977).

Schüssler, M., "Magnetic buoyancy revisited – Analytical and numerical results for rising flux tubes," *Astron. Astrophys.* **71**, 79–91 (1979).

Schüssler, M., "Flux tube dynamo approach to the solar cycle," *Nature* **288**, 150–152 (1980).

Schüssler, M., "On the structure of magnetic fields in the solar convection zone," in: *The Hydrodynamics of the Sun* (Eds. T.D. Guyenne and J.J. Hunt), ESA SP-220, pp. 67–76 (1984).

Schüssler, M., "Magnetic fields and the rotation of the solar convection zone," in: *The internal Solar Angular Velocity* (Ed. B.R. Durney), Reidel, Dordrecht, pp. 303–320 (1987).

Schüssler, M., "Flux tubes and dynamos," in: *The Cosmic Dynamo* (Eds. F. Krause, K.-H. Rädler and G. Rüdiger), Kluwer, Dordrecht, IAU-Symp. No. 157, pp. 27–39 (1993).

Schüssler, M., and Ferriz-Mas, A. "Magnetic flux tubes and the dynamo problem," this volume, Ch. 5, pp. 123–146 (2003).

Schüssler, M., Caligari, P., Ferriz-Mas, A. and Moreno-Insertis, F., "Instability and eruption of magnetic flux tubes in the solar convection zone," *Astron. Astrophys.* **281**, L69–L72 (1994).

Smith, B. T., Boyle, J. M., Dongarra, J. J., Garbow, B. S., Ikebe, Y., Klema, V. C. and Moler C. B., *Matrix Eigenstystem Routines, EISPACK Guide*, Springer (1976).

Solanki, S.K., "Smallscale solar magnetic fields – An overview," *Space Sci. Rev.* **63**, 1–188 (1993).

Soward, A.M., "Thermal and magnetically driven convection in a rapidly rotating fluid layer," *J. Fluid Mech.* **90**, 669–684 (1979).

Spiegel, E.A. and Weiss, N.O., "Magnetic activity and variations in solar luminosity," *Nature* **287**, 616–617 (1980).

Spruit, H.C. and van Ballegooijen, A.A., "Stability of toroidal flux tubes in stars," *Astron. Astrophys.* **106**, 58–66 (1982).

Steenbeck, M. and Krause, F., "Zur Dynamotheorie stellarer und planetarer Magnetfelder. I. Berechnung sonnenähnlicher Wechselfeldgeneratoren," *Astron. Nachr.* **291**, 49–84 (1969).

Steenbeck, M., Krause, F. and Rädler, K.-H., "Berechnung der mittleren Lorentz-Feldstärke $v \times B$ für ein elektrisch leitendes Medium in turbulenter, durch Coriolis-Kräfte beeinflußte Bewegung," *Z. Naturforsch.* **21a**, 369–376 (1966).

Stenflo, J.O., "Small-scale magnetic structures on the Sun," *Astron. Astrophys. Rev.* **1**, 3–48 (1989).

Taylor, J.B., "The magnetohdydrodynamics of a rotating fluid and the Earth's dynamo problem," *Proc. Roy. Soc. Lond.* A**274**, 274–283 (1963).

Taylor, R.J., "The adiabatic stability of stars containing magnetic fields - I," *Mon. Not. R. Astron. Soc.* **161**, 365–380 (1973).

Terquem, C. and Papaloizou, J.C.B. "On the stability of an accretion disc containing a toroidal magnetic field," *Mon. Not. R. Astron. Soc.* **279**, 767–784 (1996).

Thelen, J.-C., "An α-effect due to magnetic buoyancy," *Acta Astron. et Geophys. Univ. Comenianae* **XIX**, 221–234 (1997).

Tomczyk, S., Schou, J. and Thompson, M.J., "Measurement of the rotation rate in the deep solar interior," *Astrophys. J.* **448**, L57–L60 (1995).

van Ballegooijen, A.A., "The overshoot layer at the base of the solar convective zone and the problem of magnetic flux storage," *Astron. Astrophys.* **113**, 99–112 (1982).

van Ballegooijen, A.A., "On the stability of toroidal flux tubes in differentially rotating stars," *Astron. Astrophys.* **118**, 275–284 (1983).

Velikhov, E.P., "Stability of an ideally conducting liquid flowing between cylinders rotating in a magnetic field," *Sov. Phys. JETP* **36**, 1398–1404, (Vol. 9, p. 995 in English translation) (1959).

Wälder, M., Deinzer, W. and Stix, M., "Dynamo action associated with random waves in a rotating stratified fluid," *J. Fluid Mech.* **96**, 207–222 (1980).

Wilkinson, J.H., *The Algebraic Eigenvalue Problem*, Oxford (1965).

Wilkinson, J.H. and Reinsch, C., *Linear Algebra, Handbook for Automatic Computation II*, Berlin (1971).

Yoshimura, H., "Solar-cycle dynamo wave propagation," *Astrophys. J.* **201**, 740–748 (1975).

5   Magnetic flux tubes and the dynamo problem

Manfred Schüssler[1] and Antonio Ferriz-Mas[2]

[1] *Max-Planck-Institut für Aeronomie, Max-Planck-Str. 2, D-37191 Katlenburg-Lindau, Germany, E-mail: msch@linmpi.mpg.de*
[2] *Department of Physical Sciences, Astronomy Division, P.O. Box 3000, FIN-90014 University of Oulu, Finland; and Universidad de Vigo, Facultad de Ciencias de Orense, E-32004 Orense, Spain, E-mail: antonio.ferriz@oulu.fi*

The observed properties of the magnetic field in the solar photosphere and theoretical studies of magneto-convection in electrically well-conducting fluids suggest that the magnetic field in stellar convection zones is quite inhomogeneous: magnetic flux is concentrated into magnetic flux tubes embedded in significantly less magnetized plasma. Such a state of the magnetic field potentially has strong implications for stellar dynamo theory since the dynamics of an ensemble of flux tubes is rather different from that of a more uniform field and new phenomena like magnetic buoyancy appear.

If the diameter of a magnetic flux tube is much smaller than any other relevant length scale, the MHD equations governing its evolution can be considerably simplified in terms of the thin-flux-tube approximation. Studies of thin flux tubes in comparison with observed properties of sunspot groups have led to far-reaching conclusions about the nature of the dynamo-generated magnetic field in the solar interior. The storage of magnetic flux for periods comparable to the amplification time of the dynamo requires the compensation of magnetic buoyancy by a stably stratified medium, a situation realized in a layer of overshooting convection at the bottom of the convection zone. Flux tubes stored in mechanical force equilibrium in this layer become unstable with respect to an undular instability once a critical field strength is exceeded, flux loops rise through the convection zone and erupt as bipolar magnetic regions at the surface. For parameter values relevant for the solar case, the critical field strength is of the order of 10^5 G. A field of similar strength is also required to prevent the rising unstable flux loops from being strongly deflected poleward by the action of the Coriolis force and also from 'exploding' in the middle of the convection zone. The latter process is caused by the superadiabatic stratification.

The magnetic energy density of a field of 10^5 G is two orders of magnitude larger than the kinetic energy density of the convective motions in the lower solar convection zone. This raises serious doubts whether the conventional turbulent dynamo process based upon cyclonic convection can work on the basis of such a strong field. Moreover, it is unclear whether solar differential rotation is capable of generating a toroidal magnetic field of 10^5 G; it is conceivable that thermal processes like an entropy-driven outflow from exploded flux tubes leads to the large field strength required.

The instability of magnetic flux tubes stored in the overshoot region suggests an alternative dynamo mechanism based upon growing helical waves propagating along the tubes. Since this process operates only for field strengths exceeding a critical value, such a dynamo can fall into a 'grand minimum' once the field strength is globally driven below this value, for instance by magnetic flux pumped at random from the convection zone into the dynamo region in the overshoot layer. The same process may act as a (re-)starter of the dynamo operation. Other non-conventional dynamo mechanisms based upon the dynamics of magnetic flux tubes are also conceivable.

5.1. Introduction

The major part of the magnetic flux at the surface of the Sun is not distributed in a diffuse manner, but appears in discrete structures, namely magnetic flux tubes, of which sunspots are the most prominent manifestation (e.g. Zwaan, 1992). More than 90% of the magnetic flux on the solar surface outside of sunspots is concentrated into intense flux tubes with a field strength of 1–2 kG and diameters between $\simeq 500$ and less than 100 km. These small magnetic elements are surrounded by plasma permeated by much weaker field. The filamentary state of the observed magnetic field is a consequence of the very large magnetic Reynolds number of the photospheric flows and the unstable (superadiabatic) stratification below the photospheric surface. The idea that the discrete appearance of the magnetic field extends further through the convection zone was already advanced at the beginning of the eighties; it is also supported by numerical simulations of magnetoconvection (Galloway and Weiss, 1981; Parker, 1984; Nordlund *et al.*, 1992; Proctor, 1992; Brandenburg *et al.*, 1995). Simulations of the kinematics of magnetic fields in flows with chaotic streamlines also indicate the widespread formation of tube-like magnetic structures (e.g. Dorch, 2000).

The concentration of magnetic flux into flux tubes has important consequences for its storage. Since a magnetic field gives rise to an effective magnetic pressure, a flux tube surrounded by almost field-free plasma easily becomes buoyant. Buoyancy can be a very efficient mechanism for removing magnetic flux and bringing it to the solar surface (Parker, 1955a, 1975). Simplified estimates as well as detailed numerical simulations show that flux tubes with a field strength of the order of equipartition with the energy density of convective motions (or larger) rise to the solar surface within one month or less (see, e.g. Moreno-Insertis, 1992, 1994, and references therein); this time scale is much shorter than the 11-year characteristic time for field generation and amplification by the dynamo process. Such rapid buoyant flux loss is suppressed if the toroidal magnetic flux is generated and stored within a subadiabatically stratified (i.e. convectively stable) layer of overshooting convection below the convection zone proper (see, e.g. Spiegel and Weiss, 1980; Galloway and Weiss, 1981; van Ballegooijen, 1982a,b, 1983; Schüssler, 1983; Schmitt *et al.*, 1984; Tobias *et al.*, 1998). The stable stratification impedes radial motion and allows flux tubes to find an equilibrium position with vanishing buoyancy (Moreno-Insertis *et al.*, 1992; Ferriz-Mas, 1996).

In the past decade, studies of the equilibrium, stability, and dynamics of magnetic flux tubes have led to the conclusion that the field strength of the dynamo-generated magnetic field stored at the bottom of the solar convection zone is of the order of 10^5 G, an order of magnitude larger than the (equipartition) field strength following from setting the magnetic energy density equal to the kinetic energy density of convection or differential rotation (e.g. Moreno-Insertis, 1992; Schüssler, 1996; Fisher *et al.*, 2000). Apart from the obvious question concerning the origin of such a strong field, this results also raises doubts with regard to the traditional view of the 'turbulent' dynamo process. Such strong fields cannot be considered in the framework of a kinematical theory and the turbulence is probably strongly affected in the presence of a super-equipartition field.

In this contribution we give an overview of our present understanding of the magnetic field dynamics in the solar convection zone and its implications for the dynamo theory of the large-scale solar magnetic field. We begin with a short description of the commonly used thin flux tube approximation (Section 5.2) and then discuss the flux tube dynamics in the convection zone in Section 5.3. This concerns the storage of magnetic flux (Section 5.3.1), the instability and rise of magnetic flux tubes (Section 5.3.2), as well as the origin of the strong super-equipartition toroidal field (Section 5.3.3). Section 5.4 is devoted to discussing a dynamo model based on magnetic flux tubes. We consider the α-effect resulting from the undular instability of a toroidal flux tube in a rotating star (Section 5.4.1) and describe the results of a simple nonlinear dynamo model based upon these ideas. A possible consequence of such a model is the appearance of extended periods with no strong magnetic field; such periods could be related to the 'grand minima' of solar activity such as the Maunder minimum in the seventeenth century. In Section 5.5 we summarize the main results.

5.2. Thin flux tubes

The study of the structure and dynamics of magnetic fields in stars is a complex mathematical problem owing to the nonlinear nature of the magnetohydrodynamic (MHD) equations, to the large hydrodynamic and magnetic Reynolds numbers leading to strong structuring of the magnetic field, and to the stratification of the plasma. On the other hand, a filamentary nature of magnetic fields in the convection zone allows for a considerable simplification of the MHD equations if the diameter of a flux tube is small compared to all other relevant length scales (scale heights, radius of curvature, wavelengths of MHD waves, etc.). The flux tube is idealized as a bundle of magnetic field lines, which is separated from its non-magnetic environment by a tangential discontinuity. It is assumed that instantaneous lateral pressure balance is maintained between the flux tube and the exterior. Diffusion of the magnetic field due to Ohmic resistance is neglected. The formal thin flux-tube approximation has been developed with different degrees of generality by Defouw (1976); Roberts and Webb (1978); Spruit (1981) and Ferriz-Mas and Schüssler (1989, 1993). This approximation permits the reduction of the full set of MHD equations to a mathematically more easily tractable form while retaining the full effects of the compressibility of the plasma, the magnetic Lorentz force and gravity. In physical terms, the approximation amounts to describing the dynamics of a magnetic flux tube as the motion of a quasi-one-dimensional continuum in a three-dimensional environment. A magnetic tube with a flux of 3×10^{21} Mx (corresponding to a medium-size sunspot) and a field strength of 10^5 G would have a diameter of approximately 2000 km. Since for the stability and the first stages of the rise one is interested in perturbations that affect the flux tube as a whole, and given the large value of the pressure scale-height at the bottom of the convection zone (about 60,000 km), the use of the thin flux-tube approximation for the dynamics of the magnetic flux concentrations is perfectly justified.

With the aid of this approximation it has been possible to obtain a consistent model of magnetic field storage, instability, dynamics, and eruption not only for the Sun, but also for other stars with outer convection zones (Schüssler et al., 1994, 1996).

In the following, the subscript 'i' refers to quantities inside the tube (except for the magnetic field, which we simply denote **B**), while external quantities are labelled with 'e'.

The dynamics of the matter in the flux tube is governed by the momentum equation for compressible MHD that we write in a reference frame rotating with (constant) angular velocity $\mathbf{\Omega}$:

$$\rho_i \frac{D\mathbf{v}_i}{Dt} = -\nabla \left(p_i + \frac{B^2}{8\pi} \right) + \frac{(\mathbf{B} \cdot \nabla)\mathbf{B}}{4\pi} + \rho_i[\mathbf{g} - \mathbf{\Omega} \times (\mathbf{\Omega} \times \mathbf{r})] + 2\rho_i \mathbf{v}_i \times \mathbf{\Omega}. \quad (5.1)$$

Here p is gas pressure, ρ is mass density, $\mathbf{v}$ is the velocity field, $\mathbf{g}$ is the acceleration of gravity and $\mathbf{B}$ the magnetic field.

We assume that the flux tube moves within a non-magnetic medium; this external medium is in stationary – albeit not necessarily static – equilibrium and its equation of motion is

$$\rho_e(\mathbf{v}_e \cdot \nabla)\mathbf{v}_e = -\nabla p_e + \rho_e[\mathbf{g} - \boldsymbol{\Omega} \times (\boldsymbol{\Omega} \times \mathbf{r})] + 2\rho_e\mathbf{v}_e \times \boldsymbol{\Omega}, \tag{5.2}$$

where $\mathbf{v}_e$ is the external velocity relative to the rotating frame of reference.

The continuity of normal stress across the boundary between the flux tube and the surrounding fluid (idealized by a surface of discontinuity) yields the condition of instantaneous lateral balance of total pressure, which relates the internal and external gas pressures:

$$p_i + \frac{B^2}{8\pi} = p_e. \tag{5.3}$$

The path of the flux tube is described by the space curve $\mathbf{r}(s, t)$, which is parametrized by the instantaneous arc-length, s. At each point on the curve we set up an orthonormal triad (the Frenet basis) made up of the tangent, $\mathbf{e}_t$, the normal, $\mathbf{e}_n$, and the binormal, $\mathbf{e}_b$, unit vectors. The equation of motion (5.1) is projected onto this basis. By making use of (5.2) and (5.3), the projection yields:

$$\rho_i\left(\frac{D\mathbf{v}_i}{Dt}\right) \cdot \mathbf{e}_t = -\frac{\partial p_i}{\partial s} + \rho_i\,\mathbf{g} \cdot \mathbf{e}_t - \rho_i\,[\boldsymbol{\Omega} \times (\boldsymbol{\Omega} \times \mathbf{r})] \cdot \mathbf{e}_t + 2\rho_i\,(\mathbf{v}_i \times \boldsymbol{\Omega}) \cdot \mathbf{e}_t. \tag{5.4a}$$

$$\rho_i\left(\frac{D\mathbf{v}_i}{Dt}\right) \cdot \mathbf{e}_n = \frac{B^2}{4\pi}\kappa + (\rho_i - \rho_e)\,\mathbf{g} \cdot \mathbf{e}_n - \rho_i[\boldsymbol{\Omega} \times (\boldsymbol{\Omega} \times \mathbf{r})] \cdot \mathbf{e}_n + 2\rho_i(\mathbf{v}_i \times \boldsymbol{\Omega}) \cdot \mathbf{e}_n$$
$$- \rho_e\,[2\mathbf{v}_e \times \boldsymbol{\Omega} - \boldsymbol{\Omega} \times (\boldsymbol{\Omega} \times \mathbf{r}) - (\mathbf{v}_e \cdot \nabla)\mathbf{v}_e] \cdot \mathbf{e}_n. \tag{5.4b}$$

$$\rho_i\left(\frac{D\mathbf{v}_i}{Dt}\right) \cdot \mathbf{e}_b = (\rho_i - \rho_e)\,\mathbf{g} \cdot \mathbf{e}_b - \rho_i[\boldsymbol{\Omega} \times (\boldsymbol{\Omega} \times \mathbf{r})] \cdot \mathbf{e}_b + 2\rho_i\,(\mathbf{v}_i \times \boldsymbol{\Omega}) \cdot \mathbf{e}_b$$
$$- \rho_e\,[2\mathbf{v}_e \times \boldsymbol{\Omega} - \boldsymbol{\Omega} \times (\boldsymbol{\Omega} \times \mathbf{r}) - (\mathbf{v}_e \cdot \nabla)\mathbf{v}_e] \cdot \mathbf{e}_b. \tag{5.4c}$$

In these equations, $\kappa(s, t)$ denotes the curvature of the tube's axis and D/Dt is the material derivative. A possible choice for the Lagrangian coordinate is to take the arc-length along the tube's axis at a given initial time, s_0, in which case $D/Dt = (\partial/\partial t)_{s_0}$.

The forces which mainly determine the dynamics of a magnetic flux tube are the buoyancy force $(\rho_i - \rho_e)\mathbf{g}$, the magnetic curvature force (for a non-straight tube), $(B^2/4\pi)\kappa\,\mathbf{e}_n$, and the Coriolis force, $2\rho_i(\mathbf{v}_i \times \boldsymbol{\Omega})$.

In ideal MHD, the equations of continuity and induction can be combined into

$$\frac{D}{Dt}\left(\frac{\mathbf{B}}{\rho}\right) = \left(\frac{\mathbf{B}}{\rho} \cdot \nabla\right)\mathbf{v}, \tag{5.5}$$

called Walén's equation. The thin-flux-tube version of Walén's equation is

$$\frac{D}{Dt}\left(\frac{\rho_i}{B}\right) + \frac{\rho_i}{B}\left[\frac{\partial(\mathbf{v}_i \cdot \mathbf{e}_t)}{\partial s} - \kappa\,(\mathbf{v}_i \cdot \mathbf{e}_n)\right] = 0. \tag{5.6}$$

If isentropic evolution is assumed, the energy equation reads:

$$\frac{Dp_i}{Dt} = \frac{\gamma p_i}{\rho_i} \frac{D\rho_i}{Dt}. \tag{5.7}$$

Finally, we take as constitutive relation the ideal gas model, $p = (\mathcal{R}/\mu)\rho T$, where $\mathcal{R}$ is the universal gas constant, μ the mean molar mass and T the temperature.

The set of equations is completed by the geometrical equations which express the relations between the unit vectors of the Frenet basis and determine their time evolution as a consequence of the Lagrangian velocity.

5.3. Flux tube dynamics in the solar convection zone

The consequences of flux tube dynamics for the stellar dynamo problem are best studied in the case of the Sun. Flux tubes can be directly observed in the solar photosphere and the properties of sunspot groups, especially in their early phases, provide indirect evidence on the structure and dynamics of the magnetic field in the solar interior. A whole variety of observations shows that a sunspot group forms through the rapid emergence of a coherent magnetic structure, which is not passively carried by convective flows, from a source region of well-ordered magnetic flux in the solar interior. The observations are consistent with the 'rising tree' picture (Zwaan, 1978, 1992; Zwaan and Harvey, 1994) of a partially fragmented magnetic structure that ascends towards the surface and emerges in a dynamically active way. Only later, after the initial stage of flux emergence, the surface fields progressively come under the influence of convective flow patterns. Consequently, magnetic flux in the convection zone has a dynamics of its own that must be taken into account when considering the dynamo problem; a convenient starting point for such a study is to consider the dynamics of isolated magnetic flux tubes.

5.3.1. Storage of magnetic flux

Already in the 1950s, Parker (1955a) and Jensen (1955) suggested that magnetic flux is brought to the solar surface through the action of magnetic buoyancy. In fact, it turns out that the buoyancy force is in some respect too efficient in doing so, since simplified estimates (Parker, 1975) as well as detailed numerical simulations (Moreno-Insertis, 1983, 1986) and stability analyses (Spruit and van Ballegooijen, 1982; Ferriz-Mas and Schüssler, 1993, 1995) show that flux tubes with a field strength of the order of the equipartition with the energy density of the convective motion (or larger) rise to the top of the convection zone within a month or less. This time scale is much shorter than the 11-year characteristic time for field generation and amplification by the dynamo. As a consequence, magnetic flux is lost from the convection zone much faster than it can be regenerated by the dynamo process.

The strong convective motions and the unstable stratification of the convection zone (that is to say, its superadiabatic temperature gradient, i.e. specific entropy decreasing outward) prohibit a solution of this 'magnetic flux storage problem' in terms of a magnetic configuration in mechanical equilibrium within the convection zone proper. On the other hand, numerical simulations indicate that magnetic flux with a field strength not strongly exceeding the convective equipartition value is pumped below the convection zone by the sinking plumes of overshooting convection (Tobias *et al.*, 1998). Owing to the subadiabatic (stable) stratification of this layer of convective overshoot, the flux can achieve a stable equilibrium configuration there. Possible equilibria include a homogeneous layer of magnetic flux and an ensemble of magnetic flux tubes.

In the case of a layer of toroidal magnetic field in mechanical equilibrium, the magnetic Lorentz force is balanced by a combination of gas pressure gradient and Coriolis force due to a field-aligned flow (Rempel *et al.*, 2000). The relative importance of both forces for the balance of the magnetic curvature force depends on the degree of subadiabaticity of the stratification as measured by $\delta \stackrel{\text{def}}{=} \nabla - \nabla_{\text{ad}}$, where $\nabla_{\text{ad}} \stackrel{\text{def}}{=} (\mathrm{d} \ln T/\mathrm{d} \ln p)_{\text{ad}}$ is the adiabatic logarithmic temperature gradient (i.e. the logarithmic temperature gradient in a homoentropic stratification). In a strongly subadiabatic region (like the radiative core of the Sun with $\delta < -0.1$), the latitudinal pressure gradient is dominant for the magnetic equilibrium, while the contribution of the Coriolis force is rather very small. Inclusion of the energy equation leads to a slow (Eddington-Sweet-like) meridional circulation. For $\delta \simeq -10^{-3}$, the contributions of both forces are similar, while for a value of $\delta \simeq -10^{-6}$, as is probably realistic for a layer of convective overshoot, the magnetic curvature force is balanced practically by the Coriolis force alone. For a field strength of the order of 10^5 G (10 Tesla), a toroidal velocity of the order of $100 \, \mathrm{m \, s^{-1}}$ relative to the background rotation is required, so that the profile of differential rotation would be significantly modified.

Another important point is the stability of a magnetic layer. In the simplest case, the decrease of the field strength, B, at the top of the magnetic layer drives an instability of Rayleigh–Taylor type, whose nonlinear evolution leads to the formation of flux tubes (Matthews *et al.*, 1995). Assuming the temperature profile to be undisturbed by the presence of the magnetic layer, ignoring effects of diffusion and rotation, and assuming local Cartesian geometry (i.e. neglecting the field line curvature of the equilibrium field), the linear criterion for instability can be written as (cf. Hughes and Proctor, 1988)

$$\frac{1}{\gamma H_{\mathrm{p}}} \frac{\mathrm{d}}{\mathrm{d}z} (\ln B) > \frac{k_y^2 k^2}{k_x^2} + \frac{N^2}{v_{\mathrm{A}}^2}, \tag{5.8}$$

where H_{p} is the pressure scale height, k_y the perturbation wavenumber parallel to the undisturbed field, k_x the perturbation wavenumber perpendicular to both undisturbed field and gravity, k the total wavenumber perpendicular to the field, v_{A} the Alfvén speed, and $N^2 = -g\delta/H_{\mathrm{p}}$ the square of the Brunt–Väisälä, or buoyancy, frequency. Assuming very large wavelength along the field, and a length scale d for the decrease of B at the top of the layer, we obtain from (5.8) a necessary condition for instability, namely

$$B > 1.5 \left[\frac{d}{10^4 \, \mathrm{km}} \right]^{1/2} \left[\frac{|\delta|}{10^{-6}} \right]^{1/2} \times 10^4 \, \mathrm{G}. \tag{5.9}$$

The values $d = 10^4 \, \mathrm{km}$ and $\delta = -10^{-6}$ are 'typical' values for the conditions assumed in the overshoot layer below the solar convection zone. Consequently, we expect a magnetic layer contained in the overshoot region to become unstable to flux tube formation when the field strength exceeds a value of the order of the equipartition field strength with respect to the kinetic energy density of the convective velocities. Inclusion of (thermal, magnetic, and viscous) diffusion opens additional possibilities for double- or triple-diffusive instabilities (see, e.g. Hughes and Proctor, 1988), so that it seems difficult to maintain a smooth magnetic layer.

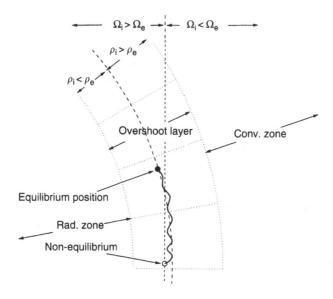

Figure 5.1 Sketch of the oscillatory adjustment of an initially buoyant flux tube (open circle) with $T_i = T_e$ and $\Omega_i = \Omega_e$ to a position of mechanical (force) equilibrium (full circle) characterized by neutral buoyancy ($\rho_i = \rho_e$) and faster internal rotation, $\Omega_i > \Omega_e$ (from Caligari *et al.*, 1998).

As long as they remain within the stably stratified overshoot region, the flux tubes formed by the instability of a magnetic layer can find a new equilibrium governed by the balance of curvature force and Coriolis force (Moreno-Insertis *et al.*, 1992), which is similar to the equilibrium of a magnetic layer in a slightly subadiabatic region (Rempel *et al.*, 2001). To obtain mechanical equilibrium in the idealized case of a toroidal flux tube, i.e. a flux ring contained in a plane parallel to the equator, the buoyancy force must vanish since its component parallel to the axis of rotation cannot be balanced by any other force. In the direction perpendicular to the axis of rotation, the magnetic curvature force is balanced by the Coriolis force due to a faster rotation of the plasma within the flux ring compared to its nonmagnetic environment.

The approach of an initially buoyant flux ring in temperature equilibrium with its environment ($T_i = T_e$) and rotating with the same angular velocity as the surrounding plasma ($\Omega_i = \Omega_e$) toward its equilibrium position is schematically illustrated in Fig. 5.1, which shows a cut through a meridional plane. As the tube rises, it loses its buoyancy because of the subadiabatic stratification in the overshoot region. The magnetic curvature force leads to a poleward displacement of the flux ring, reducing its distance from the axis of rotation. Owing to conservation of angular momentum, the rotational velocity of the plasma inside the flux ring increases during this process until the resulting Coriolis force balances the curvature force. The final result is a neutrally buoyant flux ring ($\rho_i = \rho_e$) with a somewhat higher internal rotation rate. The tube initially performs superimposed buoyancy and inertial oscillations around this equilibrium position (Moreno-Insertis *et al.*, 1992). Because of the drag force exerted on the moving flux tube, the amplitude of the oscillations decreases until the equilibrium position is reached.

The mechanical equilibrium of flux tubes implies no thermal equilibrium: neutral buoyancy requires the flux tube interior to be somewhat cooler than its surroundings. As a consequence, radiative heating of the tube perturbs the force equilibrium and leads to a slow outward drift of the flux ring. The time span for which a tube can be stored is therefore limited by the duration of this slow rise through the overshoot layer. Once the tube enters the convection zone proper, it rapidly rises to the surface due to convective buoyancy (Moreno-Insertis, 1983). Consideration of the variation of the temperature gradient in the lower convection zone (Fan and Fisher, 1996; Moreno-Insertis *et al.*, 2002) indicates that the radiative heating of a flux tube is indeed more efficient than previously thought. This places tight (and uncomfortable) limits on the possible storage times unless the overshoot layer is rather more subadiabatic than usually estimated, i.e. a value of $\delta \simeq 10^{-4}$ is required (Rempel, 2003).

5.3.2. Instability and rise of magnetic flux tubes

For a linear stability analysis one considers a toroidal flux tube symmetric with respect to the solar rotation axis and lying at an arbitrary latitude in a plane parallel to the equator. The flux tube evolves through a sequence of equilibria while being continuously amplified by radial differential rotation.

Stability against isentropic perturbations can be examined by means of a normal-mode analysis of both axisymmetric and non-axisymmetric displacements of the equilibrium path of the tube. Assume the equilibrium path to be given by the function $\mathbf{r}(s_0, t)$, where we use the unperturbed arc-length, s_0, as a Lagrangian (or material) coordinate. The equilibrium configuration is a toroidal flux tube (a flux ring) lying in a plane perpendicular to the equator and at a distance R_0 from the star's rotation axis. Now consider three-dimensional, isentropic perturbations about the equilibrium path: $\mathbf{r}(s_0, t) = \mathbf{r}_0(s_0) + \boldsymbol{\xi}(s_0, t)$. The equations governing the dynamics of the flux tube are linearized about the equilibrium configuration and all perturbed quantities are expressed in terms of the Lagrangian displacement vector, $\boldsymbol{\xi}$. The resulting linear, homogeneous system of equations with constant coefficients permits wave solutions of the form

$$\boldsymbol{\xi} = \hat{\boldsymbol{\xi}} \exp \mathrm{i}(m\phi_0 + \omega t), \qquad (5.10)$$

where ω is the (complex) frequency, m the (integer) azimuthal wavenumber, and $\phi_0 = s_0/R_0$. The dispersion relation is a sixth-order polynomial in the eigenfrequency ω with real coefficients (for details, see, Ferriz-Mas and Schüssler, 1995). The stability properties depend on the various parameters (e.g. latitude, field strength, superadiabaticity of the stratification, angular velocity and its gradients) which enter into the coefficients of the dispersion relation.

In general, the resulting six modes represent mixtures of longitudinal and transversal tube modes. For axisymmetric modes ($m = 0$), the dispersion relation reduces to the form $\omega^4 + a_2\omega^2 + a_0 = 0$, with real coefficients a_2 and a_0. A mode is unstable if $\mathcal{I}m(\omega) < 0$. Depending on the values of a_0 and a_2 the roots are either real (stable modes), purely imaginary (monotonically unstable modes), or pairs of complex conjugates (oscillatory unstable modes). For non-axisymmetric modes ($m \geq 1$), the equations for all three components of the displacement $\boldsymbol{\xi}$ are coupled and the full sixth-order equation has to be solved numerically.

We have applied the stability analysis to a model of the solar convection zone provided by Stix (cf. Skaley and Stix, 1991). The model makes use of a non-local mixing-length treatment of the convection zone following the formalism of Shaviv and Saltpeter (1973) and yields a consistently calculated overshoot layer of about 10,000 km depth. The super-adiabaticity becomes negative at $r = 5.38 \times 10^5$ km. The bottom of the convection zone proper is at $r = 5.13 \times 10^5$ km (where $\delta = -4.2 \cdot 10^{-7}$). The radiative core begins at $r = 5.02 \times 10^5$ km ($\delta = -1.4 \times 10^{-4}$). The superadiabaticity δ becomes negative already in the lower part of the convection zone proper. In the present model, the total extent of the subadiabatic layer is $\simeq$36,000 km, while the overshoot layer extends over $\simeq$10,000 km. Fig. 5.2 shows a stability diagram on the (B_0, λ_0)-plane, where B_0 is the magnetic field strength and λ_0 is the latitude of the equilibrium flux tubes; their location in depth is in the lower part of the overshoot region, i.e. 2000 km above its lower boundary where the superadiabaticity has the value $\delta = -2.6 \times 10^{-6}$. The value of 2000 km represents the approximate radius of a flux tube with a magnetic flux of 10^{22} Mx (corresponding to a large active region) and a field strength of 10^5 G. The white region corresponds to stable flux tubes, the shaded areas are domains of non-axisymmetric instability: dark shading indicates that the mode with azimuthal wave number $m = 1$ has the largest growth rate (shortest growth time), while light shading indicates dominance of the mode $m = 2$.

The magnetic field of a toroidal flux tube is probably amplified by differential rotation (Ω-effect) and by other mechanisms, such as the explosion of 'weak' flux tubes in the convection zone (see Section 5.3.3). In any case, once the field has become sufficiently strong, instabilities set in; the most unstable perturbations (i.e. with the shortest growth time) are non-axisymmetric (i.e. undular) and give rise to the formation of rising loops. The critical magnetic field strength for the onset of the instability lies around 10^5 G for conditions prevailing at the bottom of the solar convection zone (e.g. Schüssler *et al.*, 1994).

The weak instability in region II of Fig. 5.2, with large growth times, results from the combined effect of Coriolis force and buoyancy instability. In Section 5.4.1. it will be shown that this instability gives rise to helical waves of growing amplitude, which propagate along the tube and produce an inductive effect (α-effect) regenerating poloidal field from toroidal field (Ferriz-Mas *et al.*, 1994). Once the field becomes larger than $\simeq 1.2 \times 10^5$ G, a second regime of instability sets in (corresponding to region III on the diagram), with much smaller growth times. This instability drives the flux tubes into the convection zone proper on a short time scale and is ultimately responsible for the rise of unstable loops towards the solar surface within about 1 month (see, e.g. Caligari *et al.*, 1995).

A linear stability analysis provides the proper initial conditions for numerical simulations of the emergence of magnetic flux loops through the convection zone. Once an unstable loop has entered the superadiabatic part of the convection zone, the subsequent evolution becomes nonlinear and very fast, so that numerical simulations are necessary to follow its rise towards the surface. The nonlinear dynamic evolution of unstable flux tubes has been studied in detail by numerical integration of the thin flux-tube equations (see, e.g. D'Silva and Choudhuri, 1993; Schüssler *et al.*, 1994; Fan *et al.*, 1994; Caligari *et al.*, 1995, 1998; Fisher *et al.*, 2000). For initial field strengths of the order of equipartition (10^4 G) or a few times this value, the Coriolis force plays a dominant role and flux tubes starting from low latitudes are deflected polewards parallel to the solar rotation axis and emerge at unrealistically high latitudes (Choudhuri and Gilman, 1987), in contradiction with sunspot observations. With a field strength of approximately 10^5 G the flux tubes emerge at low latitudes, the deviation of the eruption latitude from the starting latitude being less than 5°. Other features that are reproduced by the numerical simulations starting from an initial field strength of the order of 10^5 G are the tilt angle (inclination with respect to the East–West direction) of sunspot groups and the asymmetry between the two legs of the rising tubes (Fig. 5.3).

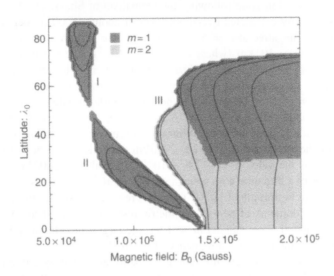

Figure 5.2 Stability diagram for non-axisymmetric perturbations of a toroidal flux tube. All parameters entering the dispersion relation are chosen as to represent the overshoot layer below the solar convection zone (approximately 2000 km above the top of the radiative region). The shaded regions denote instability; the degree of shading indicates the azimuthal wavenumber of the mode with the largest growth rate (from Ferriz-Mas and Schüssler, 1995).

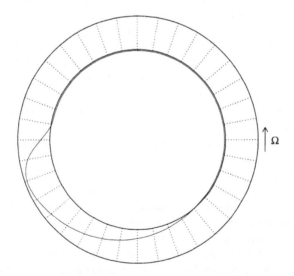

Figure 5.3 Snapshot of an unstable magnetic flux tube that started at the bottom of the convection zone. The direction of solar rotation is counterclockwise (from Caligari *et al.* 1995).

A number of two- and three-dimensional simulations of buoyantly rising magnetic flux tubes not relying on the thin flux-tube approximation has been performed in recent years (Longcope *et al.*, 1996; Moreno-Insertis and Emonet, 1996; Dorch and Nordlund, 1998; Emonet and Moreno-Insertis, 1998; Fan *et al.*, 1998a,b; Dorch *et al.*, 1999; Abett *et al.*, 2000).

While basically confirming the results of the thin-tube approach, these simulations indicate that a rising magnetic flux tube is liable to continued fragmentation into counter-rotating line vortices whose interaction eventually stops the rise altogether and destroys the flux tube as a coherent entity (see also Schüssler, 1979; Tsinganos, 1980). The fragmentation process is suppressed if the initial flux tube is sufficiently strongly twisted, i.e. has an azimuthal field component in addition to the axial field. At a later stage of the rise, the twist can lead to kink instability, the signature of which seems to be indicated by sunspot observations and X-ray data (Fan *et al.*, 1999; Matsumoto *et al.*, 2000). On the other hand, Fan (2001) has shown by means of three-dimensional simulations that the effect of solar rotation on a rising tube may be sufficient to maintain its coherence even in the case of an initially untwisted flux tube.

5.3.3. Origin of the strong field

By which mechanism is the magnetic field amplified at the bottom of the convection zone to the value of $\simeq 10^5$ G indicated by the studies of flux-tube dynamics? Flux expulsion by convection should lead to about equipartition field strength, but the magnetic energy density of a 10^5 G field is two orders of magnitude larger than the mean kinetic energy density of the convective motions. It is possible that convective flows could *locally* be much stronger (for instance, in concentrated downflows) and compress the field; however, such local concentrations correspond to large azimuthal wave numbers (certainly $m > 10$), for which the undular instability would require a field strength well in excess of 10^6 G.

While it is obvious that the differential rotation in a radial shear layer at the bottom of the convection zone (the 'tachocline') can generate a toroidal field in the first place, it is not so clear which field strength can be reached by this process. Since the magnetic pressure is much smaller than the gas pressure for all conceivable field strengths in the lower convection zone, the change in field strength of a stretched flux tube is simply proportional to its elongation, $\Delta B/B \simeq \Delta l/l \simeq \Delta v \cdot \tau/d$, where Δv is the difference in rotation speed over the thickness d of the shear layer, and τ is the elapsed time. Assuming $\Delta v = 100\,\mathrm{m\,s^{-1}}$ from helioseismology (e.g. Tomczyk *et al.*, 1995), $d = 10^4$ km (estimated thickness of the overshoot layer) and a necessary amplification factor $\Delta B/B = 1000$ (from 100 to 10^5 G) we find $\tau \simeq 3$ years, which appears to be a reasonable value for the amplification time.

However, the dynamical aspect of the problem has to be taken into account: as the field strength grows, the tension force leads to an increasing resistance of the tube against further stretching. Assume a magnetic loop with a radius of curvature equal to the thickness of the shear layer, $R_c \simeq 10^4$ km, being stretched by the velocity difference $\Delta v = 100\,\mathrm{m\,s^{-1}}$ over the layer. Further stretching is inhibited when the resisting tension force, $B^2/4\pi R_c$, balances the aerodynamic drag force, $C_D \rho (\Delta v)^2/a$, which provides the stretching (C_D is the drag coefficient, which is of order unity and a is the radius of the tube). Equating both forces and inserting values yields a relationship between the field strength that can be reached by stretching and the tube radius, viz.

$$B \simeq 3 \times 10^4 \, a_8^{-1/2}\,\mathrm{G},$$

where a_8 is the radius in units of 10^8 cm. The smaller the radius the stronger is the coupling of the tube to the shear flow by means of the drag force. In order to reach field strengths in excess of 10^5 G, the tube radius would have to be smaller than 100 km; i.e. only very thin tubes containing much less magnetic flux than a large active region can be sufficiently stretched. It is possible, however, that a larger tube may fray into smaller tubes near the apex of the stretched loop due to the action of the interchange and Kelvin–Helmholtz-like instabilities, so that the stretching could continue.

Another aspect of the problem arises from energy considerations. The kinetic energy contained in the differential rotation of the shear layer is between one (for $B \simeq 10^4$ G) and two (for $B \simeq 10^5$ G) orders of magnitude smaller than the magnetic energy of the toroidal flux generated within one 11-year cycle. Consequently, a very efficient mechanism is required that continuously restores the shear layer and feeds energy into the differential rotation. This estimate also implies that the back-reaction of the magnetic field on the differential rotation represents an important nonlinear effect in the dynamo process.

A quite different alternative mechanism for the intensification of magnetic fields stored in the overshoot region is related to the 'explosion' of flux tubes (Moreno-Insertis *et al.*, 1995). This phenomenon is caused by the superadiabatic stratification of the convection zone and the nearly adiabatic evolution of a flux tube owing to the very long thermal exchange time with its environment. As a simplified model, assume that the flux tube rises quasi-statically in the convection zone, so that hydrostatic equilibrium (along the field lines) is maintained. If the plasma in the flux tube was homentropic initially, the internal gas pressure, p_i, as a function of height, z, is given by

$$p_i(z) = p_{i,0} \left(1 - \frac{z \nabla_{ad}}{H_{i,0}} \right)^{1/\nabla_{ad}}.$$ (5.11)

where $H_{i,0}$ is the internal pressure scale height at $z = 0$ and ∇_{ad} is the adiabatic logarithmic temperature gradient. In the external medium we assume a polytropic stratification with (constant) temperature gradient ∇,

$$p_e(z) = p_{e,0} \left(1 - \frac{\nabla z}{H_{e,0}} \right)^{1/\nabla},$$ (5.12)

where $H_{e,0}$ is the external pressure scale height at $z = 0$. The temperature difference between the plasma in the flux tube and in its environment as a function of height is

$$\Delta T \stackrel{\text{def}}{=} T_i - T_e = T_{i,0} - T_{e,0} + \frac{\mu g}{\mathcal{R}}(\nabla - \nabla_{ad})z.$$ (5.13)

where μ is the mean molar mass, g the gravitational acceleration, and $\mathcal{R}$ the gas constant. Since the external stratification is superadiabatic ($\nabla > \nabla_{ad}$), ΔT grows and the internal scale height eventually becomes larger than the external scale height. Consequently, $p_i(z)$ decreases less rapidly with height than $p_e(z)$ and there is a critical height, z_c, at which we have $p_i(z_c) = p_e(z_c)$. Since the balance of total pressure between the flux tube and its surroundings is maintained, this means that $B^2/8\pi = p_e - p_i \to 0$ for $z \to z_c$ and conservation of magnetic flux formally demands that the radius of the tube becomes infinite at that point: the flux tube 'explodes' as the top of the rising loop approaches the height z_c.

The value of z_c depends on the strength of the initial field at $z = 0$: a stronger field leads to a higher initial gas pressure deficit in the tube and thus to a larger explosion height. While a field strength of 10^4 G results in explosion already in the lower half of the solar convection zone, flux tubes with an initial field strength of 10^5 G can reach the surface without exploding.

Numerical simulations of rising thin flux loops confirm the simple model laid out above. As the loop apex reaches the height z_c, the tube expands strongly and the field strength decreases accordingly; the simulations cannot be continued beyond that event since the thin flux-tube approximation breaks down and the flux tube looses its identity as a coherent object. The abrupt weakening of the magnetic field and the concomitant inflation of the upper parts of a flux loop could provide a mechanism for the intensification of the field strength in the deepest reaches of a flux tube: matter is 'sucked' up from below into the inflated summit region, the gas pressure in the submerged part of the tube decreases and the magnetic field strength there increases correspondingly. As the field strength at the top of the loop drops abruptly, the flux tube loses its coherence and becomes passive with respect to the convective motions: the remaining tube consists of two more or less inclined tubular 'stumps' connected by a large web of tangled weak field. The further evolution following the explosion may be driven by the buoyancy of the high-entropy material within the flux tubes and lead to a continuous outflow of plasma from the stumps. If this process occurs in the middle of the convection zone, then a noticeable field concentration in the deepest part of the flux tube could possibly ensue. In numerical experiments, Rempel and Schüssler (2001) indeed find a significant intensification of the remaining flux tube after an explosion, which closely follows the scenario sketched above.

In summary, although the evidence for strong field of the order of 10^5 G at the bottom of the convection zone is compelling, it is still unclear how such a marked intensification can be achieved. While field line stretching is limited by the back-reaction of the Lorentz force on the plasma motion, amplification via 'explosion' is, in principle, only limited by the external gas pressure. However, whether this mechanism really works or is just of academic interest, remains to be demonstrated.

5.4. A dynamo model on the basis of flux tubes

It is widely accepted that the solar activity cycle is the result of a hydromagnetic dynamo operating at the bottom of the solar convection zone. The principle of dynamo action was first suggested by Larmor (1919). Basically, it requires that the plasma moves in such a way as to induce electric currents capable of maintaining and amplifying a 'seed' field against Ohmic decay. In the case of stars like the Sun, one of the main ingredients for dynamo action is differential rotation, which generates the toroidal field component from the poloidal component. The generation of the poloidal field from the toroidal field is explained in terms of an electric current parallel to the mean magnetic field induced by the effect of cyclonic convection (α-effect), an idea originally put forward by Parker (1955b).

The concentration of magnetic flux into intense tubes has important consequences for the operation of the dynamo. The conventional kinematic approach – in which the back-reaction of the Lorentz force on the flow field is neglected and in which turbulent flows play a key role in regenerating the poloidal field from the toroidal component – is not applicable unless the flux tubes are an ensemble of very thin ($r \leq 100$ km) fibrils (Parker, 1982). A way out of this dilemma is the idea of a two-layer dynamo, advanced by Parker (1993). In this model, the two induction effects of an $\alpha\Omega$-type dynamo are spatially separated: the α-effect is thought to operate in the turbulent part of the convection zone, while the amplification by differential rotation takes place in the overshoot layer. Another possibility is an α-effect driven by buoyancy: although the α-effect was originally formulated for rotating, turbulent convective systems, the basic idea behind Parker's topological argument is that the generation of an α-current is due to the lack of mirror symmetry of the flow, and this can also be achieved by the combination of rotation and buoyancy instability. The picture of an α-effect working with strong (super-equipartition) fields without invoking convective turbulence is outlined in the next subsection.

5.4.1. *An α-effect due to unstable flux tubes*

The key ideas for a hydromagnetic dynamo driven by the instability of magnetic flux tubes may be sketched as follows (Ferriz-Mas *et al.*, 1994):

- *Storage*: The flux tubes are stored in mechanical force equilibrium in a layer of over-shooting convection at the bottom of the convection zone, as explained in Section 5.3.1 (see, e.g. Moreno-Insertis *et al.*, 1992).
- *Magnetic field amplification*: The magnetic field is amplified by differential rotation (Ω-effect) or by other mechanisms such as the explosion of 'weak' flux tubes in the convection zone (see Section 5.3.3).
- *Onset of instability*: When the magnetic field strength reaches a threshold value, buoyancy instabilities set in. The critical value for the onset of the instability depends on the stratification, latitude, and angular velocity distribution (Ferriz-Mas and Schüssler, 1995).
- *α-effect*: Call $\mathbf{v}'$ and $\mathbf{B}'$ the instability-generated perturbations of the fields $\mathbf{v}$ and $\mathbf{B}$, respectively, in the flux tube. We compute the azimuthal average of the ϕ-component of the electric field induced by the perturbations $\mathbf{v}'$ and $\mathbf{B}'$, and obtain that (Ferriz-Mas *et al.*, 1994)

$$\overline{\mathbf{v}' \times \mathbf{B}'} = \alpha\overline{\mathbf{B}}, \tag{5.14}$$

where the mean magnetic field $\overline{\mathbf{B}}$ is equal to the unperturbed field $\mathbf{B}_0$.

The linear stability analysis of toroidal flux tubes shows that the non-axisymmetric modes ($m = 1, 2, 3, \ldots$) of flux tubes outside the equatorial plane give rise to a mean electric current parallel to the mean magnetic field. The α-effect is non-vanishing only for non-axisymmetric and unstable modes; the requirement for instability comes from the necessary phase difference between the fields $\mathbf{v}'$ and $\mathbf{B}'$. The function $\alpha = \alpha(r, \theta, B_0)$ is antisymmetric with respect to the equatorial plane.

The non-axisymmetric perturbations of the flux tubes combined with the Coriolis force give rise to helical waves of growing amplitude along the flux tube, leading to an electric current parallel or anti-parallel to the mean (unperturbed) magnetic field. The α-effect of unstable flux tubes thus results from a combination of buoyancy instability and Coriolis force. We stress that this α-effect is 'dynamic' in the sense that the Lorentz force and its feedback on the velocity field are taken into account.

A crucial feature of this dynamic α-effect is that it operates only in a finite interval of field strengths, say (B_1, B_2). Below a critical value B_1 the flux tube is stable and, if perturbed, performs oscillations about its equilibrium path. The α-effect only appears when the tube becomes unstable at the critical field strength, B_1. The regeneration of poloidal field from toroidal field proceeds while the magnetic field strength increases, until a second critical value, B_2, is reached. Around this value of the magnetic field, the growth rate of the insta-bility becomes so large that the tube leaves the stably stratified overshoot region, enters the convection zone proper, and rapidly rises towards the surface so that it is no longer avail-able for the α-effect. The function $\alpha(B)$ is therefore non-vanishing only within an interval (B_1, B_2) of the magnetic field strength.

5.4.2. *Global picture of a dynamo with flux tubes*

We suggest the possibility of two separate, full dynamos operating in two separate layers and each one responsible for different aspects of the solar activity cycle:

(**I**) A boundary layer, *strong-field dynamo* (with $B \gtrsim 10^5$) located in the overshoot layer (depth of about 10^4 km) with strong (super-equipartition) fields concentrated in isolated flux tubes. This large-scale dynamo would ultimately be responsible for the solar activity cycle of sunspots. Grand minima would occur when this dynamo ceases to work.

(**II**) A turbulent *weak-field dynamo* (with $B < 10^4$) operating throughout the convection zone. This dynamo would generate a more irregular field from the turbulent flow, that would not disappear during a grand minimum and probably maintain a residual level of activity.

Both dynamos are coupled with each other and interchange magnetic flux (Fig. 5.4). The convection zone feeds the overshoot layer with the weak magnetic field ($<10^4$ G) dragged

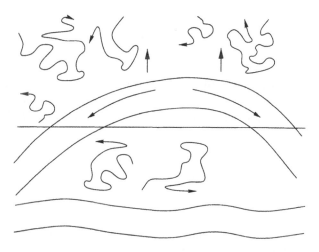

Figure 5.4 Qualitative picture of the magnetic coupling between a turbulent dynamo distributed throughout the convection zone and an interface dynamo located in the underlying overshoot layer. The strong-field dynamo operating in the overshoot layer generates superequiparti-tion fields while the turbulent, weak-field dynamo generates a more irregular field. The dynamo in the convection zone stochastically sheds poloidal flux loops into the overshoot layer. Instabilities driven by buoyancy lead to flux loss from the overshoot layer into the convection zone proper.

downwards by convective cells. The differential rotation of the boundary layer (shear layer) between the bottom of the convection zone and the radiative region intensifies the magnetic field (Ω-effect of classical dynamo theory). The overshoot layer expels to the convection zone the intense flux tubes ($\gtrsim 10^5$ G) which become unstable. The possible decoupling between both dynamos could be the explanation for activity minima such as the Maunder minimum from 1645 to 1715. Of course, the magnetic activity of fully convective stars would be qualitatively different (for instance, those with spectral type later than M5).

5.4.3. Model equations

In order to investigate how a dynamo with strong (super-equipartition) magnetic fields could operate, we shall combine the properties of an α-effect based on flux tube instability with the one-dimensional non-linear dynamo equations of Schmitt and Schüssler (1989) for a thin shell.

The isolated flux tubes are responsible for the dynamic α-effect described in Section 5.4.1, but our approach is within the framework of the mean-field dynamo theory (see, e.g. Moffatt, 1978; Krause and Rädler, 1980; Hoyng, Chapter 1, this volume), which involves a strong 'parametrization' of turbulence (α-effect and enhanced η via turbulent diffusivity).

The magnetic field $\mathbf{B}$ and the velocity field $\mathbf{u}$ are written as $\mathbf{B} = \bar{\mathbf{B}} + \mathbf{B}'$, and $\mathbf{u} = \bar{\mathbf{u}} + \mathbf{u}'$, where the averages are taken in the azimuthal direction and $\mathbf{u}'$, $\mathbf{B}'$ are the deviations from axial symmetry. With this decomposition, the magnetic induction equation splits into two equations governing the evolution of the mean field, $\bar{\mathbf{B}}$, and of the deviation, $\mathbf{B}'$, respectively.

Consider spherical coordinates (r, θ, ϕ) and split the mean fields $\bar{\mathbf{u}}$ and $\bar{\mathbf{B}}$ into poloidal and toroidal parts:

$$\bar{\mathbf{u}} = \bar{\mathbf{u}}_P + \bar{\mathbf{u}}_T = \bar{\mathbf{u}}_P + \bar{u}_\phi \mathbf{e}_\phi,$$
$$\bar{\mathbf{B}} = \bar{\mathbf{B}}_P + \bar{\mathbf{B}}_T = \bar{\mathbf{B}}_P + \bar{B}_\phi \mathbf{e}_\phi. \tag{5.15}$$

Since the mean $\mathbf{B}$-field is axisymmetric, it can be expressed in terms of two scalar functions, $A(r, \theta, t)$ and $B(r, \theta, t)$, where A is the ϕ-component of the vector potential for the mean poloidal field and B is the ϕ-component of the mean toroidal field (i.e. $B \overset{\text{def}}{=} \bar{B}_\phi$):

$$\bar{\mathbf{B}} = \underbrace{\nabla \times (0, 0, A)}_{\bar{\mathbf{B}}_P} + \underbrace{(0, 0, B)}_{\bar{\mathbf{B}}_T}. \tag{5.16}$$

We call $\Omega \overset{\text{def}}{=} \bar{u}_\phi / (r \sin \theta)$ the mean angular velocity and η_{eff} the effective diffusivity (turbulent plus molecular) and introduce the following scaling factors:

$$L = \text{characteristic length of variation,}$$
$$\alpha_0 = \text{characteristic value of } \alpha(r, \theta), \tag{5.17}$$
$$\Omega_0' = \text{characteristic value of } \|\nabla \Omega(r, \theta)\|.$$

With η_{eff} and L, a natural time scale for Ohmic diffusion of the magnetic field emerges, $\tau \overset{\text{def}}{=} L^2/\eta_{\text{eff}}$. The Reynolds numbers of the α- and the Ω-effects are defined as:

$$\mathcal{R}_\alpha \overset{\text{def}}{=} \frac{\alpha_0 L}{\eta_{\text{eff}}}, \qquad \mathcal{R}_\Omega \overset{\text{def}}{=} \frac{\Omega_0' L^3}{\eta_{\text{eff}}}. \tag{5.18}$$

$\mathcal{R}_\alpha$ is a measure of the induction effect of the helical motions, while $\mathcal{R}_\Omega$ is a measure of the induction effect of differential rotation. Their product yields the dynamo number, viz.

$$\mathcal{N} \overset{\text{def}}{=} \mathcal{R}_\alpha \cdot \mathcal{R}_\Omega = \alpha \Omega_0' L^4 / \eta_{\text{eff}}^2. \tag{5.19}$$

For conditions prevailing in the Sun and solar-like stars we have $|\mathcal{R}_\Omega| \gg |\mathcal{R}_\alpha|$, so that the α-effect is responsible mainly for regenerating the mean poloidal field from the toroidal field, while the Ω-effect – due to differential rotation – converts mean poloidal field into toroidal field ($\alpha\Omega$-dynamo).

Neglecting the α-effect in the equation for B, the mean-field equations for an $\alpha\Omega$-dynamo without mean poloidal flow (i.e. neglecting the redistributing effect of the meridional circulation) in dimensionless form are

$$\frac{\partial A}{\partial t} = \overbrace{\mathcal{R}_\alpha f(r,\theta)B}^{\alpha\text{-effect}} + \overbrace{\mathcal{D}(A)}^{\text{diffusion of }\overline{\mathbf{B}}_P},$$

$$\frac{\partial B}{\partial t} = \underbrace{\mathcal{R}_\Omega r \sin\theta\, \mathbf{B}_P \cdot \nabla\Omega}_{\Omega\text{-effect}} + \underbrace{\mathcal{D}(B)}_{\text{diffusion of }\overline{\mathbf{B}}_T}, \tag{5.20}$$

where $f(r,\theta)$ describes the spatial dependence of the α-effect and $\mathcal{D} \overset{\text{def}}{=} \eta_{\text{eff}}(\nabla^2 - r^{-2}\sin^{-2}\theta)$ is the diffusion operator (for constant η_{eff}).

As a first step towards a mean-field dynamo model based on an α-effect due to buoyancy instability we consider a one-dimensional non-linear dynamo operating in a thin shell (Schmitt and Schüssler, 1989). The thin-shell dynamo can be considered a first approach to a dynamo operating in the overshoot layer. In this simplified model, the r-dependence is neglected and the diffusion operator takes the form

$$\mathcal{D} = \frac{1}{\sin\theta} \frac{\partial}{\partial\theta} \left(\sin\theta \frac{\partial}{\partial\theta} \right) - \frac{1}{\sin^2\theta}, \tag{5.21}$$

which represents latitudinal diffusion (while radial diffusion is assumed to be suppressed by the stable stratification and flux expulsion by convection).

We introduce two new ingredients to the thin-shell dynamo equations: (1) An α-effect arising from weak flux tube instability, operating in the range $B_1 < B < B_2$, as described in Section 5.4.1, and (2) strong convective downdrafts which pump magnetic flux from the convection zone into the overshoot layer.

The function $f(\theta, B)$, which gives the dependence of the α-effect with colatitude and with B, is obtained from the linear stability analysis described in Section 5.3.2.

We assume that strong convective downdrafts transport magnetic flux from the convection zone proper into the overshoot region. This magnetic flux is generated in the convection zone by means of a turbulent dynamo. A source function $S(t, \theta)$ describing the pumping of poloidal magnetic flux has been inserted arbitrarily into the dynamo equations and is not derived rigorously. As a working hypothesis, assume that these flux loops are shed stochastically into the overshoot layer. $S(t, \theta)$ represents therefore stochastic excitation.

The model equations for a non-linear $\alpha\Omega$-shell-dynamo operating in the overshoot region are then

$$
\frac{\partial A}{\partial t} = \mathcal{R}_\alpha f(\theta, B) B + \mathcal{D}(A) + S(\theta, t),
$$

$$
\frac{\partial B}{\partial t} = \mathcal{R}_\Omega \frac{\partial (A \sin \theta)}{\partial \theta} + \mathcal{D}(B).
$$

(5.22)

5.4.4. *Numerical simulations and discussion of the results*

We have solved numerically the nonlinear system (5.22) by means of a time-stepping method (see Schmitt and Schüssler, 1989; Schmitt *et al.*, 1996) and using parameter values typical for the overshoot layer at the bottom of the solar convection zone. The α-effect is limited to $\pm 35°$ latitude about the equator and works only in the range $[B_1, B_2]$ (i.e. the interval of slowly growing instability); on the basis of the stability criteria for flux tubes in the overshoot layer (Ferriz-Mas and Schüssler, 1995; see also Fig. 5.2) we have chosen $B_1 = 8 \cdot 10^4$ G for the critical field strength at which instability sets in and $B_2 = 1.2 \cdot 10^5$ G for the field strength leading to rapid eruption. We have taken $\alpha_0 \simeq 2\,\mathrm{cm\,s}^{-1}$ (which is the order of magnitude for the α-effect arising from flux tube instability, as shown in Ferriz-Mas *et al.*, 1994). The length scale L [see (5.17)] is taken to be the distance of the dynamo layer from the center of the Sun ($L = 5 \cdot 10^{10}$ cm), and the diffusion time L^2/η_{eff} is the time scale. We have chosen $\eta_{\mathrm{eff}} = 6 \cdot 10^{11}\,\mathrm{cm}^2\,\mathrm{s}^{-1}$ so that the dynamo period becomes 22 years. With the choices of L, η_{eff} and α_0 we have $\mathcal{R}_\alpha \simeq 0.2$. The value of Ω_0' we take corresponds to a difference of rotational velocity of $50\,\mathrm{m\,s}^{-1}$ over the vertical extent of 10^4 km of the overshoot layer; this yields $\mathcal{R}_\Omega = -13{,}000$.

The most delicate point of the model is the choice of the source function $S(\theta, t)$. The threshold for the dynamo effect may lead to intermittent behaviour if the system is perturbed by random fluctuations or by deterministic variations.

Strong convective downdrafts transport magnetic flux from the overlying convection zone into the overshoot layer. We parametrize in a simple way this magnetic pumping by considering that $S(\theta, t)$ is a randomly varying source function with correlation time τ_c, correlation length λ_c, and an amplitude varying randomly in the interval $(-S_0, S_0)$ with a uniform distribution. We have taken $\tau_c \simeq 1$ month (i.e. the duration of one flux injection is of the order of the typical correlation time at the bottom of the convection zone), $\lambda_c \simeq 10^5$ km (typical horizontal length of large-scale convective flows), and $S_0 = 0.5$. If a filling factor of 0.1 for flux tubes in the overshoot layer is assumed, the fluctuation level of S_0 corresponds to a radial magnetic flux of 10^{22} Mx to be generated each month by the turbulent dynamo.

Fig. 5.5 gives the results from a numerical simulation using the above parameter values. As a rough measure of the global magnetic activity we have represented the magnetic energy of the mean field in the overshoot layer (in arbitrary units) versus time for a run over 500

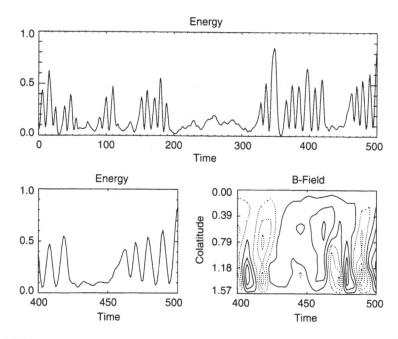

Figure 5.5 Long-term evolution of dynamo activity from a simulation with level of fluctuation $S_0 = 0.5$ and parameter values indicated in the main text. The top panel shows the total magnetic energy (in arbitrary units) as a function of time (with arbitrary zero point). The lower panel shows a detail of a 100-year interval together with the corresponding diagram $B(\theta, t)$.

years. The run exhibits 11-year cycles of varying amplitude, irregularly interrupted by grand minima. The intermittent pattern resembles the irregular behaviour of solar activity.

5.5. Concluding remarks

The magnetic field at the solar surface reflects the presence of isolated flux tubes below the photosphere. The existence of this hierarchy of concentrated magnetic structures in the observable layers is indicative of the intermittent nature of magnetic fields in the underlying convection zone. The equilibrium and dynamics of flux tubes are basically different from those of a diffuse field. This has important consequences for the storage of magnetic flux, the onset and development of instabilities, the rise of magnetic flux towards the surface to form the observed active regions, and also for the understanding of the dynamo mechanism.

Magnetic buoyancy due to the increased effective pressure inside the tubes poses a dilemma: on the one hand, it is necessary to bring the magnetic flux to the solar surface; on the other it may lead to a rapid loss of magnetic flux from the convection zone, thus preventing the storage of magnetic flux long enough for the operation of a hydromagnetic dynamo. It is nowadays widely accepted that the magnetic field emerging at the surface is stored prior to eruption within a stable layer of overshooting convection immediately below the convection zone, where helioseismology has revealed steep radial gradients in the angular velocity.

Amplification of this field by large-scale shear eventually leads to the onset of buoyancy-driven instabilities, which may give rise to an electric current (anti-)parallel to the unperturbed magnetic field, equivalent to an α-effect. The α-effect requires a non-vanishing phase difference between the velocity and magnetic field perturbations. Since only unstable modes yield a net α-effect, strong magnetic fields (well above equipartition) are required for its operation. It can be shown that toroidal flux tubes in a rotating star do not provide a net α-effect, even if they are subject to random forcing, unless they are unstable to small displacements (Ossendrijver, 2000). Once the field becomes larger than about $1.1-1.2 \times 10^5$ G, a second regime of instability is reached, which drives the flux tubes into the convection zone proper and leads to the rise of unstable loops up to the photosphere within about one month.

A dynamo based only on this α-effect and differential rotation is not self-excited in the sense that a field is generated out of an arbitrarily small seed field. Another mechanism has to be invoked in order to operate at field strengths below the threshold for instability. In Section 5.4 we have modelled a nonlinear, strong-field dynamo based on unstable flux tubes and operating in the overshoot layer by using a simple one-dimensional $\alpha\Omega$-mean-field approach (cf. Schmitt and Schüssler, 1989). We have assumed that the poloidal flux generated by a turbulent convection-zone dynamo is pumped stochastically into the overshoot layer; this effect has been modelled by introducing a stochastic source term in the equation for the evolution of the mean poloidal field. Stochastic fluctuations in the coupling between a weak-field dynamo operating throughout the convection zone and the strong-field dynamo operating in the tachocline can lead to the observed patterns of on–off intermittency. The fluctuations lead to qualitative effects if the stochastic source term S is comparable to the α-effect term in the equation for the evolution of the poloidal field. Reducing the α-effect renders the system more susceptible to stochastic fluctuations and *vice versa* (see, for details, Schmitt *et al.*, 1996). Stronger fluctuations may destroy the cyclic behaviour of the overshoot layer dynamo and lead to increased but irregular activity. Such activity is observed in fast rotating cool stars. On the other hand, stars with low and non-variable magnetic activity may be in a state with only the turbulent convection zone dynamo active.

A problem that has not been addressed here is the backreaction of the magnetic field on the angular velocity distribution in the tachocline. A reservoir of about 10^{24} Mx in toroidal flux at the bottom of the solar convection zone is required in order to account for the flux emerging in large active regions over the 11-year cycle (see, e.g. Galloway and Weiss, 1981). Assuming that these active regions are the result of the rise and emergence of large toroidal flux tubes starting their journey from the bottom of the convection zone with field strength close to 10^5 G, this means an energetic requirement of 10^{39} erg, to be supplied by the dynamo process over 11 years. Depending on the depth of the tachocline this value can be up to a factor 10 larger than the energy of differential rotation, so that an efficient mechanism is required that restores the shear layer and replenishes the kinetic energy on a time-scale less than 11 years. This means that one needs a mechanism that permanently redistributes angular momentum and feeds energy into differential rotation. On the other hand, there must exist a strong backreaction of the magnetic field on differential rotation via the Lorentz force (which is neglected in the 'classical' kinematic approach). This suggests that the modification of the differential rotation (Ω-quenching) may be an important nonlinear process for the operation of the dynamo. The nonlinear feedback of strong magnetic fields on differential rotation in the mean-field conservation law of angular momentum has been suggested by some authors to be the cause leading to grand minima in the cyclic activity (e.g. Rüdiger and Arlt, Chapter 6 and references therein).

References

Abett, W.P., Fisher, G.H. and Fan, Y., "The three-dimensional evolution of rising, twisted magnetic flux tubes in a gravitationally stratified model convection zone," *Astrophys. J.* **540**, 548–562 (2000).

Brandenburg, A., "Astrophysical large scale dynamos: Simulations of MHD turbulence," this volume, Ch. 9, pp. 269–344 (2003).

Brandenburg, A., Procaccia, I. and Segel, D., "The size and dynamics of magnetic flux structures in magnetohydrodynamic turbulence," *Phys. Plasmas*, **2**, 1148–1156 (1995).

Brandenburg, A. and Schmitt, D., "Simulations of an alpha-effect due to magnetic buoyancy," *Astron. Astrophys.* **338**, L55–L58 (1998).

Caligari, P., Moreno-Insertis, F. and Schüssler, M., "Emerging flux tubes in the solar convection zone I. Asymmetry, tilt, and emergence latitude," *Astrophys. J.* **441**, 886–902 (1995).

Caligari, P., Schüssler, M. and Moreno-Insertis, F., "Emerging flux tubes in the solar convection zone II. The influence of initial conditions," *Astrophys. J.* **502**, 481–492 (1998).

Choudhuri, A.R. and Gilman, P.A., "The influence of the Coriolis force on flux tubes rising through the solar convection zone," *Astrophys. J.* **316**, 788–800 (1987).

Defouw, R.J., "Wave propagation along a magnetic tube," *Astrophys. J.* **209**, 266–269 (1976).

D'Silva, S. and Choudhuri, A.R., "A theoretical model for tilts of bipolar magnetic regions," *Astron. Astrophys.* **272**, 621–633 (1993).

Dorch, S.B.F., "On the structure of the magnetic field in a kinematic ABC flow dynamo," *Physica Scripta* **61**, 717–722 (2000).

Dorch, S.B.F. and Nordlund, Å., "Numerical 3D simulations of buoyant magnetic flux tubes," *Astron. Astrophys.* **338**, 329–339 (1998).

Dorch, S.B.F., Archontis, V. and Nordlund, Å, "3D simulations of twisted magnetic flux ropes," *Astron. Astrophys.* **352**, L79–L82 (1999).

Emonet, T. and Moreno-Insertis, F., "The physics of twisted magnetic tubes rising in a stratified medium: two-dimensional results," *Astrophys. J.* **492**, 804–821 (1998).

Fan, Y., "Nonlinear growth of the three-dimensional undular instability of a horizontal magnetic layer and the formation of arching flux tubes," *Astrophys. J.* **546**, 509–527 (2001).

Fan, Y. and Fisher, G.H., "Radiative heating and the buoyant rise of magnetic flux tubes in the solar interior," *Solar Physics* **166**, 17–41 (1996).

Fan, Y., Fisher, G.H. and McClymont, A.N., "Dynamics of emerging active region flux loops," *Astrophys. J.* **436**, 907–928 (1994).

Fan, Y., Zweibel, E.G. and Lantz, S.R., "Two-dimensional simulations of buoyantly rising, interacting magnetic flux tubes," *Astrophys. J.* **493**, 480–493 (1998a).

Fan, Y., Zweibel, E.G., Linton, M.G. and Fisher, G.H., "The rise of kink-unstable magnetic flux tubes in the solar convection zone," *Astrophys. J.* **505**, L59–L63 (1998b).

Fan, Y., Zweibel, E.G., Linton, M.G. and Fisher, G.H., "The rise of kink-unstable magnetic flux tubes and the origin of delta-configuration sunspots," *Astrophys. J.* **521**, 460–477 (1999).

Ferriz-Mas, A., "On the storage of magnetic flux tubes at the base of the solar convection zone," *Astrophys. J.* **458**, 802–816 (1996).

Ferriz-Mas, A., Schmitt, D. and Schüssler, M., "A dynamo effect due to instability of magnetic flux tubes," *Astron. Astrophys.* **289**, 949–956 (1994).

Ferriz-Mas, A. and Schüssler, M., "Radial expansion of the magnetohydrodynamic equations for axially symmetric configurations," *Geophys. Astrophys. Fluid Dynam.* **48**, 217–234 (1989).

Ferriz-Mas, A. and Schüssler, M., "Instabilities of magnetic flux tubes in a stellar convection zone I: Equatorial flux rings in differentially rotating stars," *Geophys. Astrophys. Fluid Dynam.* **72**, 209–247 (1993).

Ferriz-Mas, A. and Schüssler, M., "Instabilities of magnetic flux tubes in a stellar convection zone II: Flux rings outside the equatorial plane," *Geophys. Astrophys. Fluid Dynam.* **81**, 233–265 (1995).

Fisher, G.H., Fan, Y., Longcope, D.W., Linton, M.G. and Pevtsov, A.A., "The solar dynamo and emerging flux," *Solar Phys.* **192**, 119–139 (2000).

Galloway, D.J. and Weiss, N.O., "Convection and magnetic fields in stars," *Astrophys. J.* **243**, 945–953 (1981).

Hoyng, P., "Mean field theory: back to the basics," this volume, Ch. 1, pp. 1–36 (2003).

Hughes, D.W. and Proctor, M.R.E., *Ann. Rev. Fluid Mech.* **20**, 187–223 (1988).

Jensen, E., "On tubes of magnetic forces embedded in stellar material," *Ann. d'Astrophys.* **18**, 127–132 (1955).

Krause, F. and Rädler, K.-H., *Mean-Field Magnetohydrodynamics and Dynamo Theory*, Pergamon, Oxford (1980).

Larmor, J., "How could a rotating body such as the Sun become a magnet?," *Rep. Brit. Assoc. Adv. Sci. 1919*, 159–160 (1919).

Longcope, D.W., Fisher, G.H. and Arendt, S., "The evolution and fragmentation of rising magnetic flux tubes," *Astrophys. J.* **464**, 999–1011 (1996).

Matsumoto, R., Tonooka, H., Tajima, T., Chou, W. and Shibata, K., "Three-dimensional MHD simulations of the emergence of twisted flux tubes," *Advances in Space Research* **26**, 543–546 (2000).

Matthews, P.C., Hughes, D.W. and Proctor, M.R.E., "Magnetic buoyancy, vorticity, and three-dimensional flux tube formation," *Astrophys. J.* **448**, 938–941 (1995).

Moffat, H.K., *Magnetic Field Generation in Electrically Conducting Fluids*, Cambridge University Press, Cambridge (1978).

Moreno-Insertis, F., "Rise times of horizontal magnetic flux tubes in the convection zone of the Sun," *Astron. Astrophys.* **122**, 241–250 (1983).

Moreno-Insertis, F., "Nonlinear time-evolution of kink-unstable magnetic flux tubes in the convective zone of the Sun," *Astron. Astrophys.* **166**, 291–305 (1986).

Moreno-Insertis, F., "The motion of magnetic flux tubes in the convection zone and the subsurface origin of active regions," in: *Sunspots: Theory and Observations*, (Eds. J.H. Thomas and N.O. Weiss), Kluwer Academic Publishers, Dordrecht, pp. 385–410 (1992).

Moreno-Insertis, F., "The magnetic field in the convection zone as a link between the active regions on the surface and the field in the solar interior," in: *Solar Magnetic Fields* (Eds. M. Schüssler and W. Schmidt), Cambridge University Press, pp. 117–135 (1994).

Moreno-Insertis, F., Caligari, P. and Schüssler, M., "*Explosion* and intensification of magnetic flux tubes," *Astrophys. J.* **452**, 894–900 (1995).

Moreno-Insertis, F. and Emonet, T., "The rise of twisted magnetic tubes in a stratified medium," *Astrophys. J.* **472**, L53–L56 (1996).

Moreno-Insertis, F., Schüssler, M. and Ferriz-Mas, A., "Storage of magnetic flux tubes in a convective overshoot region," *Astron. Astrophys.* **264**, 686–700 (1992).

Moreno-Insertis, F., Schüssler, M. and Glampedakis, C., "Radiative heating of magnetic flux tubes. I. Solution of the diffusion problem," *Astron. Astrophys.* **388**, 1022–1035 (2002).

Nordlund, Å., Brandenburg, A., Jennings, R.L., Rieutord, M., Ruokolainen, J., Stein, R.F. and Tuominen, I., "Dynamo action in stratified convection with overshoot," *Astrophys. J.* **392**, 647–652 (1992).

Ossendrijver, M.A.J.H., "The dynamo effect of magnetic flux tubes," *Astron. Astrophys.* **359**, 1205–1210 (2000).

Parker, E.N., "The formation of sunspots from the solar toroidal field," *Astrophys. J.* **121**, 491–507 (1955a).

Parker, E.N., "Hydromagnetic dynamo models," *Astrophys. J.* **122**, 293–314 (1955b).

Parker, E.N., "The generation of magnetic fields in astrophysical bodies. X. Magnetic buoyancy and the solar dynamo," *Astrophys. J.* **198**, 205–209 (1975).

Parker, E.N., "The dynamics of fibril magnetic fields. II. The mean field equations," *Astrophys. J.* **256**, 302–315 (1982).

Parker, E.N., "Stellar fibril magnetic systems. I. Reduced energy state," *Astrophys. J.* **283**, 343–348 (1984).

Parker, E.N., "A solar dynamo surface wave at the interface between convection and nonuniform rotation," *Astrophys. J.* **408**, 707–719 (1993).

Proctor, M.R.E., "Magnetoconvection," in: *Sunspots: Theory and Observations* (Eds. J.H. Thomas and N.O. Weiss), Kluwer Academic Publishers, Dordrecht, pp. 221–241 (1992).

Rempel. M., "Thermal properties of magnetic flux tubes. II. Storage of flux in the overshoot region," *Astron. Astrophys.* in press (2003).

Rempel, M., Schüssler, M., "Intensification of magnetic fields by conversion of potential energy," *Astrophys. J.* **552**, L171–L174 (2001).

Rempel, M.. Schüssler, M. and Toth, G., "Storage of magnetic flux at the bottom of the solar convection zone," *Astron. Astrophys.* 363, 789–799 (2000).

Roberts, B. and Webb, A.R., "Vertical motions in an intense magnetic flux tube", *Solar Phys.* **56**, 5–35 (1978).

Rüdiger, G. and Arlt, R., "The physics of the solar cycle," this volume, Ch. 6, pp. 147–194 (2003).

Schmitt, D., "Magnetically driven alpha-effect and stellar dynamos," this volume (2001).

Schmitt, D. and Schüssler, M., "Non-linear dynamos I. One-dimensional model of a thin layer dynamo," *Astron. Astrophys.* **223**, 343–351 (1989).

Schmitt, D., Schüssler, M. and Ferriz-Mas, A., "Intermittent solar activity by an on-off dynamo," *Astron. Astrophys.* **311**, L1–L4 (1996).

Schmitt, J.H.M.M., Rosner, R. and Bohn, H.-U., "The overshoot region at the bottom of the solar convection zone," *Astrophys. J.* **282**, 316–329 (1984).

Schüssler, M., "Magnetic buoyancy revisited: Analytical and numerical results for rising flux tubes," *Astron. Astrophys.* **71**, 79–91 (1979).

Schüssler, M., "Stellar dynamo theory", in: *Solar and Stellar Magnetic Fields: Origins and Coronal Effects* (Ed. J.O. Stenflo), IAU-Symp. No. 102, Reidel, Dordrecht, pp. 213–236 (1983).

Schüssler, M., "Magnetic flux tubes and the solar dynamo," in: *Solar and Astrophysical Magnetohydrodynamic Flows* (Ed. K. Tsinganos), Kluwer Academic Publishers, Dordrecht, pp. 17–37 (1996).

Schüssler, M., Caligari, P., Ferriz-Mas, A. and Moreno-Insertis, F., "Instability and eruption of magnetic flux tubes in the solar convection zone," *Astron. Astrophys.* **281**, L69–L72 (1994).

Schüssler, M., Caligari, P., Ferriz-Mas, A., Solanki, S.K. and Stix, M., "Distribution of starspots on cool stars I. Young and main sequence stars of 1 $M_\odot$," *Astron. Astrophys.* **314**, 503–513 (1996).

Shaviv, G. and Saltpeter, E.E., "Convective overshooting in stellar interior models," *Astrophys. J.* **184**, 191–200 (1973).

Skaley, D. and Stix, M., "The overshoot layer at the base of the solar convection zone," *Astron. Astrophys.* **241**, 227–232 (1991).

Spiegel, E.A. and Weiss, N.O., "Magnetic activity and variations in solar luminosity," *Nature* **287**, 616–617 (1980).

Spruit, H.C., "Motion of magnetic flux tubes in the solar convection zone and chromosphere," *Astron. Astrophys.* **98**, 155–160, (1981).

Spruit, H.C. and van Ballegooijen, A.A., "Stability of toroidal flux tubes in stars," *Astron. Astrophys.* **106**, 58–66 (1982).

Tobias, S.M., Brummell, N.H., Clune, T.L. and Toomre, J., "Pumping of Magnetic Fields by Turbulent Penetrative Convection," *Astrophys. J.* **502**, L177–L180 (1998).

Tomczyk, S., Schou, J. and Thompson, M.J., "Measurement of the rotation rate in the deep solar interior," *Astrophys. J.* **448**, L57–L60 (1995).

Tsinganos, K.C., "Sunspots and the physics of magnetic flux tubes. X – On the hydrodynamic instability of buoyant fields," *Astrophys. J.* **239**, 746–760 (1980).

van Ballegooijen, A.A., "The structure of the solar magnetic field below the photosphere," *Astron. Astrophys.* **106**, 43–52 (1982a).

van Ballegooijen, A.A., "The overshoot layer at the base of the solar convective zone and the problem of magnetic flux storage," *Astron. Astrophys.* **113**, 99–112 (1982b).

van Ballegooijen, A.A., "On the stability of toroidal flux tubes in differentially rotating stars," *Astron. Astrophys.* **118**, 275–284 (1983).

Zwaan, C., "On the appearance of magnetic flux in the solar photosphere," *Solar Phys.* **60** , 213–240 (1978).

Zwaan, C., "The evolution of sunspots," in: *Sunspots: Theory and Observations* (Eds. J.H. Thomas and N.O. Weiss), Kluwer Academic Publishers, Dordrecht, pp. 75–100 (1992).

Zwaan, C. and Harvey, K.L., "Patterns in the solar magnetic field," in: *Solar Magnetic Fields* (Eds. M. Schüssler and W. Schmidt), Cambridge University Press, 27–48 (1994).

6 Physics of the solar cycle

Günther Rüdiger and Rainer Arlt

*Astrophysikalisches Institut Potsdam, An der Sternwarte 16, D-14482 Potsdam,
Germany, E-mail: gruediger@aip.de* (Günther Rüdiger),
E-mail: rarlt@aip.de (Rainer Arlt)

The theory of the solar/stellar activity cycles is presented, based on the mean-field concept in magnetohydrodynamics. A new approach to the formulation of the electromotive force (EMF) as well as the theory of differential rotation and meridional circulation is described for use in dynamo theory. Activity cycles of dynamos in the overshoot layer (BL-dynamo) and distributed dynamos are compared, with the latter including the influence of meridional flow. The overshoot layer dynamo is able to reproduce the solar cycle periods and the butterfly diagram only if $\alpha = 0$ in the convection zone. The problems of too many magnetic belts and too short cycle times emerge if the overshoot layer is too thin. The distributed dynamo including meridional flows with a magnetic Reynolds number $R_m \gtrsim 20$ (low magnetic Prandtl number) reproduces the observed butterfly diagram even with a positive dynamo-α in the bulk of the convection zone.

The nonlinear feedback of strong magnetic fields on differential rotation in the mean-field conservation law of angular momentum leads to grand minima in the cyclic activity similar to those observed. The two-dimensional model described here contains the large-scale interactions as well as the small-scale feedback of magnetic fields on differential rotation and induction in terms of a mean-field formulation (Λ-quenching, α-quenching). Grand minima may also occur if a dynamo occasionally falls below its critical eigenvalue. We expressed this idea by an on-off α function which is non-zero only in a certain range of magnetic fields near the equipartition value. We never found any indication that the dynamo collapses by this effect after it had once been excited.

The full quenching of turbulence by strong magnetic fields in terms of reduced induction (α) and reduced turbulent diffusivity (η_T) is studied with a one-dimensional model. The full quenching results in a stronger dependence of cycle period on dynamo number compared with the model with α-quenching alone giving a very weak cycle period dependence.

Also the temporal fluctuations of α and η_T from a random-vortex simulation were applied to a dynamo model. Then the low 'quality' of the solar cycle can be explained with a relatively small number of giant cells acting as the dynamo-active turbulence. The simulation contains the transition from almost regular magnetic oscillations (many vortices) to a more or less chaotic time series (very few vortices).

6.1. Introduction

Explaining the characteristic period of the quasicyclic activity oscillations of stars with the Sun included is one of the challenges for stellar physics. The main period is an essential property of the dynamo mechanism. Solar dynamo theory is reviewed here in the special context of the cycle-time problem. The parameters of the convection zone turbulence do not easily provide us with the 22-year time scale for the solar dynamo. Even for the boundary layer dynamos, it is only possible if a dilution factor in the turbulent EMF smaller than unity is introduced which parameterizes the intermittent character of the magnetohydrodynamic turbulence.

A notable number of interesting phenomena have been investigated in the search for the solution of this problem: flux-tube dynamics, magnetic quenching, parity breaking, and chaos. Nevertheless, even the simplest observation – the solar cycle period of 22 years – is hard to explain (cf. DeLuca and Gilman, 1991; Stix, 1991; Gilman, 1992; Levy, 1992; Schmitt, 1993; Brandenburg, 1994a; Weiss, 1994). How can we understand the existence of the large ratio of the mean cycle period and the correlation time of the turbulence? Three main observations are basic in this respect:

- There is a factor of about 300 between the solar cycle time and the Sun's rotation period.
- This finding is confirmed by stellar observations (Fig. 6.1).
- The convective turnover time near the base of the convection zone is very similar to the solar rotation period.

The problem of the large observed ratio of cycle and correlation times,

$$\frac{\tau_{cyc}}{\tau_{corr}} \gtrsim 10^2, \tag{6.1}$$

constitutes the primary concern of dynamo models. In a thick convection shell this number reflects (the square of) the ratio of the stellar radius to the correlation length and numbers of the order 100 in (6.1) are possible. For the thin boundary layer dynamo, however, the problem becomes more dramatic and is in need of an extra hypothesis.

The activity period of the Sun varies strikingly about its average from one cycle to another. Only a nonlinear theory will be able to explain the non-sinusoidal (chaotic or not) character

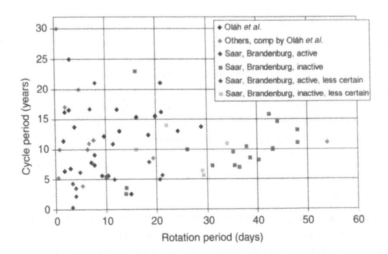

Figure 6.1 Stellar cycles: cycle time in years versus rotation period in days compiled from Saar and Brandenburg (1999) and Oláh *et al.* (2000).

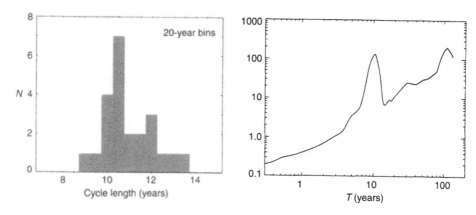

Figure 6.2 Left: The distribution of the solar cycle length does not approach a Dirac function, the 'quality' of the cycle only gives values of about 5. Right: The wavelet spectrum of the sunspot-number time series shows two peaks for both 10 and 100 years (Frick *et al.*, 1997a).

of the activity cycle (Fig. 6.2). A linear theory is only concerned with the mean value of the oscillation frequency.

6.2. Basic theory

Kinematic dynamo theory utilizes only one equation to advance the mean magnetic field in time, i.e.

$$\frac{\partial \langle \mathbf{B} \rangle}{\partial t} = \nabla \times (\langle \mathbf{u} \rangle \times \langle \mathbf{B} \rangle + \mathcal{E}). \tag{6.2}$$

Here only a non-uniform rotation will be imposed on the mean flow, $\langle \mathbf{u} \rangle$; any meridional flow shall be introduced later. The turbulent EMF, $\mathcal{E} = \langle \mathbf{u}' \times \mathbf{B}' \rangle$, contains induction α_{ij} and dissipation η_{ijk}, i.e.

$$\mathcal{E}_i = \alpha_{ij} \langle B_j \rangle + \eta_{ijk} \langle B_j \rangle_{,k} + \cdots. \tag{6.3}$$

Both tensors are pseudo-tensors. While for η_{ijk} an elementary isotropic pseudo-tensor exists (ε_{ijk}), the same is not true for α_{ij}. An odd number of Ω's is, therefore, required for the α-tensor, which is only possible with an odd number of another preferred direction, (say) **g**. The α-effect can thus only exist in stratified, rotating turbulence. The first formula reflecting this situation,

$$\alpha = c_\alpha \frac{l_{\text{corr}}^2 \Omega}{H_\rho} \cos \theta, \tag{6.4}$$

was given by Krause (1967) with Ω being the angular velocity of the basic rotation, θ the colatitude, and H_ρ the density scale height. Evidently, α is a complicated effect, where the effective α might really be very small; the unknown factor c_α in (6.4) may be much smaller than unity. The strength of this effect was computed in recent analytical and numerical simulations for both convectively unstable as well as stable stratifications. While Brandenburg *et al.* (1990) worked with a box heated from below, Brandenburg and Schmitt (1998) considered magnetic buoyancy in the transition region between the radiative solar core and the convection zone, Brandenburg (2000) probed the Balbus–Hawley instability for dynamo-α production. Ferrière (1993), Kaisig *et al.* (1993), and Ziegler *et al.* (1996) used random supernova explosions to drive the galactic turbulence. The magnitudes of the α-effects do not reach the given estimate in these cases: the dimensionless factor c_α seems really to be much smaller than unity.

6.2.1. Simple dynamos

6.2.1.1. Dynamo waves

An illuminating example of kinematic dynamo theory for the cycle period is the dynamo wave solution. In plane Cartesian geometry there is a mean magnetic field subject to a (strong) shear flow and an α-effect – all quantities vary only in the z-direction with a given wave number (Parker, 1975). The amplitude equations can then be written as

$$\dot{A} + A = C_\alpha B, \quad \dot{B} + B = iC_\Omega A, \tag{6.5}$$

with A representing the poloidal magnetic field component and B the toroidal component. C_α is the normalized α and C_Ω is the normalized shear. The eigenfrequency of the equation system is the complex number

$$\omega_{\text{cyc}} = \sqrt{\frac{D}{2}} + i\left(1 - \sqrt{\frac{D}{2}}\right). \tag{6.6}$$

A marginal solution of the magnetic field is found for a 'dynamo number' $D \equiv C_\alpha C_\Omega = 2$. In that case the field is not steady but oscillates with the (dimensionless) frequency $\omega_{\text{cyc}} = 1$. The second solution of (6.5) for $D = 2$ is $\omega_{\text{cyc}} = -1 + 2i$ which describes a decaying field mode. Nevertheless we have found that simultaneously ω_{cyc} and $-\omega_{\text{cyc}}$ belong to the same eigenvalue. The consequence is that nonaxisymmetric field modes are always drifting in the azimuth rather than oscillating. The 'flip-flop'-solutions for fast rotating stars (Tuominen *et al.*, 1999), on the other hand, might be interpreted by the (nonlinear) superposition of a stationary nonaxisymmetric dipole with an oscillating axisymmetric mode.

The (re-normalized) cycle period is thus given simply by combining the eddy diffusivity and the wave number, i.e. $\omega_{\text{cyc}} \simeq \eta_T k^2$ (which is simply the skin–effect relation). With a mixing-length expression for the eddy diffusivity, $\eta_T = c_\eta l_{\text{corr}}^2 / \tau_{\text{corr}}$, one finds the basic relation between the cycle time and the correlation time as

$$\frac{\tau_{\text{cyc}}}{\tau_{\text{corr}}} = \frac{1}{2\pi c_\eta}\left(\frac{R}{l_{\text{corr}}}\right)^2. \tag{6.7}$$

Note that $c_\eta \leq 0.3$. The ratio between the global scale, R, and the correlation length of the turbulence, l_{corr}, determines the cycle time. As this ratio has a minimum value of 10, it is thus no problem to reach a factor of 100 between the cycle period and the correlation time.[1] The rotation period does not enter – it only influences the cycle period in the nonlinear regime (for dynamo numbers exceeding 2). For such numbers the frequency increases – and the cycle time becomes shorter.

6.2.1.2. Shell dynamo

The simplest assumptions about the α-effect and the differential rotation are used in the shell dynamo, i.e.

$$\alpha = \alpha_0 \cos\theta, \quad \Omega = \text{const.} + \Omega' x \tag{6.8}$$

for $x_i < x < 1$ (Roberts, 1972; Roberts and Stix, 1972). Positive Ω' means radial super-rotation and negative Ω' means sub-rotation. The solution with the lowest eigenvalue for the sub-rotation case in a thick shell is a solution with dipolar symmetry, which is, in the vicinity of the bifurcation point, the only stable one (Krause and Meinel, 1988).

The cycle time grows with the linear thickness D of the shell. Compared with the dynamo wave, we find a reduction of (6.7) by the normalized thickness, $d \leq 0.5$. As long as convection zones of main-sequence stars are considered, that is not too dramatic, but what about rather thin layers like the solar overshoot region? As the cycle period is found to vary linearly with D in similar shell models,

$$\tau_{\text{cyc}} \simeq 0.26 \frac{RD}{\eta_{\text{T}}}, \tag{6.9}$$

the cycle time must become dramatically short in very thin boundary layers.

6.2.1.3. The dynamo dilemma

The sign of α has been presumed as positive in the northern hemisphere and the differential rotation has been considered unknown (Steenbeck and Krause, 1969) from the early years of dynamo theory. Only sub-rotation – with angular velocity increasing inwards – could then produce a suitable butterfly diagram from the latitudinal migration of the toroidal field. In the other case, e.g. where the surface rotation law is applied in the entire convection zone, one cannot reproduce the migration of the toroidal field towards the equator (Köhler, 1973).

The correct field migration according to the sunspot butterfly diagram is generally produced by a negative dynamo number only, which may be either due to positive α in the northern hemisphere and sub-rotation, or negative α and super-rotation. For all dynamos with super-rotation, however, the phase relation between the radial field component and the toroidal field component is $\langle B_r \rangle \langle B_\phi \rangle > 0$, again disagreeing with the observations (Stix, 1976). The prediction of dynamo theory for the solar differential rotation was thus a clear sub-rotation, $\partial\Omega/\partial r < 0$. This prediction contrasts with the results of helioseismology which revealed sub-rotation only near the poles, whereas near the solar equator there is a super-rotation at the base of the convection zone (Fig. 6.3).

Agreement with the butterfly diagram shall thus be satisfied by a negative α which can be only explained for the overshoot region. But the phase relation of the field components does then not agree with the known one (Yoshimura, 1976; Parker, 1987; Schlichenmaier and Stix, 1995). With the incorrect phase relation it is even problematic to derive the observed properties of the torsional oscillations (Howard and LaBonte, 1980), which are suggested to arise as a result of the backreaction of the mean-field Lorentz force (Schüssler, 1981; Yoshimura, 1981; Rüdiger et al., 1986; Küker et al., 1996).

In the light of recent developments for dynamos with meridional flow included, there could easily exist a solution of this dynamo dilemma in quite an unexpected way (see Section 6.4).

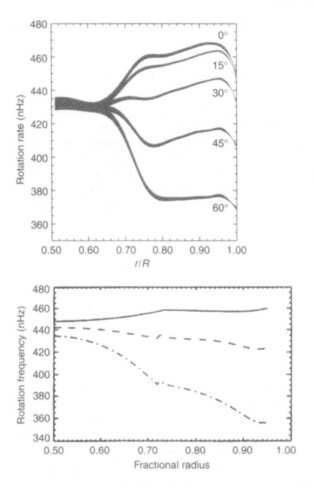

Figure 6.3 Top: The internal rotation of the Sun after the inversion of SOHO data (after Kosovichev
et al., 1997). Bottom: Theory of the solar internal rotation by Kitchatinov and Rüdiger
(1995). The rotation frequency is given for the equator (solid), mid-latitudes (dashed) and
poles (dashed-dotted).

6.2.2. *Differential rotation*

The theory of the maintenance of differential rotation in convective stellar envelopes might
be an instructive detour from dynamo theory. Differential rotation is also turbulence-induced
but without the complications due to the magnetic fields. It is certainly unrealistic to expect a
solution of the complicated problem of the solar dynamo without an understanding of mean-
field hydrodynamics. There is even no hope for the stellar dynamo concept if the internal
stellar rotation law cannot be predicted.

The main observational features of the solar differential rotation are

- surface equatorial acceleration of about 30%,
- strong polar sub-rotation and weak equatorial super-rotation,
- reduced equator-pole difference in Ω at the lower convection-zone boundary.

The characteristic Taylor–Proudman structure in the equatorial region and the characteristic disk-like structure in the polar region are comprised by the results. In the search for stellar surface differential rotation, chromospheric activity has been monitored for more than a decade. Surprisingly enough, there is not yet a very clear picture. For example, the rotation pattern of the solar-type star HD 114710 might easily be reversed compared with that of the Sun (Donahue and Baliunas, 1992).[2]

In close correspondence to dynamo theory we develop the theory of differential rotation in a mean-field formulation starting from the conservation of angular momentum,

$$\frac{\partial}{\partial t}(\rho r^2 \sin^2 \theta \, \Omega) + \frac{\partial}{\partial x_i}(\rho r \sin \theta \, Q_{i\phi}) = 0, \tag{6.10}$$

where the Reynolds stress $Q_{i\phi}$ derived from the correlation tensor

$$Q_{ij} = \langle u_i'(\mathbf{x}, t) u_j'(\mathbf{x}, t) \rangle \tag{6.11}$$

corresponds to the turbulent EMF in mean-field electrodynamics.

The correlation tensor involves both dissipation (eddy viscosity) as well as induction (Λ-effect):

$$Q_{ij} = \Lambda_{ijk} \Omega_k - \mathcal{N}_{ijkl} \Omega_{k,l}. \tag{6.12}$$

Both effects are represented by tensors and must be computed carefully. For anisotropic and rotating turbulence the zonal fluxes of angular momentum can be written as

$$Q_{r\phi} = -\nu_\perp r \sin \theta \frac{\partial \Omega}{\partial r} - (\nu_\parallel - \nu_\perp) \sin \theta \cos \theta \left(r \cos \theta \frac{\partial \Omega}{\partial r} - \sin \theta \frac{\partial \Omega}{\partial \theta} \right)$$
$$+ \nu_T \left(V^{(0)} + \sin^2 \theta V^{(1)} \right) \Omega \sin \theta, \tag{6.13}$$

$$Q_{\theta\phi} = -\nu_\perp \sin \theta \frac{\partial \Omega}{\partial \theta} - (\nu_\parallel - \nu_\perp) \sin^2 \theta \left(\sin \theta \frac{\partial \Omega}{\partial \theta} - r \cos \theta \frac{\partial \Omega}{\partial r} \right)$$
$$+ \nu_T H^{(1)} \Omega \sin^2 \theta \cos \theta \tag{6.14}$$

(see Kitchatinov, 1986; Durney, 1989; Rüdiger, 1989). All coefficients are found to be strongly dependent on the Coriolis number

$$\Omega^* = 2\tau_{\text{corr}} \Omega \tag{6.15}$$

with τ_{corr} as the convective turnover time (Gilman, 1992; Kitchatinov and Rüdiger, 1993). Moreover, the most important terms of the Λ-effect ($V^{(0)}$, $H^{(0)}$) correspond to higher orders of the Coriolis number. The Coriolis number exceeds unity almost everywhere in the convection zone except the surface layers. That is true for all stars – in this sense all main-sequence stars are rapid rotators. Theories linear in Ω are not appropriate for stellar activity physics. The α-effect will thus never run with $\cos \theta$ in its first power. The Coriolis number Ω^* is smaller than unity at the top of the convection zone and larger than unity at its bottom. At that depth we find minimal eddy viscosities (Ω-quenching) and maximal $V^{(1)} = H^{(1)}$ (Fig. 6.4). Since the latter are known as responsible for pole-equator differences in Ω, we can state that the differential rotation is produced in the deeper layers of the convection zone where the rotation must be considered as rapid ($\Omega^* \lesssim 10$).

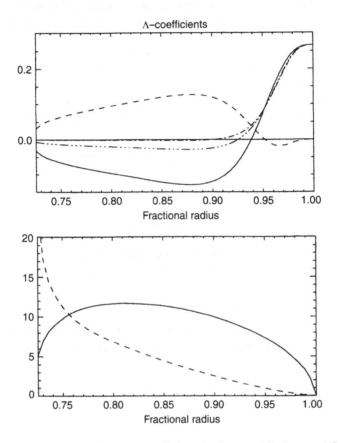

Figure 6.4 Top: The turbulence-originated coefficients in the non-diffusive zonal fluxes of angular momentum in the solar convection zone. $V^{(0)}$ (solid), $V^{(1)}$ (dashed), $V^{(0)} + V^{(1)}$ (dot-dashed). Bottom: Eddy viscosity in units of $10^{12}\,\mathrm{cm^2\,s^{-1}}$ (solid) and Coriolis number Ω^* in the solar convection zone.

The solution of (6.10) with the turbulence quantities in Fig. 6.4 is given in Küker *et al.* (1993) and Kitchatinov and Rüdiger (1995) using a mixing-length model by Stix and Skaley (1990). With a mixing-length ratio ($\alpha_{\mathrm{MLT}} = 5/3$) we find the correct equatorial acceleration of about 30%. There is a clear radial sub-rotation ($\partial\Omega/\partial r < 0$) below the poles while in mid-latitudes and below the equator the rotation is basically rigid (bottom figure of Fig. 6.3). In this way the bottom value of the pole-equator difference is reduced and the resulting profiles of the internal angular velocity are close to the observed ones, Fig. 6.5 presents the results of an extension of the theory to a sample of main-sequence stellar models given in Kitchatinov and Rüdiger (1999). Simple scalings like

$$\frac{\delta\Omega}{\Omega} \propto \Omega^\kappa \tag{6.16}$$

may be introduced for both the latitudinal as well as the radial differences of the angular velocity. The κ-exponents prove to be negative and of order -1 (bottom figure of Fig. 6.5).

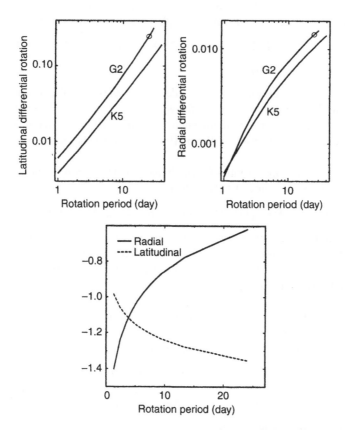

Figure 6.5 The rotation differences $\delta\Omega/\Omega$ for both a solar model and a K5 main-sequence star versus basic rotation after Kitchatinov and Rüdiger (1999). Top left: Normalized equator-pole difference of surface rotation. Top right: The normalized radial differential rotation. Bottom: The scaling exponent κ in (6.16) for the solar model. In both cases $\kappa \simeq -1$ is approximated.

So we find that for one and the same spectral type the approximation

$$\delta\Omega \simeq \text{const.} \tag{6.17}$$

should be not too rough. Recent observations of AB Dor (Donati and Cameron, 1997) and PZ Tel (Barnes *et al.*, 2000) seem to confirm this surprising and unexpected result where in all three cases the constant value approaches 0.06 day^{-1}.

6.2.3. The EMF for turbulence

We now step forward to the turbulent EMF using the same turbulence model as in Section 6.2.2. We are working with the same quasilinear approximation and the same convection zone model. The additional equation for the magnetic fluctuations is

$$\frac{\partial \mathbf{B}'}{\partial t} - \eta_t \Delta \mathbf{B}' = \nabla \times (\mathbf{u}' \times \langle \mathbf{B} \rangle), \tag{6.18}$$

where $\eta_t \simeq \nu_t$ is the small-scale diffusivity. In this configuration we deal with an α-tensor being highly anisotropic, even for the most simple case of slow rotation ($\Omega^* \ll 1$),

$$\alpha_{ij} = \gamma \varepsilon_{ijk} G_k - \alpha_1 (\mathbf{G} \cdot \mathbf{U}) \delta_{ij} - \alpha_2 (G_i \Omega_j + G_j \Omega_i) + \hat{\gamma} \varepsilon_{ijk} U_k - \hat{\alpha}_1 (\mathbf{U} \cdot \boldsymbol{\Omega}) \delta_{ij}$$
$$- \hat{\alpha}_2 (U_i \Omega_j + U_j \Omega_i) \tag{6.19}$$

(Moffatt, 1978; Krause and Rädler, 1980; Wälder *et al.*, 1980) with the stratification vectors $\mathbf{G} = \nabla \log \rho$ and $\mathbf{U} = \nabla \log u_T$ with the rms velocity $u_T = \sqrt{\langle \mathbf{u}'^2 \rangle}$. The γ-terms describe advection effects, such as turbulent diamagnetism (Rädler, 1968; Kitchatinov and Rüdiger, 1992) and buoyancy (Kitchatinov and Pipin, 1993) and cover the off-diagonal elements of the α-tensor.

The diagonal elements are essential for the induction. The mixing-length approximation used in Rüdiger and Kitchatinov (1993) leads to

$$\alpha_{rr} = \hat{\alpha} \left(\mathbf{U} + \tfrac{1}{4} \mathbf{G} \right) \cdot \boldsymbol{\Omega}, \quad \alpha_{\phi\phi} = \alpha_{\theta\theta} = -\hat{\alpha} \left(\mathbf{U} + \tfrac{3}{2} \mathbf{G} \right) \cdot \mathbf{U}. \tag{6.20}$$

While the most important component $\alpha_{\phi\phi}$ becomes positive (if density stratification dominates), the component α_{rr} becomes negative in the northern hemisphere. In contrast to the standard formulation, the α_2-components in (6.19) are dominant as confirmed by numerical simulations by Brandenburg *et al.* (1990) and Ossendrijoer *et al.* (2001). Our quasilinear theory of the α-effect provides $\hat{\alpha} = 8/15 \tau_{\mathrm{corr}}^2 u_T^2$.

The general case of an arbitrary rotation rate is not easy to present. The overshoot region, however, is of particular interest and is characterized by $|\mathbf{U}| > |\mathbf{G}|$ and $\Omega^* \gg 1$. In cylindrical coordinates (s, ϕ, z) and for very fast rotation one gets

$$\boldsymbol{\alpha} = c_\alpha \begin{bmatrix} -(3\pi/8) \cos \theta & (3\pi/8\Omega^*) \cos \theta & 0 \\ -(3\pi/8\Omega^*) \cos \theta & -(3\pi/8) \cos \theta & 0 \\ 0 & -(3\pi/8\Omega^*) \sin \theta & 0 \end{bmatrix} \frac{\mathrm{d}}{\mathrm{d}r} \left(\tau_{\mathrm{corr}} u_T^2 \right) \tag{6.21}$$

(Rüdiger and Kitchatinov, 1993). Note that

- all components with index z disappear,
- the remaining diagonal terms (the α-effect) do not vanish for very rapid rotation,
- the α-terms are negative (in the northern hemisphere) for outward increase of the turbulence intensity (like in the overshoot region),
- the advection terms tend to vanish for rapid rotation; their relation to the α-effect is given by the Coriolis number Ω^*.

It might seem that the α-effect is now in our hands. There are, however, shortcomings beyond the use of the quasilinear approximation. The main point is the absence of terms higher than first order in $\mathbf{G}$ or $\mathbf{U}$, respectively. In particular, all terms with $(\mathbf{G} \cdot \mathbf{U})^3 \sim \cos^3 \theta$ are missing so that we have no information about the α-effect at the poles where for slow rotation $\alpha_{\phi\phi}$ is largest. However, this need not be true for 'rapid' rotation. Simulations of $\alpha_{\phi\phi}$ with its latitudinal profile by Brandenburg (1994b) indeed showed that the maximum occurs at mid-latitudes (Fig. 6.6). Schmitt (1985, 1987) derived a similar profile for a different type of instability.

A new discussion of the α-effect in shear flows recently emerged (cf. Brandenburg and Schmitt, 1998). A quasilinear computation of the influence of the differential rotation on the α-effect leads to

$$\alpha \simeq -l_{\mathrm{corr}}^2 \frac{\partial \Omega}{\partial \theta} \frac{\mathrm{d} \log(\rho u_T)}{\mathrm{d}r} \sin \theta \tag{6.22}$$

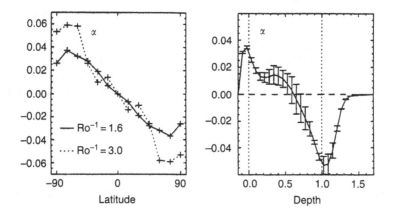

Figure 6.6 Simulation of the α-effect after Brandenburg (1994b) versus latitude and depth. Note the negative values at the bottom of the convection zone (right) and the maximum for high latitudes (left).

for a sphere or

$$\alpha \simeq -l_{\mathrm{corr}}^2 s \frac{d\Omega}{ds} \frac{d\log\rho}{dz} \qquad (6.23)$$

for a disk (Rüdiger and Pipin, 2000). The latter relation for accretion disks with $\partial\Omega/\partial s < 0$ yields negative α-values in the northern hemisphere and positive values in the southern hemisphere (see Brandenburg and Donner, 1997). According to (6.21) or (6.22), negative values for the northern $\alpha_{\phi\phi}$ in the solar overshoot region are only achievable with a very steep decrease of the turbulence intensity with depth.

There are very interesting observational issues concerning the current helicity

$$\mathcal{H}_{\mathrm{curr}} = \langle \mathbf{j}' \cdot \mathbf{B}' \rangle \qquad (6.24)$$

with the electric current $\mathbf{j} = \nabla \times \mathbf{B}/\mu_0$. The current helicity has the same kind of equatorial (anti-) symmetry as the dynamo-α. For homogeneous global magnetic fields, the dynamo-α is related to the turbulent EMF according to $\langle B_\phi \rangle^2 \alpha_{ij} \langle B_i \rangle \langle B_j \rangle = \mathcal{E} \cdot \langle \mathbf{B} \rangle$. After Rädler and Seehafer (1990) we read this equation as $\alpha_{\phi\phi} = \mathcal{E} \cdot \langle \mathbf{B} \rangle / \langle B_\phi \rangle^2$, where $\alpha_{\phi\phi}$ is the dominant component of the α-tensor. We are, in particular, interested in checking their and Keinigs' (1983) antiphase relation,

$$\frac{\alpha_{\phi\phi} \langle B \rangle^2}{\mu_0 \mathcal{H}_{\mathrm{curr}}} = -\eta < 0, \qquad (6.25)$$

between α-effect and current helicity. In addition, we have recently shown that indeed the negativity of the left-hand side of (6.25) is preserved for magnetic-driven turbulence fields which do not fulfill the conditions for which the Keinigs relation (6.25) originally has been derived (Rüdiger *et al.*, 2001).

An increasing number of papers present observations of the current helicity at the solar surface, all showing that it is negative in the northern hemisphere and positive in the southern hemisphere (Hale, 1927; Seehafer, 1990; Pevtsov *et al.*, 1995; Abramenko *et al.*, 1996; Bao and Zhang, 1998; see Low (1996) for a review). There is thus a strong empirical evidence that the α-effect is positive in the bulk of the solar convection zone in the northern hemisphere.

For a given field of magnetic fluctuations the dynamo-α as well as the kinetic and current helicities, have been computed by Rüdiger *et al.* (2001) assuming that the turbulence is subject to magnetic buoyancy and global rotation. In particular, the role of magnetic buoyancy appears quite important for the generation of α-effect. So far, only the role of density stratification has been discussed for both α-effect and Λ-effect, and their relation to kinetic helicity and turbulence anisotropy. If density fluctuations shall be included then it makes no sense here to adopt the anelastic approximation $\nabla \cdot (\rho \mathbf{u}') = 0$ but one has to work with the mass conservation law in the form

$$\frac{\partial \rho'}{\partial t} + \rho \nabla \cdot \mathbf{u}' = 0. \tag{6.26}$$

For the turbulent energy equation one can simply adopt the polytropic relation $p' = c_{ac}^2 \rho'$, where c_{ac} is the isothermal speed of sound. The turbulence is thus assumed to be driven by the Lorentz force and it is subject to a global rotation. The result is an α-effect of the form

$$\alpha_{\phi\phi} = -\frac{1}{5} \frac{\tau_{corr}^2}{c_{ac}^2} \frac{\langle B^{(0)^2} \rangle}{\mu_0 \rho} (\mathbf{g} \cdot \mathbf{\Omega}). \tag{6.27}$$

Here $\mathbf{g}$ is the gravitional acceleration, and the energy of the magnetic fluctuations forcing the turbulence is also given. For rigid rotation the α-effect proves thus to be positive in the northern hemisphere and negative in the southern hemisphere. Again we do not find any possibility to present a negative α-effect in the northern hemisphere.

The kinetic helicity

$$\mathcal{H}_{kin} = \langle \mathbf{u}' \cdot \nabla \times \mathbf{u}' \rangle \tag{6.28}$$

has just the same latitudinal distribution as the α-effect, but the magnetic model does not provide the minus sign between the dynamo-α and the kinetic helicity which is characteristic within the conventional framework, $\alpha \propto -\tau_{corr} \mathcal{H}_{kin}$. If a rising eddy can expand in density-reduced surroundings, then a negative value of the kinetic helicity is expected. The magnetic-buoyancy model, however, leads to another result.

If the real convection zone turbulence is formed by a mixture of both dynamic-driven and magnetic-driven turbulence, the kinetic helicity should also be a mixture of positive parts and negative parts so that it should be a rather small quantity. The opposite is true for the current helicity which also changes its sign at the equator. Again it turns out to be negative in the northern hemisphere and positive in the southern hemisphere. The current helicity (due to fluctuations) and the α-effect are thus always out of phase. The current helicity at the solar surface might thus be considered a much more robust observational feature than the kinetic helicity (6.28).

Strong magnetic fields are known to suppress turbulence and the α-effect is also expected to decrease. The field strength at which the α-effect reduces significantly is assumed, in energy, to be comparable with the kinetic energy of the velocity fluctuations, and is called the equipartition field strength $B_{eq} = (\mu_0 \rho u_T^2)^{1/2}$. The magnetic feedback on the turbulence was studied by Rüdiger and Kitchatinov (1993) using a δ-function spectral distribution of velocity fluctuations. The $\alpha_{\phi\phi}$-component of the α-tensor acting in an $\alpha\Omega$-dynamo is found to be suppressed by the magnetic field with the α-quenching function

$$\Psi = \frac{15}{32\beta^4} \left(1 - \frac{4\beta^2}{3(1+\beta^2)^2} - \frac{1-\beta^2}{\beta} \arctan \beta \right) \tag{6.29}$$

with

$$\beta = |\langle B \rangle| / B_{\text{eq}}. \tag{6.30}$$

The quenching function decreases as β^{-3} for strong magnetic fields. The function is similar to the heuristic approach $(1 + \beta^2)^{-1}$ which is artificial but nevertheless frequently used in α-effect dynamos. For small fields, Ψ runs like $1 - (12/7)\beta^2$, the mentioned heuristic approximation runs like $1 - \beta^2$.

We can also compute the eddy diffusivity using the same turbulence theory as in the previous section. The simplest expression is

$$\eta_{\text{T}} = c_\eta \tau_{\text{corr}} u_{\text{T}}^2 \tag{6.31}$$

with $c_\eta \lesssim 0.3$. This expression, however, fails to include rotation and stratification. Even a non-uniform but weak magnetic field induces a turbulent EMF, without rotation and stratification in accordance to $\eta_{ijk} = \eta_{\text{T}} \varepsilon_{ijk}$. A much stronger magnetic quenching of such an eddy diffusivity has been discussed by Vainshtein and Cattaneo (1992) and Brandenburg (1994a) and is still a matter of debate. A more conventional theory of the η_{T}-quenching and its consequences for the stellar activity cycle theory is given in Section 6.6.

6.3. A boundary-layer dynamo for the Sun

The spatial location of the dynamo action is unknown unless helioseismology will reveal the exact position of the magnetic toroidal belts beneath the solar surface (cf. Dziembowski and Goode, 1991). There are, however, one or two arguments in favour of locating it deep within or below the convection zone:

- Hale's law of sunspot parities can only be fulfilled if the toroidal magnetic field belts are very strong (10^5 G, see Moreno-Insertis, 1983; Choudhuri, 1989; Fan et al., 1993; Caligari et al., 1995).
- The dynamo field strength is approximately the equipartition value $B_{\text{eq}} = \left(\mu_0 \rho u_{\text{T}}^2 \right)^{1/2}$. Using mixing-length theory arguments $\rho u_{\text{T}}^3 \simeq$ const., hence B_{eq} is increasing inwards with $\rho^{1/6}$ (one order of magnitude, $\lesssim 10^4$ G).
- The radial gradient of Ω is maximal below the convection zone.

As a consequence of these arguments high field amplitudes might be generated only in the layer between the convection zone and the radiative interior (Boundary layer (BL), 'tachocline', see van Ballegooijen, 1982). On the other hand, if there is some form of turbulence in this layer, it will be hard to understand the present-day finite value of lithium in the solar convection zone. The lithium burning only starts 40,000 km below the bottom of the convection zone. Any turbulence in this domain would lead to a rapid and complete depletion of the lithium in the convection zone, which is not observed. Moreover, as shown by Rüdiger

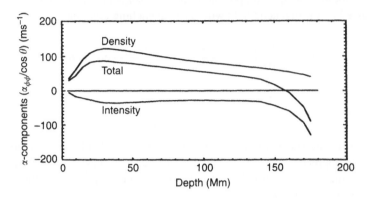

Figure 6.7 The depth-dependence of the α-effect with $c_\alpha = 1$. Note the different contributions by density gradient and turbulence intensity distribution. The α-effect becomes negative only at the bottom of the convection zone (Krivodubskij and Schultz, 1993).

and Kitchatinov (1997), it is also not easy to generate very strong toroidal magnetic fields in the solar tachocline as the Maxwell stress tends to deform the rotation profile to more and more smooth functions. In our computation with molecular diffusion coefficient values the toroidal field amplitude never did exceed 1 kG, independent of the radial magnetic field applied.

Krivodubskij and Schultz (1993) (with the inclusion of the depth-dependence of the Coriolis number Ω^* and using a mixing-length model) derived a profile of the α-effect (Fig. 6.7) with a magnitude of $100\,\mathrm{m\,s^{-1}}$, positive in the convection zone and negative in the overshoot layer. The negativity of α there can only be relevant for the dynamo if the bulk of the convection zone is free from α. It has been argued that the short rise-times of the flux tubes in the convection zone prevent the formation of the α-effect (Spiegel and Weiss, 1980; Schüssler, 1987; Stix, 1991). However, the rise-times may be much longer in the overshoot region (Ferriz-Mas and Schüssler, 1993, 1995; van Ballegooijen, 1998). Anyway, as $c_\alpha = 1$ is used in Fig. 6.7, the given values can only be considered as maximal values.

There is a serious shortcoming of the BL concept. The characteristic scales of the magnetic fields are no longer much larger than the scales of the turbulence. The validity of the local formulations of the mean-field electrodynamics is not ensured for such thin layers. Nevertheless, there is an increasing number of corresponding quantitative models (Choudhuri, 1990; Belvedere *et al.*, 1991; Prautzsch, 1993; Markiel and Thomas, 1999). Even in this case a couple of new questions will arise. For a demonstration of this puzzling situation a model has been established by Rüdiger and Brandenburg (1995) under the following assumptions:

- Intermittency (or the weakness of the turbulence in the BL) is introduced by a dilution factor ϵ in $\mathcal{E}_i = \epsilon(\alpha_{ij}\langle B_j\rangle + \eta_{ijk}\langle B_j\rangle_{,k})$. The factor ϵ controls the weight of the turbulent EMF in relation to the differential rotation.
- The ignorance of the α-effect in the polar regions is parameterized with α_u with $\alpha \sim (1 - \alpha_u \cos^2\theta)\cos\theta$.
- α exists only in the overshoot region, η_T only in the convection zone.
- The correlation time is taken from $\tau_{\mathrm{corr}} \simeq D/u_T \simeq 10^6\,\mathrm{s}$ so that $\Omega^* \simeq 5$ results.
- The tensors α and η_T are computed for a rms velocity profile by Stix (1991), the rotation law is directly taken from Christensen-Dalsgaard and Schou (1988).

The main results from these models are:

- The cycle period has just the same sensitivity to the BL's thickness as the shell dynamo, i.e. $\tau_{cyc}[\text{yr}] \sim D/\epsilon$ for D in Mm ($D \simeq 15 \cdots 35$ Mm).
- Due to the rotational η_T-quenching the linear model yields the correct cycle time. Generally the 22-year cycle can only exist for a dilution of $\epsilon \simeq 0.5$ (cf. Fig. 6.8).
- For $\alpha_u \simeq 0$ the magnetic activity is concentrated near the poles, for $\alpha_u \simeq 1$ it moves to the equator.
- For too thin BL's there are too many toroidal magnetic belts in each hemisphere (Fig. 6.9).

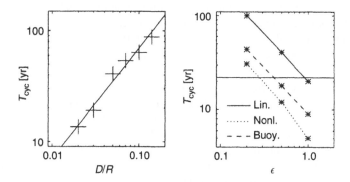

Figure 6.8 Cycle period τ_{cyc} as a function of the BL thickness D/R and and the dilution factor ϵ for $\alpha_u = 0$. Left: $\epsilon = 0.5$, the results are well represented by $\tau_{cyc} \propto D$. Right: $D = 35$ Mm. The horizontal line indicates the solar cycle period of 22 years.

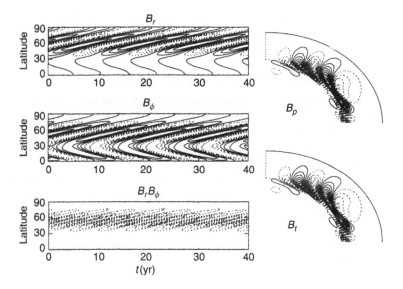

Figure 6.9 The magnetic field geometry for the nonlinear (α-quenched) solution without buoyancy and $\epsilon = 0.5$, $D = 35$ Mm, $\alpha_u = 1$. An obvious problem of the BL-dynamo is the large number of toroidal field belts.

The dependence of the cycle period on ϵ is shown in Fig. 6.8. For $\epsilon \approx 1$ the linear solutions have the correct 22-year solar cycle period (right panel). The cycle period increases linearly with D:

$$\tau_{\text{cyc}} \propto D, \tag{6.32}$$

(see left panel of Fig. 6.8), which is in agreement with the linear relation (6.9) for spherical shell dynamos. For too thin boundary layers the cycle time will thus become very short. Nonlinearity further shortens the cycle period and this effect has to be compensated for by taking a smaller value of ϵ, depending on whether or not magnetic buoyancy is included. In those cases where α-quenching and magnetic buoyancy are included, the rms-values of the poloidal and toroidal fields are 11 G and 5 kG. Poloidal and toroidal fields as well as their mean and fluctuating parts are difficult to disentangle observationally. A toroidal field of 5 kG is, however, consistent with the observed total flux of 10^{24} Mx, distributed over 50 Mm in depth and 400 Mm in latitude. On the other hand, a poloidal field of 11 G is larger than the observed value.

In all cases the magnetic field exceeds the equipartition value which is here 6 kG. This is partly due to the turbulent diamagnetism leading to an accumulation of magnetic fields at the bottom of the convection zone, and partly due to too optimistic estimates for α-quenching.

The dilution factor ϵ is the main free parameter that must be chosen such that the 22-year magnetic cycle period is obtained. The thickness of the overshoot layer must be chosen to match the correct number of toroidal field belts of the Sun. Our results suggest that the thickness should not be much smaller than $D \approx 35\,\text{Mm} \approx H_p/2$.

Nonlinear effects typically lead to a reduction of the magnetic cycle period. Unfortunately, there are no consistent quenching expressions that are valid for rapid rotation. Magnetic buoyancy also provides a nonlinear feedback (e.g. Durney and Robinson, 1982; Moss *et al.*, 1990), but computations by Kitchatinov and Pipin (1993) indicate that this only leads to a mean field advection of a few m s^{-1}, i.e. less than the advection due to the diamagnetic effect. However, magnetic buoyancy is important in the upper layers (where the diamagnetic effect is weak), and may have significant effects on the field geometry and the cycle period.

6.4. Distributed dynamos with meridional circulation

Not only differential rotation but also meridional flow u^m will influence the mean-field dynamo. This influence can be expected to be only a small modification if and only if its characteristic time scale τ_{drift} exceeds the cycle time τ_{cyc} of about 11 years. With $\tau_{\text{drift}} \simeq R/u^m$ we find $u^m \simeq 2\,\text{m s}^{-1}$ as a critical value. If the flow is faster (Fig. 6.10) the modification can be drastic. In particular, if the flow counteracts the diffusion wave, i.e. if the drift at the bottom of the convection zone is polewards, the dynamo might come into trouble. The behaviour of such a solar-type dynamo (with superrotation at the bottom of the convection zone and perfect conduction beneath the convection zone) is demonstrated in the present section.

In the following the dynamo equations are given with the inclusion of the induction by meridional circulation. For axisymmetry the mean flow in spherical coordinates is given by $\mathbf{u} = (u_r, u_\theta, r \sin\theta \Omega)$.

With a similar notation the magnetic field is

$$\mathbf{B} = \left(\frac{1}{r^2 \sin\theta} \frac{\partial A}{\partial \theta}, \; -\frac{1}{r \sin\theta} \frac{\partial A}{\partial r}, \; B \right) \tag{6.33}$$

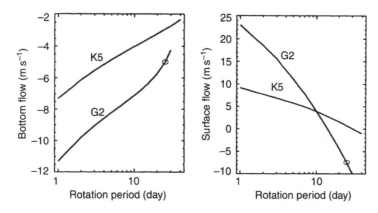

Figure 6.10 The meridional circulation at mid-latitudes (45°) at the bottom (left) and the top (right) of the convection zone for both a solar model and a K5 main-sequence star as a function of rotation period (Kitchatinov and Rüdiger, 1999). The circle represents the solar case. The flow at the bottom is equatorwards (as it is the flow at the surface, there are two cells in radius).

with A as the poloidal-field potential and B as the toroidal field. Their evolution is described by

$$\frac{\partial A}{\partial t} + (\mathbf{u} \cdot \nabla)A = \alpha s B + \eta_T \left[\frac{\partial^2 A}{\partial r^2} + \frac{\sin\theta}{r^2} \frac{\partial}{\partial\theta} \left(\frac{1}{\sin\theta} \frac{\partial A}{\partial\theta} \right) \right], \tag{6.34}$$

$$\frac{\partial B}{\partial t} + s\rho(\mathbf{u} \cdot \nabla)\frac{B}{s\rho} = \frac{1}{r} \left(\frac{\partial\Omega}{\partial r}\frac{\partial A}{\partial\theta} - \frac{\partial\Omega}{\partial\theta}\frac{\partial A}{\partial r} \right) - \frac{1}{s}\frac{\partial}{\partial r}\left(\alpha\frac{\partial A}{\partial r} \right) - \frac{1}{r^3}\frac{\partial}{\partial\theta}\left(\frac{\alpha}{\sin\theta}\frac{\partial A}{\partial\theta} \right)$$

$$+ \frac{\eta_T}{s} \left[\frac{\partial^2(sB)}{\partial r^2} + \frac{\sin\theta}{r}\frac{\partial}{\partial\theta}\left(\frac{1}{s}\frac{\partial(sB)}{\partial\theta} \right) \right] \tag{6.35}$$

for uniform η_T and with $s = r\sin\theta$ (cf. Choudhuri *et al.*, 1995). In order to produce here a solar-type butterfly diagram with $\mathbf{u} = 0$ the α-effect must be taken as negative in the northern hemisphere (Steenbeck and Krause, 1969; Parker, 1987). Our rotation law and the radial meridional flow profile at 45° are given in Fig. 6.11. A one-cell flow pattern is adopted in each hemisphere. The u^m approximates the latitudinal drift close to the convection zone bottom, the circulation is counterclockwise (in the first quadrant) for positive u^m, i.e. towards the equator at the bottom of the convection zone. Note that the cycle period becomes shorter for clockwise flow (Roberts and Stix, 1972) and it becomes longer for the more realistic counterclockwise flow. The latter forms a poleward flow at the surface in agreement with observations and theory (Fig. 6.10). For too strong meridional circulation ($u^m \geq 50\,\mathrm{m\,s^{-1}}$) our dynamo stops operation. The same happens already for $u^m \approx -10\,\mathrm{m\,s^{-1}}$ if the flow at the bottom is opposite to the magnetic drift wave (Fig. 6.12).

However, it is hard to explain why the convection zone itself is producing the negative α-values used in the model. More easy to understand are the positive values adopted in Fig. 6.13 (see Durney, 1995, 1996). Of course, now without circulation the butterfly diagram becomes wrong, with a poleward migration of the toroidal magnetic belts. One can also understand

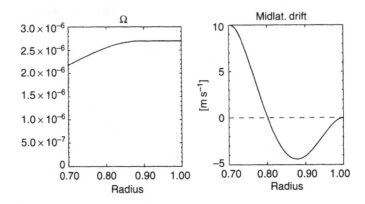

Figure 6.11 Rotation law and the pattern of a meridional flow at 45° for the Steenbeck–Krause dynamo model in Fig. 6.12. The meridional circulation is prescribed (Rüdiger, 1989) and not yet a result of differential rotation theory. The rotation profile always has a positive slope.

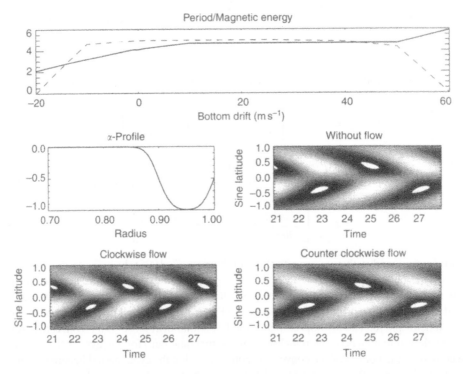

Figure 6.12 Operation of a Steenbeck–Krause dynamo with negative α-effect (in northern hemisphere) with and without meridional circulation. Eddy diffusivity is $5 \times 10^{12} \, \mathrm{cm^2 \, s^{-1}}$. Positive bottom drift means counterclockwise flow (equatorwards at the bottom of the convection zone, polewards at the surface). Top: The cycle period (solid, in years) and the magnetic energy (dashed) versus the latitudinal drift at the bottom of the convection zone. Middle: α-profile in radius and butterfly diagram of the circulation-free dynamo. The maximal toroidal field is plotted versus the time in years. Bottom: Butterfly diagrams for clockwise and counterclockwise circulations.

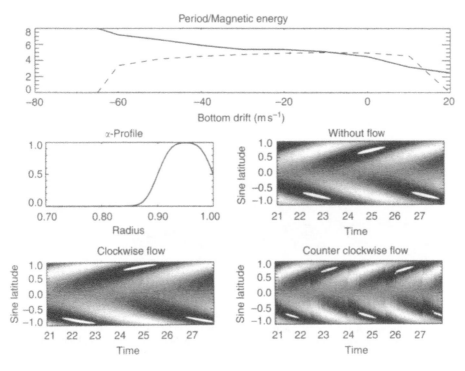

Figure 6.13 The same as in Fig. 6.12 but for positive α-effect (in northern hemisphere). Note the decay of the dynamo already for $10 \cdots 20 \text{ m s}^{-1}$ (equatorwards at the bottom, polewards at the top of the convection zone).

that the counterclockwise flow (equatorwards at the convection zone bottom) rapidly destroys the dynamo operation, here already for 10 m s^{-1} amplitude. This amplitude is so small that it cannot be a surprise that such a dynamo might not be realized in the Sun. For the eddy diffusivity $5 \times 10^{12} \text{ cm}^2 \text{ s}^{-1}$ is used here, so the magnetic Reynolds number $R_m = u^m R / \eta_T$ is

$$R_m = 14 \frac{u^m}{10 \text{ m s}^{-1}}, \qquad (6.36)$$

i.e. $R_m = 14$ for $u^m = 10 \text{ m s}^{-1}$. In their paper Choudhuri *et al.* (1995) apply Reynolds numbers of order 500 while Dikpati and Charbonneau (1999) even take $R_m = 1400$. Indeed, their dynamos are far beyond the critical values of $R_m \simeq 10$ which are allowed for the weak-circulation $\alpha\Omega$-dynamo. Their dynamos exhibit the same period reduction for increasing u^m shown in Fig. 6.13 but the butterfly diagram is opposite. It seems that quite another branch of dynamos would exist for $R_m \gg 10$. This is indeed true.

The butterfly diagram of such a dynamo with positive α-effect (3 m s^{-1}) and a magnetic Reynolds number of 450 is given in Fig. 6.14. It works with low eddy diffusivity of $10^{11} \text{ cm}^2 \text{ s}^{-1}$ and the flow pattern of Fig. 6.11. In the lower panel the phase relation of poloidal and toroidal magnetic fields is shown. In lower latitudes the observed phase lag between the field components is obtained. In Table 6.1 the cycle periods in years are given for positive-α dynamos. Bold numbers indicate the solar-type butterfly diagram. For fast meridional flow

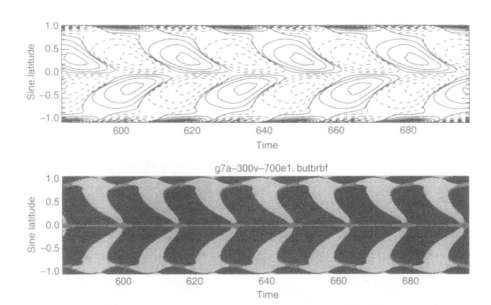

Figure 6.14 Top: Butterfly diagram of a marginal dynamo with *positive* α-effect (3 m s^{-1}), with eddy diffusivity of 10^{11} cm^2 s^{-1} and a bottom flow of 7 m s^{-1}. Bottom: $\langle B_r \rangle \langle B_\phi \rangle$, black colour indicates negative sign, i.e. the observed phase lay between poloidal and toroidal field.

Table 6.1 Cycle periods in years of solar-type-butterfly dynamos (bold) with positive α-effect ($=3$ m s^{-1}) and meridional flow of u^m at the bottom of the convection zone, the eddy diffusivity given in the headline

u^m [m s^{-1}]	1×10^{11} cm^2 s^{-1}	2×10^{11} cm^2 s^{-1}	5×10^{11} cm^2 s^{-1}
1	−120	−70	−36
3	**60**	−45	−24
5	**45**	**40**	Decay
7	**31**	**26**	Decay
11	**20**	**20**	Decay

Signature '−' indicates polewards migration, 'decay' means decaying fields. Note the good representation of the η-dependence of the cycle time of (6.9) in the first line of the table.

and small eddy diffusivity we indeed find oscillating dynamos with $\omega_{\text{cyc}} \propto u^m$. The eddy diffusivity is the same as that known from the sunspot decay, i.e. 10^{11} cm^2 s^{-1} (see Brandenburg, 1993).

6.5. The Maunder minimum

The period of the solar cycle and its amplitude are far from constant. The most prominent activity drop was the Maunder minimum between 1670 and 1715 (Spörer, 1887). The variability of the cycle period can be expressed by the 'quality', $\omega_{\text{cyc}}/\Delta\omega_{\text{cyc}}$, which is as low as 5 for the Sun (Hoyng, 1993, cf. Fig. 6.2).

Measurements of ^{14}C abundances in sediments and long-lived trees provide much longer time series than sunspot counts. In agreement with Schwarz (1994), Voss *et al.* (1996, 1997) found a secular periodicity of 80–90 years as well as a long-duration period of about 210 years. The measurements of atmospheric ^{14}C abundances by Hood and Jirikowic (1990) suggested a periodicity of 2400 years which is also associated with a long-term variation of solar activity. The variety of frequencies found in solar activity may even indicate chaotic behaviour as discussed by Rozelot (1995), Kurths *et al.* (1993, 1997) and Knobloch and Landsberg (1996).

The activity cycle of the Sun is not exceptional: the observation of chromospheric Ca-emission of solar-type stars yields activity periods between 3 and 20 years (Noyes *et al.*, 1984; Baliunas and Vaughan, 1985; Saar and Baliunas, 1992a,b). A few of these stars do not show any significant activity. This suggests that even the existence of the grand minima is a typical property of cool main-sequence stars like the Sun. From ROSAT X-ray data Hempelmann *et al.* (1996) find that up to 70% of the stars with a constant level of activity exhibit a rather low level of coronal X-ray emission. HD 142373 with its X-ray luminosity of only log $F_X = 3.8$ is a typical candidate. We conclude that during a grand minimum not only the magnetic field in the activity belts is weaker than usual but the total magnetic field energy is also reduced.

The short-term cycle period appears to decrease at the end of a grand minimum according to a wavelet analysis of sunspot data by Frick *et al.* (1997). There is even empirical evidence for a very weak but persistent cycle throughout the solar Maunder minimum as found by Wittmann (1978) and recently Beer *et al.* (1998). The latitudinal distribution of the few sunspots observed during the Maunder minimum was highly asymmetric (Spörer, 1887; Ribes and Nesme-Ribes, 1993; Nesme-Ribes *et al.*, 1994). Short-term deviations from the north–south symmetry in regular solar activity are readily observable (Verma, 1993), yet a 30-year period of asymmetry in sunspot positions as seen during the Maunder minimum remains a unique property of grand minima and should be associated with a parity change of the internal magnetic fields (see, however, Pulkkinen *et al.*, 1999).

6.5.1. *Mean-field magnetohydrodynamics*

The explanation of grand minima in the magnetic activity cycle by a dynamo action has been approached by two concepts. The first one considers the stochastic character of the turbulence and studies its consequences for the variations of the α-effect and all related phenomena with time (see Section 6.8). The alternative concept includes the magnetic feedback to the internal solar rotation (Weiss *et al.*, 1984; Jennings and Weiss, 1991). Kitchatinov *et al.* (1994a) and Tobias (1996, 1997) even introduced the conservation law of angular momentum in the convection zone including magnetic feedback in order to simulate the intermittency of the dynamo cycle.

A theory of differential rotation based on the Λ-effect concept is coupled with the induction equation in a spherical two-dimensional mean-field model. The mean-field equations for the convection zone include the effects of diffusion, α-effect, toroidal field production by differential rotation and the Lorentz force. They are

$$\frac{\partial A}{\partial t} = \eta_T \frac{\partial^2 A}{\partial r^2} + \eta_T \frac{\sin\theta}{r^2} \frac{\partial}{\partial \theta} \left(\frac{1}{\sin\theta} \frac{\partial A}{\partial \theta} \right) + \alpha r \sin\theta B, \tag{6.37}$$

$$\frac{\partial B}{\partial t} = \frac{1}{r}\frac{\partial}{\partial r}\left(\eta_T\frac{\partial(Br)}{\partial r}\right) + \frac{\eta_T}{r^2}\frac{\partial}{\partial\theta}\left(\frac{1}{\sin\theta}\frac{\partial(B\sin\theta)}{\partial\theta}\right)$$

$$+ \frac{1}{r}\frac{\partial\Omega}{\partial r}\frac{\partial A}{\partial\theta} - \frac{1}{r}\frac{\partial\Omega}{\partial\theta}\frac{\partial A}{\partial r} - \frac{1}{r\sin\theta}\frac{\partial}{\partial r}\left(\alpha\frac{\partial A}{\partial r}\right) - \frac{1}{r^3}\frac{\partial}{\partial\theta}\left(\frac{\alpha}{\sin\theta}\frac{\partial A}{\partial\theta}\right), \tag{6.38}$$

$$\rho r \sin\theta\frac{\partial\Omega}{\partial t} = -\frac{1}{r^3}\frac{\partial}{\partial r}\left(r^3\rho Q_{r\phi}\right) - \frac{1}{r\sin^2\theta}\frac{\partial}{\partial\theta}\left(\sin^2\theta\rho Q_{\theta\phi}\right)$$

$$+ \frac{1}{\mu_0 r^2 \sin\theta}\left(\frac{1}{r}\frac{\partial A}{\partial\theta}\frac{\partial(Br)}{\partial r} - \frac{1}{\sin\theta}\frac{\partial A}{\partial r}\frac{\partial(B\sin\theta)}{\partial\theta}\right). \tag{6.39}$$

Since the importance of the large-scale Lorentz force in the momentum equation was discussed by Malkus and Proctor (1975), we call this term 'Malkus–Proctor' effect.

The computational domain is a spherical shell covering the outer parts of the Sun down to a fractional solar radius, $x = r/R$, of 0.5. The convection zone extends from $x = 0.7$ to $x = 1$. The α-effect works only in the lower part from $x = 0.7$ to $x = 0.8$ while turbulent diffusion of the magnetic field, turbulent viscosity, and the Λ-effect are present in the entire convection zone. Below $x = 0.7$ both the magnetic diffusivity and the viscosity are two orders of magnitude smaller than in the convection zone. The boundary conditions are specified as $Q_{r\phi} = \partial A/\partial r = B = 0$ at $r = R$, and $Q_{r\phi} = A = B = 0$ at the inner boundary. The angular momentum flux is given by

$$Q_{r\phi} = \nu_T \sin\theta\left(-r\frac{\partial\Omega}{\partial r} + V^{(0)}\Omega\right), \quad Q_{\theta\phi} = -\nu_T \sin\theta\frac{\partial\Omega}{\partial\theta}. \tag{6.40}$$

$V^{(0)}$ determines the radial rotation law without magnetic field.

Five dimensionless numbers define the model, namely the magnetic Reynolds numbers of both the differential rotation and the α-effect, $C_\Omega = \Omega_0 R^2/\eta_T$, $C_\alpha = \alpha_0 R/\eta_T$; the magnetic Prandtl number $P_m = \nu_T/\eta_T$; the Elsasser number

$$E = \frac{B_{eq}^2}{\mu_0 \rho \eta_T \Omega_0}; \tag{6.41}$$

and the Λ-effect amplitude, $V^{(0)}$. In the α-effect,

$$\alpha = \alpha_0 \cos\theta \sin^2\theta, \tag{6.42}$$

the factor $\sin^2\theta$ has been introduced to restrict magnetic activity to low latitudes and $\alpha_0 \simeq l_{corr}\Omega_0$, so that $C_\Omega/|C_\alpha| \simeq R/l_{corr}$, hence in general C_Ω exceeds C_α ($\alpha\Omega$ dynamo). Our dynamo works with $C_\alpha = -10$ and $C_\Omega = 10^5$. $V^{(0)}$ is positive in order to produce the required super-rotation, and its amplitude is 0.37. With the eddy diffusivity (6.31) the Elsasser number reads $E = 2/c_\eta\Omega^*$ and is set to unity here. (See also Küker *et al.*, 1999; Pipin, 1999.)

6.5.2. Results of the interplay of dynamo and angular velocity

Figures 6.15 and 6.16 demonstrate the action of different effects and show the variation of the toroidal magnetic field at a fixed point ($x = 0.75, \theta = 30°$), the total magnetic energy, the variation of the cycle period, and the parity $P = (E_S - E_A)/(E_S + E_A)$, derived from

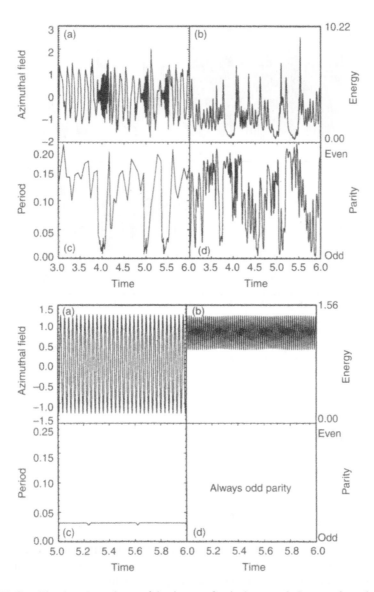

Figure 6.15 Top: The time dependence of the dynamo for the large-scale Lorentz-force feedback only ($P_m = 0.1$, $E = 1$). (a): Toroidal magnetic field and (b): magnetic energy. (c): Cycle time and (d): magnetic parity. Bottom: The same with α-quenching but still without Λ-quenching ($\lambda = 0$).

the decomposition of the magnetic energy into symmetric and antisymmetric components (Brandenburg *et al.*, 1989). All times and periods are given in units of a diffusion time R^2/η_T. Field strengths are measured in units of B_{eq}.

If the large-scale Lorentz force (Malkus–Proctor effect) is the only feedback on rotation, the time series given in the upper panel of Fig. 6.15 may be compared with the results in Tobias (1996) although a number of assumptions are different between Tobias' Cartesian

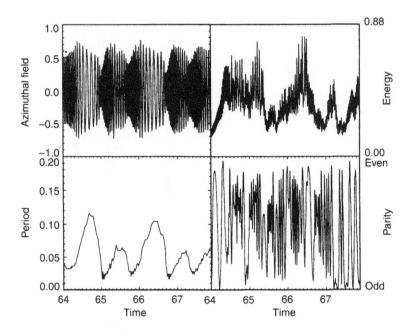

Figure 6.16 The same as in Fig. 6.15 but for strong Λ-quenching ($\lambda = 25$). Top: Toroidal magnetic field (left) and magnetic energy (right). Bottom: Cycle time (left) and magnetic parity (right).

approach and our spherical model. Fig. 6.15 shows a quasi-periodic behaviour with activity interruptions like grand minima. This model, however, neglects the feedback of strong magnetic fields on the α-effect and the differential rotation.

The lower panel of Fig. 6.15 shows the result of the same model but with a local α-quenching,

$$\alpha \propto \frac{1}{1 + (B_{tot}/B_{eq})^2},\tag{6.43}$$

where B_{tot} is the absolute value of the magnetic field. The variability of the cycles turns into a solution with only one period. Similar to the suppression of dynamo action, a quenching of the Λ-effect causing the differential rotation is according to

$$V^{(0)} \propto \frac{1}{1 + \lambda(B_{tot}/B_{eq})^2}.\tag{6.44}$$

If λ is near unity, the maximum field strength and total magnetic energy decrease slightly, but the periodic behaviour remains the same, i.e. the effect of the Λ-quenching is too small to alter the differential rotation significantly. However, an increase of λ leads to grand minima – an example for $\lambda = 25$ is given in Fig. 6.16. Minima in cycle period occur shortly after a grand activity minimum in agreement with the analysis of sunspot data by Frick *et al.* (1997a,b). The amplitude of the period fluctuations in Fig. 6.16 is much lower than in the Malkus–Proctor model but is still stronger than that observed. The magnetic Prandtl number

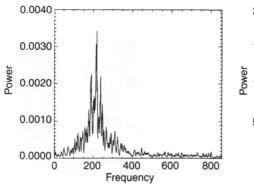 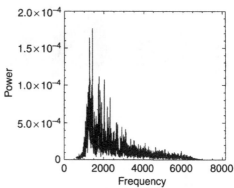

Figure 6.17 Left: Power spectrum of the magnetic-field amplitude variations for the Malkus–Proctor model of Fig. 6.15. The frequency is given in arbitrary units. Right: The same but for the model with strong Λ-quenching. The highest peaks are the main cycle frequency, whereas the difference between two peak neighbours indicates the secular cycle.

used for the solutions in Figs. 6.15 and 6.16 was $P_m = 0.1$ and it is noteworthy that grand minima do not appear for $P_m = 1$.

The strong variations of the parity between symmetric and antisymmetric states appearing in the nonperiodic solutions can be explained by the numerous (usually $5 \cdots 6$) magnetic field belts migrating towards the equator. Slight shifts of this belt-structure against the equator result in strong variations in the parity. Averaged over time, dipolar and quadrupolar components of the fields have roughly the same strength; all periodic solutions have strict dipolar structure.

Spectra of long-time series of the toroidal magnetic field are given in Fig. 6.17 for both the Malkus–Proctor model and the strong Λ-quenching model. The long-term variations of the field will be represented by a set of close frequencies whose difference is the frequency of the grand minima. The Malkus–Proctor model shows a number of lines close to the main cycle frequency. The difference between the two highest peaks can be interpreted as the occurrence rate of grand minima. However, the shape of the spectrum indicates that the magnetic field appears rather irregularly. The spectrum of the model with all feedback terms and strong Λ-quenching is also given in Fig. 6.17 and shows a similar behaviour with highest amplitudes near the main cycle frequency of the magnetic field. The average frequency of the grand minima is represented by the distance between the two highest peaks.

The interplay of magnetic fields and differential rotation is demonstrated in Fig. 6.18 showing the variation of the radial rotational shear averaged over latitude, $\langle (\partial\Omega/\partial r)^2 \rangle$, versus time, compared with the toroidal field. A minimum in differential rotation is accompanied by a decay of the magnetic field and followed by a grand minimum. The differential rotation is being restored during the grand minimum since the suppressing effect of magnetic Λ-quenching is reduced. The behaviour of the system can be evaluated in a phase diagram of the dipolar component E_A of the magnetic field energy, the quadrupolar component E_S, and the mean angular velocity gradient, as also shown in Fig. 6.18. The dots in the interior of the diagram represent the actual time series; projections of the trajectory are given at the sides.

The trajectory resides at strong differential rotation during normal cyclic activity. The magnetic field oscillates in a wide range of energies. The trajectory moves down as the

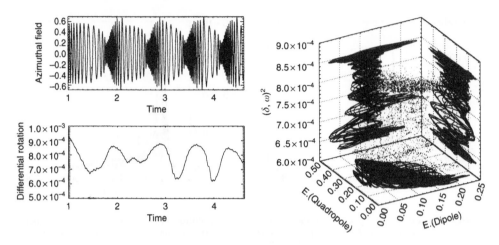

Figure 6.18 Left: Correlation between magnetic field oscillations and variations of the differential rotation measure $(\partial\Omega/\partial r)^2$, averaged over the latitude θ. Right: Correlations between magnetic field energies for both parities (dipolar and quadrupolar energy component) and the differential rotation. For most of the time the orbit resides in the state of regular cycles in the upper part of the box. Differential rotation is suppressed during grand minima.

field starts to suppress the differential rotation, with oscillation amplitudes diminishing. The actual grand minimum is expressed by a loop at very low dipolar energies; the quadrupolar component, however, remains present throughout the minimum. The differential-rotation measure is already growing at that time. At first glance, the phase graph may be associated with Type 1 modulation as classified in Knobloch *et al.* (1998). The effect of large-scale Lorentz forces on the differential rotation as the only feedback of strong magnetic fields produces irregular grand minima with strong variations in cycle period. The complex time series turns into a single-period solution, if the suppression of dynamo action (α-quenching) is included. If a strong feedback of small-scale flows on the generation of Reynolds stress (Λ-quenching) is added, grand minima occur at a reasonable rate between 10 and 20 cycle times. The cycle period varies by a factor of 3 or 4. The northern and southern hemispheres differ slightly in their temporal behaviour. This is a general characteristic of mixed-mode dynamo explanations of grand minima. The magnetic Prandtl number directs the intermittency of the activity cycle. Values smaller than unity are required for the existence of grand minima, whereas the occurrence of grand minima again becomes more and more exceptional for very small values of P_m (Fig. 6.19).

6.6. Stellar cycles

The cycle period may be considered as a main test of stellar dynamo theory because it reflects an essential property of the dynamo mechanism. A realistic solar dynamo model should provide the correct 22-year cycle period and, in the case of stellar cycles, the observed dependence of the cycle period on the rotation period.

The following sections will deal with a number of applications of one-dimensional models illuminating some effects of more complicated EMF. This will include the additional

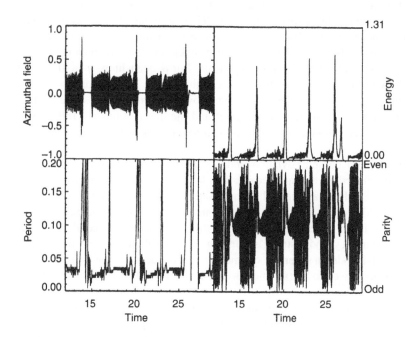

Figure 6.19 The same as in Fig. 6.16 but for $P_m = 0.01$ showing much rarer occurrences of grand minima than for $P_m = 0.1$.

consideration of the magnetic-diffusivity quenching, a dynamo-induced α-function, and the influence of temporal fluctuations of α and η_T.

The relations between stellar parameters and the amplitude and duration of magnetic activity cycles are fundamental for our knowledge of stellar physics. Observational results cover a variety of findings which depend on how stars are grouped and how unknown parameters must be chosen. Generally, relations for the magnetic field and the cycle frequency are expressed by

$$\langle B \rangle_{\max} \propto \Omega^{*m}, \quad \omega_{\mathrm{cyc}} \propto \Omega^{*n}, \tag{6.45}$$

with different m and n. The value discussed by Noyes *et al.* (1984) was $n = 1.28$. Saar (1996) derives a very small m (Saar, fig. 3), while Baliunas *et al.* (1996) find $n \simeq 0.47$ for young stars and $n \simeq 1.97$ for old stars. Brandenburg *et al.* (1998) derive $n_I = 1.46$ and $n_A = 1.48$ for inactive (I) and active (A) stars. Saar and Brandenburg (1999) find $n = 0.6$ for the young (super-active) stars and $n = 1.5$ for the old ones. Higher values are reported by Ossendrijver (1997) consistent with $n = 2 \cdots 2.5$. All the reported exponents so far are positive. The relations in (6.45) can be reformulated with the dynamo number $\mathcal{D}$ as

$$\langle B \rangle_{\max} = B_{\mathrm{eq}} \mathcal{D}^{m'}, \quad \omega_{\mathrm{cyc}} = \mathcal{D}^{n'}/\tau_{\mathrm{diff}}, \tag{6.46}$$

where B_{eq} means the equipartition field and $\tau_{\mathrm{diff}} = H^2/\eta_0$ the diffusion time. Instead of the normalization used in (6.46)$_2$, Soon *et al.* (1993) proposed the use of the basic rotation rate Ω_0; a procedure we do not follow here to retain consistency with conventional definitions of the dynamo number.

The main issues of stellar activity and the relations to dynamo theory concerning cycle times were formulated by Noyes *et al.* (1984). Their argumentation concerns a zeroth-order $\alpha\Omega$-dynamo model for which the coefficient n in the scaling $\omega_{\mathrm{cyc}} \propto \mathcal{D}^{n/2}$ is determined. The linear case yields $n = 4/3$ for the most unstable mode (Tuominen *et al.*, 1988). Different nonlinearities lead to different exponents. For α-quenching, $n = 0$ is found while for models with flux loss (rather than α-quenching) $n = 1$ results or $n = 2/3$ – if only the toroidal field is quenched by magnetic buoyancy. Note the absence of finite n for simple α-quenching. If the theory is correct and the observations are following relations such as (6.45), then the nonlinear dynamo can never work with α-quenching as the basic nonlinearity.

A number of more complex dynamo models confirming the findings of Noyes *et al.* (1984) have been developed in the last decade. For large dynamo numbers Moss *et al.* (1990) also derived $n = 1$ from a spherical dynamo saturated by magnetic buoyancy. Schmitt and Schüssler (1989) showed with a one-dimensional model (Schmitt and Schüsler, fig. 6) that there is practically no dependence of the cycle frequency on the (large) dynamo number for α-quenching, but they find a strong scaling ($n \simeq 2$) for their flux-loss models. Also for a two-dimensional model in spherical symmetry the dependence of the cycle frequency on the dynamo number proved to be extremely weak, $n \lesssim 0.1$ (Rüdiger *et al.*, 1994, dashed line in their fig. 4).

The latter application, however, requires further consideration. One finds a finite value for n if the magnetic feedback is not only considered acting upon the α-effect, but also for the eddy diffusivity tensor. What we assume here is that the magnetic field always suppresses and deforms the turbulence field, and this has consequences for both the α-effect and the eddy diffusivity. The theory for a second-order correlation-approximation is given in Kitchatinov *et al.* (1994b), applications are summarized in Rüdiger *et al.* (1994).

The turbulent-diffusivity quenching concept was also described in Noyes *et al.* (1984). As the magnetic fields become super-equipartitioned, however, a series expansion such as used cannot be adopted. Tobias (1998), with a two-dimensional global model in Cartesian coordinates, finds n' varying between 0.38 and 0.67 depending on the nonlinearity. He also finds the exponent growing with increasing effect of η-quenching. The results for the Malkus–Proctor effect alone yield the weakest dependence for the cycle period with $\mathcal{D}$.

Since the paper by Brandenburg *et al.* (1998), anti-quenching expressions such as

$$\alpha \propto \langle B \rangle^p, \quad \eta_{\mathrm{T}} \propto \langle B \rangle^q \tag{6.47}$$

are also under consideration, with $p > 0$ and $q > 0$. The idea is to find the consequences of EMF quantities being induced by the magnetic field itself (see Section 6.7). Saturation of such dynamos requires $q > p$. All their global models, however, lead to exponents n of order 0.5. It concerns both one- and two-dimensional models, characteristic values for q were numbers up to 6.

It is shown in the following that a special one-dimensional slab dynamo, as a generalization of Parker's zero-dimensional wave dynamo without η-quenching, has a very low n while n exceeds unity if η-quenching is included. Buoyancy effects are therefore not the only possible addition to dynamo theory to explain observations as given in (6.45); the magnetic suppression of the turbulent magnetic diffusivity also yields an explanation of the observations.

6.6.1. The model

Here we are not working with the simplest possibility, which would read $\alpha_{ij} = \alpha \delta_{ij}$ and $\eta_{ijk} = \eta_T \varepsilon_{ijk}$. The full feedback of the induced magnetic field on the turbulent EMF is included, i.e. the magnetic suppression and deformation of both the tensors α and η. In particular, the influence of the magnetic field on the magnetic diffusivity is often ignored in dynamo computations, and we shall demonstrate the differences in the gross properties of the solutions with and without η-quenching. The main consequence of the inclusion of η-quenching is the appearance of a nonlinear 'magnetic velocity' in $\mathcal{E}$:

$$\mathcal{E} = \cdots + \mathbf{U}^{\mathrm{mag}} \times \langle \mathbf{B} \rangle \tag{6.48}$$

with

$$\mathbf{U}^{\mathrm{mag}} = \hat{\eta} \nabla \log\langle B \rangle + \eta_z \frac{(\nabla \times \langle \mathbf{B} \rangle) \times \langle \mathbf{B} \rangle}{\langle B \rangle^2} \tag{6.49}$$

(Kitchatinov *et al.*, 1994b) and

$$\eta_T = \eta_0 \varphi, \quad \eta_z = \eta_0 \varphi_z, \quad \hat{\eta} = \eta_0 \hat{\varphi}. \tag{6.50}$$

The coefficient functions $\varphi(\beta)$ are

$$\varphi = \frac{3}{2\beta^2} \left(-\frac{1}{1 + \beta^2} + \frac{1}{\beta} \arctan \beta \right), \tag{6.51}$$

$$\varphi_z = \frac{3}{8\beta^2} \left(1 + \frac{2}{1 + \beta^2} + \frac{\beta^2 - 3}{\beta} \arctan \beta \right), \tag{6.52}$$

$$\hat{\varphi} = \frac{3}{8\beta^2} \left(-\frac{5\beta^2 + 3}{(1 + \beta^2)^2} + \frac{3}{\beta} \arctan \beta \right) \tag{6.53}$$

with

$$\beta = |\langle \mathbf{B} \rangle| / B_{\mathrm{eq}}. \tag{6.54}$$

The components $\alpha_{\phi\phi}$ and α_{ss} of the α-tensor are taken from Rüdiger and Schultz (1997). They are equal in our approximation and read

$$\alpha_{ss} = \alpha_{\phi\phi} = -\frac{2}{5} \Omega^* \frac{d(\log \rho)}{dz} u_T^2 \tau_{\mathrm{corr}} \Psi \tag{6.55}$$

with the α-quenching function Ψ given in (6.29). These quenching functions are approximated by $\Psi(\beta) \simeq 1 - (12/7)\beta^2$, $\varphi(\beta) \simeq 1 - (6/5)\beta^2$, $\varphi_z(\beta) \simeq (2/5)\beta^2$ and $\hat{\varphi}(\beta) \simeq (3/5)\beta^2$ for weak magnetic fields.

The only coordinate about which our quantities are allowed to vary is the direction of the rotation axis, i.e. z. The z-dependencies in (6.55) are summarized in the form $\alpha = -\hat{\alpha} \sin 2z$ where the lower and upper boundary are located at $z = 0$ and $z = \pi$. The plane has infinite extent in the x and y directions and is restricted by boundaries in the z-direction. The normalized dynamo equations then are

$$\frac{\partial A}{\partial t} = C_\alpha \hat{\alpha}(z) \Psi B + \varphi \frac{\partial^2 A}{\partial z^2} - w \frac{\partial A}{\partial z},$$

$$\frac{\partial B}{\partial t} = -C_\alpha \frac{\partial}{\partial z} \left(\hat{\alpha}(z) \Psi \frac{\partial A}{\partial z} \right) - C_\Omega \frac{\partial A}{\partial z} + \frac{\partial}{\partial z} \left(\varphi \frac{\partial B}{\partial z} \right) - \frac{\partial}{\partial z} (w B) \tag{6.56}$$

with $w = (2\hat{\varphi} - \varphi_z)\partial \log B_{\text{tot}}/\partial z$ and $B_{\text{tot}}^2 = B^2 + (\partial A/\partial z)^2$. The boundary conditions are $B = \partial A/\partial z = 0$ at $z = 0$ and $z = \pi$ which limits the magnetic field to the dynamo layer. The dimensionless parameters

$$C_\alpha = \frac{\alpha_0 H}{\eta_{\text{T}}}, \quad C_\Omega = \frac{\partial u_y}{\partial x} \frac{H^2}{\eta_{\text{T}}} \tag{6.57}$$

represent the strength and sign of the α-effect and the shear flow. C_α is fixed to a value of 5 here. A positive C_α describes a positive α in the upper half of the layer and vice versa. We chose to vary C_Ω with positive values which represent a positive shear.

6.6.2. The results

All the magnetic fields of the nonlinear one-dimensional model are symmetric with respect to the equatorial plane. In our domain of positive shear the solutions of both the α^2-dynamo (small C_Ω) and the $\alpha\Omega$-dynamo (large C_Ω) are oscillatory. The transition between these regimes occurs at $C_\Omega \simeq 50$ (cf. Rüdiger and Arlt, 1996). The cycle periods are always constant over time. Fig. 6.20 shows the resulting cycle frequencies for the conventional α-quenching expression. The cycle period is strikingly independent of the dynamo number (see also Noyes et al., 1984; Schmitt and Schüssler, 1989; Jennings and Weiss, 1991). On the other hand, Fig. 6.21 gives the resulting cycle frequencies of the oscillatory solutions for the $\alpha^2\Omega$-dynamo with the η-quenching concept. The cycle periods are given for models with and without η-quenching. The exponent n' is variable and positive for the model with η-quenching whereas it is much more complex for the model without η-quenching. The resulting exponent $n' \simeq 0.5$ for large dynamo numbers in Fig. 6.21 is remarkably similar to the behaviour of linear dynamos. In order to compare this value with the observed value of n in (6.45) we have to introduce the scaling of the physical quantities with Ω^*. We adopt the scaling $\eta_0 \propto 1/\Omega^*$ from Kitchatinov et al. (1994b) and use the parameterization

$$\frac{\partial \log \Omega}{\partial \log r} \propto \Omega^{*\kappa}. \tag{6.58}$$

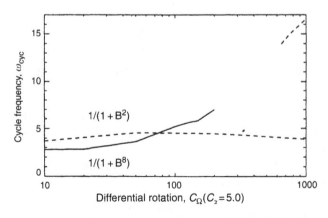

Figure 6.20 The cycle frequency of oscillatory solutions of the $\alpha^2\Omega$-dynamo versus C_Ω with different α-quenching functions. A second solution can be excited for the slower cut-off function (dashed) for high C_Ω. The steeper cut-off function (solid) delivers chaotic solutions above $C_\Omega \sim 200$.

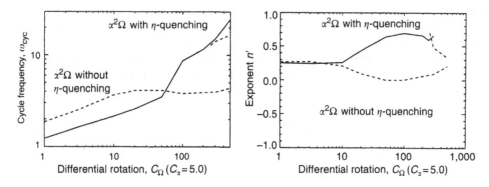

Figure 6.21 Left: The cycle frequency of oscillatory solutions of the $\alpha^2\Omega$-dynamo versus C_Ω with (solid) and without (dashed) η-quenching. Right: The exponent n' for the scaling (6.46). Only diffusivity quenching yields exponents of order 0.5. The model without n-quenching contains a second solution which can be excited at high C_Ω.

Negative κ describe a decrease of the unknown normalized differential rotation in stellar convection zones with increasing basic rotation. Negative κ are indeed produced by theoretical models of the differential rotation, though only if meridional flow is included. The results in Fig. 6.5 lead to $\kappa \simeq -1$. The dynamo number scales as

$$\mathcal{D} \propto \Omega^{*\mu+\kappa+3}, \tag{6.59}$$

if the α-effect is assumed as scaling with $\Omega^{*\mu}$. Then the relations (6.45) and (6.46) yield

$$n = (\mu + 3 + \kappa)n' - 1. \tag{6.60}$$

We apply the exponent $n' \simeq 0.5$ from Fig. 6.21 for large dynamo numbers and get $n = (\mu + \kappa)/2 + 0.5$, so that with $\kappa = -1$ we find $\mu \simeq 2n$. Consequently, the observed exponents $n = 0.5 \cdots 1.5$ would lead to

$$\mu = 1 \cdots 3 \tag{6.61}$$

for the α-effect running with the Coriolis number which does not seem too unreasonable.

6.7. Dynamo-induced on-off α-effect

A two-component model for the solar dynamo has been suggested by Ferriz-Mas *et al.* (1994). Toroidal magnetic fields are generated by differential rotation in the overshoot layer whereas a dynamo acts in the upper part of the convection zone. The thin overshoot layer is assumed to be stably stratified up to magnetic field strengths of 10^5 Gauss (Ferriz-Mas and Schüssler, 1995). The dynamo may operate in the turbulent convective zone and impose stochastic perturbations to the overshoot layer.

Schmitt *et al.* (1996) discuss whether the interaction of the weak-field convection zone dynamo and the magnetic fields in the overshoot layer might explain the grand-minimum behaviour of the solar cycle. They assumed a mean-field dynamo with an α-effect working only when the magnetic field exceeds a certain threshold. Grand minima are expected when the perturbations from the convection zone do not keep the overshoot layer instabilities above this threshold. Here we test whether a dynamo, once excited, will really turn off in this sense.

The dynamo equations are studied in both one- and two-dimensional domains. In the one-dimensional approach the integration region extends only in the z-direction whence the remaining radial and azimuthal components of the magnetic field depend on z only. Normalization of times with the diffusion time $\tau_{\text{diff}} = H^2/\eta_T$ and vertical distances z with H yields

$$\frac{\partial A}{\partial t} = C_\alpha \hat{\alpha}(z)\Psi(B_{\text{tot}})B + \frac{\partial^2 A}{\partial z^2}, \qquad (6.62)$$

$$\frac{\partial B}{\partial t} = -C_\alpha \frac{\partial}{\partial z}\left(\hat{\alpha}(z)\Psi(B_{\text{tot}})\frac{\partial A}{\partial z}\right) - C_\Omega \frac{\partial A}{\partial z} + \frac{\partial^2 B}{\partial z^2}. \qquad (6.63)$$

The α depends on the magnetic field as well as on the location in the object. Hence we have two different dependencies in $\alpha = \hat{\alpha}(z)\alpha_0\Psi(B_{\text{tot}})$. Again, the first factor is expressed by $\hat{\alpha}(z) = -\sin 2z$.

The α-quenching by magnetic fields has usually been expressed by functions such as $1/[1 + (B_{\text{tot}}/B_{\text{max}})^2]$, with a cut-off field strength B_{max} related to the energy of velocity fluctuations or the gas pressure. In the present approach α acts only in a certain range of B_{tot} and it is zero otherwise, i.e.

$$\Psi(B_{\text{tot}}) = \begin{cases} 1 & \text{for } B_{\text{min}} < B_{\text{tot}} < B_{\text{max}}, \\ 0 & \text{otherwise.} \end{cases} \qquad (6.64)$$

The lower threshold B_{min} represents the onset of instability in the strong-field dynamo layer at the bottom of the convection zone, and the upper threshold B_{max} is related to the buoyancy-driven disappearance of flux tubes from the overshoot layer. Again a vacuum surrounds the one-dimensional model.

The normalized induction equations of the two-dimensional model read

$$\frac{\partial B}{\partial t} = \frac{1}{r}\frac{\partial}{\partial r}\left(\eta_T\frac{\partial(Br)}{\partial r}\right) + \frac{\eta_T}{r^2}\frac{\partial}{\partial\theta}\left(\frac{1}{\sin\theta}\frac{\partial(B\sin\theta)}{\partial\theta}\right)$$
$$+ \frac{1}{r}\frac{\partial\Omega}{\partial r}\frac{\partial A}{\partial\theta} - \frac{1}{r\sin\theta}\frac{\partial}{\partial r}\left(\alpha\frac{\partial A}{\partial r}\right) - \frac{1}{r^3}\frac{\partial}{\partial\theta}\left(\frac{\alpha}{\sin\theta}\frac{\partial A}{\partial\theta}\right), \qquad (6.65)$$

$$\frac{\partial A}{\partial t} = \eta_T\frac{\partial^2 A}{\partial r^2} + \eta_T\frac{\sin\theta}{r^2}\frac{\partial}{\partial\theta}\left(\frac{1}{\sin\theta}\frac{\partial A}{\partial\theta}\right) + \alpha r\sin\theta\, B. \qquad (6.66)$$

The computational domain extends from $r = 0.5$ to 1.0 measured in solar radii, and from $\theta = 0$ to π in colatitude. The dynamo effect is assumed to be placed between $r = 0.7$ and 0.8 neighboured by a weakly dissipative layer with $\eta_T = 0.01$ below and a zone with strong turbulent diffusion ($\eta_T = 1$) above the dynamo layer; η_T is also unity in the dynamo shell. Here we neglect the θ-dependence of the angular velocity and assume a constant $\partial\Omega/\partial r$ for simplicity.

An $\alpha^2\Omega$-dynamo provides oscillatory and steady-state solutions. The first test in the one-dimensional model applies $C_\Omega = 100$ and $C_\alpha = 5$ and two on-off functions with both $B_{\text{min}} = 0.5$ and 0.9. The dynamo was excited with an initial magnetic field strong enough to reach the 'on'-range of α. Fig. 6.22 shows the resulting oscillatory solutions. The dynamo

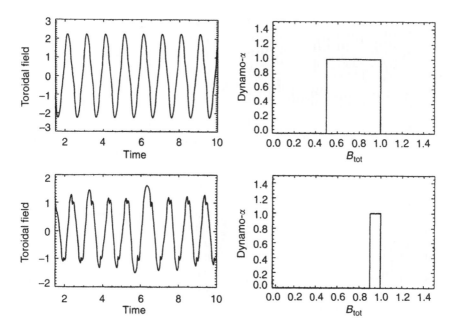

Figure 6.22 The toroidal component of the magnetic field (left) at a fixed point of the domain for two different on-off α-functions (right) in a one-dimensional model. The initial conditions imply a field strong enough to cover partly the 'on'-range of α.

never dies since the α-effect acts before the magnetic fields completely 'dives' through the on-off function.

The same is found in a two-dimensional model as shown in Fig. 6.23. We used the parameters $C_\Omega = 10^5$ and $C_\alpha = -10$ and two on-off functions with $B_{min} = 0.38$ and 0.5, respectively If the on-off function is too narrow the solutions change from oscillatory to steady-state, although we never find the dynamo stopping its operation. The parity of the oscillatory solution is antisymmetric with respect to the equator but equator symmetry is found for the steady-state solution.

The results indicate that the dynamo, once initiated by sufficiently strong magnetic fields, will not switch off on its own and, hence, does not impose grand minima to its cyclic behaviour. The dynamo will only cease operating if the flux tubes migrate into the upper convection zone on a time scale much shorter than the magnetic diffusion time.

6.8. Random α

Two different types of magnetic dynamos appear to be acting in the Sun and the Earth. In contrast to the distinct oscillatory solar magnetic activity, the terrestrial magnetic field is almost permanent. Indeed, two different sorts of dynamos in spherical geometry have been constructed with such properties. The α^2-dynamo yields stationary solutions while the $\alpha\Omega$-dynamo yields oscillatory solutions. Observation and theory seem to be rather in agreement. There are, however, differences in the time behaviour: solar activity is not strictly periodic and the Earth's dynamo is not strictly permanent. The Earth's magnetic field also varies.

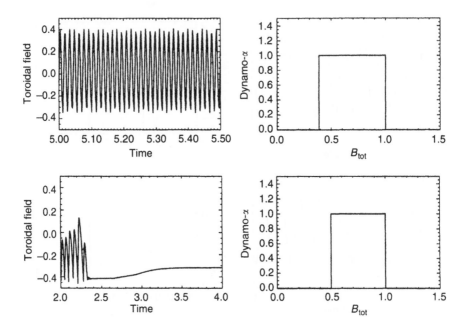

Figure 6.23 The toroidal component of the magnetic field (left) for two different on-off α-functions (right) in a two-dimensional model. $B_{\mathrm{max}} = 1$ in both cases.

There are irregular changes of the field strength, the shortest period of stability being 40,000 years. The average length of the intervals of constant field is almost 10 times larger than this minimum value (cf. Krause and Schmidt, 1988). Such a reversal was simulated numerically with a three-dimensional MHD code by Glatzmaier and Roberts (1995).

In Section 6.5 we explained grand minima such as the Maunder minimum by various non-linearities in the mean-field equations. The averages are taken over an ensemble, i.e. a great number of identical examples. The other possibility to explain the temporal irregularities is to consider the characteristic turbulence values as a time series. The idea is that the averaging procedure concerns only a periodic spatial coordinate, e.g. the azimuthal angle ϕ. In other words, when expanding in Fourier series such as $e^{im\phi}$ the mode $m = 0$ is considered as the mean value. Again, if the time scale of this mode does not vary significantly during the correlation time, local formulations such as equation (6.69) below are reasonable. Nevertheless, the turbulence intensity, the α-effect, and the eddy diffusivity become time-dependent quantities (Hoyng, 1988; Choudhuri, 1992; Moss *et al.*, 1992; Hoyng, 1993; Hoyng *et al.*, 1994; Otmianowska-Mazur *et al.*, 1997; Vishniac and Brandenburg, 1997).

However, questions concerning the amplitude, the time scales, and the phase relation between (say) helicity and eddy diffusivity need to be studied. Are the fluctuations strong enough to influence the dynamo significantly? We construct here a random turbulence model and evaluate the complete turbulent EMF as a time series. The consequences of this concept are then computed for a simple plane-layer dynamo with shear. The main parameter which we vary is the number of cells in the unit length.

One can define mean values of a field F by integration over (say) space. Of particular interest here is an averaging procedure over longitude, i.e.

$$\langle F \rangle = \frac{1}{2\pi} \int_0^{2\pi} F \, d\phi \qquad (6.67)$$

(Braginsky, 1964; Hoyng, 1993). The turbulent EMF for a given position in the meridional plane forms a time series with the correlation time τ_{corr} as a characteristic scale. The peak-to-peak variations in the time series should depend on the number of cells. They remain finite if the number of cells is restricted as it is in reality. For an infinite number of the turbulence cells the peak-to-peak variation in the time series goes to zero but it will grow for a finite number of cells along the unit length.

The tensors constituting the local mean-field EMF in (6.3), α and η, must be calculated from one and the same turbulence field. We propose to define a helical turbulence existing in a rectangular parcel and to compute simultaneously their related components.

We restrict ourselves to the computation of the turbulent EMF in the high-conductivity limit. Then the second-order correlation approximation yields

$$\mathcal{E} = \int_0^\infty \left\langle \mathbf{u}'(\mathbf{x}, t) \times \nabla \times \left(\mathbf{u}'(\mathbf{x}, t - \tau) \times \langle \mathbf{B}(\mathbf{x}, t) \rangle \right) \right\rangle d\tau, \qquad (6.68)$$

which for short correlation times can be written in the form (6.3). A Cartesian frame is used where y represents the azimuthal direction over which the average is performed. For the dynamo only the components $\mathcal{E}_x$ and $\mathcal{E}_y$ are relevant.

It would be tempting to apply (6.68) as it stands. In components it reads

$$\mathcal{E}_x = \alpha_{xx} \langle B_x \rangle + \eta_T \frac{\partial \langle B_y \rangle}{\partial z},$$

$$\mathcal{E}_y = \alpha_{yy} \langle B_y \rangle - \eta_T \frac{\partial \langle B_x \rangle}{\partial z}. \qquad (6.69)$$

η_T plays the role of a common eddy diffusivity. From (6.68) one can read

$$\alpha_{xx} = \int_0^\infty \left\langle u'_y(t) \frac{\partial u'_z(t - \tau)}{\partial x} - u'_z(t) \frac{\partial u'_y(t - \tau)}{\partial x} \right\rangle d\tau,$$

$$\alpha_{yy} = \int_0^\infty \left\langle u'_z(t) \frac{\partial u'_x(t - \tau)}{\partial y} - u'_x(t) \frac{\partial u'_z(t - \tau)}{\partial y} \right\rangle d\tau, \qquad (6.70)$$

$$\eta_T = \int_0^\infty \langle u'_z(t) u'_z(t - \tau) \rangle \, d\tau$$

(Krause and Rädler, 1980).

6.8.1. The turbulence model

We study the time evolution of the dynamo coefficients (6.70) generated by turbulent gas motions. We analyse a parcel of the solar gas situated in the convective zone. The rectangular coordinate system has the xy-plane parallel to the solar equator, the x-axis pointing parallel to the solar radius, the y-axis tangential to the azimuthal direction and the z-axis directed to the south pole. The parcel is permanently perturbed by vortices of the form of rotating columns of gas randomly distributed at all angles in three-dimensional space. At every time step the resulting velocity field is used to calculate the EMF-coefficients after (6.70).

A single vortex rotates with the velocity ω around its axis of rotation and moves with the velocity (w) up and down along this axis. The eddy itself has also its own lifetime growing linearly with subsequent time steps. The vortex velocity ω as well as the vertical drift w decrease with the distance r from the axis and with the lifetime t according to

$$(\omega, w) = (\omega_0, w_0) \exp\left\{-\tfrac{1}{2}\left[(r/l_{\mathrm{corr}})^2 + (z/z_{\mathrm{corr}})^2\right]\right\} e^{-t/\tau_{\mathrm{corr}}}. \tag{6.71}$$

where l_{corr} is the adopted vortex radius (Otmianowska-Mazur et al., 1992), z_{corr} is its length scale along the axis of rotation and τ_{corr} is the eddy decay time scale. All velocities are truncated at $3l_{\mathrm{corr}}$ and $3z_{\mathrm{corr}}$. In the model all vortices are orientated like right-handed screws. The resulting motion has maximum helicity. A uniform distribution of eddies in three-dimensional space is approximated with 12 possible inclination angles of their rotational axes to the xy-plane.

The initial state is a number N of moving turbulent cells having random inclinations and positions in the xy-plane. After an assumed period of time (one or more time steps), a fraction of them, R_{tur} (the ratio of the new-eddies number to N), is changed to new ones with different randomly given position, inclination and with lifetime starting from zero. The situation repeats itself continuously with time.

The vortex radius l_{corr} as well as the decay time τ_{corr} and the fraction of the new turbulent cells R_{tur} are varied for four cases given in Tables 6.2 and 6.3 in normalized units (cgs units for time, velocity and diffusivity result after multiplication with 2.5×10^4 s, 10^5 cm s^{-1} and 2.5×10^{14} cm^2 s^{-1}).

We made numerical experiments with the following values for the vortex radius l_{corr} and the length scale z_{corr} in the z-direction in normalized units: 1, 2, 4 and 8. For successive

Table 6.2 Input and output for the turbulence models $\mathcal{A}, \mathcal{B}, \mathcal{C}$ and $\mathcal{D}$. N is the eddy population of the equator, R_{tur} the eddy birth rate, other quantities, and normalizations as explained in the text

	l_{corr}	τ_{corr}	N	R_{tur}	η_T	$\sigma(\eta_T)$	α_{xx}	$\sigma(\alpha_{xx})$	α_{yy}	$\sigma(\alpha_{yy})$
$\mathcal{A}$	1	10	200	1.0	0.191	0.086	−0.046	0.048	−0.077	0.055
$\mathcal{B}$	2	20	100	0.7	0.336	0.208	−0.029	0.063	−0.048	0.059
$\mathcal{C}$	4	40	50	0.005	1.045	0.860	−0.029	0.119	−0.039	0.110
$\mathcal{D}$	8	80	25	0.0003	1.793	2.278	−0.015	0.136	−0.017	0.110

Table 6.3 The ratio S of fluctuations and mean values

Model	S_{η_T}	$S_{\alpha_{xx}}$	$S_{\alpha_{yy}}$
$\mathcal{A}$	0.45	1.03	0.72
$\mathcal{B}$	0.62	2.13	1.24
$\mathcal{C}$	0.82	4.10	2.79
$\mathcal{D}$	1.27	8.94	6.62

length scales we choose the time scales 10, 20, 40 and 80. The ratio of both scales is always $l_{corr}/\tau_{corr} \cong 0.1$. In order to obtain the expected mean turbulent velocity value of 0.1, the fraction of new vortices R_{tur} is also varied (cf. Otmianowska-Mazur *et al.*, 1997).

6.8.2. Numerical experiments

The simulations deliver time series of the turbulence intensity, the eddy diffusivity and the α-coefficients, and a standard deviation σ from their temporal averages is

$$\sigma = \sqrt{E\{[X - E(X)]^2\}}, \tag{6.72}$$

where $E(X)$ is the time average of a random variable X. The standard deviation measures the amplitude of the fluctuations compared with the temporal mean of a given quantity. Table 6.2 lists the input model parameters l_{corr}, τ_{corr} and R_{tur} and the coefficients obtained. The ratios of the standard deviations to their mean values,

$$S_\eta = \sigma(\eta)/|\eta|, \tag{6.73}$$

are given in Table 6.3.

The results for a model $\mathcal{A}$ with $l_{corr} = 1$, and $\tau_{corr} = 10$, which is the sample of shortest lifetimes and smallest spatial dimensions of individual vortices in our simulation sequence is given in Fig. 6.24. All computed quantities are drawn as a time series, each in its own normalized units according to their definitions. The three EMF-coefficients as well as the turbulence intensity show fluctuations around their mean values in time. The diffusion coefficient η_T possesses positive values during most of the time. There are also short periods with a negative η_T. The negative values are not significant because the resulting mean η_T is always positive yielding 0.191 (Table 6.2) in normalized units. The ratio S_{η_T} is 0.45 (Table 6.3). Indeed, the value 0.191 corresponds to $\eta_T \simeq \tau_{corr} u_T^2$, which gives 0.1 for this case. In contrast to η_T, the coefficients α_{xx} and α_{yy} possess negative values during the major part of the time; for short periods, however, positive values are also present. The negative α results from the assumed non-zero, right-handed helicity of the vortices and is in agreement with the expectations. Averaged in time α_{xx} for model $\mathcal{A}$ is -4.6×10^3 cm s^{-1}. For α_{yy} we get -7.7×10^3 cm s^{-1} resulting in a dynamo number of $C_\alpha = 10$. It can be seen that α_{xx} is always slightly smaller than α_{yy}. This fact is connected with the assumed averaging only along the y-axis, which certainly influences the value of α_{xx}. Model $\mathcal{A}$ yields the ratios $\sigma(\alpha_{xx})/|\alpha_{xx}| = 1.03$ and $\sigma(\alpha_{yy})/|\alpha_{yy}| = 0.72$.

The second experiment, $\mathcal{B}$, uses doubled correlation lengths and times, $l_{corr} = 2$, $\tau_{corr} = 20$. A fraction of 0.7 new vortices at every time step was applied. The resulting fluctuations of η_T and α-coefficients are higher than in the case $\mathcal{A}$ (Table 6.2). The averaged value of η_T increases to 0.336, yielding 8.5×10^{13} cm^2 s^{-1}. The ratios $S_{\alpha_{xx}}$ and $S_{\alpha_{yy}}$ increase up to 2.13 and 1.24, respectively (Table 6.3). It means that the fluctuations of α are higher for larger eddies with longer lifetimes – as it should.

The experiment $\mathcal{D}$ works with largest correlation times and longest lifetimes, $l_{corr} = 8$ and $\tau_{corr} = 80$. The birth rate R_{tur} is 0.0003 per time step – really low in comparison with the previous cases. Fig. 6.25 presents the time series of studied quantities. The resulting fluctuations of the alpha coefficients are extremely high. The mean value of η_T is 1.793 (4.5×10^{14} cm^2 s^{-1}). The ratio $S_{\eta_T} = 1.27$ indicates the significant growth of the fluctuations compared with the mean value. The obtained ratios are $S_{\alpha_{xx}} = 8.94$ and $S_{\alpha_{yy}} = 6.62$. The

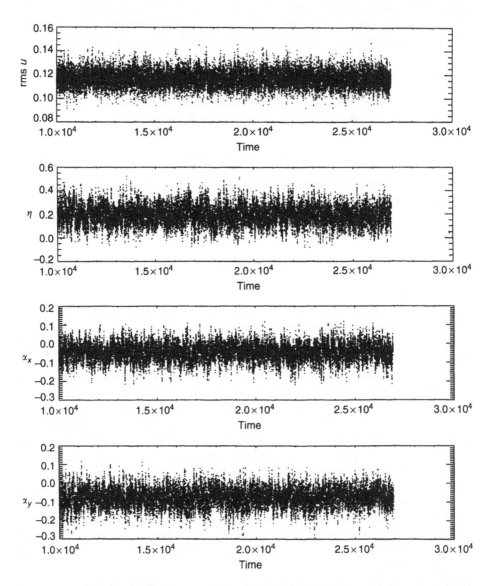

Figure 6.24 Time series for turbulence intensity, eddy diffusivity and α-tensor components α_{xx} and α_{yy} for turbulence model $\mathcal{A}$. For time, velocity (u and α) and diffusivity cgs units result after multiplication with 2.5×10^4 s, 10^5 cm s^{-1} and 2.5×10^{14} cm^2 s^{-1}.

averaged in time values of α_{xx} and α_{yy} decrease again to -0.015 (-1.5×10^3 cm^2 s^{-1}) and -0.017 (-1.7×10^3 cm^2 s^{-1}).

Table 6.3 summarizes the main results of our simulations. The fluctuations in the time series become more and more dominant with the decreasing number of eddies together with the increasing vortex size. The extreme case is plotted in Fig. 6.25. For such models the fluctuations of both the eddy diffusivity as well as the α-effect are much higher than the averages. Even short-lived changes of the sign of the quantities were found.

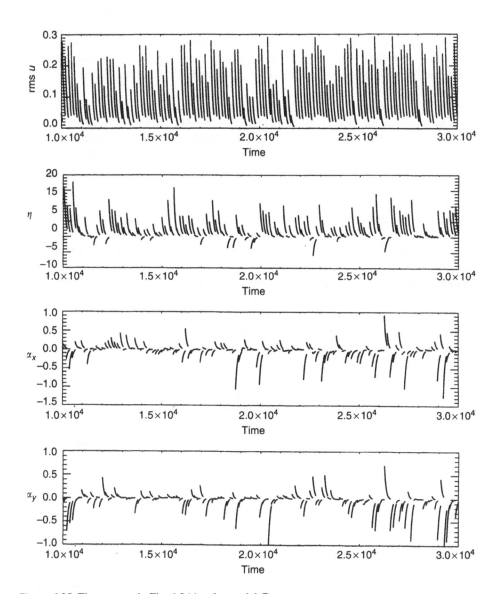

Figure 6.25 The same as in Fig. 6.24 but for model $\mathcal{D}$.

The α-effect fluctuations exceed those of the eddy diffusivity in all our models. The latter proves to be more stable than the α-effect against dilution of the turbulence. This is a confirmation, indeed, for those studies in which an α-effect time series is exclusively used in dynamo computations.

6.8.3. A plane-layer dynamo

Quite a few publications deal with the influence of stochastic α-fluctuations upon dynamo-generated magnetic fields for various models. Choudhuri (1992) discussed a simple plane-wave dynamo basically in the linear regime. The fluctuations adopted are as weak as about 10%. While in the $\alpha\Omega$-regime the oscillations are hardly influenced, the opposite is true for the α^2-dynamo. In the latter regime the solution suffers dramatic and chaotic changes even for rather weak disturbances. In the region between the both regimes the remaining irregular variations are suppressed in a model with nonlinear feedback formulated as a traditional α-quenching.

A plane one-dimensional $\alpha^2\Omega$-model similar to that of Section 6.6.1 is used to illustrate the influence of the fluctuating turbulence coefficients on a mean-field dynamo. It is not an overshoot dynamo (cf. Ossendrijver and Hoyng, 1996). The plane has infinite extent in the radial (x) and the azimuthal (y) direction, the boundaries are in the z-direction. Hence, we assume that the magnetic field components in both azimuthal direction, (B), and radial direction, ($\partial A/\partial z$), depend on z only. Note that the z-axis opposite to the colatitude θ.

The α-tensor has only one component (given in Section 6.2.3) vanishing at the equator, and it is assumed here to vanish also at the poles. The magnetic feedback is considered to be conventional α-quenching: $\alpha = \hat{\alpha}\alpha_0\Psi(B_{\text{tot}})$ with $\Psi(B_{\text{tot}}) \propto B_{\text{tot}}^{-2}$. The diffusivity is spatially uniform. The normalized dynamo equations are given in (6.62) and (6.63).

Positive C_α describes a positive α in the northern hemisphere and a negative one in the southern hemisphere. C_Ω defines the amplitude of the differential rotation, positive C_Ω represents positive shear $\partial\Omega/\partial r$. The dynamo operates with a positive α-effect in the northern hemisphere. Our dynamo numbers are $C_\alpha = 5$ and $C_\Omega = 200$. The turbulence models $\mathcal{A}$, $\mathcal{B}$ and $\mathcal{D}$ are applied, flow patterns with small ($\mathcal{A}$), medium ($\mathcal{B}$) and with very large ($\mathcal{D}$) eddies are used.

A magnetic dipole field oscillating with a (normalized) period of 1.46 (Fig. 6.26(a)) is induced by a dynamo model without EMF-fluctuations ($N \to \infty$, called $\mathcal{O}$). The period corresponds to an activity cycle of 8 years in physical units. Also the turbulence models $\mathcal{B}$ produce an oscillating dipole but with a more complicated temporal behaviour (Fig. 6.26(b) and (c)). It is not a single oscillation, the power spectrum forms a rather broad line with substructures. The quality,

$$Q = \frac{\omega_{\text{cyc}}}{\Delta\omega_{\text{cyc}}}, \tag{6.74}$$

of this line (with $\Delta\omega_{\text{cyc}}$ as its half-width) proves to be about 2.9. This value close to the observed quality of the solar cycle is produced here by a turbulence model with about 100 eddies along the equator (see also Ossendrijver et al., 1996). There are also remarkable variations of the magnetic cycle amplitude.

The turbulence field $\mathcal{D}$ leads, not surprisingly, to a highly irregular temporal behaviour in the magnetic quantities. Its power spectrum peaks at several periods (Fig. 6.26(d)). The overall shape of the spectrum, however, does no longer suggest any oscillations. The power of the lower frequencies is strongly increased, the high-frequency power decreases as $\omega_{\text{cyc}}^{-5/3}$ like a Kolmogoroff spectrum, indicating the existence of chaos.

Our experiments basically show how the dilution of turbulence is able to transform a single-mode oscillation (for very many cells) to an oscillation with low quality (for moderate eddy population) and finally to a temporal behaviour close to chaos (for very few cells). It is possible that the observed low quality of the solar cycle indicates the finite number of the giant cells driving the large-scale solar dynamo. Implications for the temporal evolution of the

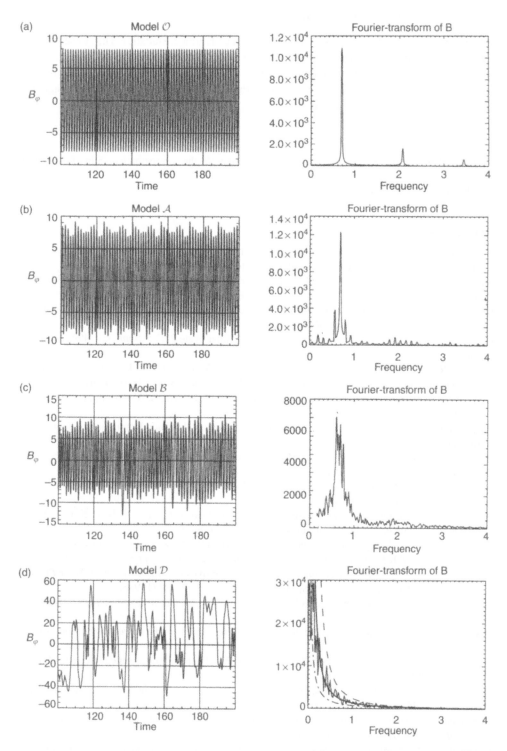

Figure 6.26 Dynamo-induced magnetic toroidal fields for the turbulence models (a): $\mathcal{O}$, (b): $\mathcal{A}$, (c): $\mathcal{B}$, and (d): $\mathcal{D}$. Left: Time series. Right: Power spectra. The real time unit is 2.7 years.

solar rotation law should be another output of such a cell number statistics. The consequences for rotation will be an independent test of the presented theory leading to the same number of eddies contributing to the turbulent angular momentum transport. For only occasional turbulence a nontrivial time series for both the turbulence EMF-coefficients and the magnetic field is an unavoidable consequence.

Notes

1 On the other hand, in order to reproduce a factor of order 10^2 the correlation length must not be too small.
2 If the same butterfly diagram is applied.

References

Abramenko, V.I., Wang, T. and Yurchishin, V.B., "Analysis of electric current helicity in active regions on the basis of vector magnetograms," *Solar Phys.* **168**, 75–89 (1996).

Baliunas, S.L. and Vaughan, A.H., "Stellar activity cycles," *Ann. Rev. Astron. Astrophys.* **23**, 379–412 (1985).

Baliunas, S.L., Nesme-Ribes, E., Sokoloff, D. and Soon, W.H., "A dynamo interpretation of stellar activity cycles," *Astrophys. J.* **460**, 848–854 (1996).

Bao, S. and Zhang, H., "Patterns of current helicity for solar cycle 22," *Astrophys. J.* **496**, L43–L46 (1998).

Barnes, J.R., Cameron, A.C., James, D.J., Watson, C.A., Vincent, F. and Donati, J.-F., "Starspot coverage and differential rotation on PZ Tel," in: *Stellar Clusters and Associations: Convection, Rotation, and Dynamos* (Eds. R. Pallavicini, G. Micela and S. Sciortino), ASP Conf. Ser. 198, p. 345 (2000).

Beer, J., Tobias, S.M. and Weiss, N.O., "An active Sun throughout the Maunder minimum," *Solar Phys.* **181**, 237–249 (1998).

Belvedere, G., Lanzafame, G. and Proctor, M.R.E., "The latitude belts of solar activity as a consequence of a boundary-layer dynamo," *Nature* **350**, 481–483 (1991).

Braginsky, S.I., "Self excitation of a magnetic field during the motion of a highly conducting fluid," *Sov. Phys. JETP* **20**, 726 (1964).

Brandenburg, A., "Simulating the solar dynamo," in: *The Cosmic Dynamo* (Eds. F. Krause, K.-H. Rädler and G. Rüdiger), Kluwer, Dordrecht, pp. 111–121 (1993).

Brandenburg, A., "Solar dynamos: computational background," in: *Lectures on Solar and Planetary Dynamos* (Eds. M.R.E. Proctor and A.D. Gilbert), Cambridge University Press, pp. 117–159 (1994a).

Brandenburg, A., "Hydrodynamical simulations of the solar dynamo," in: *Advances in Solar Physics* (Eds. G. Belvedere and M. Rodono), Springer, Heidelberg, pp. 73–84 (1994b).

Brandenburg, A., "Dynamo-generated turbulence and outflows from accretion discs," *Phil. Trans. R. Soc. Lond.* A **358**, 759 (2000).

Brandenburg, A. and Donner, K.J., "The dependence of the dynamo alpha on vorticity," *Mon. Not. R. Astr. Soc.* **288**, L29–L33 (1997).

Brandenburg, A. and Schmitt, D., "Simulations of an alpha-effect due to magnetic buoyancy," *Astron. Astrophys.* **338**, L55–L58 (1998).

Brandenburg, A., Saar, S.H. and Turpin, C.R., "Time evolution of the magnetic activity cycle period," *Astrophys. J.* **498**, 51–54 (1998).

Brandenburg, A., Tuominen, I. and Moss, D., "On the nonlinear stability of dynamo models," *Geophys. Astrophys. Fluid Dynam.* **49**, 129–142 (1989).

Brandenburg, A., Nordlund, Å., Pulkkinen, P., Stein, R.F. and Tuominen, I., "3-D simulation of turbulent cyclonic magneto-convection," *Astron. Astrophys.* **232**, 277–291 (1990).

Caligari, P., Moreno-Insertis, F. and Schüssler, M., "Emerging flux tubes in the solar convection zone. I. Asymmetry, tilt and emergence latitude," *Astrophys. J.* **441**, 886–902 (1995).

Choudhuri, A.R., "The evolution of loop structures in flux rings within the solar convection zone," *Solar Phys.* **123**, 217–239 (1989).

Choudhuri, A.R., "On the possibility of an $\alpha^2\omega$-type dynamo in a thin layer inside the Sun," *Astrophys. J.* **355**, 733–744 (1990).

Choudhuri, A.R., "Stochastic fluctuations of the solar dynamo," *Astron. Astrophys.* **253**, 277–285 (1992).

Choudhuri, A.R., Schüssler, M. and Dikpati, M., "The solar dynamo with meridional circulation," *Astron. Astrophys.* **303**, L29–L32 (1995).

Christensen-Dalsgaard, J. and Schou, J., "Differential rotation in the solar interior," in: *Seismology of the Sun and Sun-like stars* (Eds. V. Domingo and E.J. Rolfe), ESA SP-286, pp. 149–153 (1988).

DeLuca, E.E. and Gilman, P.A., "The solar dynamo," in: *Solar Interior and Atmosphere* (Eds. A.N. Cox, W.C. Livingston and M.S. Matthews), Arizona University Press, Tucson, pp. 275–303 (1991).

Dikpati, M. and Charbonneau, P., "A Babcock–Leighton flux transport dynamo with solar-like differential rotation," *Astrophys. J.* **518**, 508–520 (1999).

Donahue, R.A. and Baliunas, S.L., "Evidence of differential surface rotation in the solar-type star HD 114710," *Astrophys. J.* **393**, L63–L66 (1992).

Donati, J.-F. and Cameron, A.C., "Differential rotation and magnetic polarity patterns on AB Doradus," *Mon. Not. R. Astr. Soc.* **291**, 1–19 (1997).

Durney, B.R., "On the behavior of the angular velocity in the lower part of the solar convection zone," *Astrophys. J.* **338**, 509–527 (1989).

Durney, B.R., "On a Babcock–Leighton dynamo model with a deep-seated generating layer for the toroidal magnetic field," *Solar Phys.* **160**, 213–235 (1995).

Durney, B.R., "On a Babcock–Leighton dynamo model with a deep-seated generating layer for the toroidal magnetic field, II," *Solar Phys.* **166**, 231–260 (1996).

Durney, B.R. and Robinson, R.D., "On an estimate of the dynamo-generated magnetic fields in late-type stars," *Astrophys. J.* **253**, 290–297 (1982).

Dziembowski, W.A. and Goode, P.R., "Seismology for the fine structure in the Sun's oscillations varying with its activity cycle," *Astrophys. J.* **376**, 782–786 (1991).

Fan, Y., Fisher, G.H. and DeLuca, E.E., "The origin of morphological asymmetries in bipolar active regions," *Astrophys. J.* **405**, 390–401 (1993).

Ferrière, K., "The full alpha-tensor due to supernova explosions and superbubbles in the galactic disk," *Astrophys. J.* **404**, 162–184 (1993).

Ferriz-Mas, A. and Schüssler, M., "Instabilities of magnetic flux tubes in a stellar convection zone. I. Equatorial flux rings in differentially rotating stars," *Geophys. Astrophys. Fluid Dynam.* **72**, 209–248 (1993).

Ferriz-Mas, A. and Schüssler, M., "Instabilities of magnetic flux tubes in a stellar convection zone. II. Flux rings outside the equatorial plane," *Geophys. Astrophys. Fluid Dynam.* **81**, 233–266 (1995).

Ferriz-Mas, A., Schmitt, D. and Schüssler, M., "A dynamo effect due to instability of magnetic flux tubes," *Astron. Astrophys.* **289**, 949–956 (1994).

Frick, P., Galyagin, D., Hoyt, D.V., Nesme-Ribes, E., Schatten, K.H., Sokoloff, D. and Zakharov, V., "Wavelet analysis of solar activity recorded by sunspot groups," *Astron. Astrophys.* **328**, 670–681 (1997a).

Frick, P., Nesme-Ribes, E. and Sokoloff, D., "Wavelet analysis of solar activity recorded by sunspot groups and solar diameter data," in: *Stellar and Planetary Magnetoconvection* (Eds. J. Brestenský and S. Ševčik), Acta Astron. et Geophys. Univ. Comenianae XIX, p. 113 (1997b).

Gilman, P.A., "What can we learn about solar cycle mechanisms from observed velocity fields," in: *The Solar Cycle* (Ed. K.L. Harvey), ASP Conf. Ser. 27, pp. 241–255 (1992).

Glatzmaier, G.A. and Roberts, R., "A three-dimensional convective dynamo solution with rotating and finitely conducting inner core and mantle," *Phys. Earth Planet. Inter.* **91**, 63 (1995).

Hale, G.E., "The fields of force in the atmosphere of the Sun," *Nature* **119**, 708–714 (1927).

Hempelmann, A., Schmitt, J. and Stępień, K., "Coronal X-ray emission of cool stars in relation to chromospheric activity and magnetic cycles," *Astron. Astrophys.* **305**, 284–295 (1996).

Hood, L.L. and Jirikowic, J.L., "A probable approx. 2400 year solar quasi-cycle in atmospheric delta C-14," in: *Climate Impact of Solar Variability*, NASA, 98–105 (1990).

Howard, R.F. and LaBonte, B.J., "The sun is observed to be a torsional oscillator with a period of 11 years," *Astrophys. J.* **239**, L33–L36 (1980).

Hoyng, P., "Turbulent transport of magnetic fields. III. Stochastic excitation of global magnetic modes," *Astrophys. J.* **332**, 857–871 (1988).

Hoyng, P., "Helicity fluctuations in mean field theory: an explanation for the variability of the solar cycle," *Astron. Astrophys.* **272**, 321–339 (1993).

Hoyng, P., Schmitt, D. and Teuben, L.J.W., "The effect of random alpha-fluctuations and the global properties of the solar magnetic field," *Astron. Astrophys.* **289**, 265–278 (1994).

Jennings, R.L. and Weiss, N.O., "Symmetry breaking in stellar dynamos," *Mon. Not. R. Astr. Soc.* **252**, 249–260 (1991).

Kaisig, M., Rüdiger, G. and Yorke, H.W., "The alpha-effect due to supernova explosions," *Astron. Astrophys.* **274**, 757–764 (1993).

Keinigs, R.K., "A new interpretation of the alpha effect," *Phys. Fluids* **76**, 2558–2560 (1983).

Kitchatinov, L.L., "Turbulent transport of angular momentum and differential rotation," *Geophys. Astrophys. Fluid Dynam.* **35**, 93–110 (1986).

Kitchatinov, L.L., "Turbulent transport of magnetic fields and the solar dynamo," in: *The Cosmic Dynamo* (Eds. F. Krause, K.-H. Rädler and G. Rüdiger), Kluwer, Dordrecht, pp. 13–17 (1993).

Kitchatinov, L.L. and Pipin, V.V., "Mean-field buoyancy," *Astron. Astrophys.* **274**, 647–652 (1993).

Kitchatinov, L.L. and Rüdiger, G., "Magnetic-field advection in inhomogeneous turbulence," *Astron. Astrophys.* **260**, 494–498 (1992).

Kitchatinov, L.L. and Rüdiger, G., "Lambda-effect and differential rotation in stellar convection zones," *Astron. Astrophys.* **276**, 96–102 (1993).

Kitchatinov, L.L. and Rüdiger, G., "Differential rotation in solar-type stars: revisiting the Taylor number puzzle," *Astron. Astrophys.* **299**, 446–452 (1995).

Kitchatinov, L.L. and Rüdiger, G., "Differential rotation models for late-type dwarfs and giants," *Astron. Astrophys.* **344**, 911–917 (1999).

Kitchatinov, L.L., Rüdiger, G. and Küker, M., "Λ-quenching as the nonlinearity in stellar turbulence dynamos," *Astron. Astrophys.* **292**, 125–132 (1994a).

Kitchatinov, L.L., Pipin, V.V. and Rüdiger, G., "Turbulent viscosity, magnetic diffusivity, and heat conductivity under the influence of rotation and magnetic field," *Astron. Nachr.* **315**, 157–170 (1994b).

Knobloch, E. and Landsberg, A.S., "A new mode of the solar cycle," *Mon. Not. R. Astr. Soc.* **278**, 294–302 (1996).

Knobloch, E., Tobias, S.M. and Weiss, N.O., "Modulation and symmetry changes in stellar dynamos," *Mon. Not. R. Astr. Soc.* **297**, 1123–1138 (1998).

Köhler, H., "The solar dynamo and estimates of the magnetic diffusivity and the α-effect," *Astron. Astrophys.* **25**, 467–476 (1973).

Kosovichev, A.G., Schou, J., Scherrer, P.H., *et al.*, "Structure and rotation of the solar interior: initial results from the MDI medium-1 program," *Solar Phys.* **170**, 43–61 (1997).

Krause, F., "Eine Lösung des Dynamoproblems auf der Grundlage einer linearen Theorie der magnetohydrodynamischen Turbulenz," *Thesis*, Universität Jena (1967).

Krause, F. and Meinel, R., "Stability of simple nonlinear α^2-dynamos," *Geophys. Astrophys. Fluid Dynam.* **43**, 95–117 (1988).

Krause, F. and Rädler, K.-H., *Mean-field Magnetohydrodynamics and Dynamo Theory*, Pergamon Press, Oxford (1980).

Krause, F. and Schmidt, H.-J., "A low-dimensional attractor for modeling the reversals of the Earth's magnetic field," *Phys. Earth Planet. Inter.* **52**, 23 (1988).

Krivodubskij, V.N. and Schultz, M., "Complete alpha-tensor for solar dynamo," in: *The Cosmic Dynamo* (Eds. F. Krause, K.-H. Rädler and G. Rüdiger), Kluwer, Dordrecht, pp. 25–26 (1993).

Küker, M., Arlt, R. and Rüdiger, G., "The Maunder minimum as due to magnetic Λ-quenching," *Astron. Astrophys.* **343**, 977–982 (1999).

Küker, M., Rüdiger, G. and Kitchatinov, L.L., "An alpha-omega model of the solar differential rotation," *Astron. Astrophys.* **279**, L1–L4 (1993).

Küker, M., Rüdiger, G. and Pipin, V.V., "Solar torsional oscillations as due to the magnetic quenching of the Reynolds stress," *Astron. Astrophys.* **312**, 615–623 (1996).

Kurths, J., Brandenburg, A., Feudel, U. and Jansen, W., "Chaos in nonlinear dynamo models," in: *The Cosmic Dynamo* (Eds. F. Krause, K.-H. Rädler and G. Rüdiger), Kluwer, Dordrecht, pp. 83–89 (1993).

Kurths, J., Feudel, U., Jansen, W., Schwarz, U. and Voss, H., "Solar variability: simple models and proxy data," in: *Proc. Int. School Physics Enrico Fermi, Course CXXXIII* (Eds. G. Castagnoli and A. Provenzale), IOS Press, Amsterdam, p. 247 (1997).

Levy, E.H., "Physical assessment of stellar dynamo theory," in: *Cool Stars, Stellar Systems, and the Sun* (Eds. M.S. Giampapa and J.A. Bookbinder), ASP Conf. Ser. 26, p. 223 (1992).

Low, B.C., "Solar activity and the corona," *Solar Phys.* **167**, 217–265 (1996).

Malkus, W.V.R. and Proctor, M.R.E., "The macrodynamics of α-effect dynamos in rotating fluids," *J. Fluid Mech.* **67**, 417–443 (1975).

Markiel, J.A. and Thomas, J.H., "Solar interface dynamo models with a realistic rotation profile," *Astrophys. J.* **523**, 827–837 (1999).

Moffatt, K.H., *Magnetic Field Generation in Electrically Conducting Fluids*, Cambridge University Press (1978).

Moreno-Insertis, R., "Rise times of horizontal magnetic flux tubes in the convection zone of the sun," *Astron. Astrophys.* **122**, 241–250 (1983).

Moss, D., Tuominen, I. and Brandenburg, A., "Nonlinear dynamos with magnetic buoyancy in spherical geometry," *Astron. Astrophys.* **228**, 284–294 (1990).

Moss, D., Brandenburg, A., Tavakol, R. and Tuominen, I., "Stochastic effects in mean field dynamos," *Astron. Astrophys.* **265**, 843–849 (1992).

Nesme-Ribes, E., Sokoloff, D., Ribes, J.C. and Kremliovsky, M., "The Maunder minimum and the solar dynamo," in: *The Solar Engine and its Influence on Terrestrial Atmosphere and Climate* (Ed. E. Nesme-Ribes), Springer-Verlag, Berlin, p. 71 (1994).

Noyes, R.W., Weiss, N.O. and Vaughan, A.H., "The relation between stellar rotation rate and activity cycle periods," *Astrophys J.* **287**, 769–773 (1984).

Oláh, K., Kolláth, Z. and Strassmeier, K.G., "Multiperiodic light variations of active stars," *Astron. Astrophys.* **356**, 643–653 (2000).

Ossendrijver, A.J.H., "On the cycle periods of stellar dynamos," *Astron. Astrophys.* **323**, 151–157 (1997).

Ossendrijver, A.J.H. and Hoyng, P., "Stochastic and nonlinear fluctuations in a mean field dynamo," *Astron. Astrophys.* **313**, 959–970 (1996).

Ossendrijver, A.J.H., Hoyng, P. and Schmitt, D., "Stochastic excitation and memory of the solar dynamo," *Astron. Astrophys.* **313**, 938–948 (1996).

Ossendrijver, M., Stix, M. and Brandenburg, A., "Magnetoconvection and dynamo coefficients: Dependence of the alpha effect on rotation and magnetic field," *Astron. Astrophys.* **376**, 713–726 (2001).

Otmianowska-Mazur, K., Urbanik, M. and Terech, A., "Magnetic field in a turbulent galactic disk," *Geophys. Astrophys. Fluid Dynam.* **66**, 209–222 (1992).

Otmianowska-Mazur, K., Rüdiger, G., Elstner, D. and Arlt, R., "The turbulent EMF as a time series and the 'quality' of dynamo cycles," *Geophys. Astrophys. Fluid Dynam.* **86**, 229–248 (1997).

Parker, E.N., "The generation of magnetic fields in astrophysical bodies. X-Magnetic buoyancy and the solar dynamo," *Astrophys. J.* **198**, 205–209 (1975).

Parker, E.N., "The dynamo dilemma," *Solar Phys.* **110**, 11–21 (1987).

Pevtsov, A.A., Canfield, R.C. and Metcalf, T.R., "Latitudinal variation of helicity of photospheric magnetic fields," *Astrophys. J.* **440**, L109–L112 (1995).

Pipin, V.V., "The Gleissberg cycle by a nonlinear $\alpha\Lambda$ dynamo," *Astron. Astrophys.* **346**, 295–302 (1999).

Prautzsch, T., "The dynamo mechanism in the deep convection zone of the Sun," in: *Theory of Solar and Planetary Dynamos* (Eds. PC. Matthews and A.M. Rucklidge), Cambridge University Press, p. 249 (1993).

Pulkkinen, P.J., Brooke, J., Pelt, J. and Tuominen, I., "Long-term variation of sunspot lattitudes," *Astron. Astrophys.* **341**, L43–L46 (1999).

Rädler, K.-H., "Zur Elektrodynamik turbulent bewegter leitender Medien. Teil II," *Zeitschr. f. Naturforschg.* **23a**, 1851–1860 (1968).

Rädler, K.-H. and Seehafer, N., "Relations between helicities in mean-field dynamo models," in: *Topological Fluid Mechanics* (Eds. H.K. Moffatt and A. Tsinober), Cambridge University Press, p. 157 (1990).

Ribes, J.C. and Nesme-Ribes, E., "The solar sunspot cycle in the Maunder minimum AD1645," *Astron. Astrophys.* **276**, 549–563 (1993).

Roberts, P.H., "Kinematic dynamo models," *Phil. Trans. R. Soc. Lond.* A. **272**, 663 (1972).

Roberts, P.H. and Stix, M., "α-Effect dynamos by the Bullard-Gellman formalism," *Astron. Astrophys.* **18**, 453–466 (1972).

Rozelot, J.P., "On the chaotic behaviour of the solar activity," *Astron. Astrophys.* **297**, L45–L48 (1995).

Rüdiger, G., Tuominen, I., Krause, F. and Virtanen, H., "Dynamo-generated flows in the Sun," *Astron. Astrophys.* **166**, 306–318 (1986).

Rüdiger, G., *Differential Rotation and Stellar Convection: Sun and Solar-type Stars*, Gordon and Breach Science Publishers, New York (1989).

Rüdiger, G. and Arlt, R., "Cycle times and magnetic amplitudes in nonlinear 1D $\alpha^2\Omega$-dynamos," *Astron. Astrophys.* **316**, L17–L20 (1996).

Rüdiger, G. and Brandenburg, A., "A solar dynamo in the overshoot layer: cycle period and butterfly diagram," *Astron. Astrophys.* **296**, 557–566 (1995).

Rüdiger, G. and Kitchatinov, L.L., "α-effect and α-quenching," *Astron. Astrophys.* **269**, 581–588 (1993).

Rüdiger, G. and Kitchatinov, L.L., "The slender solar tachocline: a magnetic model," *Astron. Nachr.* **318**, 273–279 (1997).

Rüdiger, G. and Pipin, V.V., "Viscosity alpha and dynamo-alpha for magnetic-driven turbulence in density-stratified Kepler disks," *Astron. Astrophys.* **362**, 756–761 (2000).

Rüdiger, G. and Schultz, M., "Nonlinear galactic dynamo models with magnetic-supported interstellar gas-density stratification," *Astron. Astrophys.* **319**, 781–787 (1997).

Rüdiger, G., Pipin, V.V. and Belvedere, G., "Alpha-effect, helicity and angular momentum transport for a rotating magnetically driven turbulence in the solar convection zone," *Solar Phys.* **198**, 241–251 (2001).

Rüdiger, G., Kitchatinov, L.L., Küker, M. and Schultz, M., "Dynamo models with magnetic diffusivity-quenching," *Geophys. Astrophys. Fluid Dynam.* **78**, 247–260 (1994).

Saar, S.H., "Recent measurements of stellar magnetic fields," in: *Stellar Surface Structure* (Eds. K.G. Strassmeier and J.L. Linsky), Kluwer, Dordrecht, pp. 237–243 (1996).

Saar, S.H. and Brandenburg, A., "Time evolution of the magnetic activity cycle period. II. Results for an expanded stellar sample," *Astrophys J.* **524**, 295–310 (1999).

Saar, S.H. and Baliunas, S.L., "Recent advances in stellar cycle research," in: *The solar cycle* (Ed. K.L. Harvey), *ASP Conf. Ser.* **27**, 150–167 (1992a).

Saar, S.H. and Baliunas, S.L., "The magnetic cycle of κ Ceti (G5V)," in: *The Solar Cycle* (Ed. K.L. Harvey), ASP Conf. Ser. 27, pp. 197–202 (1992b).

Schlichenmaier, R. and Stix, M., "The phase of the radial mean field in the solar dynamo," *Astron. Astrophys.* **302**, 264–270 (1995).

Schmitt, D., *Thesis*, Universität Göttingen (1985).

Schmitt, D., "An $\alpha\omega$-dynamo with an α-effect due to magnetostrophic waves," *Astron. Astrophys.* **174**, 281–287 (1987).

Schmitt, D., "The solar dynamo," in: *The Cosmic Dynamo* (Eds. F. Krause, K.-H. Rädler and G. Rüdiger), Kluwer, Dordrecht, pp. 1–12 (1993).

Schmitt, D. and Schüssler, M., "Non-linear dynamos. I. One-dimensional model of a thin layer dynamo," *Astron. Astrophys.* **223**, 343–351 (1989).

Schmitt, D., Schüssler, M. and Ferriz-Mas, A., "Intermittent solar activity by an on-off dynamo," *Astron. Astrophys.* **311**, L1–L4 (1996).

Schüssler, M., "The solar torsional oscillation and dynamo models of the solar cycle," *Astron. Astrophys.* **94**, L17–L18 (1981).

Schüssler, M., "Magnetic fields and the rotation of the solar convection zone," in: *The Internal Solar Angular Velocity* (Ed. R. Durney), Reidel, Dordrecht, pp. 303–320 (1987).

Schwarz, U., "Zeitreihenanalyse astrophysikalischer Aktivitätsphänomene," *Thesis*, Universität Potsdam (1994).

Seehafer, N., "Electric current helicity in the solar atmosphere," *Solar Phys.* **125**, 219–323 (1990).

Soon, W.H., Baliunas, S.L. and Zhang, Q., "An interpretation of cycle periods of stellar chromospheric activity," *Astrophys. J.* **414**, L33–L36 (1993).

Spiegel, E.A. and Weiss, N.O., "Magnetic activity and variations in solar luminosity," *Nature* **287**, 616–617 (1980).

Spörer, G., "Über die Periodizität der Sonnenflecken seit dem Jahre 1618," *Vierteljahresschrift der Astron. Ges.* **22**, 323 (1887).

Steenbeck, M. and Krause, F., "Zur Dynamotheorie stellarer und planetarer Magnetfelder. I. Berechnung sonnenähnlicher Wechselfeldgeneratoren," *Astron. Nachr.* **291**, 49–84 (1969).

Stix, M., "Differential rotation and the solar dynamo," *Astron. Astrophys.* **47**, 243–254 (1976).

Stix, M., "The solar dynamo," *Geophys. Astrophys. Fluid Dynam.* **62**, 211–228 (1991).

Stix, M. and Skaley, D., "The equation of state and the frequencies of solar p modes," *Astron. Astrophys.* **232**, 234–238 (1990).

Tobias, S.M., "Grand minima in nonlinear dynamos," *Astron. Astrophys.* **307**, L21–L24 (1996).

Tobias, S.M., "The solar cycle: parity interactions and amplitude modulation," *Astron. Astrophys.* **322**, 1007–1017 (1997).

Tobias, S.M., "Relating stellar cycle periods to dynamo calculations," *Mon. Not. R. Astr. Soc.* **296**, 653–661 (1998).

Tuominen, I., Rüdiger, G. and Brandenburg, A., "Observational constraints for solar-type dynamos," in: *Activity in Cool Star Envelopes* (Eds. Havnes *et al.*), Kluwer, Dordrecht, p. 13 (1988).

Tuominen, I., Berdyugina, S.V., Korpi, M.J. and Rönty, T., "Nonaxisymmetric stellar dynamos," in: *Stellar Dynamos: Nonlinearity and Chaotic Flows* (Eds. M. Núñez, A. Ferriz-Mas), *ASP Conf. Ser.* **178**, 195 (1999).

Vainshtein, S.I. and Cattaneo, F., "Nonlinear restriction on dynamo action," *Astrophys. J.* **393**, 165–171 (1992).

van Ballegooijen, A.A., "The overshoot layer at the base of the solar convective zone and the problem of magnetic flux storage," *Astron. Astrophys.* **113**, 99–112 (1982).

van Ballegooijen, A.A., "Understanding the solar cycle," in: *Synoptic Solar Physics* (Eds. K.S. Balasubramaniam, J. Harvey and D. Rabin), ASP Conf. Ser. 140, p. 17 (1998).

Verma, V.K., "On the north-south asymmetry of solar activity cycles," *Astrophys. J.* **403**, 797–800 (1993).

Vishniac, E.T. and Brandenburg, A., "An incoherent $\alpha\omega$-dynamo in accretion disks," *Astrophys. J.* **475**, 263–274 (1997).

Voss, H., Kurths, J. and Schwarz, U., "Reconstruction of grand minima of solar activity from radiocarbon data – linear and nonlinear signal analysis," *J. Geophys. Res.* A **101**, 15637–15644 (1996).

Voss, H., Sanchez, A., Zolitschka, B., Brauer, A. and Negendank, J.F.W., "Solar activity variations recorded in varved sediments from the crater lake of Holzmaar – A maar lake in the Westeifel volcanic field, Germany," *Surv. Geophys.* **18**, 163 (1997).

Wälder, M., Deinzer, W. and Stix, M., "Dynamo action associated with random waves in a rotating stratified fluid," *J. Fluid Mech.* **96**, 207–222 (1980).

Weiss, N.O., "Solar and stellar dynamos," in: *Lectures on Solar and Planetaty Dynamos* (Eds. M.R.E. Proctor and A.D. Gilbert), Cambridge University Press, pp. 59–95 (1994).

Weiss, N.O., Cattaneo, F. and Jones, C.A., "Periodic and aperiodic dynamo waves," *Geophys. Astrophys. Fluid Dynam.* **30**, 305–342 (1984).

Wittmann, A., "The sunspot cycle before the Maunder minimum," *Astron. Astrophys.* **66**, 93–97 (1978).

Yoshimura, H., "Phase relation between the poloidal and toroidal solar-cycle general magnetic fields and location of the origin of the surface magnetic fields," *Solar Phys.* **50**, 3–23 (1976).

Yoshimura, H., "Solar cycle Lorentz force waves and the torsional oscillations of the sun," *Astrophys. J.* **247**, 1102–1112 (1981).

Ziegler, U., Yorke, H.W. and Kaisig, M., "The role of supernovae for the galactic dynamo: the full alpha-tensor," *Astron. Astrophys.* **305**, 114–124 (1996).

7 Highly supercritical convection in strong magnetic fields

Keith Julien[1], Edgar Knobloch[2] and Steve Tobias[3]

[1]*Department of Applied Mathematics, University of Colorado, Boulder,*
  CO 80309-0526, USA, E-mail: julien@colorado.edu
[2]*Department of Physics, University of California, Berkeley, CA 94720, USA,*
  E-mail: knobloch@physics.berkeley.edu
[3]*Department of Applied Mathematics, University of Leeds, Leeds, LS2 9JT, UK,*
  E-mail: smt@amsta.leeds.ac.uk

Fully nonlinear convection in a strong imposed magnetic field is studied in the regime in which the convective velocities are not strong enough to distort the magnetic field substantially. Motivated by convection in sunspots both vertical and inclined imposed fields are considered. In this regime the leading order nonlinearity is provided by the distortion of the horizontally averaged temperature profile. For overstable convection this profile is determined from the solution of a nonlinear eigenvalue problem for the (time-averaged) Nusselt number and oscillation frequency, and evolves towards an isothermal profile with increasing Rayleigh number. In the presence of variable magnetic Prandtl number $\zeta(z)$ the profile is asymmetric with respect to midlevel, but nonetheless develops an isothermal core in the highly supercritical regime. A hysteretic transition between two distinct convection regimes is identified in the inclined case, and used to suggest an explanation for the sharp boundary between the sunspot umbra and penumbra. These results are obtained via an asymptotic expansion in inverse powers of the Chandrasekhar number, and generalize readily to a polytropic atmosphere.

7.1. Introduction

The study of convection in an imposed magnetic field is motivated primarily by astrophysical applications, particularly by the observed magnetic field dynamics in the solar convection zone (Hughes and Proctor, 1988). Applications to sunspots (Thomas and Weiss, 1992) have led several authors to investigate the suppression of convection by strong magnetic fields. The linear and weakly nonlinear theory describing this suppression is summarized by Chandrasekhar (1961) and Proctor and Weiss (1982).

In this paper, we summarize a recent development that allows us to extend these strong-field results, semi-analytically, into the fully nonlinear regime and generalize it to include background stratification. The resulting solutions are valid at Rayleigh numbers arbitrarily far above onset. These solutions are constructed via an asymptotic expansion in inverse powers of the Chandrasekhar number Q. This dimensionless number measures the strength of the imposed magnetic field, and in the large Q limit leads to a simplified set of dynamical equations. In these equations the dominant nonlinearity arises from the nonlinear distortion of the mean temperature profile; the strong magnetic field resists distortion by the velocity field, and the Lorentz force arising from the distortion of the magnetic field remains small. As a result our solutions are characterized by a single wavenumber in the horizontal with the vertical structure given by the solution of a nonlinear eigenvalue problem for the Nusselt number. The derivation of this nonlinear eigenvalue problem can be performed analytically although the problem itself must be solved numerically, and can be performed for both steady and oscillatory magnetoconvection and for both vertical and tilted imposed fields. Although these results are formally obtained for stress-free boundary conditions, in the strong magnetic field limit they describe the dynamics outside of narrow boundary layers for other types of boundary conditions as well.

Of particular interest is the fact that our approach applies equally to the case in which the magnetic Prandtl number ζ depends nontrivially on the depth z within the layer. Not only does this allow us to explore more realistic profiles of the magnetic and thermal diffusivities but it also makes accessible the interesting case in which ζ passes through one somewhere in the layer. Recall that near the surface of the solar convection zone ($1500\,\text{km} < z < 20{,}000\,\text{km}$) the thermal diffusivity is reduced owing to the increase in opacity caused by ionization and in this region $\zeta > 1$, favouring steady convection. Both above and below this region $\zeta < 1$, favouring overstable convection. It has been suggested (Knobloch and Weiss, 1984; Weiss et al., 1990) that these changes in ζ are responsible for the presence of umbral dots with the more efficient steady convection penetrating into the overlying regions of less efficient overstable convection and forming the intermittent bright spots observed in sunspot umbrae. Our solutions in this regime lend further support to this idea. We find strongly nonlinear two- and three-dimensional solutions in which overturning convection in the lower part of the layer is coupled to overstable convection in the upper part. The resulting solution is periodic in time but the oscillation amplitude is small near the bottom and large near the top. Moreover, the oscillation period becomes independent of the applied Rayleigh number at high Rayleigh numbers indicating that the oscillation is of magnetic origin. In the case of tilted magnetic field our methods permit us to construct two-dimensional solutions only. Although not as general, these solutions reveal an unexpected and important transition with increasing Rayleigh number to a novel form of convection in which the input heat is converted into magnetic energy instead of being transported across the layer. This transition occurs only for sufficiently large tilt angles and is hysteretic, and may be related to the abrupt transition from umbra to penumbra observed in sunspots. As a first step towards a realistic sunspot model based on these ideas we also describe how our methods generalize to a stratified atmosphere.

7.2. Formulation of the problem

We begin with Boussinesq magnetoconvection in a plane horizontal layer described by the dimensionless equations

$$\frac{1}{\sigma}\frac{D\mathbf{v}}{Dt} = -\nabla\pi + \zeta Q\mathbf{B}\cdot\nabla\mathbf{B} + RaT\hat{\mathbf{z}} + \nabla^2\mathbf{v}, \tag{7.1}$$

$$\frac{DT}{Dt} = \nabla^2 T, \tag{7.2}$$

$$\frac{D\mathbf{B}}{Dt} = \mathbf{B}\cdot\nabla\mathbf{v} - \nabla\times(\zeta\nabla\times\mathbf{B}), \tag{7.3}$$

$$\nabla\cdot\mathbf{v} = 0, \quad \nabla\cdot\mathbf{B} = 0. \tag{7.4}$$

Here $\mathbf{v} = (u, v, w)$ is the velocity field in Cartesian coordinates (x, y, z) with z vertically upwards. The symbol T denotes the temperature, while π is the total (thermal and magnetic) pressure. The velocity field is written in the form $\mathbf{v} = \bar{\mathbf{U}} + \mathbf{u}$, where $\bar{\mathbf{U}}(z)$ is a possible mean flow and $\mathbf{u}(x, y, z, t)$ is the convective flow. Likewise, the dimensionless magnetic field is assumed to be the superposition $\mathbf{B} = \hat{\mathbf{r}} + \bar{\mathbf{B}} + \mathbf{b}$ of an imposed oblique field of unit strength, a mean field $\bar{\mathbf{B}}(z)$, and a three-dimensional field $\mathbf{b}(x, y, z, t)$ both due to the presence of convection. The oblique field is denoted by $\hat{\mathbf{r}} = (\sin \vartheta, 0, \cos \vartheta)$, where ϑ denotes the angle with respect to the vertical in the (x, z) plane. The equations have been nondimensionalized with respect to the thermal diffusion time in the vertical. The resulting dimensionless parameters

$$Q = \frac{B_0^2 d^2}{\mu_0 \rho \eta \nu}, \quad Ra = \frac{g \alpha \Delta T d^3}{\nu \kappa}, \quad \sigma = \frac{\nu}{\kappa}, \quad \zeta = \frac{\eta}{\kappa}, \tag{7.5}$$

are the Chandrasekhar, Rayleigh, and thermal and magnetic Prandtl numbers, respectively. We write

$$\begin{aligned}
\mathbf{u}(x, y, z, t) &= \nabla \times \phi(x, y, z, t)\hat{\mathbf{z}} + \nabla \times \nabla \times \psi(x, y, z, t)\hat{\mathbf{z}}, \\
\mathbf{b}(x, y, z, t) &= \nabla \times A(x, y, z, t)\hat{\mathbf{z}} + \nabla \times \nabla \times B(x, y, z, t)\hat{\mathbf{z}}.
\end{aligned} \tag{7.6}$$

Equations (7.1)–(7.4) are solved for a fluid confined between boundaries at fixed temperatures,

$$T(0) = 1, \quad T(1) = 0, \tag{7.7}$$

which are impenetrable and either stress-free or no-slip at the top and bottom. The simplest magnetic boundary conditions, employed by Matthews *et al.* (1992), require that the field be tilted by the same angle ϑ to the vertical everywhere on the top and bottom boundaries. However, in the large Q limit the detailed nature of the boundary conditions becomes unimportant: the solutions for different magnetic or velocity boundary conditions differ in thin boundary layers at top and bottom only. Periodic boundary conditions in the horizontal are assumed.

For large values of the Chandrasekhar number Q simplified equations describing the nonlinear problem can be obtained using the scaling (cf. Julien *et al.*, 1999, 2000)

$$(x, y) = Q^{-1/4}(x', y'), \quad t = Q^{-1/2}t'. \tag{7.8}$$

With this scaling we focus on small horizontal scales (and high frequency oscillations in the case of overstable convection). For an alternative scaling see Matthews (1999). Because the strong magnetic field tends to align the cells with the tilt we expect the solutions to manifest small scale oscillations in the vertical with an $O(Q^{1/4})$ vertical wavenumber, in addition to variation on the usual $O(1)$ scale corresponding to the layer depth. We denote these scales by z' and Z, respectively, and write

$$\partial_x, \partial_y = Q^{1/4}(\partial_{x'}, \partial_{y'}), \quad \partial_z = Q^{1/4}\partial_{z'} + D, \quad \partial_t = Q^{1/2}\partial_{t'}, \tag{7.9}$$

where $D \equiv \partial_Z$. Throughout we focus on $O(Q)$ Rayleigh numbers, i.e. we also write $Ra = QRa'$. The Prandtl numbers σ and ζ are not scaled and remain of order one. However, the latter is allowed to vary with depth. The resulting expansion reflects the tendency towards small scale motion aligned with the inclined magnetic field and describes correctly not only the linear and nonlinear properties of solutions with these wavenumbers but also those with the $O(Q^{1/6})$ wavenumbers selected by linear theory (Chandrasekhar, 1961). In addition the analysis with $O(Q^{1/4})$ wavenumbers captures the transition from overstable convection preferred at small ζ to steady overturning convection preferred for $\zeta > 1$. Next, we scale the fluid variables such that all three components of the velocity and magnetic field perturbations are comparable, with $\mathbf{u} \approx O(Q^{1/4})$ and $\mathbf{b} \approx O(Q^{-1/4})$, i.e. we write

$$\phi = \phi'(x', y', z', Z, t'; Q), \tag{7.10}$$

$$\psi = Q^{-1/4}\psi'(x', y', z', Z, t'; Q), \tag{7.11}$$

$$A = Q^{-1/2}A'(x', y', z', Z, t'; Q), \tag{7.12}$$

$$B = Q^{-3/4}B'(x', y', z', Z, t'; Q). \tag{7.13}$$

Finally, to allow substantial deformation of the mean temperature gradient, we write

$$T = \theta_0(Z; Q) + Q^{-1/4}\theta'(x', y', z', Z, t'; Q), \tag{7.14}$$

where θ' denotes the fluctuations from the mean. In the following we drop all primes.

The above scaling works for both stress-free and no-slip boundary conditions at top and bottom since it describes the dynamics in the bulk, i.e. outside of thin boundary layers required by specific velocity and magnetic field boundary conditions (Julien *et al.*, 1999, 2000). The resulting equations are solved by an asymptotic expansion in powers of $Q^{-1/4}$ of the form

$$\mathbf{\Psi} = \mathbf{\Psi}_1 + Q^{-1/4}\mathbf{\Psi}_2 + Q^{-1/2}\mathbf{\Psi}_3 + \cdots, \tag{7.15}$$

where $\mathbf{\Psi} \equiv (\phi, \psi, \theta, A, B)^{\mathrm{T}}$. At $O(Q^0)$ one obtains

$$\hat{\mathbf{r}} \cdot \nabla_0\phi_1 = 0, \quad \hat{\mathbf{r}} \cdot \nabla_0\psi_1 = 0, \quad \hat{\mathbf{r}} \cdot \nabla_0 A_1 = 0, \quad \hat{\mathbf{r}} \cdot \nabla_0 B_1 = 0, \tag{7.16}$$

where $\nabla_0 \equiv (\partial_x, \partial_y, \partial_z)$.

Thus on small scales all perturbations align with the imposed magnetic field. Solutions of this type take the form

$$\mathbf{\Psi}_1(\mathbf{x}, Z, t) = \int \mathbf{\Psi}_1(\mathbf{k}_0, Z, t)e^{i\mathbf{k}_0 \cdot \mathbf{x}}d\mathbf{k}_0 + \text{c.c.}, \tag{7.17}$$

where $\mathbf{k}_0 \cdot \hat{\mathbf{r}} = 0$. Since $\hat{\mathbf{r}} = (\sin\vartheta, 0, \cos\vartheta)$ and $(k_{0x}, k_{0y}) \equiv k_{0\perp}(\cos\chi, \sin\chi)$ it follows that $k_{0z} = -k_{0x}\tan\vartheta = -k_{0\perp}\cos\chi\tan\vartheta$. The solvability condition at $O(Q^{-1/4})$ yields evolution equations for the amplitudes $\mathbf{\Psi}_1(\mathbf{k}_0, Z, t)$:

$$\frac{1}{\sigma}\left(\partial_t\phi_1 - \frac{1}{k_{0\perp}^2}PN_\phi(\mathbf{\Psi}_1)\right) = \zeta\left(\hat{r}_z DA_1 - \frac{1}{k_{0\perp}^2}PM_\phi(\mathbf{\Psi}_1)\right) - k_0^2\phi_1, \tag{7.18}$$

$$\frac{1}{\sigma}\left(\partial_t\psi_1 + \frac{1}{k_0^2 k_{0\perp}^2}PN_\psi(\mathbf{\Psi}_1)\right) = \frac{Ra}{k_0^2}\theta_1 + \zeta\left(\hat{r}_z DB_1 + \frac{1}{k_0^2 k_{0\perp}^2}PM_\psi(\mathbf{\Psi}_1)\right) - k_0^2\psi_1, \tag{7.19}$$

$$\partial_t A_1 - \frac{1}{k_{0\perp}^2}PM_A(\mathbf{\Psi}_1) = \hat{r}_z D\phi_1 - \zeta k_0^2 A_1, \tag{7.20}$$

$$\partial_t B_1 - \frac{1}{k_{0\perp}^2}PM_B(\mathbf{\Psi}_1) = \hat{r}_z D\psi_1 - \zeta k_0^2 B_1, \tag{7.21}$$

where P is a projection operator that filters out the fast spatial variation, defined by

$$Pf(\boldsymbol{\Psi}_1) \equiv \frac{1}{(2\pi)^3} \int e^{-i\mathbf{k_0}\cdot\mathbf{x}} f(\boldsymbol{\Psi}_1)\,d\mathbf{x}, \tag{7.22}$$

and the quantities M, N denote nonlinear terms. Explicit expressions for these terms can be found in Julien *et al.* (1999).

The temperature equation yields the following equations at $O(Q^{1/4})$ and $O(Q^0)$, respectively,

$$\partial_t\theta_1 - J[\phi_1,\theta_1] + \left(\nabla_{0\perp}\partial_z\psi_1 \cdot \nabla_{0\perp}\theta_1 - \nabla_{0\perp}^2\psi_1\partial_z\theta_1\right) - \nabla_{0\perp}^2\psi_1 D\theta_0 = \nabla_0^2\theta_1, \tag{7.23}$$

$$\partial_t\theta_2 - J[\phi_1,\theta_2] - J[\phi_2,\theta_1] + \left(\nabla_{0\perp}\partial_z\psi_1 \cdot \nabla_{0\perp}\theta_2 - \nabla_{0\perp}^2\psi_1\partial_z\theta_2\right)$$

$$+ \left(\nabla_{0\perp}\partial_z\psi_2 \cdot \nabla_{0\perp}\theta_1 - \nabla_{0\perp}^2\psi_2\partial_z\theta_1\right) + \left(\nabla_{0\perp}D\psi_1 \cdot \nabla_{0\perp}\theta_1 - \nabla_{0\perp}^2\psi_1 D\theta_1\right)$$

$$- \nabla_{0\perp}^2\psi_2 D\theta_0 = \nabla_0^2\theta_2 + 2\partial_z D\theta_1 + D^2\theta_0. \tag{7.24}$$

Here the symbol $J(f,g)$ denotes the horizontal Jacobian $f_x g_y - f_y g_x$. Equation (7.23) can be solved for θ_1. Once this is done the solvability condition for the mean part of θ_2 yields

$$D^2\theta_0 + D(\overline{\nabla_\perp^2\psi_1\,\theta_1}) = 0, \tag{7.25}$$

which can be integrated once, obtaining

$$D\theta_0 + \overline{\nabla_\perp^2\psi_1\,\theta_1} = -K. \tag{7.26}$$

For steady patterns the constant K is identified with the Nusselt number; for oscillatory patterns we extend the meaning of the overbar to indicate an average over time as well. For such patterns K represents the time-averaged Nusselt number.

Equations (7.18)–(7.21), (7.23) and (7.26) constitute a closed set of evolution equations for the vertical profiles of the different fields. The simplest case of such evolution is offered by (tilted) two-dimensional solutions of the form

$$\boldsymbol{\Psi}_1 = \boldsymbol{\Psi}_L(Z)\exp(i\omega t + i\mathbf{k_0}\cdot\mathbf{x}) + \boldsymbol{\Psi}_R(Z)\exp(i\omega t - i\mathbf{k_0}\cdot\mathbf{x}) + \text{c.c.} \tag{7.27}$$

For these solutions the nonlinear terms in (7.18)–(7.21) vanish and consequently the toroidal fields ϕ_1, A_1 also vanish (Julien *et al.*, 1999). The remaining quantities ψ_1, B_1 and θ_1 satisfy linear equations:

$$\left(\frac{i\omega}{\sigma} + k_0^2\right) k_0^2\psi_{(L,R)} = Ra\theta_{(L,R)} + \zeta k_0^2\hat{r}_z DB_{(L,R)}, \tag{7.28}$$

$$(i\omega + \zeta k_0^2)B_{(L,R)} = \hat{r}_z D\psi_{(L,R)}, \tag{7.29}$$

$$(i\omega + k_0^2)\theta_{(L,R)} = -k_{0\perp}^2\psi_{(L,R)}\,D\theta_0. \tag{7.30}$$

These coupled equations may be collapsed using the transformation

$$\boldsymbol{\Psi}_{1L} = \frac{\boldsymbol{\Psi}_1}{\sqrt{1+|c|^2}}, \qquad \boldsymbol{\Psi}_{1R} = \frac{c\boldsymbol{\Psi}_1}{\sqrt{1+|c|^2}}, \tag{7.31}$$

where $c = O(1)$ for travelling (standing) waves. Here $\Psi_1(Z) = (\Psi(Z), B(Z), \Theta(Z))$. From (7.26) we now obtain

$$D\theta_0\left[1 + \frac{2k_0^2 k_{0\perp}^4}{\omega^2 + k_0^4}|\Psi|^2\right] = -K, \tag{7.32}$$

with K given by the requirement that $\theta_0(0) = 1, \theta_0(1) = 0$:

$$K = \left[\int_0^1 \frac{\omega^2 + k_0^4}{\omega^2 + k_0^4 + 2k_0^2 k_{0\perp}^4 |\Psi|^2}\,dZ\right]^{-1}. \tag{7.33}$$

while (7.28)–(7.30) yield the nonlinear eigenvalue problem

$$D^2\Psi - \frac{(D\zeta)k_0^2}{i\omega + \zeta k_0^2}D\Psi - \frac{1}{\hat{r}_\zeta^2 \zeta}\left(\frac{i\omega}{\sigma} + k_0^2\right)(i\omega + \zeta k_0^2)\Psi$$

$$+ \frac{RaK}{\hat{r}_\zeta^2 \zeta} \frac{(i\omega + \zeta k_0^2)(-i\omega + k_0^2)}{\omega^2 + k_0^4 + 2k_0^2 k_{0\perp}^4 |\Psi|^2} \frac{k_{0\perp}^2}{k_0^2}\Psi = 0. \tag{7.34}$$

The solutions of this problem depend on the prescribed function $\zeta(Z)$ as well as the parameters Ra, k_0, $k_{0\perp}$ and σ. The corresponding results for a vertical magnetic field are recovered on setting $\hat{r}_\zeta = 1$ (Julien *et al.*, 1999). Note that steady solutions ($\omega = 0$) are independent of both σ and ζ, at least if ζ is depth-independent. The latter is not generally true and for finite Q the steady solutions do depend on ζ.

7.2.1. *Dynamics and symmetry*

The solutions obtained by solving the nonlinear eigenvalue problem (7.34) represent a special type of simple, stationary, spatially periodic solutions. Despite their simplicity these solutions are important. For example, Julien *et al.* (1999) show that in the case of an imposed vertical field all spatially periodic three-dimensional solutions possess the same (time-averaged) Nusselt number and frequency. The selection among these planforms is due to subdominant eigenvalues describing their relative stability but absent from the theory. In addition to this lack of planform selection there is one other effect that disappears in the large Q limit. Boussinesq convection in a tilted field has a special property: the equations (and boundary conditions) are equivariant (i.e. symmetric) under a reflection in a vertical plane followed by a reflection in the midplane of the layer. As a result if a left-travelling wave is a solution so is a right-travelling wave, and consequently so are standing waves. Moreover, steady solutions in the form of tilted convection cells are also possible. However, as noted by Matthews *et al.* (1992) and discussed in more detail by Knobloch (1994), when the properties of the layer depend on depth the midplane symmetry is usually lost. In this case there are generically no steady state bifurcations from the conduction state and the primary instability is a Hopf bifurcation to travelling waves with a preferred direction of propagation selected by the tilt together with the depth-dependence. There are also no bifurcations to standing waves, the counterparts of which are quasiperiodic waves that appear in a secondary bifurcation from the primary travelling wave branch. These effects are absent in our limit because they rely on frequency splitting due to the depth-dependence and this remains of order one, i.e. on the $O(Q^{1/2})$ timescale the drift of our steady solutions is negligible and so is the second frequency accompanying our standing waves, or the frequency difference between left- and right-travelling waves. Likewise, the splitting in the critical Rayleigh numbers for the onset of left- and right-travelling waves remains $O(1)$ and hence small compared to the $O(Q)$ Rayleigh numbers considered. It is important to bear these facts in mind when interpreting the solutions described below.

These shortcomings notwithstanding, the asymptotic equations (7.18)–(7.23) and (7.26) provide a much simplified description of the dynamics of the system in the high Q limit. The resulting equations for three-dimensional magnetoconvection in a vertical field,

$$\frac{1}{\sigma}\partial_t\phi_1 = \zeta(Z)DA_1 + \nabla_\perp^2\phi_1, \tag{7.35}$$

$$\frac{1}{\sigma}\partial_t\nabla_\perp^2\psi_1 = -Ra\theta_1 + \zeta(Z)D\nabla_\perp^2 B_1 + \nabla_\perp^4\psi_1, \tag{7.36}$$

$$\partial_t A_1 = D\phi_1 + \zeta(Z)\nabla_\perp^2 A_1, \tag{7.37}$$

$$\partial_t B_1 = D\psi_1 + \zeta(Z)\nabla_\perp^2 B_1, \tag{7.38}$$

$$\partial_t\theta_1 - \nabla_\perp^2\theta_1 = \nabla_\perp^2\psi_1 D\theta_0, \tag{7.39}$$

coupled via (7.26), bear a considerable similarity to those derived Babin *et al.* (1997) for rapidly rotating turbulence. These equations have an invariant subspace $\phi_1 = A_1 = 0$. The dynamics in this subspace are described by (7.36), (7.38), (7.39) and (7.26); at finite amplitude vortical motion may be excited by secondary instabilities, resulting in the spontaneous generation of helicity, both fluid and magnetic. However, even aside from this interesting possibility, it is clear that the dynamics in the zero-helicity invariant subspace can be turbulent, much as in the related problem of rapidly rotating convection (Julien *et al.*, 1998), i.e. that the solutions described by the nonlinear eigenvalue problem (7.34) need not be stable, particularly if three-dimensional periodic boxes with spatial period larger than $2\pi/k_{0\perp}$ are employed. Simulations of these equations (and of equations (7.18)–(7.23), (7.26)) in these circumstances are therefore of fundamental interest and will be described elsewhere.

7.3. Results

In this section we summarize the results obtained thus far from the nonlinear eigenvalue problem (7.34). This problem is to be solved subject to the boundary conditions

$$\Psi(0) = \Psi(1) = 0, \tag{7.40}$$

imposing impermeability of the boundaries. Once this is done the mean (i.e. averaged horizontally and in time) temperature profile θ_0 can be found from the relation (7.32). The calculations are performed on a discretised one-dimensional mesh using an iterative Newton–Raphson–Kantorovich scheme with $O(10^{-10})$ accuracy in the L_2 norm of $\Psi(Z)$ and the corresponding eigenvalues. For each Ra the solution determines K and ω as eigenvalues of equation (7.34) and the associated eigenfunction $\Psi(Z)$.

In the following we present results first for the case of a vertical magnetic field ($\vartheta = 0$), and then discuss the case of an inclined field ($\vartheta \neq 0$) in the two cases $\chi = 0$, $\chi = \pi/2$ (see Fig. 7.1). All results are obtained with the horizontal wavenumber $k_{0\perp} = 1$ and Prandtl number $\sigma = 1.1$. (In Julien *et al.* (1999) solutions are obtained for both $k_{0\perp} = 1$ and $k_0 = 1$.)

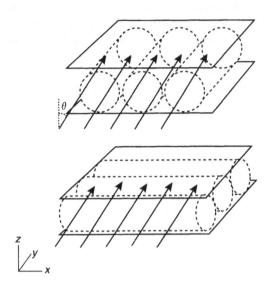

Figure 7.1 A sketch indicating the orientation of perpendicular ($\chi = 0$) and parallel ($\chi = \pi/2$) rolls.

7.3.1. Linear theory

We first summarize the linear stability properties of the conduction solution $\mathbf{v} = \mathbf{0}$, $T = 1 - z$, $\mathbf{B} = \hat{\mathbf{r}}$ in the two cases $\chi = 0$, $\chi = \pi/2$ obtained from the full equations with stress-free boundaries and fixed field inclination. Fig. 7.2 shows the critical Rayleigh number Ra_c and the corresponding wavenumber k_c as a function of Q for steady and oscillatory convection for several different values of the tilt angle ϑ when $\chi = 0$. As expected, when $\chi = 0$ the critical Rayleigh number increases rapidly with increasing Q, i.e. the inclined magnetic field has a stabilizing effect. This effect depends only weakly on the tilt angle but decreases as this angle increases. Fig. 7.2c shows the approach of $Ra_c(Q)$ to its asymptotic behaviour, and confirms that in this regime $Ra_c = O(Q)$ for the tilt angles considered. However, for large Q one expects $k_c = O(Q^{1/6})$ but this behaviour is found only for tilt angles that are small enough; for tilt angles exceeding approximately 30° an O(1) wavenumber is selected instead. These results reflect the dominant component of the field: for steady convection in an imposed vertical field, $k_c \sim (\frac{1}{2}\pi^4 Q)^{1/6}$ while for an imposed horizontal field, $k_c \sim \pi^{3/2} Q^{-1/4}$. In both cases $Ra_c \sim \pi^2 Q$. These two types of behaviour indicate that as the tilt angle increases the system undergoes a transition in which its behaviour changes from one in which the vertical field dominates to one in which the horizontal field dominates. We shall see that such a transition occurs in the nonlinear regime as well. In contrast, when $\chi = \pi/2$ $k_c \sim (\frac{1}{2}\pi^4 Q \cos^2 \vartheta)^{1/6}$, $Ra_c \sim \pi^2 \cos^2 \vartheta \, Q$, provided that $Q \cos^2 \vartheta \gg 1$, and the transition to the horizontal field behaviour occurs only for tilt angles $\vartheta = \pi/2 - O(Q^{-1/2})$, i.e. for horizontal fields. In this case the magnetic field has no effect on the onset of convection and the selected wavenumber is therefore O(1). Thus in either case large wavenumbers are selected for tilt angles that are not too large.

These results are to be compared with those obtained from the linearized eigenvalue problem (7.34). For constant ζ we obtain

$$Ra^{(s)} = \left(1 + \cos^2 \chi \tan^2 \vartheta\right)\left[\pi^2 \cos^2 \vartheta + (1 + \cos^2 \chi \tan^2 \vartheta)^2 k_{0\perp}^4\right] Q, \qquad (7.41)$$

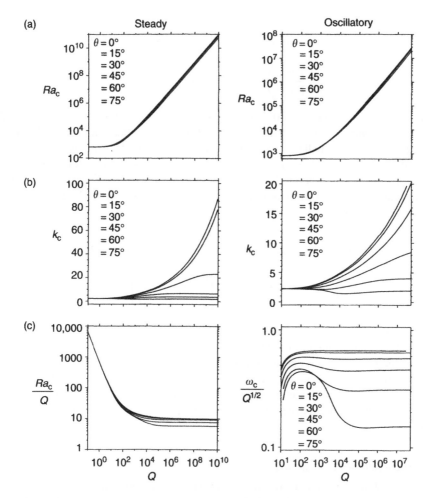

Figure 7.2 The critical Rayleigh number Ra_c (a) and wavenumber k_c (b) for onset of steady and oscillatory perpendicular rolls ($\chi = 0$) as functions of Q for several values of ϑ and $\zeta = 0.1$. (c) Approach of $Ra_c(Q)$ to its asymptotic behaviour at large Q.

where $k_0^2 = k_{0\perp}^2(1 + \cos^2 \chi \tan^2 \vartheta)$. The minimum occurs at $k_{0\perp} = 0$:

$$Ra_c^{(s)} = (1 + \cos^2 \chi \tan^2 \vartheta)\pi^2 \cos^2 \vartheta\, Q. \tag{7.42}$$

This equation predicts that for $\chi = 0$, Ra_c is independent of ϑ, in good agreement with the results of Fig. 7.2. Likewise, for $\zeta < 1$, the onset of overstable oscillations occurs at

$$Ra^{(o)} = (1 + \cos^2 \chi \tan^2 \vartheta)\left[\frac{\sigma + \zeta}{\sigma(1 + \sigma)}\right]$$

$$\times \left[\pi^2 \cos^2 \vartheta\, \sigma\zeta + (1 + \sigma)(1 + \zeta)(1 + \cos^2 \chi \tan^2 \vartheta)^2 k_{0\perp}^4\right] Q, \tag{7.43}$$

$$\omega^{(o)2} = \frac{\zeta}{(1 + \sigma)}\left[(1 - \zeta)\pi^2 \cos^2 \vartheta\, \sigma - \zeta(1 + \sigma)(1 + \zeta)(1 + \cos^2 \chi \tan^2 \vartheta)^2 k_{0\perp}^4\right] Q. \tag{7.44}$$

Again, the minimum occurs at $k_{0\perp} = 0$:

$$Ra_c^{(0)} = \left(1 + \cos^2 \chi \tan^2 \vartheta\right)\left(\frac{\sigma + \zeta}{1 + \sigma}\right)\pi^2 \cos^2 \vartheta \; \zeta Q,$$

$$\omega_c^{(0)2} = \left(\frac{1 - \zeta}{1 + \sigma}\right)\pi^2 \cos^2 \vartheta \; \sigma\zeta Q \tag{7.45}$$

and is independent of ϑ when $\chi = 0$. In these expressions the wavenumber $k_{0\perp}$ is the scaled horizontal wavenumber. The selection of $k_{0\perp} = 0$ is thus a reflection of the selection (in the exact problem) of a smaller wavenumber, viz. $k_{0\perp} = O(Q^{1/6})$ in unscaled variables. However, the $Q^{1/4}$ scaling determines correctly the resulting minimum Rayleigh numbers and frequency. Moreover, as pointed out by Chandrasekhar (1961), the $O(Q^{1/4})$ wavenumber scaling also captures the transition from steady to oscillatory convection. This transition takes place when $\omega_c^{(0)} = 0$, i.e. at

$$Ra_{TB} = \left(1 + \cos^2 \chi \tan^2 \vartheta\right)\left[\frac{\sigma + \zeta}{\zeta(1 + \sigma)}\right]\pi^2 \cos^2 \vartheta \, Q, \tag{7.46}$$

$$k_{\perp TB} = \left[\frac{\pi \cos \vartheta}{1 + \cos^2 \chi \tan^2 \vartheta}\right]^{1/2}\left[\frac{\sigma(1 - \zeta)}{\zeta(1 + \sigma)}\right]^{1/4} Q^{1/4}, \tag{7.47}$$

and defines the Takens–Bogdanov (TB) point. Consequently the $O(Q^{1/4})$ scaling describes correctly not only the vicinity of the TB point but also the behaviour for onset wavenumbers far from this point, i.e. it allows us to retain the full wavenumber dependence of the problem.

For future reference we note that when $\vartheta = 0$ all χ dependence necessarily drops out. However, in contrast to the moderate Q results of Matthews *et al.* (1992) $Ra_c(\chi = \pi/2) < Ra_c(\chi = 0)$ for all $\vartheta \neq 0$ and hence away from the TB point parallel rolls are always selected at onset.

7.3.2. *Vertical field with constant ζ*

In Fig. 7.3 we show the (time-averaged) Nusselt number K and frequency for both steady and overstable convection in a vertical field as a function of the scaled Rayleigh number Ra.

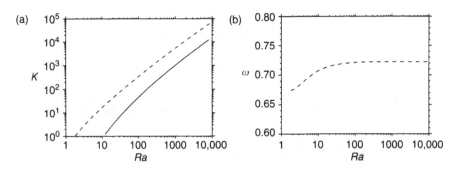

Figure 7.3 The (time-averaged) Nusselt number K for (a) steady (solid line) and oscillatory (dashed line) convection in a vertical field as a function of the scaled Rayleigh number Ra when $\zeta = 0.1$ and $\sigma = 1.1$. (b) The corresponding (scaled) oscillation frequency ω.

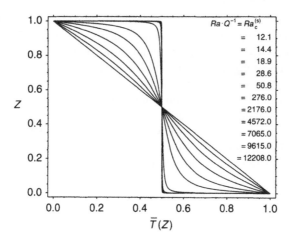

Figure 7.4 Mean temperature profiles $\overline{T}(Z)$ for steady convection at several values of the (scaled) Rayleigh number showing the development of an isothermal core with increasing Ra when $\zeta = 0.1$.

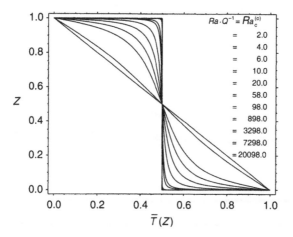

Figure 7.5 Same as Fig. 7.4 but for oscillatory convection when $\zeta = 0.1, \sigma = 1.1$.

Observe that solutions can be obtained for highly supercritical Rayleigh numbers and that K increases monotonically with increasing Ra while the frequency ω appears to saturate, indicating that the oscillations are of magnetic origin. Figs. 7.4 and 7.5 show the corresponding mean temperature profiles $\theta_0 \equiv \overline{T}(Z)$ for several values of Ra. For both steady and oscillatory convection the temperature gradients are confined to thinner and thinner boundary layers at the top and bottom as Ra increases. At the same time the bulk of the layer becomes more and more isothermal. Note that these boundary layers are symmetrical with respect to $Z = 1/2$ and that the isothermal interior has temperature $T = 1/2$, i.e. a temperature that is exactly half way between the temperatures at the top and bottom boundaries.

7.3.3. Vertical field with depth-dependent ζ

The situation changes dramatically when $\zeta \equiv \zeta(Z)$. We follow Weiss *et al.* (1990, 1996) and Julien *et al.* (1999, 2000) and choose a linear dependence on depth, $\zeta(Z) = \zeta_0 + \epsilon(1 - Z)$, focusing on the case in which ζ changes from favouring oscillatory convection at the top of the layer ($\zeta < 1$) to favouring steady convection at the bottom of the layer ($\zeta > 1$).

Fig. 7.6(a) shows the Nusselt number $K(Ra)$ for several values of ϵ when $\zeta_0 = 1$. Fig. 7.6(b) shows the corresponding results for the more interesting case $\zeta_0 = 0.2$. In this case the convective instability is oscillatory and consequently we also show the oscillation frequency ω (Fig. 7.6(c)). These figures show that the differences between the effects of variable ζ on steady and oscillatory convection extend into the nonlinear regime. In the steady case (Fig. 7.6(a)) the Nusselt number is quite insensitive to ϵ and one has to go to Ra values in

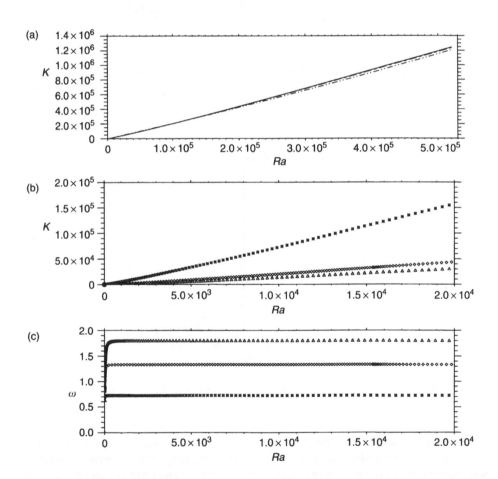

Figure 7.6 The Nusselt number $K(Ra)$ for $\zeta(Z) = \zeta_0 + \epsilon(1 - Z)$ at several values of ϵ when $\sigma = 1.1$. (a) Steady convection when $\zeta_0 = 1.0$ and $\epsilon = 0.0$ (solid), $\epsilon = 0.5$ (dashed), $\epsilon = 1.0$ (dot-dashed) and $\epsilon = 5.0$ (dot-dot-dot-dashed). (b) Oscillatory convection when $\zeta_0 = 0.2$ and $\epsilon = 0.0$ (asterisks), $\epsilon = 2.0$ (diamonds) and $\epsilon = 5.0$ (triangles). (c) The oscillation frequency ω corresponding to (b). Notice that for steady convection $K(Ra)$ remains largely independent of ϵ, but for oscillatory convection both K and the asymptotic frequency ω_∞ depend strongly on ϵ.

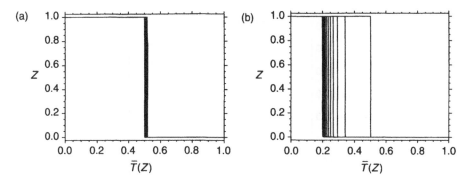

Figure 7.7 Mean temperature profiles $\overline{T}(Z)$ for (a) steady convection and $\epsilon = 0(1)5$ at $Ra = 52100$ and (b) oscillatory convection as a function of $\epsilon = 0(0.5)5$ at $Ra = 20106$, $\sigma = 1.1$. The core temperature shifts monotonically from $T_{\text{core}} = 1/2$ as ϵ increases.

excess of 10^5 to see a real difference. Nonetheless the trend is clear: at each Rayleigh number the Nusselt number decreases monotonically with increasing ϵ. For oscillatory convection the ϵ dependence is much stronger. As expected the Nusselt number is largest for small ϵ. These solutions also have the smallest asymptotic frequency at large Ra; not surprisingly this makes convective transport more efficient and hence increases the Nusselt number. In Fig. 7.7 we show the dependence of the mean temperature profile on ϵ at a high Rayleigh number in each of these two cases. Observe that as ϵ increases away from zero the mean temperature profile acquires asymmetry with respect to the midplane $Z = 1/2$. For large Rayleigh numbers the layer still develops an isothermal core but now its temperature T_{core} differs from $T = 1/2$. For steady convection with $\epsilon > 0$ $T_{\text{core}} > 1/2$ while the opposite is the case when $\epsilon < 0$. Consequently the temperature jump in the upper boundary layer is larger than that in the lower one when $\epsilon > 0$ and conversely. Once again the local decrease in the thermal diffusivity near the lower boundary makes convection easier near the bottom and hence we expect its maximum amplitude to fall below $Z = 1/2$ as ϵ increases. In addition lower thermal diffusivity implies that at a given Rayleigh number the temperature jump across the lower thermal boundary layer also falls below $1/2$, as seen in Fig. 7.7(a). This is ultimately why T_{core} increases with ϵ. Despite this appealing picture for steady convection the results for oscillatory convection reveal an opposite trend. Thus, as shown in Fig. 7.7(b), the core temperature moves towards lower values with increasing ϵ, indicating that despite the decrease in thermal diffusivity near the bottom boundary the time-averaged temperature drop across the lower boundary layer increases. This is presumably because the repeated flow reversals do, on average, increase the boundary layer thickness, but what is unexpected is the magnitude of the resulting shift in the core temperature. Fig. 7.8 summarizes the shifts with ϵ in the core temperatures in the two cases. Fig. 7.9 shows the profiles of the square of the eigenfunction $|\psi_1(Z)|$ in the oscillatory regime when $\epsilon = 0.0$ and $\epsilon = 1.0$, both at onset and at $Ra = 20106$. When $\epsilon = 0.0$ the profiles are symmetric, with largest amplitude at $Z = 1/2$. When $\epsilon > 0$ both profiles become asymmetric and peak above midheight; the effect is enhanced by nonlinearity.

As shown in Fig. 7.10 nonlinear oscillations may be present even when linear theory predicts steady onset. In this case the oscillation frequency decreases to zero at finite amplitude where the branch of oscillatory solutions bifurcates from the steady branch (cf. Julien *et al.*, 1999).

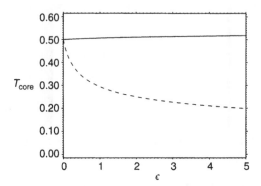

Figure 7.8 The core temperature T_{core}, determined from high Rayleigh number computations, as a function of ϵ for steady convection (solid line) when $\zeta_0 = 1.0$ and oscillatory convection (dashed line) when $\zeta_0 = 0.2$, $\sigma = 1.1$. The isothermal core temperature increases slightly with increasing ϵ in the steady case but decreases in the oscillatory case.

7.3.4. Tilted field with constant ζ

Calculations show that these results are not changed qualitatively when the magnetic field is tilted, provided that the tilt angle ϑ is not too large. However, with increasing tilt both steady and oscillatory convection become less efficient at transporting heat, and the Rayleigh number dependence of the Nusselt number becomes weaker (see Fig. 7.11). The increase in tilt angle leads to a larger Lorentz force, which in turn leads to a suppression of the heat transport. For the steady case the dependence on tilt angle is much weaker. This is to be expected since in the oscillatory regime ohmic diffusion has only a finite time to reduce the Lorentz force due to field distortion before the flow reverses and the Lorentz force depends strongly on the tilt angle. In contrast in the steady case the Lorentz force exerts a much weaker effect and the reduction of the Nusselt number is largely due to a geometrical effect: the strong oblique magnetic field inclines the convection cells relative to the vertical allowing them more time to lose their upward buoyancy to adjacent descending fluid.

Fig. 7.12 shows the corresponding results for the oscillatory mode when $\vartheta = \pi/4$. The figure reveals a remarkable behaviour: the Nusselt number K initially increases rapidly with Ra as in the vertical magnetic field case, but then undergoes a hysteretic transition to a new state characterized by a small Nusselt number that decreases slowly with increasing Ra. Since the conductive flux increases with Ra the convective flux in this regime must decrease even more rapidly. As this state is followed to larger Rayleigh numbers we see that the mean temperature becomes almost piecewise linear (Fig. 7.13), with a limited isothermal core. The extent of this core quickly saturates, in contrast to the case of a vertical field for which the isothermal core grows continuously with Ra as the temperature gradients are compressed into ever thinner thermal boundary layers (as in Fig. 7.5). Evidently, in this state increasing the heat input does not result in increased heat transport across the layer. Instead, as discussed further below, the added energy is all stored in the magnetic field perturbations (since the field strength is large this is achieved with small deformation of the field); moreover, the perturbation magnetic field suppresses the convective motion in the boundary layers near the top and bottom (see Fig. 7.13(a)) thereby reducing the transport of heat across the layer. These conclusions are supported by detailed computations of energy fluxes (Julien *et al.*, 2000) and

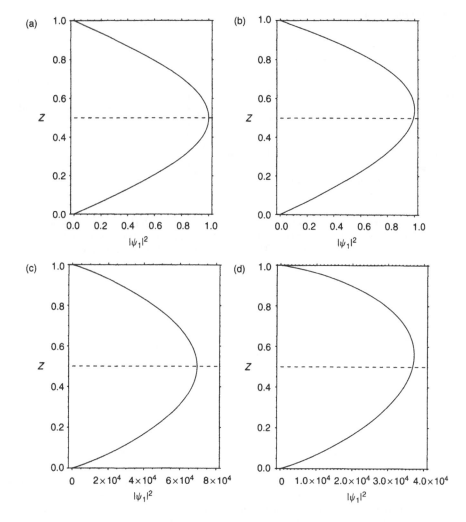

Figure 7.9 The square of the eigenfunction $|\psi_1(Z)|$ for oscillatory convection at onset for (a) $\epsilon = 0.0$ and (b) $\epsilon = 1.0$, showing the gradual development of localized oscillations in the upper part of the layer. (c, d) The corresponding $|\psi_1(Z)|^2$ at $Ra = 20106$. The Prandtl number $\sigma = 1.1$.

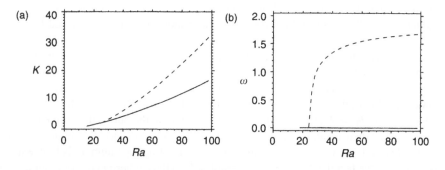

Figure 7.10 (a) The time-averaged Nusselt number K and (b) oscillation frequency ω as functions of Ra for $\epsilon = 5.0$, $\sigma = 1.1$.

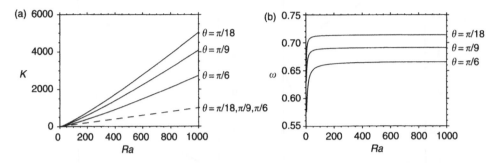

Figure 7.11 (a) The (time-averaged) Nusselt number K for (a) steady (solid line) and oscillatory (dashed line) perpendicular rolls and $\vartheta = 10°, 20°, 30°$ as a function of the scaled Rayleigh number Ra when $\zeta = 0.1$, $\sigma = 1.1$. (b) The corresponding oscillation frequency ω.

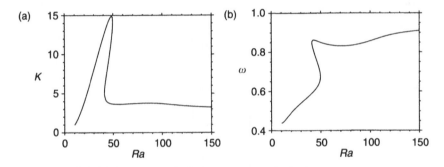

Figure 7.12 (a) The (time-averaged) Nusselt number K for oscillatory perpendicular rolls as a function of the scaled Rayleigh number Ra for $\vartheta = \pi/4$ and $\zeta = 0.1$, $\sigma = 1.1$. (b) The corresponding frequency ω. Note the hysteretic transition from the 'vertical' convection mode to the 'horizontal' convection mode with increasing Ra.

are a consequence of the fact that the flow must always cross magnetic field lines. In this regime (i.e. on the branch where the Nusselt number remains low as Ra is increased) the system of perpendicular rolls therefore behaves much more like one with an imposed horizontal field.

Fig. 7.13(b) indicates that the new regime (hereafter the 'horizontal' regime) is characterized by broad thermal boundary layers. This is a simple consequence of the suppression of all flow in these layers (see Fig. 7.13(a)) by the perturbation magnetic field. As a result the temperature profile in these boundary layers is linear. For example, (7.32) shows that in the top boundary layer $\theta_0 = K(1 - Z)$. Since the temperature at the outer edge of the boundary layers is $\theta_0 = 1/2$ their width is approximately $1/2K$. Moreover, since K is almost independent of the Rayleigh number so is their structure once the Rayleigh number is high enough. This is so despite the fact that the convective amplitude in the isothermal interior continues to increase as $Ra^{1/2}$ in this regime. Note that the Rayleigh number must exceed a critical value before the 'horizontal' convection mode sets in. This is because the flow in the interior must

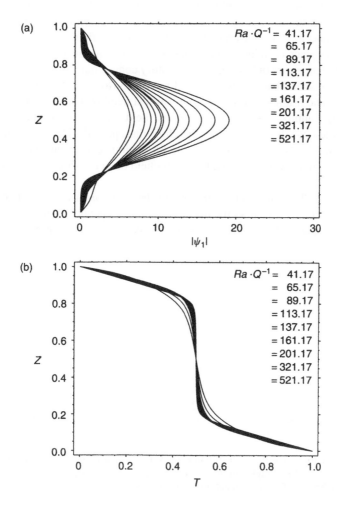

Figure 7.13 (a) Convection amplitude as measured by $|\psi_1(Z)|$ and (b) mean temperature profiles $\theta_0(Z)$ for oscillatory perpendicular rolls at $\vartheta = \pi/4$ at several values of the (scaled) Rayleigh number showing the development of broad boundary layers and small isothermal core with increasing Ra when $\zeta = 0.1$, $\sigma = 1.1$. These properties are characteristic of the 'horizontal' convection mode.

be strong enough to expell the magnetic field perturbation into the boundary layers at the top and bottom; this expulsion occurs primarily *along* the imposed magnetic field. In steady convection the resulting boundary layer thickness is determined by magnetic Reynolds number and is narrow if this is large. In contrast in oscillatory flow the flow reversals prevent the formation of such narrow boundary layers and the boundary layer thickness is determined by the perturbation Lorentz force rather than ohmic diffusion. The resulting boundary layers are therefore wider than in the case of steady convection. A number of conclusions follow immediately from the above considerations. First, the transition between the two regimes occurs at lower Rayleigh numbers when ϑ is larger. Indeed for small values of tilt angle ϑ we have shown that this transition to the lower 'horizontal branch' does not occur (for this value of ζ) but that solutions stay on the efficient 'vertical branch'. Moreover, since the ability of the magnetic field to suppress oscillatory convection increases with decreasing ζ the value of the Rayleigh number at which the transition from the 'vertical field' regime to the 'horizontal field' regime takes place is an increasing function of ζ. This argument also explains why the two convection regimes are only found in oscillatory convection.

In Fig. 7.14 we show that if ζ is increased too much (to $\zeta = 0.15$ for this value of ϑ) the situation becomes radically different. The 'vertical field' branch now extends to arbitrarily large Rayleigh numbers while the 'horizontal field' branch has become disconnected. The upper saddle-node bifurcation has therefore disappeared. The disconnected branch moves away from the 'vertical branch' as the tilt angle is decreased; moreover, the critical value of ζ at which the vertical and horizontal field branches disconnect is an increasing function of ϑ.

When $\chi = \pi/2$ the behaviour is similar but the 'horizontal branch' appears only for larger values of ϑ. Fig. 7.15 shows the Rayleigh number dependence of the Nusselt number and oscillation frequency for steady and oscillatory parallel rolls when $\vartheta = 1, \zeta = 0.1$ and

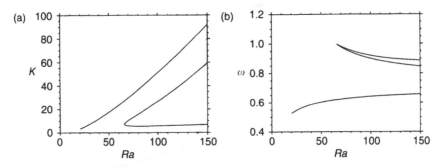

Figure 7.14 (a) The (time-averaged) Nusselt number K for oscillatory perpendicular rolls as a function of the scaled Rayleigh number Ra for $\vartheta = \pi/4$ and $\zeta = 0.15, \sigma = 1.1$, showing multiple branches. (b) The corresponding frequencies ω.

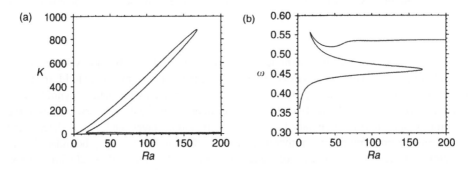

Figure 7.15 (a) The (time-averaged) Nusselt number K for steady (dashed line) and oscillatory (solid line) parallel rolls as a function of the scaled Rayleigh number Ra for $\vartheta = 1$ when $\zeta = 0.1$, $\sigma = 1.1$. (b) The corresponding (scaled) oscillation frequency ω.

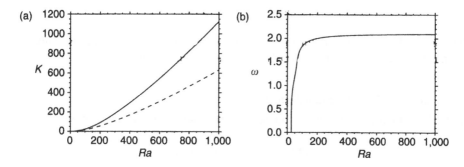

Figure 7.16 (a) The (time-averaged) Nusselt number for oscillatory perpendicular rolls when $\vartheta = \pi/4$ and $\zeta(Z) = \zeta_0 + \epsilon(1 - Z)$ with $\zeta_0 = 0.1$ and $\epsilon = 2.0$. (b) The corresponding frequency $\omega(Ra)$. The primary instability is a steady state one but the resulting steady convection loses stability almost immediately to oscillations.

$\sigma = 1.1$. Both convection modes are present at this tilt angle, in contrast to $\vartheta = \pi/4$ for which the solutions remain on the 'vertical branch' for all values of Ra, cf. Fig. 7.12 for $\chi = 0$.

7.3.5. Tilted field with depth-dependent ζ

The results for perpendicular rolls with $\zeta(Z) = \zeta_0 + \epsilon(1 - Z)$ and $\vartheta = \pi/4$ are shown in Fig. 7.16. The behaviour of both the Nusselt number and frequency with increasing Ra is very similar to that in the vertical case. For the present parameter values the primary bifurcation is a steady state one but the resulting steady convection loses stability almost immediately in a TB bifurcation to an oscillatory mode. As Ra is increased further the oscillatory mode behaves in the usual manner for the 'vertical branch' with K increasing monotonically with Ra and the frequency ω saturating. Because of the up–down asymmetry introduced by the depth dependence of ζ, the solution is no longer symmetric with respect to $Z = 1/2$. For these parameter values the depth dependence postpones the appearance of the 'horizontal branch' to larger tilt angles.

7.4. Generalizations

The procedure outlined above can be generalized to astrophysically more realistic situations. We describe here some of these generalizations.

7.4.1. Newton's law of cooling at the top

If instead of a fixed temperature boundary condition at the top we use the more realistic boundary condition

$$D\theta_0 + \beta\theta_0 = 0 \quad \text{on } Z = 1, \quad \beta > 0, \tag{7.48}$$

integration of equation (7.32) yields

$$K^{-1} = \int_0^1 \frac{\omega^2 + k_0^4}{\omega^2 + k_0^4 + 2k_0^2 k_{0\perp}^4 |\Psi|^2} \, dZ + \beta^{-1}. \tag{7.49}$$

Thus

$$K^{-1} = K'^{-1} + \beta^{-1}, \tag{7.50}$$

where K' is the Nusselt number for fixed temperature boundary conditions. This result demonstrates that Newton's law of cooling *reduces* the heat transport.

7.4.2. *Fixed heat flux at top and bottom*

If instead we impose fixed heat flux boundary conditions at both top and bottom

$$D\theta_0 = -F \quad \text{on } Z = 0, 1, \tag{7.51}$$

then $K = 1$ regardless of the value of the Rayleigh number. This simply states that whatever flux goes into the layer must also leave the layer. In this case the Nusselt number is clearly not a useful measure of the amplitude of convection. Instead the eigenvalue problem (7.34) becomes a nonlinear eigenvalue problem for Ra and ω and we may use the kinetic energy

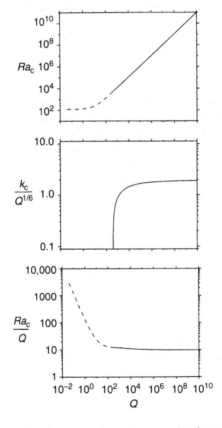

Figure 7.17 Linear theory results for onset of steady convection in a vertical magnetic field with fixed heat flux boundary conditions, showing that $O(Q^{1/6})$ wavenumbers are selected for sufficiently large Q. The selected wavenumber k_c vanishes for $Q < 376.5$ (theory predicts that the selected wavenumber vanishes at $Q = Q_\infty = 376$).

$E \equiv (1/2V) \int |\mathbf{u}|^2 \, dV$ as a suitable measure of the amplitude of convection. Thus, no new calculations are necessary to solve the fixed flux problem. In general, an important feature of fixed flux convection is the selection of large scales at onset (Chapman and Proctor, 1980; Depassier and Spiegel, 1981). If this were the case for magnetoconvection in the large Q limit the scaling employed in the derivation would be violated. Fig. 7.17 shows that this concern is unfounded. Although the selected wavenumber remains zero for sufficiently small Q it becomes finite for larger Q and rapidly approaches $O(Q^{1/6})$ values as in the case of fixed temperature boundaries. Moreover, in this regime the critical Rayleigh number increases as Q, as assumed in the theory.

The transition from zero to finite wavenumber with increasing Q is of interest in its own right and can be analyzed as in Knobloch (1989).

7.4.3. *Magnetoconvection in a stratified atmosphere*

Background stratification is not only important for astrophysical applications but also provides another source of vertical asymmetry. In such circumstances we expect qualitatively similar results to those found for depth-dependent ζ. In this section we describe an analogous derivation for this case but confine attention to the simpler two-dimensional problem in an imposed vertical field. We eliminate sound waves using the magneto-anelastic approximation (Gilman and Glatzmaier, 1981; Lantz and Sudan, 1995) and thereby focus on dynamics on timescales long compared with the sound travel time. The basic state is described by the equations

$$\frac{dp_0}{dz} = -g\rho_0, \quad p_0 = \mathcal{R}\rho_0 T_0, \quad S_0 = \text{const.} \tag{7.52}$$

Here $\mathcal{R}$ is the gas constant divided by the mean molar mass. It follows from the first law of thermodynamics that

$$\frac{dT_0}{dz} = -\frac{g}{c_p} \tag{7.53}$$

and hence that

$$T_0 = T_s\left(1 - \frac{z}{h}\right), \quad \rho_0 = \rho_s\left(1 - \frac{z}{h}\right)^m, \quad p_0 = p_s\left(1 - \frac{z}{h}\right)^{m+1}, \tag{7.54}$$

where the subscript s denotes a reference value (at $z = 0$), m is the polytropic index and h is the (temperature) scale height:

$$m = \frac{1}{\gamma - 1}, \quad h = \frac{c_p T_s}{g}. \tag{7.55}$$

Here $\gamma = c_p/c_v$.

In the following we discuss strongly nonlinear convection in such an atmosphere. The convection arises as a result of a buoyancy force due to a density perturbation ρ_1 of the basic state $\rho_0(z)$ assumed to be small compared to ρ_0 but still large enough to drive fully nonlinear convection. The basic equations are

$$\rho_0 \frac{D\mathbf{u}}{Dt} = -\nabla p_1 - \rho_1 g\hat{\mathbf{z}} + \frac{1}{\mu_0}(\nabla \times \mathbf{B}) \times \mathbf{B} + \nabla \cdot \tau, \tag{7.56}$$

together with the continuity equation

$$\nabla \cdot \rho_0 \mathbf{u} = 0, \tag{7.57}$$

the entropy equation

$$\rho_0 T_0 \frac{DS_1}{Dt} = \nabla \cdot (\kappa \nabla T_1) + \frac{\eta}{\mu_0} |\nabla \times \mathbf{B}|^2 + \tau_{ij} \partial_j u_i, \tag{7.58}$$

and the induction equation

$$\frac{\partial \mathbf{B}}{\partial t} = \nabla \times (\mathbf{u} \times \mathbf{B}) - \nabla \times (\zeta \nabla \times \mathbf{B}), \quad \nabla \cdot \mathbf{B} = 0. \tag{7.59}$$

Here the subscripts 0, 1 denote the basic state and the perturbation about it, respectively, and $\mathbf{u}$, p, ρ, T and S are the velocity field, pressure, density, temperature and entropy per unit mass. In addition $\mathbf{B}$ is the magnetic field while τ is the viscous part of the stress tensor: $\tau_{ij} = \mu(\partial_i u_j + \partial_j u_i - \frac{2}{3} \nabla \cdot \mathbf{u} \delta_{ij})$, where μ is the (possibly depth-dependent) coefficient of dynamic viscosity; bulk viscosity is neglected. The equations are complemented by the perturbed equation of state

$$\frac{\rho_1}{\rho_0} = \frac{p_1}{p_0} - \frac{T_1}{T_0}, \tag{7.60}$$

and the thermodynamic relation

$$S_1 = c_p \frac{T_1}{T_0} - \mathcal{R} \frac{p_1}{p_0}, \tag{7.61}$$

valid for $p_1 \ll p_s$, $\rho_1 \ll \rho_s$, $T_1 \ll T_s$.

To obtain results analogous to (7.34) it suffices to consider two-dimensional motions (see Section 7.2). We therefore write

$$\rho_0 \mathbf{u} = \left(\frac{\partial \psi}{\partial z}, 0, -\frac{\partial \psi}{\partial x} \right), \quad \mathbf{B} = \left(\frac{\partial A}{\partial z}, 0, -\frac{\partial A}{\partial x} \right). \tag{7.62}$$

The resulting equations are nondimensionalized using a characteristic thermal diffusivity K_s (cm^2 s^{-1}) to define units of time and velocity. We choose the layer depth d as unit of length and B_0 as the unit of the magnetic field strength. The background pressure, density and temperature are scaled with $p_s \equiv \mathcal{R} \rho_s T_s$, ρ_s and T_s. The thermodynamic relations become

$$\frac{\rho_1}{\rho_0} = -\frac{\Delta T \rho_s}{T_s \Delta \rho} \left(\frac{T_1}{T_0} - \frac{\Delta p}{\Delta T \mathcal{R} \rho_s} \frac{p_1}{p_0} \right), \quad S_1 = \frac{T_1}{T_0} - \left(1 - \frac{1}{\gamma} \right) \frac{\Delta p}{\Delta T \mathcal{R} \rho_s} \frac{p_1}{p_0}. \tag{7.63}$$

where Δp, $\Delta \rho$ and ΔT are the units used to nondimensionalize p_1, ρ_1 and T_1. The entropy S_1 has been expressed in units of $c_p \Delta T / T_s$. Since the equation of motion indicates that the appropriate unit of p_1 is the pressure $\Delta p \equiv \rho_s K_s^2 / d^2$ we see that

$$\frac{\Delta p}{\Delta T \mathcal{R} \rho_s} = \frac{d}{h} \frac{\gamma}{\gamma - 1} Ra^{-1} \tag{7.64}$$

where

$$Ra = \frac{g \Delta T d^3}{K_s^2 T_s}, \quad Q = \frac{B_0^2 d^2}{\mu_0 \rho_s K_s^2}. \tag{7.65}$$

These definitions should be compared with (7.5). In the following we shall be interested in $d/h = O(1)$, $Ra = O(Q)$. Consequently

$$s_1 = \frac{T_1}{T_0} + O(Q^{-1}), \quad \frac{\rho_1}{\rho_0} = -\frac{\Delta T \rho_s}{T_s \Delta \rho} \frac{T_1}{T_0} + O(Q^{-1}) \tag{7.66}$$

and we do not have to calculate the pressure perturbation p_1. This offers a considerable simplification of the resulting equations which are

$$\nabla^2 \psi_t + \frac{1}{\rho_0} \psi_z \nabla^2 \psi_x - \psi_x \nabla^2 \frac{\psi_z}{\rho_0} + 2 \frac{d}{h} \frac{\rho_{0z}}{\rho_0^2} \psi_x \psi_{xx}$$

$$= -Ra \frac{\rho_0}{T_0} T_{1x} + Q(A_z \nabla^2 A_x - A_x \nabla^2 A_z) + \left[2\sigma \left(\frac{\psi_{xz}}{\rho_0} + \frac{\partial}{\partial z} \frac{\psi_x}{\rho_0} \right) \right]_{xz}$$

$$- (\partial_{xx} - \partial_{zz}) \sigma \left(\frac{\partial}{\partial z} \frac{\psi_z}{\rho_0} - \frac{\psi_{xx}}{\rho_0} \right), \tag{7.67}$$

$$\rho_0 T_{1t} + \psi_z T_{1x} - \psi_x T_0 \frac{\partial}{\partial z} \left(\frac{T_1}{T_0} \right)$$

$$= \nabla \cdot \kappa \nabla T_1 + \frac{\zeta Q}{Ra} \frac{d}{h} (\nabla^2 A)^2 + \frac{\sigma}{Ra \rho_0^2} \frac{d}{h} \left[\left(2\psi_{xz} - \frac{\rho_{0z}}{\rho_0^2} \psi_x \right)^2 \right.$$

$$\left. + \left(\psi_{xx} - \psi_{zz} + \frac{\rho_{0z}}{\rho_0^2} \psi_z \right)^2 + \frac{1}{3} \left(\frac{\rho_{0z}}{\rho_0^2} \psi_x \right)^2 \right], \tag{7.68}$$

$$A_t + \frac{1}{\rho_0} \psi_z A_x - \frac{1}{\rho_0} \psi_x A_z = \zeta \nabla^2 A. \tag{7.69}$$

Here $\zeta(z) = \eta(z)/K_s$, $\sigma(z) = \mu/\rho_s K_s$ and $\kappa(z)$ has been expressed in units of $\rho_s c_p K_s$.

Equations (7.67)–(7.69) form the basis for the subsequent study. We observe that these equations admit an equilibrium with a uniform vertical magnetic field and therefore write $A \to -x + A$ so that nonzero A now indicates departure from the uniform vertical field. Since the imposed field is vertical the solutions vary on an O(1) vertical scale only. Consequently, we set (cf. (7.9)):

$$\partial_x = Q^{1/4} \partial_{x'}, \quad \partial_z = D, \quad \partial_t = Q^{1/2} \partial_{t'}, \quad A = Q^{-1/2} A', \quad Ra = Q Ra' \tag{7.70}$$

and do not scale ψ. In addition we expand (dropping primes) the temperature in the form

$$T_1(x, z, t) = T_{10}(z) + Q^{-1/4} T_{11}(x, z, t) + Q^{-1/2} T_{12}(x, z, t) + \cdots \tag{7.71}$$

Proceeding as in Section 7.2 we obtain at leading order

$$\psi_{xt} = -Ra \frac{\rho_0}{T_0} T_{11} + DA_x + \frac{\sigma}{\rho_0} \psi_{xxx}, \tag{7.72}$$

$$A_t = \frac{\mathrm{D}\psi}{\rho_0} + \zeta A_{xx} \tag{7.73}$$

$$\rho_0 T_{11t} - \psi_x T_0 \mathrm{D}\left(\frac{T_{10}}{T_0}\right) = \kappa T_{11xx}. \tag{7.74}$$

The solvability condition for T_{12} yields

$$\mathrm{D}(\kappa \mathrm{D}T_{10}) = -\mathrm{D}(\overline{\psi_x T_{11}}) + \overline{\psi_x T_{11}}\frac{\mathrm{D}T_0}{T_0} - \frac{\zeta}{Ra}\frac{d}{h}\overline{A_{xx}^2} - \frac{\sigma}{Ra\rho_0^2}\frac{d}{h}\overline{\psi_{xx}^2}. \tag{7.75}$$

For solutions of the form (7.27) we find that $\Psi(z)$ satisfies the nonlinear eigenvalue problem

$$\mathrm{D}\left[\frac{\mathrm{D}\Psi}{\rho_0(i\omega + \zeta k^2)}\right] - \left(i\omega + \frac{\sigma k^2}{\rho_0}\right)\Psi - Ra\frac{\mathrm{D}(T_{10}/T_0)}{i\omega + (\kappa/\rho_0)k^2}\Psi = 0, \tag{7.76}$$

$$\mathrm{D}(\kappa \mathrm{D}T_{10}) = -\frac{1}{2}\kappa k^4\left\{\frac{|\Psi|^2 T_0 \mathrm{D}(T_{10}/T_0)}{\rho_0^2\omega^2 + \kappa^2 k^4}\right\} + \frac{1}{2}\kappa k^4\frac{|\Psi|^2 \mathrm{D}T_0 \mathrm{D}(T_{10}/T_0)}{\rho_0^2\omega^2 + \kappa^2 k^4}$$

$$- \frac{1}{2}\frac{\zeta}{Ra}\frac{d}{h}\frac{k^4|\mathrm{D}\Psi|^2}{\rho_0^2(\omega^2 + \zeta^2 k^4)} - \frac{1}{2}\frac{\sigma}{Ra\rho_0^2}\frac{d}{h}k^4|\Psi|^2. \tag{7.77}$$

In these expressions

$$T_0 = T_s\left(1 - \frac{d}{h}z\right), \quad \rho_0 = \rho_s\left(1 - \frac{d}{h}z\right)^m, \tag{7.78}$$

and D indicates derivatives with respect to z. Note that the mean temperature equation no longer has an integral; an integral exists only when $d \ll h$. In this limit one recovers the Boussinesq formulation.

The linear problem in the stratified case is recovered on setting $\Psi = 0$ in the temperature equation. Then

$$\mathrm{D}T_{10} = -N/\kappa, \quad N^{-1} = \int_0^1 \frac{dz}{\kappa}, \tag{7.79}$$

and the eigenvalue problem becomes *linear*. This problem is to be solved for the critical Rayleigh number Ra_c for the onset of instability and ω its frequency for given profiles of κ, ζ, ρ_0 and T_0. The nonlinear problem then describes the evolution of the convection amplitude Ψ and the accompanying deformation of the mean temperature profile as Ra increases above Ra_c much as in Section 7.2.

7.5. Implications for sunspots

In this article we have summarized some of the fully nonlinear results that can be obtained by reformulating the problem of convection in an imposed magnetic field as a nonlinear eigenvalue problem. Such a reformulation is possible when the field strength is large and the distortion of the field by the flow remains small. This reformulation promises to have significant applications in astrophysics and in particular for convection in sunspots, because realistic profiles of density and of the various diffusivities can be readily incorporated. We have considered here only the case where ζ varies linearly with height, decreasing upwards, but considered both vertical and oblique magnetic fields. This choice of ζ variation was motivated by the numerical simulations of two-dimensional compressible magnetoconvection in $m = 1$ polytropic atmospheres with imposed vertical field by Weiss and colleagues. We have presented explicit results for incompressible convection, but have indicated how our procedure generalizes to the more realistic anelastic case. We have found that at large Rayleigh numbers the incompressible system develops an isothermal core just as in the case of constant ζ except that the core temperature shifts away from $T = 1/2$, and that the maximum amplitude of convection is displaced from midlevel. We have also seen that in overstable convection the dependence on the nonuniformity in ζ, as parametrized by the parameter ϵ, is nontrivial due to two competing effects. First, the assumed ζ variation tends to localize the oscillations towards the upper boundary. At the same time, however, it tends to raise the value of ζ at midlevel and hence to suppress oscillations altogether. We found, moreover, that when the field is inclined the behaviour of the overstable system falls into one of two possible regimes. For small tilt angles the magnetic field plays a minor role in inhibiting convection and the Nusselt number is an increasing function of the Rayleigh number. If the tilt angle is increased past a threshold value (which depends on the value of ζ and on the roll orientation χ) a hysteretic transition takes place with increasing Rayleigh number from this 'vertical field' regime to a 'horizontal field' regime in which the field plays a major role in inhibiting the heat transport. Detailed results demonstrating this behaviour were presented for both perpendicular and parallel rolls. Although we have been unable to verify that this novel convection regime persists for the (smaller) wavenumbers selected by linear theory for perpendicular rolls at large Q at these tilt angles, for parallel rolls the selected wavenumbers remain large and the asymptotic theory captures correctly both convection regimes. Continuity arguments suggest that in the absence of secondary Hopf and parity-breaking bifurcations (and of sideband instabilities) the 'vertical field' branch will be stable up to the first saddle-node bifurcation (if present), as will the 'horizontal field' branch beyond the second saddle-node bifurcation; the branch between the two saddle-nodes is expected to be unstable. These expectations will be checked using direct numerical simulations of the reduced equations.

It is tempting to speculate about the possible role of the fully nonlinear solutions discovered here for the structure of a sunspot. Sunspots consist of a dark central region, the umbra, surrounded by a non-axisymmetric filamentary penumbra. The penumbra is characterised by radial striations of alternating bright and dark filaments. The reason for the sudden transition between the umbra and the penumbra is poorly understood – but the nonlinear results discussed above suggest a possible mechanism. Observations have shown (e.g. Title *et al.*, 1992) that the magnetic field strength does not change significantly across the spot, although the tilt angle does. In the sunspot umbra the magnetic field is nearly vertical, while in the penumbra it is tilted, with the tilt angle away from the vertical an increasing function of distance from the centre. Danielson (1961) in his study of tilted magnetoconvection speculated that the gradual change in tilt angle would lead to a change in the nature of convection from three-dimensional to two-dimensional but did not explain why the transition should be so abrupt. The results described in this paper suggest a possible scenario (see Figs. 7.18 and 7.19). For small tilt angles the system shows little preference between parallel and perpendicular rolls and the convection is expected to be fully three-dimensional. (For vertical magnetic fields the results of Clune and Knobloch (1993) indicate that, in the weakly nonlinear regime, three-dimensional structures are preferred at large Q if the onset of convection is oscillatory.) Moreover, both parallel and perpendicular rolls remain on the vertical field branch as the tilt

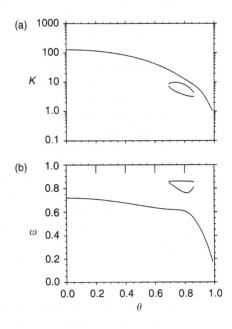

Figure 7.18 (a) The (time-averaged) Nusselt number for oscillatory perpendicular rolls as a function
of the tilt angle ϑ when $Ra = 45$, $\zeta = 0.1$, $\sigma = 1.1$. (b) The corresponding frequency
$\omega(Ra)$. Note the presence of an isola of solutions; of these those above and between the
two saddle-node bifurcations are expected to be stable.

angle is increased (at fixed supercritical Ra) and both transport energy efficiently. We have
seen, however, that there is a critical tilt angle ϑ_c at which there is a saddle-node bifurcation
beyond which the system settles onto the horizontal field branch (Fig. 7.19). Although such a
saddle-node bifurcation is present for both perpendicular and parallel rolls it is encountered
first for the perpendicular rolls as ϑ (or, equivalently, the radial distance from the spot centre)
increases. As described above, convection on the horizontal field branch is very inefficient
and for $\vartheta > \vartheta_c$ the Nusselt number drops to small values. This argument suggests that for
$\vartheta > \vartheta_c$ heat will be transported only by parallel (i.e. radial) rolls which continue to be efficient transporters of heat; we suppose these to be in the form of *standing* waves. Although
this argument ignores the nonlinear nature of the solutions at supercritical Rayleigh numbers
it does provide a natural and promising explanation for the observed sharp transition from
three-dimensional to two-dimensional radial structures observed in sunspots. Moreover, it
has much in common with the sunspot model put forward recently by Rucklidge *et al.* (1995).
In particular Figs. 7.18 and 7.19 suggest that when $Ra = 100$ the transition to a penumbra
occurs at $\vartheta \approx 0.66$ but is absent for $Ra = 45$, i.e. the penumbra develops with increasing
Ra. The same argument suggests that the second transition, at $\vartheta \approx 1$, should be associated
with the appearance of a new mode of convection, presumably field-free convection outside
the spot. It is of interest that the tilt angle of the field at the outer edge of the penumbra of a
typical sunspot is $\approx 70°$.

The scenario outlined above competes with an additional effect due to the variation of
the effective Rayleigh number across the spot. If we suppose that the temperature difference
between some reference depth and the surface remains constant across the spot and that the

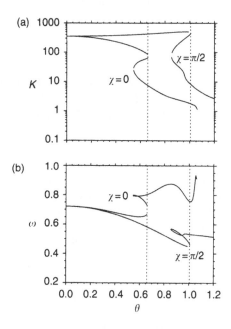

Figure 7.19 (a) Same as Fig. 7.18(a) but for both perpendicular ($\chi = 0$) and parallel ($\chi = \pi/2$) rolls when $Ra = 100$. (b) The corresponding frequency $\omega(Ra)$. The transition to the inefficient 'horizontal mode' occurs first for the perpendicular rolls and defines the critical tilt angle ϑ_c.

temperature at depth is uniform, the imposed Rayleigh number Ra will be constant across the spot. In this case Nusselt number variations across the spot (see Figs. 18 and 19) do not translate into variations in temperature at the visible surface ($\tau = 1$ depth). However, linear theory shows that at large Q the critical Rayleigh number Ra_c steadily decreases as the tilt angle increases (Fig. 7.2), so that the supercriticality, or *effective* Rayleigh number, increases with radial distance from the spot centre. Although this property of the exact linear problem is absent in the asymptotic regime considered here it suggests that we identify our low Rayleigh number 'vertical mode' results with convection in the umbra, and our higher Rayleigh number 'horizontal mode' results with convection in the penumbra. In this scenario the saddle-node bifurcation is again responsible for the abrupt change in the properties of the umbra and penumbra even though the effective Rayleigh number may vary gradually across the spot.

The $K(Ra)$ plots in Figs. 7.12, 7.14 and the $K(\vartheta)$ plots in Fig. 7.19 suggest what might happen if we instead imagine fixing the heat flux K (cf. Busse, 1967) rather than fixing Ra. Fig. 7.12 shows that there may be as many as three values of Ra at which the (time-averaged) heat transport takes the desired value K, a stable one at low Ra, an unstable one in the middle, and another stable one at very large Ra; the latter may not be present in the figure which examines only O(Q) values of Ra, but must appear at larger Ra. This argument also argues in favour of a low Ra vertical field regime and a large Ra horizontal field regime, with an abrupt transition between the two; Fig. 7.19 indicates that in a spot with a gradually varying angle of tilt these regimes would be spatially disjoint. Suitably reformulated, an argument of this type would allow an abrupt temperature change at the $\tau = 1$ depth with increasing tilt angle, and could, combined with the first argument, describe a filamentary penumbra of the observed type.

Acknowledgements

This work reported here was supported by NASA under SR&T grants NAG5-4918 and MASW-99026 (KJ), DOE under grant DE-FG03-95ER-25251 (EK), NSF under grant DMS-9703684 (EK) and NASA under SPTP grant NAG5-2256 (SMT).

References

Babin, A.V., Mahalov, A., Nicolaenko, B. and Zhou, Y., "On the asymptotic regimes and the strongly stratified limit of rotating Boussinesq equations," *Theor. Comp. Fluid Dynam.* **9**, 223–251 (1997).

Busse, F.H., "Non-stationary finite amplitude convection," *J. Fluid Mech.* **28**, 223–239 (1967).

Chandrasekhar, S., *Hydrodynamic and Hydromagnetic Stability*, Oxford University Press (1961).

Chapman, C.J. and Proctor, M.R.E., "Nonlinear Rayleigh–Bénard convection between poorly conducting boundaries," *J. Fluid Mech.* **101**, 759–782 (1980).

Clune, T. and Knobloch, E., "Pattern selection in three-dimensional magnetoconvection," *Physica D* **74**, 151–176 (1993).

Danielson, R.E., "The structure of sunspot penumbras. II. Theoretical," *Astrophys. J.* **134**, 289–311 (1961).

Depassier, M.C. and Spiegel, E.A., "The large-scale structure of compressible convection," *Astron. J.* **86**, 496–512 (1981).

Gilman, P.A. and Glatzmaier, G.A., "Compressible convection in a rotating spherical shell. I. Anelastic equations," *Astrophys. J. Suppl.* **45**, 335–349 (1981).

Hughes, D.W. and Proctor, M.R.E., "Magnetic fields in the solar convection zone," *Ann. Rev. Fluid Mech.* **20**, 187–223 (1998).

Julien, K. and Knobloch, E., "Strongly nonlinear convection cells in a rapidly rotating fluid layer: the tilted f-plane," *J. Fluid Mech.* **360**, 141–178 (1998).

Julien, K., Knobloch, E. and Tobias, S.M., "Strongly nonlinear magnetoconvection in three dimensions," *Physica D* **128**, 105–129 (1999).

Julien, K., Knobloch, E. and Tobias, S.M., "Nonlinear magnetoconvection in the presence of strong oblique fields," *J. Fluid Mech.* **410**, 285–322 (2000).

Julien, K., Knobloch, E. and Werne, J., "A new class of equations for rotationally constrained flows," *Theor. Comp. Fluid Dynam.* **11**, 251–261 (1998).

Knobloch, E., "Pattern selection in binary-fluid convection at positive separation ratios," *Phys. Rev. A* **40**, 1549–1559 (1989).

Knobloch, E., "Bifurcations in rotating systems," in: *Lectures on Solar and Planetary Dynamos* (Eds. M.R.E. Proctor and A.D. Gilbert), Cambridge University Press, pp. 331–372 (1994).

Knobloch, E. and Weiss, N.O., "Convection in sunspots and the origin of umbral dots," *Mon. Not. R. Astr. Soc.* **207**, 203–214 (1984).

Lantz, S.R. and Sudan, R.N., "Magnetoconvection dynamics in a stratified layer. I. Two-dimensional simulations and visualization," *Astrophys. J.* **441**, 903–924 (1995).

Matthews, P.C., "Asymptotic solutions for nonlinear magnetoconvection," *J. Fluid Mech.* **387**, 397–409 (1999).

Matthews, P.C., Hurlburt, N.E., Proctor, M.R.E. and Brownjohn, D.P., "Compressible magnetoconvection in oblique fields: linearized theory and simple nonlinear models," *J. Fluid Mech.* **240**, 559–569 (1992).

Proctor, M.R.E. and Weiss, N.O., "Magnetoconvection," *Rep. Prog. Phys.* **45**, 1317–1379 (1982).

Rucklidge, A.M., Schmidt, H.U. and Weiss, N.O., "The abrupt development of penumbrae in sunspots," *Mon. Not. R. Astr. Soc.* **273**, 491–498 (1995).

Thomas, J.H. and Weiss, N.O., "The theory of sunspots," in: *Sunspots: Theory and Observations* (Eds. J.H. Thomas and N.O. Weiss), Kluwer, pp. 3–59 (1992).

Title, A.M., Frank, Z.A., Shine, R.A., Tarbell, T.D., Topka, K.P., Scharmer, G. and Schmidt, W., "High resolution observations of the magnetic and velocity field of simple sunspots," in: *Sunspots: Theory and Observations* (Eds. J.H. Thomas and N.O. Weiss), Kluwer, pp. 195–219 (1992).

Weiss, N.O., Brownjohn, D.P., Hurlburt, N.E. and Proctor, M.R.E., "Oscillatory convection in sunspot umbrae," *Mon. Not. R. Astr. Soc.* **245**, 434–452 (1990).

Weiss, N.O., Brownjohn, D.P., Matthews, P.C. and Proctor, M.R.E., "Photospheric convection in strong magnetic fields," *Mon. Not. R. Astr. Soc.* **283**, 1153–1164 (1996).

8   Thin aspect ratio $\alpha\Omega$-dynamos in galactic discs and stellar shells

Andrew M. Soward

School of Mathematical Sciences, University of Exeter, Laver Building, North Park Road, Exeter, EX4 4QE, UK, E-mail: A.M.Soward@ex.ac.uk

When an astrophysical dynamo is confined to a region with small aspect ratio, ε, asymptotic approximations may lead to analytic solutions.

In the case of galactic discs, for which ε is the ratio of the disc width $2H_0$ to its diameter $2L$, analytic progress is possible when both the α- and Ω-effects vary on the long radial length scale L. As a prototype example, we will consider Stix's (1975, 1978) oblate spheroid, which is characterised by a local dynamo number $\mathcal{D}(s)$ that increases from zero at the symmetry axis $s = 0$ to a maximum (extremum) D at some radius $s = s_E$ and decreases to zero at the disc's edge $s = L$. Steady dynamo modes have a radial length scale large compared to H_0 but short compared to L. Consequently, the effect of radial-s diffusion inside the disc is negligible when compared to the potential field coupling outside. Both the steady quadrupole and 'forgotten' dipole modes lead to integral equations, which may be solved by Fourier transform methods. The oscillatory quadrupole mode, which has a radial scale comparable to the disc height, is resolved by WKBJ methods, which highlight the importance of global stability criteria, namely the vanishing of both the group velocity and phase mixing.

In the case of stellar shells, ε is the ratio of the shell thickness to radius. Here the radial structure across the disc is averaged leading to dependence on the latitude θ alone. The resulting one-dimensional $\alpha\Omega$-dynamo is characterised by a local dynamo number $\mathcal{D}(\theta)$ that increases from zero at the Equator $\theta = 0$ to a maximum D at some latitude $\theta = \theta_E$ and decreases to zero at the North Pole $\theta = \pi/2$. The ensuing latitudinally modulated kinematic dynamo wave is again resolved by WKBJ methods. In the case of quenching, its weakly nonlinear development signals a complicated and almost spontaneous (in terms of the maximum dynamo number D) transition to full nonlinearity. That state is characterised by a localised finite amplitude Parker wave, which emerges smoothly at a low latitude and terminates abruptly across a front at a high latitude. At lowest order the frequency and front location are fixed by the Dee and Langer (1983) criterion that the group velocity vanishes ahead of the front. We remark on other recent developments.

8.1. Introduction

8.1.1. Simple $\alpha\Omega$-dynamo models

The underlying theme of this chapter is to develop the asymptotic methods necessary for the investigation of mean field dynamos (Krause and Rädler, 1980) in astrophysical objects with small aspect ratio ε. The two particular geometries considered here are thin discs appropriate to galaxies and accretion discs, and thin shells which model the dynamo active regions in stars, e.g. the convection zone or possibly the tachocline in the Sun. In both cases, we restrict attention to axisymmetric $\alpha\Omega$-dynamos, for which relative to cylindrical polar coordinates (s, ϕ, z) the magnetic field **B** and steady flow **u** take the simple forms

$$\mathbf{B} := B(s, z, t)\hat{\boldsymbol{\phi}} + \nabla \times \left(A(s, z, t)\hat{\boldsymbol{\phi}} \right), \quad \mathbf{u} := s\Omega(s, z)\hat{\boldsymbol{\phi}}. \tag{8.1}$$

Here the only motion considered is the steady azimuthal velocity $s\Omega$ responsible for the Ω-effect. To complete the dynamo coupling between the azimuthal and meridional magnetic field components, we include an axisymmetric α-effect. Inside the electrically conducting fluid the meridional and azimuthal components of the magnetic induction equation are

$$\frac{\partial A}{\partial t} = \alpha B + \eta \Delta A, \tag{8.2a}$$

$$\frac{\partial B}{\partial t} = [\nabla\Omega \times \nabla (sA)] \cdot \hat{\boldsymbol{\phi}} + \eta \Delta B \tag{8.2b}$$

(see standard texts, e.g. Parker, 1979; Krause and Rädler, 1980; Zeldovich *et al.*, 1980); Moffatt, 1987), where t is time, $\Delta \equiv \nabla^2 - s^{-2}$ and η is the magnetic diffusivity. When the region exterior to the fluid is an insulator, (8.2) is solved subject to $B = 0$ on the boundary and with A and ∇A continuous across it, where the external magnetic field is potential satisfying $\Delta A = 0$.

We will assume that the flow velocity and α-effect have the reflectional symmetries

$$\Omega(s, -z, t) = \Omega(s, z, t), \qquad \alpha(s, -z, t) = -\alpha(s, z, t), \tag{8.3a}$$

about the equatorial plane $z = 0$. In consequence, dynamo solutions are possible of either dipole

$$A(s, -z, t) = A(s, z, t), \qquad B(s, -z, t) = -B(s, z, t), \tag{8.3b}$$

or quadrupole

$$A(s, -z, t) = -A(s, z, t), \qquad B(s, -z, t) = B(s, z, t) \tag{8.3c}$$

parity. In the case of galactic dynamos discussed in Section 8.2, only kinematic dynamos are considered with both α and Ω taking prescribed steady values. For the shellular dynamos discussed in Section 8.3, we will also consider the possibility that the α obeys a simple quenching law as pioneered, e.g. by Braginsky (1970) and Stix (1972). For simplicity, we choose $\alpha = \alpha_0(s, z)/(1+|B|^2)$, where α_0 is steady and the magnetic field **B** has been suitably normalised.

8.1.2. The WKBJ approach

8.1.2.1. Global stability criteria

The essential idea which forms the cornerstone of our approach is the necessity of two length scales imposed by the small aspect ratio $\varepsilon := H_0/L$ of the geometry. For disc dynamos, the short and long lengths are the disc height $2H_0$ and diameter $2L := 2s_0$; while, for the shellular dynamo, they are the shell depth $H_0 := \varepsilon r_0$ and radius $L := r_0$. Suppose that we introduce local cartesian coordinates (x, y), where x measures distance along the long side and y measures distance across the short side both in units of L. Then, in both cases, we assume that the local dynamo generating properties are functions of x and $\varepsilon^{-1}y$. So whereas, the dynamo length scale is ε, its character only changes on the $O(1)$ length scale of the long side. This suggests that we should seek separable solutions of the form

$$A := \text{Re}\left\{ a(x)\hat{A}(x, \varepsilon^{-1}y)\exp\left[i\left(\frac{1}{\varepsilon}\int_{\bar{x}}^{x} k\,dx - \omega t\right)\right]\right\}, \tag{8.4}$$

in which ω is the constant complex frequency and eigenvalue of the problem, $k(x)$ is a slowly varying complex wave number, $\bar{x}$ is a constant to be chosen later at our convenience, the amplitude $\hat{A}(x, \varepsilon^{-1}y)$ has been suitably normalised, while any further amplitude modulation is taken up by the amplitude $a(x)$.

Suppose that our dynamo problem is characterised by the dynamo number D, which provides a dimensionless measure of the product of the α and Ω-effects. Then we may substitute (8.4) into the dynamo equations (8.2), apply appropriate boundary conditions on the long sides and so obtain the local dispersion relation

$$\omega = \omega(k, x, D). \tag{8.5}$$

If we characterise local spatially wave-like disturbances by real wavenumbers k, we say that local dynamo action occurs at some position x, whenever, $\text{Im}\{\omega\} > 0$. So if we minimise $|D|$ over all real k and admissible x subject $\text{Im}\{\omega\} = 0$, we may determine a critical dynamo number D_{loc}, frequency ω_{loc} corresponding to local dynamo action with wavenumber k_{loc} at position x_{loc}, where

$$\text{Im}\{\omega_{,k}\} = 0, \quad \text{Im}\{\omega_{,x}\} = 0 \quad \text{with } \text{Im}\{\omega\} = 0 \tag{8.6}$$

and $,\bullet$ denotes the partial derivative with respect to $\bullet$.

The objective in this section is to outline sufficient conditions for dynamo action by constructing solutions which are not simply local in character but are valid over the entire spatial extent $x_l < x < x_r$ (say) of the dynamo. The complete statement of the eigenvalue problem, for which D is specified and ω is the complex eigenvalue, requires the inclusion of the homogeneous boundary conditions to be applied at the left and right endpoints x_l and x_r, respectively. In the subsections, we consider criteria for two types of spatially localised modes that commonly occur. Interior modes are considered in Section 8.1.2.2, for which the boundaries may be regarded as remote; while wall modes are considered in Section 8.1.2.3 for which one of the boundaries is regarded as close. The terms remote and close are notions that we can only make precise following our detailed discussion of them.

8.1.2.2. Interior modes

The conditions (8.6) for local dynamo action do not guarantee global dynamo action, in the sense that (8.4) should provide a true eigenfunction of the complete eigenvalue problem. This is only guaranteed if the local solution valid on the $O(\varepsilon)$ length scale in the neighbourhood of x_{loc} can be extended throughout the dynamo region on the longer $O(1)$ length scale of the problem and meet appropriate end point boundary conditions. Generally, the local result fails to meet these global requirements, when either the group velocity $\omega_{,k}$ or the phase mixing $\omega_{,x}$ are non-zero at $(k_{\text{loc}}, x_{\text{loc}})$. The physical pictures that are usually appealed to in order to explain the shortcomings of the local results are the following. In the former case, locally unstable wave packets are convected away from the point x_{loc} at the group velocity and subsequently decay. In the latter case (see, e.g. Heyvaerts and Priest, 1983), the wave beats at slightly different frequencies at neighbouring locations. The local wave number suffers a secular change advecting it away from k_{loc} in Fourier space by phase mixing, again leading to decay; typically the length scale shortens leading to enhanced dissipation. The conditions generally necessary for global instability are

$$\omega_{,k} = 0, \quad \omega_{,x} = 0 \qquad \text{with } \text{Im}\{\omega\} = 0 \tag{8.7}$$

at some (k_S, x_S). The location x_S provides the natural value for the lower limit $\bar{x}$ of the phase integral in (8.4). In general, these global conditions (see, e.g. Huerre and Monkewitz, 1990) are met at complex values of $(k_S, x_S) \neq (k_{loc}, x_{loc})$ with corresponding values $\omega_S \neq \omega_{loc}$ and $|D_S| > |D_{loc}|$. There is a further limitation to the applicability of these results, concerning the location of the so-called anti-Stokes lines. That is a technical matter that we will develop, but the appropriate condition is stated in (8.18).

The physical arguments appealed to above are temporal in nature and linked to the notion of an initial value problem. To appreciate the significance of the global stability criteria (and for that matter to determine when they are appropriate), we need to approach the problem from the steady state boundary value point of view with ω like D regarded as fixed. Then (8.5) becomes an equation for $k(x)$ possibly possessing multiple roots

$$k := k_i(x) \qquad (i = 1, 2, \ldots, n), \tag{8.8}$$

where, e.g. $n = 4$ for our stellar shell $\alpha\Omega$-dynamo problems discussed in Section 8.3. Of course, all four of the WKBJ-modes corresponding to each k_i are involved in the complete solution and must be combined so as to meet the end point boundary conditions. Nevertheless, a typical interior mode localised in the neighbourhood of $x = x_M$ (say) inside the dynamo domain is dominated by a single mode $k = k_1(x)$, whose imaginary part increases monotonically with x over the entire range $[x_l, x_r]$ and vanishes at, $x = x_M$, where $k = k_M$ (say, real). Using the Taylor series expansion $k = k_M + k'_M(x - x_M) + \cdots$ (prime $'$ denotes derivative), we see that the solution is a wave with wavelength $O(\varepsilon)$ and lies under the wider Gaussian envelope of width $O(\varepsilon^{1/2})$. Its amplitude and phase modulation are proportional to

$$\exp\left[i\left(k_M \frac{x - x_M}{\varepsilon} - \omega t\right)\right] \exp\left[ik'_M \frac{(x - x_M)^2}{2\varepsilon}\right] \qquad \text{with } \text{Im}\{k'_M\} > 0. \tag{8.9}$$

Evidently, if (k_S, x_S) coincides with (k_M, x_M), then the global and local theory yield the same results: $(k_{loc}, x_{loc}) = (k_S, x_S) = (k_M, x_M)$. This is the most readily understood case and the more complicated situation $(k_S, x_S) \neq (k_M, x_M)$ must be regarded as a generalisation of it.

A deeper understanding of the role of the global criteria is gained from the equation

$$k'(x) = -\omega_{,x} / \omega_{,k} \tag{8.10}$$

governing the spatial derivative of $k(x)$. Evidently, it is undefined at (k_S, x_S), where both the numerator and denominator vanish. So to evaluate it we utilise the lowest order terms in the Taylor series expansion

$$\omega(k, x, D) - \omega_S = \tfrac{1}{2}\omega_{,kk}(k - k_S)^2 + \omega_{,kx}(k - k_S)(x - x_S)$$

$$+ \tfrac{1}{2}\omega_{,xx}(x - x_S)^2 + \omega_{,D}(D - D_S), \tag{8.11}$$

where all partial derivatives are evaluated at (k_S, x_S, D_S). Solving (8.11) for $(k - k_S)/(x - x_S)$ with $D = D_S$ and $\omega = \omega_S$ yields

$$k'_S := \left[-\frac{\omega_{,kx}}{\omega_{,kk}} \pm \frac{(\omega_{,kx}^2 - \omega_{,kk}\,\omega_{,xx})^{1/2}}{\omega_{,kk}} \right]_S. \tag{8.12}$$

Significantly, there are two roots identified by the $\pm$ sign. One of these corresponds to the k_1 that defines our WKBJ solution, while the other corresponds to a second mode with $k = k_2$ (say) which will not satisfy the remote boundary conditions and so is unwanted.

In the light of the above remarks the WKBJ solution of interest has the local behaviour

$$\hat{a}(X) \exp\left(ik_S \frac{x - x_S}{\varepsilon} \right) \qquad \left(X := \frac{x - x_S}{\varepsilon^{1/2}} \right), \tag{8.13a}$$

where

$$\hat{a}(X) := \bar{a} \exp\left(ik'_S \frac{X^2}{2} \right) \tag{8.13b}$$

and $\bar{a}$ is an arbitrary constant. On the intermediate distance $1 \ll |X| \ll \varepsilon^{-1/2}$, large compared to the localisation length scale $|x - x_S| = O(\varepsilon^{1/2})$ but small compared to the O(1) scale of the dynamo, the WKBJ method shows that $\hat{a}(X)$ satisfies

$$-\frac{\omega_{,kk}}{2}\frac{d^2\hat{a}}{dX^2} - i\omega_{,kx} X \frac{d\hat{a}}{dX} + \left(\frac{\omega_{,xx} X^2}{2} + C + \omega_{,D}\frac{D - D_S}{\varepsilon} - \frac{\omega - \omega_S}{\varepsilon} \right)\hat{a} = 0 \tag{8.14}$$

(cf. Connor *et al.*, 1979, eq. (36); Taylor, 1979, eq. (7)), where C is an O(1) constant whose value has negligible effect at large $|X|$.

On the localisation scale $|X| = O(1)$, we may return to the governing equations with the ansatz (8.13a). On the basis that $D - D_S = O(\varepsilon)$ and $\omega - \omega_S = O(\varepsilon)$, systematic asymptotic analysis again yields the amplitude equation (8.14) together with the value of the constant C, which is the only coefficient not determined by ω and its derivatives. The reason for the failure of the WKBJ operational procedure to determine C can be traced to the fact that the dispersion relation $\omega = \omega(k, x, D)$ does not distinguish terms of the form $d(X\hat{a})/dX$ and $X d\hat{a}/dX$, which differ by $\hat{a}$.

It is instructive to substitute (8.13b) into (8.14) and to discover that it is satisfied exactly when

$$\omega_{,D}\frac{D - D_S}{\varepsilon} - \frac{\omega - \omega_S}{\varepsilon} = \frac{i\omega_{,kk} k'_S}{2} - C. \tag{8.15}$$

Equating the real and imaginary parts determines the O(ε) corrections $D - D_S$ and $\omega - \omega_S$ to the critical dynamo number and frequency. This simple statement hides the more complicated fact that (8.15) only isolates the lowest eigenvalue and corresponding eigenfunction (8.13b). Higher order modes may be generated from Parabolic Cylinder functions, as in (8.81).

On the assumption that the two WKBJ solutions identified by k_1 and k_2 are of comparable size at $x = x_S$, we can determine their relative sizes elsewhere by the value of

$$\text{Im}\{\Psi\}, \quad \text{where} \quad \Psi := \int_{x_S}^{x} (k_1 - k_2) \, dx; \tag{8.16a}$$

elsewhere they are of comparable size on the anti-Stokes lines $\mathcal{A}_+$ and $\mathcal{A}_-$ (say) defined by $\text{Im}\{\Psi\} = 0$ (see Heading, 1962). Unfortunately, the terminology is not universal and sometimes they are referred to as Stokes lines (see Fedoryuk, 1983); Heading (1962), whose terminology we adopt, defines the Stokes lines by $\text{Re}\{\Psi\} = 0$. Now $\mathcal{A}_+$ and $\mathcal{A}_-$ intersect orthogonally at $x = x_S$, where their tangents make angles

$$\arg(x - x_S) = \pm\frac{\pi}{4} - \frac{1}{2} \arg \left[\frac{(\omega_{,kk}\, \omega_{,xx} - \omega_{,kx}^2)^{1/2}}{\omega_{,kk}} \right]_S \tag{8.16b}$$

with the real x-axis. In the neighbourhood of x_S, the contours $\text{Im}\{\Psi\} = \text{constant}$ are rectangular hyperbolae defining a saddle point structure. The anti-Stokes lines $\mathcal{A}_+$ and $\mathcal{A}_-$ divide the complex x-plane into four sectors (see Fig. 8.1), where one or other of the two WKBJ solutions is dominant, $\text{Im}\{\Psi\} < \text{or} > 0$. Of some importance are the locations of the intersections $x = x_\pm$ of the lines $\mathcal{A}_\pm$ and the real x-axis, where the signs in (8.16b) are chosen so that $x_+ \geq x_-$. Restricting attention to real x, we require that the k_1 solution dominates ($\text{Im}\{\Psi\} < 0$) in the interval $\mathcal{I} = [x_-, x_+]$, while the k_2 solution dominates ($\text{Im}\{\Psi\} > 0$) outside the interval. Those exterior line segments identify the two sectors, say $\mathcal{S}_\pm$, in which the k_1 solution is subdominant ($\text{Im}\{\Psi\} > 0$) (see also fig. 2 of Soward and Jones, 1983 or fig. 7 of Huerre and Monkewitz, 1990). In turn that provides the boundary condition,

$$\hat{a} \to 0 \quad \text{as} \quad |X| \to \infty \quad \text{in sectors} \quad \mathcal{S}_\pm, \tag{8.17}$$

on the solution of (8.14) and so completes the formal specification of the eigenvalue problem. There is one further condition for the applicability of the global stability criteria (8.7) which has emerged from our construction. It is that the interval $\mathcal{I}$, on which the k_1 solution is

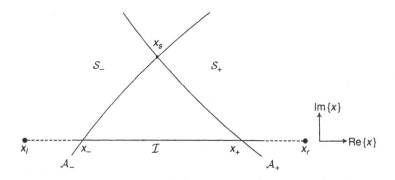

Figure 8.1 The sector structure of the complex x-plane, when $x_l \in \mathcal{S}_-$ and $x_r \in \mathcal{S}_+$. The anti-Stokes lines $\mathcal{A}_\pm$ are the continuous curves. The physical problem is defined on the interval $x_l \leq \text{Re}\{x\} \leq x_r$ of the real axis $\text{Im}\{x\} = 0$ and shown by dashed lines except in the sub-interval $\mathcal{I}$, which is drawn continuous.

dominant, lies within the dynamo domain. More precisely, the left- and right-end point boundary conditions at say x_l and x_r must be applied outside $\mathcal{I}$:

$$\mathcal{I} \subset [x_l, x_r]. \tag{8.18}$$

Thus (8.7) and (8.18) provide the global stability criteria for interior modes. It should be stressed that dynamo action might occur at lower dynamo number for some other type of mode such as the wall modes which we discuss next.

8.1.2.3. Wall modes

Suppose that, as above, we have isolated the globally marginal mode. Consider how it is effected when the left-hand end point x_l is moved to the right. The solution remains unaltered until x_l first reaches the anti-Stokes line at the left-hand end x_- of the interval $\mathcal{I}$. There, the precise form of the boundary conditions can be met by finely tuning $D - D_S$ and $\omega - \omega_S$ so as to adjust the contribution from the k_2-WKBJ solution, which is generally comparable there but negligible for all x to the right in $\mathcal{I}$. A consequence of this property is that, once x_l lies inside $\mathcal{I}$, our single dominant k_1-WKBJ solution cannot generally be adjusted by the addition of the subdominant k_2-WKBJ contribution in a manner that enables the complete solution to meet a homogeneous boundary condition at x_l, such as $a(x_l) = 0$. The way out of this dilemma is to demand the x_l remains on the anti-Stokes line but that, in turn, means that the zero phase mixing requirement must be dropped. In this case, we relax the global conditions (8.7) and require instead that the absolute stability conditions

$$\omega_{,k} = 0 \qquad \text{with } \text{Im}\{\omega\} = 0 \tag{8.19}$$

are met at some (k_T, x_T) with corresponding D_T and ω_T. In addition, we require that an appropriate anti-Stokes line from x_T passes through x_l. By this device we may construct solutions of the differential equations that satisfy the boundary conditions. Unfortunately, the recipe does not uniquely determine D and ω. A further minimisation of $|D|$ is required over those admissible solutions. Since the solutions constructed are sensative to the boundary location, we call them wall modes.

Though the application of vanishing phase mixing essential to the interior mode criteria is well documented even for fully complex locations x_S (see, e.g. Killworth, 1980, p. 161; Soward and Jones, 1983; Yano, 1992; Soward, 1992b; Lan et al., 1993; Jones et al., 2000), the interior mode criteria involving both the concept of absolute stability and boundary location does not appear to have been clearly identified except in certain special cases. These include the case of constant local dynamo number (Worledge et al., 1997; see Section 8.2.2) and nonlinear applications to frontal discontinuities (Dee and Langer, 1983; Chomaz, 1992; Meunier et al., 1997; Tobias et al., 1997; see Sections 8.3.4 and 8.3.5), as well as an application to Bénard convection (Dormy et al., 2002). Of course, the importance of the zero group velocity requirement for absolute instability is well known; it is only the anti-Stokes line location which provides the special feature of the wall mode criteria.

Of course, phase mixing may still vanish for an important class of wall modes. Then the local analysis dependent on the governing equation (8.14) remains relevant but with a local rather than remote boundary condition (see Section 8.3.5.1 and particularly (8.102a)). In the generic case of nonzero phase mixing ($\omega_{,x} \neq 0$), however, the solution corresponding to (8.13a) in the neighbourhood of x_T (say), where only the absolute stability conditions (8.19) are met, exists on a longer length scale with the structure

$$\hat{a}(X) \exp\left(ik_T \frac{x - x_T}{\varepsilon}\right) \qquad \left(X := \frac{x - x_T}{\varepsilon^{2/3}}\right). \tag{8.20a}$$

The amplitude modulation (8.13b) is replaced by the Airy function

$$\hat{a}(X) := \bar{a} \, \mathrm{Ai}\left(\left(\frac{2\omega_{,x}}{\omega_{,kk}}\right)^{1/3}(X - \lambda)\right), \tag{8.20b}$$

which vanishes as $X \uparrow \infty$ and solves

$$-\frac{\omega_{,kk}}{2}\frac{d^2\hat{a}}{dX^2} + \left(\omega_{,x} X + \omega_{,D}\frac{D - D_T}{\varepsilon^{2/3}} - \frac{\omega - \omega_T}{\varepsilon^{2/3}}\right)\hat{a} = 0 \tag{8.20c}$$

(cf. Hastie *et al.*, 1979, eq. (A.17)) provided that the constant λ satisfies

$$\omega_{,D}\frac{D - D_T}{\varepsilon^{2/3}} - \frac{\omega - \omega_T}{\varepsilon^{2/3}} = -\omega_{,x}\,\lambda. \tag{8.20d}$$

The associated anti-Stokes lines radiate out from x_T along the three lines

$$\arg(x - x_T) = \pi + \frac{2n\pi}{3} - \arg\left[\left(\frac{\omega_{,x}}{\omega_{,kk}}\right)^{1/3}\right]_S \qquad (n = 0, \pm 1). \tag{8.21}$$

The Airy function (8.20b) is approximated by the exponentially small subdominant WKBJ solution in the domain between the $n = \pm 1$ anti-Stokes lines. The exponentially large WKBJ solution dominate its representation elsewhere but both WKBJ solutions balance on the other anti-Stokes line $n = 0$. That is the anti-Stokes line, upon which the endpoint x_l must lie. Note particularly that the zeros of the Airy function lie on the $n = 0$ line (8.21) (see Section 8.3.5.1 and particularly (8.111)).

The points at which the entire coefficient of $\hat{a}$ in (8.14) and (8.20c) vanish are called turning points and play an important role in the structure of the solutions (see, e.g. Wasow, 1985; Fedoryuk, 1991). In (8.20c) the expression linear in X defines only one turning point, whereas in (8.14) the expression quadratic in X defines two turning points. Now the arrival at a turning point generally triggers an unwanted mode. We either eliminate it by fine tuning the location of the second turning point, as proposed for the interior modes, or we locate the boundary on the appropriate anti-Stokes line as proposed for our wall modes. The latter possibility has been considered in detail by Bassom and Soward (2001).

We should stress that meeting the wall mode criteria (8.19) does not guarantee that an eigenfunction can be constructed over the entire dynamo domain and so it may not provide a valid solution. The difficulty stems from the usual issue of keeping track of the dominance or otherwise of the WKBJ solution. Though this is always an issue for both interior and wall modes, it is particularly acute for the wall modes. For in that case we have only concerned ourselves with one turning point. There is likely to be second turning point lurking with its own anti-Stokes lines whose locations play a key role. These are issues that must be considered for each individual problem addressed.

Bearing in mind the limitations of the wall mode approach, the most likely way the conditions are met, while simultaneously minimising $|D|$, is by demanding that x_T is close to an end point. So, e.g. if $\hat{a}$ vanishes at the left-hand end point x_l, the eigenvalues of the problem are determined by the zeros of the Airy function. Then the first zero will generally minimise $|D|$. This is exactly what happens for the case considered in Section 8.3.5, where x_l is regarded as the front location terminating the linear region ahead of finite amplitude nonlinear waves. That specification removes the lack of uniqueness inherent in (8.19) and so renders an unambiguous solution.

8.1.2.4. MEGA

A related approach, which has had some success, is the asymptotic investigation of rapidly growing kinematic dynamo modes at large dynamo number. Even in geometries without small aspect ratios, these modes often have very short length scale and are localised about $\mathbf{x}_E$ (say), where the local dynamo number takes its maximum (extremum) value. The concept is not restricted to the one-dimensional models considered in this chapter. Indeed, in most applications configurations with two essential space dimensions have been investigated.

MEGA builds on the more general eikonal representation

$$\sim \exp[i(\varepsilon^{-1}\varphi - \omega t)] \tag{8.22}$$

of geometric optics (Bruns, 1895), in which $\mathbf{k}(\mathbf{x}) = \nabla\varphi$ is the local wave vector. Its application to an α^2-dynamo model was pioneered by Sokoloff *et al.* (1983); for that case, the analysis resembles (8.11)–(8.14) with $\mathbf{k}_S = \mathbf{0}$ and $\mathbf{x}_S = \mathbf{x}_E$ real but both two-dimensional vectors. The method was subsequently applied to $\alpha\Omega$-dynamo models (Ruzmaikin and Starchenko, 1988; Ruzmaikin *et al.*, 1988b) and later the term MEGA was coined by Ruzmaikin *et al.* (1990).

In the more general application to $\alpha\Omega$-dynamos, the magnetic field concentrated at $\mathbf{x}_M$ is not necessarily located at the maximum $\mathbf{x}_E$ of the local dynamo number. This was illustrated explicitly in Soward's (1992b) galactic field model, but anticipated by Roberts and Soward (1992) who appraised the MEGA method from the general point of view adopted in this chapter. Interestingly, in the geophysical fluid dynamics context a model is described in p. 161 of Killworth (1980) which implies that x_E and x_M are distinct and separate from a fully complex saddle point x_S. In the solar context, however, Kuzanyan and Sokoloff's (1995) model, which we discuss here in Section 8.3 clearly illustrates the separation of x_E and x_M; though it has the simplifying feature, in the linear case, that the saddle point x_S coincides with x_E. Nevertheless, more recent applications of the MEGA method (Starchenko, 1994; Starchenko and Kono, 1996) have taken account of possible separations in the two-dimensional context.

Many of the ideas of multiple scale wave theory that we have introduced have a long history in a variety of subject areas. The link of the WKBJ method to turning point problems such as (8.14) and (8.20) is well known in quantum mechanics (Schiff, 1955). The eikonal representation (8.22) is central to the study of plasma stability (Stix, 1962), while the separated length scale representation in orthogonal directions (8.4) is particularly pertinent to applications to confined plasmas (see, e.g. Connor *et al.*, 1979, eq. (1); Taylor, 1979). The fluid mechanic applications are numerous (see, e.g. Lighthill, 1978; while we itemise more recent studies directly relevant to our development where appropriate).

8.2. Galactic dynamos

8.2.1. Background

Following Parker's (1971) pioneering application of the αΩ-equations (8.2) to the galactic magnetic field, there has been continuing interest as evinced by the conference proceedings (Beck *et al.*, 1990; Krause, 1990a), book (Ruzmaikin *et al.*, 1988a) and a recent review (Beck *et al.*, 1996), in which many recent references may be found. Stix (1975) was the first to solve numerically an αΩ-dynamo in an oblate spheroidal geometry relevant to a galactic disc and match the internally generated field to an external potential field. An analytic asymptotic study of the model in the small aspect ratio limit ($\varepsilon \ll 1$) was subsequently undertaken by Soward (1978); while further numerical results obtained by Walker and Barenghi (1994) at smaller values of ε, than previously obtained by Stix (1975), largely confirm the analytic predictions. A key feature of the model, which has important implications for the nature of the analytic solution described in the following sections, is the coupling of the magnetic field at widely separated boundary locations by the external potential magnetic field. This effect was ignored in the later analytic WKBJ studies of models for particular galaxies by Baryshnikova *et al.* (1987) and Krasheninnikova *et al.* (1989). Instead they only considered the local internal coupling caused by magnetic diffusion. This leads to very different results and, in particular, fails to capture the algebraic decay of the magnetic field far from the current sources.

Throughout this section we restrict attention to axisymmetric magnetic fields and will not address interesting issues such as bisymmetric magnetic structures (see, e.g. Bykov *et al.*, 1997) and barred galaxies (Moss *et al.*, 2001). The analysis will also be entirely kinematic which means that we are unable to consider the evolution of the ensuing finite amplitude states. Nevertheless, our study in Section 8.3 suggests that interesting nonlinear problems in galactic dynamo theory, which rely e.g. on relatively simple processes such as α-quenching (also magnetic buoyancy; Moss *et al.*, 1999), can be addressed by the asymptotic methods employed there. This might include parity selection (see, e.g. Walker and Barenghi, 1995), propagating magnetic fronts (Poezd *et al.*, 1993; Moss *et al.*, 1998) and linking of dynamo generation to density wave theory (see, e.g. Mestel and Subramanian, 1991).

We should mention briefly the related problem concerning dynamo action in accretion discs (Pudritz, 1981), about which there is ongoing interest (see, e.g. Stepenski and Levy, 1990; Brandenburg and Campbell, 1997; Campbell, 1999).

8.2.2. Mathematical formulation of Stix's model

Stix (1975) proposed a galactic dynamo model in an oblate spheroid with principal axes length L and H_0. Its half height H is given by

$$\frac{H}{H_0} = h := \sqrt{1 - x^2} \quad \text{at radius} \ \ x = \frac{s}{L}. \tag{8.23}$$

The Ω and α-effects (8.3a) adopted have the simple structure

$$\Omega := \Omega(x), \quad \alpha := \alpha_0' h f(\zeta) \quad \text{with} \ \ \zeta = \frac{z}{H}. \tag{8.24a}$$

The actual forms, that he employed, are

$$\Omega(x) := \frac{1}{2}\Omega_0'' x^2, \qquad f(\zeta) := \zeta. \tag{8.24b}$$

From them we can define local magnetic Reynolds numbers

$$R_\Omega := \frac{H^2 x}{\eta} \frac{d\Omega}{dx} = \frac{H_0^2 \Omega_0''}{\eta} x^2 h^2, \qquad R_\alpha := \frac{H_0 \alpha_0'}{\eta} h^2, \tag{8.25a}$$

which in turn define the local dynamo number

$$\mathcal{D}(x) := R_\alpha R_\Omega = Q x^2 (1 - x^2)^2, \qquad \text{where } Q := H_0^3 \alpha_0' \Omega_0'' / \eta^2. \tag{8.25b}$$

Here $\mathcal{D}$ increases monotonically from zero on the symmetry axis at $x = 0$ to a maximum

$$D := \frac{4}{27} Q \quad \text{at } x_E := \sqrt{\frac{1}{3}}, \qquad \text{where } h_E := \sqrt{\frac{2}{3}} \tag{8.26a}$$

(the extremum of $\mathcal{D}$) and then decreases monotonically to zero at the disc rim $x = 1$. For later use, we note that

$$\mathcal{D}_E' = 0, \qquad -\frac{h_E^2 \mathcal{D}_E''}{2D} = 6. \tag{8.26b}$$

Following the numerical investigations by Stix (1975, 1978) and White (1977) for moderately small aspect ratios, Soward (1978) investigated the steady dipole and quadrupole modes analytically the limit

$$\varepsilon := H_0 / L \ll 1. \tag{8.27}$$

A higher order analysis of these steady modes together with a study of the oscillatory dipole and quadrupole modes was undertaken by Soward (1992a,b). A subsequent numerical study by Walker and Barenghi (1994) took Stix's results to smaller ε, at which a tighter comparison with the asymptotics could be made.

We now take H_0 as our unit of axial length and H_0^2 / η as our unit of time so that

$$z \to H_0 z, \qquad t \to (H_0^2 / \eta) t. \tag{8.28a}$$

With the additional change of variables

$$A \to HA, \qquad B \to R_\Omega B, \tag{8.28b}$$

the governing equations (8.2) inside the galactic disc $|z| < h$ are given correct to lowest order by

$$\frac{\partial A}{\partial t} = \frac{\mathcal{D} f}{h^2} B + \left(\varepsilon^2 \frac{\partial^2 A}{\partial x^2} + \frac{\partial^2 A}{\partial z^2} \right), \tag{8.29a}$$

$$\frac{\partial B}{\partial t} = -\frac{1}{h} \frac{\partial A}{\partial z} + \left(\varepsilon^2 \frac{\partial^2 B}{\partial x^2} + \frac{\partial^2 B}{\partial z^2} \right), \tag{8.29b}$$

while in the exterior vacuum $|z| > h$:

$$\varepsilon^2 \frac{\partial^2 A}{\partial x^2} + \frac{\partial^2 A}{\partial z^2} = 0, \qquad B = 0. \tag{8.30}$$

At their interface $|z| = h(x) = \sqrt{1 - x^2}$:

$$A, \partial A/\partial z \text{ and } B \text{ are continuous.} \tag{8.31}$$

There are higher order terms in (8.29) and (8.30) that have been neglected here but influence some of the higher order results reported by Soward (1992a,b). Nevertheless, to the order of accuracy attempted in our outline development here, they are generally unimportant but, when they influence qualitative comparisons, as in Figs. 8.2, 8.4, and Table 8.1 (using (8.66)), we will draw attention to the fact.

8.2.3. Local theory

Parker (1971) was the first to take advantage of the local properties of the disc. He ignored any radial variations of the disc and assumed that $h = $ constant. He sought wave-like solutions of the form

$$[A, B] := \text{Re}\{[\mathcal{A}(\zeta), \mathcal{B}(\zeta)]\exp[i(\varepsilon^{-1}kx - \omega t)]\}, \tag{8.32}$$

in which, according to (8.29), $\mathcal{A}(\zeta)$ and $\mathcal{B}(\zeta)$ satisfy

$$\mathcal{A}'' - (K^2 - i\Omega)\mathcal{A} = -\mathcal{D}f(\zeta)\mathcal{B}, \qquad \mathcal{B}'' - (K^2 - i\Omega)\mathcal{B} = \mathcal{A}', \tag{8.33a,b}$$

where $\zeta = z/h$ (see (8.24a) and (8.28a)) and

$$K := hk, \qquad \Omega := h^2\omega. \tag{8.33c}$$

The symmetry conditions at the mid-plane $z = 0$ are met by

$$\mathcal{A}'(0) = \mathcal{B}(0) = 0 \quad \text{for Dipole parity;} \tag{8.34a}$$

$$\mathcal{A}(0) = \mathcal{B}'(0) = 0 \quad \text{for Quadrupole parity,} \tag{8.34b}$$

while a continuous match with the external potential magnetic field

$$\mathcal{A} = \mathcal{A}(1)\,\exp[-|K|(\zeta - 1)], \qquad \mathcal{B} = 0 \quad \text{in } z > h$$

is achieved when

$$\mathcal{A}'(1) + |K|\mathcal{A}(1) = 0, \qquad \mathcal{B}(1) = 0, \tag{8.35}$$

where $|K|$ means $(|\text{Re}\{K\}|/\text{Re}\{K\})K$.

The solution of (8.33) on the interval $0 < \zeta < 1$ subject to the boundary conditions (8.34) and (8.35) at fixed k and $\mathcal{D}$ poses an eigenvalue problem with complex eigenvalue ω:

$$\omega(k, x, Q) := h^{-2}\Omega(K, \mathcal{D}), \tag{8.36}$$

where $K := hk$ and $\mathcal{D}(x, Q)$ is given by (8.25b). The k-symmetries imply that $\Omega(-K, \mathcal{D}) = \Omega(K, \mathcal{D})$. Furthermore we may determine from (8.36) the formula

$$\omega_{,x} - \frac{h'}{h}k\omega_{,k} = -2\frac{h'}{h}\omega + \frac{\mathcal{D}'}{h^2}\Omega_{,\mathcal{D}}, \qquad \omega_{,k} = \frac{1}{h}\Omega_{,K}, \qquad \omega_{,Q} = \frac{\mathcal{D}}{h^2 Q}\Omega_{,\mathcal{D}}. \tag{8.37a,b,c}$$

This gives the useful result that

$$\frac{D'}{D}\omega, Q = 2\frac{h'\omega}{hQ} \qquad \text{whenever} \qquad \omega, x - \frac{h'}{h}k\omega, k = 0, \qquad (8.38a)$$

where

$$\frac{hD'}{2h'D} = \frac{3x^2 - 1}{x^2} \qquad \text{for Stix's model.} \qquad (8.38b)$$

In the case of Stix's α-effect $f(\zeta) := \zeta$, the onset of dynamo action for the steady modes (direct current) is characterised by the values

$$D_{dc} \approx -12.56, \quad \Omega = 0, \quad K \to 0 \quad \text{for Quadrupole parity,} \qquad (8.39a)$$

$$D_{dc} \approx 69.10, \quad \Omega = 0, \quad K \to 0 \quad \text{for Dipole parity.} \qquad (8.39b)$$

The onset of the oscillatory mode (alternating current) sets in at larger values of D characterised by

$$D_{ac} \approx 396.59, \quad \Omega \approx -16.6, \quad K \to 0 \qquad \text{for Quadrupole parity;} \qquad (8.40a)$$

$$D_{ac} \approx -544.6, \quad \Omega \approx -11.8, \quad K \approx \pm 0.75 \quad \text{for Dipole parity.} \qquad (8.40b)$$

The final result for the oscillatory dipole is interesting, because it defines two modes; they propagate inwards ($K > 0$) and outwards ($K < 0$) but are otherwise identical. Since local theory does not distinguish between them, they can be combined in any proportion. Indeed in equal proportion they define a standing wave, which, at first sight, appears to be the natural symmetric solution. When we embed them into our WKBJ theory only one of the two modes is acceptable. For suppose one solution $k(x)$ provides the correct exponential decay at infinity, the other with $-k(x)$ necessarily grows exponentially at infinity and so fails to meet the conditions on the required solution. This is borne out by the numerical results of Walker and Barenghi (1994) illustrated in Fig. 8.6, which exhibits inward propagation corresponding to positive K and negative Ω.

A significant feature of the mathematical model is the appearance of $|K|$, which emerges through matching with the external potential magnetic field via the boundary condition (8.35). As a direct consequence, the dispersion relation (8.36) is not analytic at $k = 0$. Indeed, if we consider only stationary modes, we obtain the expansion

$$D := D_0 + D_1|K| + D_2 K^2 + \cdots \qquad \text{from } \Omega(K, D) = 0 \qquad (8.41)$$

for small K. This means that the WKBJ approach outlined in Section 8.1 will need to be modified when we embed these local long wavelength modes into a disc of slowly varying size. The presence of the term $D_1|K|$ in (8.41) means that amplitude modulation is no longer governed locally by a differential equation as in the conventional WKBJ method. Instead the process is nonlocal and governed by an integral equation. The integral equation approach has been adopted recently by Priklonsky *et al.* (2000) and in other numerical studies of dynamos by Dobler and Rädler (1998).

The steady dipole mode has a degenerate character that is not captured by simply setting $K = 0$, which is why we have been careful to identify the long length scale modes with

$K \rightarrow 0$ rather than $K = 0$. In order to systematically construct a solution, the magnetic vector potential A must be described by an expansion of the form

$$A := \overline{A}_{-1}|K|^{-1} + A_0(\zeta) + O(K), \tag{8.42}$$

where $\overline{A}_{-1}$ is a constant. The difficulty arises because with $K = \Omega = 0$ in (8.33a), the reduced system admits the trivial solution $A = \overline{A}_{-1}|K|^{-1}$, which is needed to meet the boundary condition (8.35). Without it the reduced boundary condition $A'(1) = 0$ admits no solutions, as then the equations and boundary conditions are inconsistent. This failure was first noted by Rädler and Bräuer (1987) in the context of α^2-disc dynamos; Krause (1990b) discusses the matter and calls the issue the problem of the 'forgotten' modes.

8.2.4. Radial structure

The long length scale modes do not fit any of the WKBJ classifications of Section 8.1.2 because of the lack of analyticity at $k = 0$. Nevertheless, the ideas developed there still apply. From a purely formal point of view the stability criteria for them employed by Soward (1992a) are equivalent to

$$\omega_{,x} = 0, \qquad \text{Im}\{\omega\} = 0 \tag{8.43a}$$

at some $(k, x) = (0, \bar{x})$. The formalism is totally analogous to the wall mode idea described in Section 8.1.2.3 but applied to a 'wall' $k = 0$ in Fourier space. Unlike the WKBJ procedure outlined in Section 8.1.2, the objective is only to construct a solution locally in the neighbourhood of $\bar{x}$. Elsewhere, it is argued that the solution is small but with only algebraic – rather than exponential – decay away from $\bar{x}$. Accordingly, the far field behaviour is not controlled locally, as it is in the WKBJ case, but by the solution in the neighbourhood of $\bar{x}$. This difference is striking and gives these problems with non-analytic k their special feature.

According to the results (8.37), the conditions $\omega_{,x} = 0$ and $k = 0$ are met by (8.38a) or, equivalently,

$$\mathcal{D}'\Omega_{,\mathcal{D}} = 2hh'\omega. \tag{8.43b}$$

For the steady modes $\omega = 0$, this condition is satisfied at the local dynamo number maximum (extremum) $\bar{x} = x_E$, where $\mathcal{D}' = 0$. The situation for oscillatory modes $\text{Re}\{\omega\} \neq 0$ is less straightforward, because $\Omega_{,\mathcal{D}}$ is generally complex and any $\bar{x}$ satisfying (8.43b) is likely to be complex also. The idea is that the solution so constructed in the neighbourhood of complex $\bar{x}$ can be continued analytically to provide the solution for real x. Though this appears to be rational for sufficiently small $\text{Im}\{\bar{x}\}$ it is not clear how applicable the method is when $\bar{x}$ is moved far off the real axis. Indeed this complication arose in the case of the oscillatory quadrupole, and for the relevant parameter range Soward (1992a) was unable even to locate $\bar{x}$. He considered a related problem with small, $\text{Im}\{\bar{x}\}$. Since its range of validity, did not coincide with that of the actual problem, it is not surprising to learn that the comparison with the Walker and Barenghi's (1994) full numerical solutions was unsatisfactory. We will discuss that case no further.

8.2.4.1. Steady quadrupole

We attempt solutions localised in the neighbourhood of x_E on the $O(\varepsilon^{1/3})$ length scale of the form

$$A := a(\varpi)\mathcal{A}(\zeta) + \varepsilon^{2/3} A_1(\varpi, \zeta), \quad B := a(\varpi)\mathcal{B}(\zeta) + \varepsilon^{2/3} B_1(\varpi, \zeta) \qquad (8.44a)$$

$$\text{with } \varpi := \frac{x - x_E}{\varepsilon^{1/3} l h_E}, \qquad (8.44b)$$

where the constant l is to be chosen later at our convenience. There the local dynamo number deviates from the local critical value by the $O(\varepsilon^{2/3})$ amount

$$\mathcal{D} - \mathcal{D}_{\text{dc}} = (D - \mathcal{D}_{\text{dc}}) + \tfrac{1}{2}\varepsilon^{2/3} l^2 h_E^2 \mathcal{D}_E'' \varpi^2. \qquad (8.45)$$

The functions $\mathcal{A}$ and $\mathcal{B}$ satisfy (8.33) with $K = \Omega = 0$. After an integration which takes account of the boundary conditions (8.34b), the $O(1)$ problem reduces to solving

$$\mathcal{A}'' + \mathcal{D}_{\text{dc}} f \mathcal{B} = 0, \qquad \mathcal{B}' - \mathcal{A} = 0 \quad \text{on } 0 < \zeta < 1 \qquad (8.46a)$$

$$\text{subject to} \quad \mathcal{A}(0) = 0, \quad \mathcal{A}'(1) = 0, \quad \mathcal{B}(1) = 0. \qquad (8.46b)$$

The adjoint problem is: solve

$$\mathcal{A}_A'' + \mathcal{D}_{\text{dc}} f \mathcal{B}_A = 0, \qquad \mathcal{B}_A' + \mathcal{A}_A = 0 \quad \text{on } 0 < \zeta < 1 \qquad (8.47a)$$

$$\text{subject to} \quad \mathcal{B}_A(0) = 0, \quad \mathcal{A}_A'(0) = 0, \quad \mathcal{A}_A(1) = 0. \qquad (8.47b)$$

The consistency condition, that the $O(\varepsilon^{2/3})$ problem for A_1 and B_1 has a solution, is

$$\mathcal{B}_A(1) \left. \frac{\partial A_1}{\partial \zeta} \right|_{z=1} = -\frac{\mathcal{D} - \mathcal{D}_{\text{dc}}}{\varepsilon^{2/3}} \langle f \mathcal{B} \mathcal{B}_A \rangle a, \qquad (8.48a)$$

where $\langle \bullet \rangle = \int_0^1 \bullet \, d\zeta$. The only terms to contribute stem from variations of the local dynamo number and the boundary coupling with the external potential field, which is determined from

$$\left. \frac{\partial A_1}{\partial \zeta} \right|_{\zeta=1} = \frac{\mathcal{A}(1)}{l} \mathcal{L}\left[\frac{da}{d\varpi} \right], \qquad (8.48b)$$

where the integral operator $\mathcal{L}$ is defined by

$$\mathcal{L}[\Phi] = \frac{1}{\pi} \fint_{-\infty}^{\infty} \frac{\Phi(\varpi')}{\varpi' - \varpi} d\varpi'; \qquad (8.49)$$

here the slashed integral sign denotes that the principal value of the integral is taken. Elimination of A_1 yields

$$\frac{1}{l} \mathcal{L}\left[\frac{da}{d\varpi} \right] + \left(\frac{\mathcal{D} - \mathcal{D}_{\text{dc}}}{\varepsilon^{2/3} \mathcal{D}_{\text{dc}}} \right) v^3 a = 0, \qquad (8.50a)$$

where after some integrations by parts ν^3 may be cast in the homogeneous form

$$\nu^3 := \frac{\langle \mathcal{B}'_A \mathcal{A}' \rangle}{\langle \mathcal{B}'_A \rangle \langle \mathcal{A}' \rangle}; \quad \nu \approx 1.062 \quad \text{for Stix's model} \tag{8.50b}$$

and $\mathcal{D} - \mathcal{D}_{dc}$ has the form (8.45).

We note that the eigenvalue problem

$$\mathcal{L}\left[\frac{da}{d\varpi}\right] - (\varpi^2 + \lambda)a = 0 \quad \text{with } |a| \to 0 \text{ as } |\varpi| \to \infty \tag{8.51}$$

has eigenfunctions

$$a_n(\varpi) = \text{Re}\{\Phi_n(\varpi)\}, \quad \text{in which } \Phi_n(\varpi) := \int_0^\infty i^n \text{Ai}(k + \lambda_n)e^{ik\varpi} \, dk, \tag{8.52a}$$

where λ_n are the zeros of Ai$'$ for n even and the zeros of Ai for n odd. They form the descending sequence

$$\lambda_0 \approx -1.0188, \quad \lambda_1 \approx -2.3381, \ldots \tag{8.52b}$$

and define even and odd modes

$$\Phi_n(-\varpi) = (-1)^n \Phi_n(\varpi) \tag{8.52c}$$

much like the bounded Parabolic Cylinder functions (8.81a,b) that we require to solve (8.14). Unlike those which decay exponentially, the eigenfunctions (8.52) exhibit algebraic decay determined from the large ϖ expansions

$$\Phi_n(\varpi) \sim \begin{cases} (-1)^{n/2}\text{Ai}(\lambda_n)\left[i\left(\dfrac{1}{\varpi} - \dfrac{\lambda_n}{\varpi^3}\right) + \dfrac{1}{\varpi^4} + \cdots\right] & \text{for } n \text{ even;} \\[4mm] (-1)^{(n-1)/2}\text{Ai}'(\lambda_n)\left[i\left(-\dfrac{1}{\varpi^2} + \dfrac{\lambda_n}{\varpi^4}\right) - \dfrac{2}{\varpi^5} + \cdots\right] & \text{for } n \text{ odd.} \end{cases} \tag{8.53}$$

Comparison with (8.51) shows that the eigenfunctions of (8.50a) are the $a_n(\varpi)$ of (8.52) when

$$\frac{Q - Q_{dc}}{\varepsilon^{2/3}Q_{dc}} = \frac{D - \mathcal{D}_{dc}}{\varepsilon^{2/3}Q_{dc}} = -\frac{\lambda_n}{\nu^3 l} \quad \text{with} \quad \frac{1}{l} := \nu\left(-\frac{h_E^2 \mathcal{D}''_E}{2D}\right)^{1/3}, \tag{8.54}$$

where use has been made of (8.26a): $Q_{dc} = (27/4)\mathcal{D}_{dc}$. The parameters for Stix's model are given by (8.27), (8.39), (8.50b) and (8.52b). Soward (1992a) determined additional $O(\varepsilon^{4/3})$ corrections; the values of $Q(\varepsilon)$, which result, are illustrated in Fig. 8.2 for the lowest mode $n = 0$ together with the numerical results of Walker and Barenghi (1994).

Interestingly, since A is a potential function outside the disc, it is obtained simply from

$$A(x, z) = \text{Re}\{\Phi_n(Z)\}, \quad \text{in which } Z := \frac{(x - x_E) + i\varepsilon(z - h_E)}{\varepsilon^{1/3}lh_E} = O(1). \tag{8.55a,b}$$

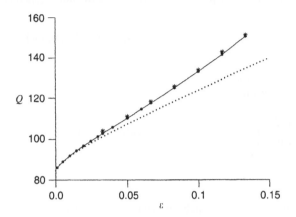

Figure 8.2 Q versus ε for the steady quadrupole. The solid line connecting the bullets ($\bullet$) indicates the numerical results of Walker and Barenghi, while the stars $\star$ correspond to those of Stix (1975). The dashed line portrays the analytic result (8.54) together with a further $\varepsilon^{4/3}$ correction (Soward, 1992a, eq. (3.21)). (Courtesy of Walker and Barenghi, 1994.)

Figure 8.3 Walker and Barenghi's numerical steady quadrupole solution at $\varepsilon = 7/60$. Contours of $B = $ constant are plotted on the right; contours of $sA = $ constant are plotted on the left. (Courtesy of Walker and Barenghi, 1994.)

Using (8.53) we see that, for n even, at intermediate distances far from the field source but still near compared to the disc radius ($1 \ll |Z| \varepsilon^{-1/3}$) the magnetic field is a two-dimensional dipole. When the source is wrapped in a circle about the symmetry axis, it describes an axisymmetric quadrupolar field far from the disc ($\varepsilon^{-1/3} \ll |Z|$). An example of Walker and Barenghi's (1994) lowest quadrupole mode $n = 0$ is illustrated in Fig. 8.3. Returning to (8.53) we see that the odd n modes are two-dimensional quadrupoles on intermediate distances but far from the disc that field largely cancels out leaving again an axial quadrupole.

8.2.4.2. *Steady dipole*

We proceed as before for the quadrupole case, but modify our approach in the light of the exceptional nature of the 'forgotten' dipole mode. In particular, we include a large axial z-independent field, which varies slowly on the $O(\varepsilon^{1/3})$ length scale, to correspond to the constant $\bar{A}_{-1}$ isolated in the expansion (8.42). Accordingly we set

$$A := \varepsilon^{-2/3} a_{-1}(\varpi) + a(\varpi)\mathcal{A}(\zeta) + \varepsilon^{2/3} A_1(\varpi, \zeta), \quad B := a(\varpi)\mathcal{B}(\zeta) + \varepsilon^{2/3} B_1(\varpi, \zeta)$$

$$(8.56a)$$

and normalise a_{-1} by the choice

$$A = \varepsilon^{-2/3} a_{-1}(\varpi) \quad \text{on the disc boundary } z = h. \tag{8.56b}$$

The same O(1) terms continue to dominate in the magnetic induction equation (8.29) and so $\mathcal{A}$ and $\mathcal{B}$ satisfy (8.33) as before. In the light of the different boundary conditions, a constant of integration emerges from integration of (8.33b). Thus, instead of (8.46), we solve

$$\mathcal{A}'' + \mathcal{D}_{\mathrm{dc}} f \mathcal{B} = 0, \qquad \mathcal{B}' - \mathcal{A} = \mathcal{B}'(1) \quad \text{on } 0 < \zeta < 1 \tag{8.57a}$$

$$\text{subject to } \mathcal{A}'(0) = 0, \quad \mathcal{B}(0) = 0, \quad \mathcal{B}(1) = 0. \tag{8.57b}$$

The adjoint problem, replacing (8.47), is: solve

$$\mathcal{A}''_A + \mathcal{D}_{\mathrm{dc}} f \mathcal{B}_A = 0, \qquad \mathcal{B}'_A + \mathcal{A}_A = 0 \quad \text{on } 0 < \zeta < 1 \tag{8.58a}$$

$$\text{subject to} \quad \mathcal{A}_A(0) = 0, \quad \mathcal{A}_A(1) = 0, \quad \mathcal{B}_A(1) = 0. \tag{8.58b}$$

Upon application of the boundary condition (8.56b) the consistency condition that the $O(\varepsilon^{2/3})$ problem for A_1 and B_1 has a solution is

$$\frac{\langle \mathcal{B}_A \rangle}{l^2} \frac{d^2 a_{-1}}{d\varpi^2} = \frac{\mathcal{D}_{\mathrm{dc}} - \mathcal{D}}{\varepsilon^{2/3}} \langle f \mathcal{B} \mathcal{B}_A \rangle a, \tag{8.59a}$$

which identifies the influence of lateral diffusion of poloidal magnetic field within the disc. It matches the external potential field when

$$A'(1)a = \frac{1}{l} \mathcal{L} \left[\frac{da_{-1}}{d\varpi} \right]. \tag{8.59b}$$

Together with a property of the integral operator $\mathcal{L}$ it yields

$$\frac{d^2 a_{-1}}{d\varpi^2} = -\mathcal{L} \left[\left(\frac{d}{d\varpi} \mathcal{L} \left[\frac{da_{-1}}{d\varpi} \right] \right) \right] = -l A'(1) \mathcal{L} \left[\frac{da}{d\varpi} \right]. \tag{8.59c}$$

In this way (8.59a) reduces to the integral equation (8.50a), as before, but with ν replaced by

$$\nu^3 := \frac{\langle f \mathcal{B} \mathcal{B}_A \rangle}{\langle f \mathcal{B} \rangle \langle \mathcal{B}_A \rangle}; \qquad \nu \approx 0.8366 \quad \text{for Stix's model.} \tag{8.60}$$

The problem posed is solved by the eigenfunctions (8.52), (8.53) and eigenvalue (8.54). As before, higher order $O(\varepsilon^{4/3})$ corrections were attempted by Soward (1992a). They had little effect on the results but those results are used in the comparison of $Q(\varepsilon)$ with the results of Walker and Barenghi (1994) in Fig. 8.4.

Though the reduced mathematical problem for the dipole is the same as that for the previous quadrupole case, the physical interpretation of the result is different. Consider the x and z-components of magnetic field

$$b_s \approx -h \frac{\partial A}{\partial z}, \qquad b_z \approx \varepsilon h \frac{\partial A}{\partial x}, \tag{8.61a}$$

in the vacuum region exterior to the disc. Since, within the framework of our two-dimensional approximation, it is potential (see (8.30)), the radial component of magnetic field on the disc

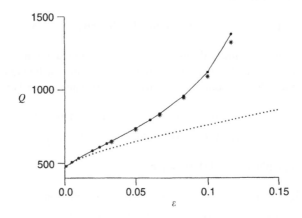

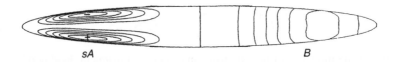

Figure 8.4 As in Fig. 8.2 but for the steady dipole. (Courtesy of Walker and Barenghi, 1994.)

Figure 8.5 As in Fig. 8.3 but for the steady dipole. (Courtesy of Walker and Barenghi, 1994.)

surface $b_s(x, h) = -\mathcal{A}'(1)a(\varpi)$ determines (b_s, b_z) everywhere as the real and imaginary parts of complex magnetic field

$$b_s - ib_z := -\mathcal{A}'(1)\Phi_n(Z), \tag{8.61b}$$

where the complex coordinate Z is again defined by (8.55b). Returning to (8.53), the interpretation of the field on intermediate distances is that due to a one-dimensional line current for n even and a two-dimensional dipole for n odd. Far from the disc, the axisymmetric field for n even is an axial dipole due to a ring current. For n odd the field is virtually annihilated but remains an axial dipole. The actual $n = 0$ fields computed numerically by Walker and Barenghi (1994), are illustrated in Fig. 8.5.

8.2.4.3. Oscillatory dipole

In view of the preliminary result (8.40b), the short length scale oscillatory dipole mode, whose radial disc scale is comparable to its transverse width, has the WKBJ representation

$$[A, B] := \mathrm{Re}\left\{a(x)[\mathcal{A}(\zeta)\mathcal{B}(\zeta)]\exp\left[i\left(\frac{1}{\varepsilon}\int_{x_S}^{x} k\,dx - \omega t\right)\right]\right\}, \tag{8.62}$$

where $\mathcal{A}, \mathcal{B}$ solve (8.33), (8.34b) and (8.35). The objective is to construct an interior global mode (see Section 8.1.2.2), which correctly crosses the saddle point x_S at a location to be determined.

Since the local dispersion relation is determined from (8.36), we see from (8.37) that the vanishing of both $\omega_{,k}$ and $\omega_{,x}$ only occurs when the condition $\mathcal{D}'\Omega_{,\mathcal{D}} = 2hh'\omega$ (see (8.43b)) is met. In a disc, where $h' = 0$ when $\mathcal{D}' = 0$, the saddle point conditions can be met simply at the real location $x_S = x_M$ of the dynamo number maximum. This type of solution is obtained trivially in the shellular dynamo model in Section 8.3, for which h is essentially a constant, i.e. there are no rescalings of the wave number of the type (8.33c) which involve a spatially varying h.

Since $\Omega_{,\mathcal{D}}$ is generally complex for real x, we expect (8.43b) and, of course, (8.7) to be satisfied by complex (k_S, x_S) for Stix's model. The numerical solution of (8.33) determines

$$k_S \approx K_S/h_S, \quad x_S \approx 0.677 + \text{i}\,0.061, \quad h_S \approx 0.741 - \text{i}\,0.056 \tag{8.63a}$$

with

$$Q_S \approx -4466, \quad \omega_S \approx -23.1, \quad K_S \approx 0.662 + \text{i}\,0.434. \tag{8.63b}$$

From (8.38), we obtain immediately that

$$(\omega_{,Q})_S = \frac{x_S^2}{3x_S^2 - 1}\frac{\omega_S}{Q_S} \approx 0.004976 - \text{i}\,0.002214. \tag{8.64}$$

Upon setting

$$Q_1 := (Q - Q_S)/\varepsilon, \quad \omega_1 := (\omega - \omega_S)/\varepsilon, \tag{8.65}$$

Soward (1992b) attempted to determine the remaining coefficients of the higher order theory that appear on the right-hand side of (8.15). The analysis led to the result

$$(\omega_{,Q})_S Q_1 - \omega_1 \approx -14.5 + \text{i}19.5, \tag{8.66a}$$

whose real and imaginary parts determine

$$Q_1 \approx -8815, \quad \omega_1 \approx -29.4. \tag{8.66b}$$

Though the lowest order results (8.63) and (8.64) are likely to be reliable, the accuracy of the $O(\varepsilon)$ correction is highly questionable. In view of the extensive algebra and computations involved in calculating the coefficient on the right-hand side of (8.66a), there is plenty of scope for errors to occur!

In spite of the reservations about the reliability of the results, we summarise the values of Q_1 and ω_1 as calculated from the results of table 4 of Walker and Barenghi (1994) for their full numerical computations for various aspect ratios. The fact that the values remain roughly constant is encouraging and lends support to the zero order asymptotics. Note, e.g. that the numerically computed value of Q for the smallest case $\varepsilon = 0.05$ differs from Q_S by about 5% and is still converging to it (Table 8.1). That should be contrasted with the local theory which from (8.40b) gives the widely different values

$$Q_{ac} \approx -3676, \quad \omega_{ac} \approx -17.7, \quad K_{ac} \approx 0.75. \tag{8.67}$$

The solution on the real x-axis is a travelling wave localised about

$$x_M \approx 0.588 \quad \text{with } k_M \approx 1.86, \quad k_M' \approx -3.27 + \text{i}\,9.85. \tag{8.68a,b,c}$$

Table 8.1 The values of $Q(= \varepsilon^3 R_{mc}^2)$ and $\omega(= \varepsilon^2 \Omega_c)$ for $\varepsilon \neq 0$ are determined from Walker and Barenghi's (1994) table 4 values for R_{mc} and Ω_c. In the bottom row $\varepsilon = 0$, the values $Q = Q_S$ and $\omega = \omega_S$ together with Q_1 and ω_1 stem from the analytic results. The remaining Q_1 and ω_1 values for $\varepsilon \neq 0$ are determined from the numerical values of Q and ω using the formulae (8.65)

ε	$-Q$	$-Q_1$	$-\omega$	$-\omega_1$
8/60	5038	4290	25.3	16
7/60	4959	4230	25.0	16
6/60	4886	4202	24.7	16
5/60	4820	4249	24.4	16
4/60	4766	4493	24.2	17
3/60	4732	5326	24.1	20
0	4466	8815	23.1	29

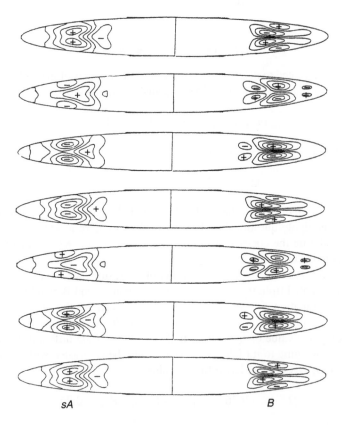

sA   B

Figure 8.6 As in Fig. 8.3 but for the oscillatory dipole. Starting from the top, each pair of plots illustrates the configuration at equal intervals of one sixth of the wave period; the bottom figure at the end of the period coincides with the top figure. (Courtesy of Walker and Barenghi, 1994.)

These characteristics define a short wavelength modulated travelling wave of the form (8.9). It is well illustrated in Fig. 8.6 taken from Walker and Barenghi (1994) for the case $\varepsilon = 7/60$. Two features of the cycle portrayed are interesting. First, the wave propagates axially towards the mid-plane from both above and below. This is compatible with the generally accepted argument for a uniform medium that propagation is in the direction $\nabla\Omega \times \alpha\hat{\phi}$. In addition, it is clear that the mode is propagating radially inwards towards the centre compatible with the result (8.68b).

From a technical point of view, we note that the anti-Stokes lines $\mathcal{A}_-$ and $\mathcal{A}_+$ intersect the real x-axis at

$$x_- \approx 0.558 \quad \text{and} \quad x_+ \approx 0.708, \qquad (8.69\text{a,b})$$

respectively. The fact, that the interval $\mathcal{I} = [x_-, x_+]$ lies inside the spatial extent of the dynamo $[0,1]$, confirms that the interior mode global criteria are met. There may, however, be possible repercussions due to x_+ being significantly far to the right of the dynamo number maximum x_M and relatively close to the end $x = 1$ of the dynamo range. That closeness could have a larger influence on the dynamo than the size of the field there suggests. In other words, the high order corrections Q_1 and ω_1 in the table might still be influenced by conditions near $x = 1$, which our asymptotic analysis fails to capture.

Despite the reservations about the higher order theory, it is encouraging to find that the lowest order results are a considerable improvement on the local theory. They correctly predict the dynamo number and frequency and the direction of wave propagation.

8.3. Stellar shell dynamos

8.3.1. Model equations and α-quenching

Parker (1955) was the first to explain the nature of the sunspot cycle and activity cycles of other stars in terms of a kinematic $\alpha\Omega$-dynamo wave. Since then considerable effort has been devoted to developing his idea, studying more detailed mathematical models, collecting observational data (see, e.g. Baliunas *et al.*, 1995), constraining (see, e.g. Dicke, 1988; Tuominen *et al.*, 1988) and interpreting the results (see, e.g. Noyes *et al.*, 1984, Baliunas *et al.*, 1996; Brandenburg *et al.*, 1998; Tobias, 1998). Essentially, the solar dynamo operates in the convective spherical shell above the stable radiative interior but may involve the convective overshoot region at their interface (see, e.g. Weiss, 1994). On the one hand, the complete system of partial differential equations governing the hydromagnetic mean field $\alpha\Omega$-model throughout the shell has been integrated (see e.g. Brandenburg *et al.*, 1990; Brandenburg, 1994; Brandenburg, Chapter 9, this volume; Rüdiger and Arlt, Chapter 6, this volume; Rüdiger and Brandenburg, 1995). On the other, the equations have been averaged radially (Parker, 1955; but see also Stix, 1989); this leads to a one-dimensional system is characterised by a local dynamo number $\mathcal{D}$ dependent on the latitude θ.

There are essentially two limiting cases of thick and thin shells. In the thick shell limit, solutions depend on the pole and equator boundary conditions; there is coupling between the Northern and Southern hemispheres and issues such as symmetry (dipole, quadrupole and mixed parity) can be addressed – sometimes within the framework of a simplified cartesian geometry (see, e.g. Jennings, 1991; Jennings and Weiss, 1991; Tobias, 1996, 1997a,b; Weiss and Tobias, 1997; Phillips *et al.*, 2002). Recent developments focus particularly on chaos and intermittency (see, e.g. Covas *et al.*, 1998; Tworkowski *et al.*, 1998). In the thin shell limit $\varepsilon \ll 1$, numerical integrations (Moss *et al.*, 1990) indicate that there is a short latitudinal length scale. This feature has been exploited along two distinct tracks. On the one hand, Worledge *et al.* (1997), Tobias *et al.* (1997, 1998) have investigated the special case of uniform spatial conditions with $\mathcal{D} = $ constant. On the other, Fioc *et al.* (1995), Kuzanyan and Sokolof (1995, 1996, 1997), Sokoloff *et al.* (1995a,b), Meunier *et al.* (1996, 1997), Galitsky and Sokoloff (1997, 1998, 1999), Sokoloff and Soward (1998) and Bassom *et al.* (1999) considered a spatially varying $\mathcal{D}$ with its maximum size at mid-latitudes. In other respects, both groups of authors make simplifying assumptions, which lead to the same one-dimensional $\alpha\Omega$-equations in the linear and nonlinear (α quenching) regimes. Further abstractions of the problem are possible which reduce the complexity of the one-dimensional system to variants of the Complex–Ginsburg–Landau (CGL) equation (see, e.g. Spiegel, 1994) or simply to time dependent ordinary differential equations (see, e.g. Weiss *et al.*, 1984; Knobloch *et al.*, 1997).

For the thin shell limit, we assume a radial shear for which the local radially averaged local Reynolds numbers have the form

$$R_\Omega := \frac{H^2 r}{\eta} \frac{\partial \Omega}{\partial r} \propto \cos\theta, \qquad R_\alpha := \frac{H\alpha}{\eta} \propto \sin\theta \qquad (8.70a)$$

from which we construct the local dynamo number

$$\mathcal{D} := -R_\Omega R_\alpha = D \sin(2\theta) \qquad (8.70b)$$

in terms of the dynamo number D. Note that to simplify the development the signs of the dynamo numbers are opposite to those usually employed. Following the change of variables (8.28), our model equations become

$$\frac{\partial A}{\partial t} = \frac{\mathcal{D}(\theta)}{(1 + |B|^2)} B + \left(\varepsilon^2 \frac{\partial^2 A}{\partial \theta^2} - A \right), \qquad (8.71a)$$

$$\frac{\partial B}{\partial t} = \varepsilon \frac{\partial A}{\partial \theta} + \left(\varepsilon^2 \frac{\partial^2 B}{\partial \theta^2} - B \right) \qquad (8.71b)$$

(Proctor and Spiegel, 1991; see also Schüssler and Ferriz-Mas, Chapter 5, this volume) similar to (8.29) except that we have included a simple form of α-quenching and under the radial averaging the second radial derivative in the diffusion operator has been replaced by -1.

Meunier *et al.* (1997) employed a harmonic expansion with respect to time but retained only the first harmonic

$$[A, B] := \mathrm{Re}\{[\mathcal{A}(\theta), \mathcal{B}(\theta)] \exp(i\omega t)\}. \qquad (8.72a)$$

Projecting the governing equations on to that harmonic yields

$$i\omega\mathcal{A} = \mathcal{D}_{\mathrm{eff}}\mathcal{B} + \left(\varepsilon^2 \frac{\partial^2 \mathcal{A}}{\partial \theta^2} - \mathcal{A} \right), \qquad (8.72b)$$

$$i\omega\mathcal{B} = \varepsilon \frac{\partial \mathcal{A}}{\partial \theta} + \left(\varepsilon^2 \frac{\partial^2 \mathcal{B}}{\partial \theta^2} - \mathcal{B} \right). \qquad (8.72c)$$

where

$$\mathcal{D}_{\mathrm{eff}}(\theta, |\mathcal{B}|) := \frac{2\mathcal{D}(\theta)}{\sqrt{1 + |\mathcal{B}|^2}\left(1 + \sqrt{1 + |\mathcal{B}|^2} \right)} \qquad (8.72d)$$

may be regarded as an effective dynamo number for the nonlinear problem. Though this approximation does not retain the accuracy of the original model equations (8.71), it does capture the important physics of the system.

In the following sections we begin by outlining the linear results and continue with the nonlinear α-quenching extension. The main thrust of the nonlinear development pivots about the appearance of fronts which separate a fully developed nonlinear region on one side from a linear region on the other. There is an extensive literature on front structure in the context of the CGL equation (see, e.g. Chomaz, 1992; Couairon, 1997; Couairon and Chomaz, 1997a). The key condition, upon which such frontal solutions hinge, is that the group velocity in the linear region immediately ahead of the front must vanish; for the stationary fronts concerning us here, that is the wall mode criterion (8.19). The whole idea, of course, has its origins in the work of Dee and Langer (1983).

The fundamental concerns are frequency selection and front location. The matter is generally resolved in a straightforward way using the Dee and Langer (1983) criterion. For a uniform background, the front is blocked by a boundary (Couairon and Chomaz, 1996, 1997b; Tobias *et al.*, 1997), while in the case of a spatially varying background blocking may be possible at an internal location (Meunier *et al.*, 1997; Pier *et al.*, 1998). In our dynamo problem, subtler issues emerge concerning its location in the transitional stages from the weakly nonlinear to the fully developed nonlinear state, which we outline in Section 8.3.5. Tobias *et al.* (1997, 1998) identified the problem and resolved the matter for the constant local dynamo number case while Bassom *et al.* (1999) provided the extension to the case of spatially varying local dynamo number (8.70b). Here there are tenuous links with Couairon and Chomaz (1999a) on some overlapping issues.

For pedagogical purposes, the nature of our nonlinear dynamo results is best understood within the framework of the single mode truncation and the concept of an effective dynamo number (8.72d). Of course, the entire notion of an α-quenched dynamo is ad hoc and so we are at liberty to advocate (8.72) as our model system. Perhaps a more natural approach in this spirit, however, is to postulate that quenching results after time averaging. In other words, we assume that the magnetic field has a wave-like character, e.g. of the form (8.72a), which is readily and unambiguously averaged to yield the quenched α-effect. Accordingly, the simplest quenched model system is again (8.71) but with the previously real amplitudes A and B now complex. This approach has been adopted by Tobias *et al.* (1997, 1998) to investigate secondary instabilities of finite amplitude states (but see also the related study Couairon and Chomaz, 1999b).

8.3.2. Linear theory

8.3.2.1. The Kuzanyan–Sokoloff model
We begin with the linear assumption that $\mathcal{D}_{\mathrm{eff}}(\theta, 0) = \mathcal{D}(\theta)$ and following Kuzanyan and Sokoloff (1995) make the WKBJ ansatz

$$[\mathcal{A}, \mathcal{B}] := [a(\theta), b(\theta)] \exp\left(\frac{\mathrm{i}}{\varepsilon} \int_{\theta_s}^{\theta} k \, \mathrm{d}\theta\right) \tag{8.73}$$

in the Northern hemisphere $0 \le \theta \le \pi/2$. It leads to the local dispersion relation

$$(\mathrm{i}\omega + 1 + k^2)^2 - \mathrm{i}\mathcal{D}k = 0. \tag{8.74}$$

For a marginal mode $\mathrm{Im}\{\omega\} = 0$, the absolute stability condition (8.19) of vanishing group velocity $\omega_{,k} = 0$ is met by the threshold values

$$k_T := \sqrt{2/3}\,e^{i\pi/6}, \quad \mathcal{D}_T := \frac{16k_T^3}{i}, \quad \omega_T := \sqrt{3}, \quad (\omega,_\mathcal{D})_T = \frac{1}{8k_T}. \tag{8.75}$$

Since ω only depends on colatitude θ through the local dynamo number $\mathcal{D}$, the vanishing of phase mixing $\omega_{,\theta} = 0$ coincides with the local extremum of D of $\mathcal{D}(\theta)$ at θ_E.

Correct to lowest order, the above results combine to satisfy the global conditions (8.7) for a marginal dynamo solution, when $D = \mathcal{D}_T$ and $\omega = \omega_T$. It only remains to construct the WKBJ solution itself from

$$k := k(\theta), \quad \text{which solves } \omega(k, \mathcal{D}(\theta)) = \omega_T. \tag{8.76}$$

The main feature that influences our choice of k is that there are two latitudes θ and $\theta_A(= \frac{1}{2}\pi - \theta$ for (8.70b)) with the same local dynamo number $\mathcal{D}(\theta_A) = \mathcal{D}(\theta)$. So if $k(\theta)$ is the required root of (8.76), whose imaginary part increases monotonically on the range $0 \le \theta \le \pi/2$, there is a second root $k_A(\theta) := k(\theta_A)$. Notice, however, that the monotonic increase envisaged is only possible because of the double root property at $\theta = \theta_E$, where $k_A = k = k_T$. The required switching of roots is not possible for the remaining two roots $k_\pm$ of (8.76) which have the generic property $k_\pm(\theta_A) = k_\pm(\theta)$; the mode switching property $k(\theta_A) \ne k(\theta)$ is exceptional but necessary for the construction of our solution.

At $\theta = \theta_E$, the roots of (8.76) are

$$k_{T\pm} := \left(1 \pm i\sqrt{2}\right)^2 k_T, \quad k_A = k = k_T, \quad (\mathrm{Im}\{k_T\} > 0). \tag{8.77}$$

Since $\mathrm{Im}\{k_T\}$ is positive, $\mathrm{Im}\{k\}$ passes through zero at a lower latitude

$$\theta_P = \frac{1}{2}\sin^{-1}\left(\frac{\mathcal{D}_P}{\mathcal{D}_T}\right), \quad \text{in which } \frac{\mathcal{D}_P}{\mathcal{D}_T} = \frac{27|k_T|}{32k_P}, \quad k_P := \left(\sqrt{3} - 1\right)^{1/2}. \tag{8.78a}$$

Here $\mathcal{D}_P := \mathcal{D}(\theta_P)$ and $k_P := k(\theta_P)$, while the remaining three roots of the dispersion relation are

$$k_{P\pm} := \left[-\sqrt{3} \pm i\left(13 + 8\sqrt{3}\right)^{1/2}\right] k_P e^{i\pi/6}, \quad k_A = k_P e^{i\pi/3}. \tag{8.78b}$$

Our mode of interest, characterised by $k(\theta)$, decays both towards the equator $\theta \downarrow 0$ and the North Pole $\theta \uparrow \pi/2$. In stark contrast, a mode defined by $k_A(\theta)$ blows up in both directions. Furthermore, $\mathrm{Im}\{k_+\}$ is positive and $\mathrm{Im}\{k_-\}$ is negative throughout the Northern hemisphere. So $k_-(\theta)$ defines a mode growing monotonically from the equator, where $k_-(0) = k(0)$, to the pole, whereas $k_+(\theta)$ defines a mode decaying monotonically from the equator up to the pole, where $k_+(\pi/2) = k(\pi/2)$.

As we explained in Section 8.1, our WKBJ solution is an interior global mode with frequency ω_T. It is manifest here as a Parker Dynamo wave localised in the vicinity of the low latitude θ_P, where the local wave number is $\varepsilon^{-1}k_P$. It is modulated under a Gaussian hump ($\mathrm{Im}\{k_P'\} > 0$, where the prime now denotes θ-derivatives) on the longer θ-length scale $O(\varepsilon^{1/2})$ (see Fig. 8.7).

The saddle point criteria are met at θ_E, where

$$i\omega_{,kk} = -3, \quad i\omega_{,kx} = 0, \quad i\omega_{,xx} = 2k_T^2 \mathcal{D}_T''/\mathcal{D}_T, \tag{8.79a}$$

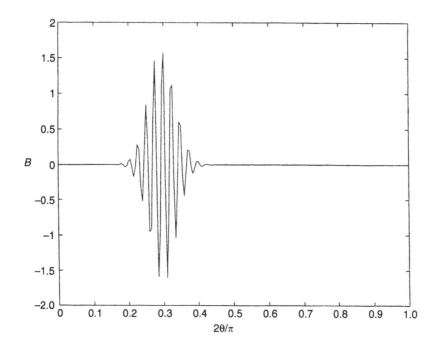

Figure 8.7 The linear solution B versus $2\theta/\pi$ at some fixed time obtained numerically for $\varepsilon = \pi/600$. The modulated Parker wave is localise close to $2\theta_p/\pi \approx 0.3$. (Courtesy of Meunier *et al.*, 1997.)

while use of (8.12) gives

$$k_E' := \left(\frac{dk}{d\theta}\right)_E = ik_T\sqrt{-\frac{2D_T''}{3D}} \qquad (\text{Im}\{k_E'\} > 0). \qquad (8.79b)$$

These values are necessary for the higher order corrections to the theory. In particular with the parameter values (8.75) and (8.79), (8.15) gives

$$\frac{4k_T^2}{3}\frac{D - D_T}{\varepsilon^2 D_T} - i\frac{2}{3}\frac{\omega - \omega_T}{\varepsilon^2} = -2ik_E'\left(\Delta + \frac{1}{2}\right) \qquad (8.80a)$$

with $\Delta = 0$. Though the constant C vanishes for the system (8.71), the slightly modified form used by Kuzanyan and Sokoloff (1995) has θ-dependence in the coefficients of both (8.71a) and (8.71b) resulting in a non zero C. In any event, the frequency $\omega - \omega_T$ and dynamo number $D - D_T$ corrections are now readily obtained by equating real and imaginary parts.

From a more general point of view, we note that (8.14) may be cast in the form

$$\frac{d^2\hat{b}}{dX^2} - ik_E'(1 + 2\Delta + ik_E'X^2)\hat{b} = 0 \qquad \left(B := \hat{b}(X)\exp\left[ik_T\frac{\theta - \theta_E}{\varepsilon}\right]\right), \qquad (8.80b)$$

in which $X := (\theta - \theta_E)/\varepsilon^{1/2}$ and where (8.80a) is now regarded as the definition of Δ. The solution which satisfies $\hat{b} \to 0$ as $X \uparrow \infty$ is

$$\hat{b}(X) := \bar{b}\, U\left(-\tfrac{1}{2} - \Delta, e^{i\pi/12}\sqrt{2|k'_E|X}\right),$$ (8.81a)

where U is the standard Weber Parabolic Cylinder function. The solution that satisfies $\hat{b} \to 0$ as $X \downarrow -\infty$ determines the eigenvalues

$$\Delta = 0, 1, 2, \ldots.$$ (8.81b)

The special case $\Delta = 0$ is simply the lowest mode.

8.3.2.2. The Worledge–Knobloch–Tobias–Proctor model

For the constant local dynamo number case of Worledge *et al.* (1997), the above analysis can be adapted with some minor modifications. Crucially, the four WKBJ solutions are described by four constant values of k. The lowest order theory that isolates $\mathcal{D}_T$ and ω_T applies. The only real difference is how the subtleties of the higher order theory are resolved; that is instructive for the treatment of the nonlinear problem that we address in the following sections.

When D is close to but not equal to $\mathcal{D}_T$, two of the complex wave numbers are close to but not equal to k_T. We may as we did above in (8.80) consider a modulated mode satisfying

$$\frac{d^2 b}{d\theta^2} + \Lambda^2 b = O(\varepsilon) \qquad \left(\mathcal{B} := \frac{b(\theta)}{\varepsilon}\exp\left[i\frac{k_T\theta}{\varepsilon}\right]\right),$$ (8.82a)

where

$$\frac{4k_T^2}{3}\frac{D - \mathcal{D}_T}{\varepsilon^2 \mathcal{D}_T} - i\frac{2}{3}\frac{\omega - \omega_T}{\varepsilon^2} = \Lambda^2.$$ (8.82b)

Hence the complete solution obtained from the four WKBJ contributions (8.77) is

$$\mathcal{B} := \frac{b(\theta)}{\varepsilon}\exp\left(i\frac{k_T\theta}{\varepsilon}\right) + \bar{b}_{T+}\exp\left(i\frac{k_{T+}\theta}{\varepsilon}\right) + \bar{b}_{T-}\exp\left(i\frac{k_{T-}(\theta - \pi/2) + k_T\pi/2}{\varepsilon}\right)$$ (8.83a)

with a similar expression for $\mathcal{A}$. Since the result (8.77) shows that $\mathrm{Im}\{k_{T+}\} > \mathrm{Im}\{k_T\} > 0 > \mathrm{Im}\{k_{T-}\}$, we may regard $\varepsilon^{-1}b(\theta)$ as characterising the mainstream solution, relative to which the terms proportional to $\bar{b}_{T+}$ and $\bar{b}_{T-}$ define boundary layers on the left $\theta = 0$ and right $\theta = \pi/2$ boundaries, respectively.

The mainstream equation must then be solved subject to reduced homogeneous boundary conditions obtained by considering the jump conditions across each of the two boundary layers. Ignoring exponentially small terms, the left-hand boundary layer solution is

$$\mathcal{B} \approx \frac{b(0) + b'(0)\theta}{\varepsilon}\exp\left(i\frac{k_T\theta}{\varepsilon}\right) + \bar{b}_{T+}\exp\left(i\frac{k_{T+}\theta}{\varepsilon}\right)$$ (8.83b)

for $\theta = O(\varepsilon)$. While the right-hand boundary layer solution is

$$\exp\left(-i\frac{k_T\pi}{2\varepsilon}\right)\mathcal{B} \approx \frac{b(\pi/2) - b'(\pi/2)\theta_A}{\varepsilon}\exp\left(-i\frac{k_T\theta_A}{\varepsilon}\right) + \bar{b}_{T-}\exp\left(-i\frac{k_{T-}\theta_A}{\varepsilon}\right)$$

(8.83c)

for $\theta_A := \frac{1}{2}\pi - \theta = O(\varepsilon)$. In view of the long length scale dependence of $b(\theta)$ on θ, we note that the θ-derivative of the corresponding modulated term in (8.83a) has two contributions:

$$\frac{ik_T}{\varepsilon^2}b + \frac{b'}{\varepsilon}$$

with a similar expression for the derivatives of $\mathcal{A}$ (defined by (8.83a) with the b's replaced by a's). Moreover, in view of the governing equations, we also have a proportional to b correct to lowest order. That means that our reduced homogenous boundary conditions involve b alone. The only such homogeneous equation is $b = 0$! Accordingly the boundary conditions on (8.82a), assuming that $b(\theta) = O(1)$, are simply

$$b(0) = O(\varepsilon), \qquad b(\pi/2) = O(\varepsilon)$$

(8.84a)

and in this way all terms in (8.83b,c) are of the same size in their respective boundary layers. The precise form of the $O(\varepsilon)$ corrections to (8.84a), that they imply, is sensitive to the actual boundary conditions adopted at the two end points.

The eigensolutions of (8.82) subject to (8.84a) are

$$b := \bar{b}\sin\Lambda\theta, \qquad \Lambda/2 = 1, 2, \ldots.$$

(8.84b)

The solution (8.83a) so constructed is localised in the equatorial $\theta = O(\varepsilon)$ boundary layer (8.83b), where both the mainstream and boundary layer term are $O(1)$.

Though we have presented the above result as simply a variant of the saddle point idea, it does emphasise slightly different issues. On the one hand, the saddle point layer has an $O(\varepsilon^{1/2})$ θ-length scale and the boundary conditions on that problem are local in the sense that the correct WKBJ solution has to be picked up on leaving the saddle point layer. On the other, when $\mathcal{D}$ is everywhere constant that layer fills the entire space and the problem must be solved subject to boundary conditions at both its ends. In view of the fact that the solution is localised close to one boundary, the conclusion that the solution depends on the other distant boundary, where the amplitude is exponentially small is counter intuitive and a triumph of Worledge *et al.*'s (1997) investigation. They reinforce the unusual character of their result by pointing out that, in the absence of a right hand boundary and with simply the requirement that the amplitudes remain bounded as $\theta \uparrow \infty$, the marginal mode is the same as that predicted by local theory, i.e. the marginal mode is wave-like and characterised by

$$\mathcal{D}_{\text{loc}} := 32/3\sqrt{3}(< \mathcal{D}_T), \qquad \omega_{\text{loc}} := 4/3, \quad k_{\text{loc}} := 1/\sqrt{3}.$$

(8.85)

To establish this result, it is necessary to consider the three other contributions to the solution linked to the remaining k's which solve $\omega(k, \mathcal{D}_{\text{loc}}) = \omega_{\text{loc}}$. One root with $\text{Im}\{k\} < 0$ gives an unacceptable contribution growing as $\theta \uparrow \infty$, which is rejected. The two remaining roots with $\text{Im}\{k\} > 0$ give spatially decaying contributions, which constitute the boundary layer structure necessary to meet appropriate boundary conditions at $\theta = 0$.

8.3.3. *Weakly nonlinear theory*

The weakly nonlinear theory for the two models discussed above follows the conventional route (Bräuer, 1979). In both cases, it is supposed that the dynamo number D and frequency ω are close to the marginal values for linear kinematic dynamos, namely $D_L \approx D_T$ and $\omega_L \approx \omega_T$. In both cases, however, the amplitude equations, which result, have unusual features worth noting.

8.3.3.1. *The Kuzanyan–Sokoloff model*

Since the lowest mode $\Delta = 0$ is described on the entire range $0 < \theta < \pi/2$ by the WKBJ solution (8.73) we may express the adjoint in the similar form

$$[\mathcal{A}_A, \mathcal{B}_A] := [a_A(\theta), b_A(\theta)] \exp\left(-\frac{i}{\varepsilon} \int_{\theta_S}^{\theta} k_A \, d\theta\right), \tag{8.86a}$$

$$\text{with } k_A(\theta) := k(\theta_A), \text{ and } \theta_A := \frac{\pi}{2} - \theta \tag{8.86b,c}$$

for our symmetric choice (8.70b) of $\mathcal{D}$. For our simplified system (8.72) it can be shown that $[a_A(\theta), b_A(\theta)] := [a(\theta_A), b(\theta_A)]$. Routine analysis yields

$$\int_0^{\pi/2} \left\{[-(D - D_L)bb_A + i(\omega - \omega_L)(ab_A + ba_A)] \exp\left(\frac{i}{\varepsilon} \int_{\theta_S}^{\theta} (k - k_A) \, d\theta\right)\right\} d\theta$$

$$= -\int_0^{\pi/2} \left\{\frac{3}{4}\mathcal{D}(\theta)|b|^2 bb_A \exp\left(\frac{i}{\varepsilon} \int_{\theta_S}^{\theta} (2k - k_A - k^*) \, d\theta\right)\right\} d\theta, \tag{8.87}$$

where $k^*(\theta)$ for complex θ is the analytic extension of the complex conjugate of $k(\theta)$ defined for real θ.

The integrals are evaluated by the method of steepest descents. The dominant contributions stem from the locations $\theta = \theta_E$, where $k - k_A$ vanishes, and $\theta = \theta_N$ (say), where $2k - k_A - k^* = 0$. In this way, we obtain a result of the form

$$\frac{4k_T^2}{3} \frac{D - D_L}{D_T} - i\frac{2}{3}(\omega - \omega_L) = \cdots |b_{\max}|^2 \exp\left(-\frac{p + iq}{\varepsilon}\right), \tag{8.88a}$$

where $|b_{\max}|$ is the maximum amplitude of b at θ_P (see (8.78a)) and

$$p + iq := \int_{\theta_N}^{\theta_P} (k - k^*) \, d\theta + \int_{\theta_N}^{\theta_E} (k - k_A) \, d\theta. \tag{8.88b}$$

The omitted constant coefficient on the right-hand side of (8.88a) has algebraic dependence on ε and is of no particular interest.

Now the realised value of the constant p is

$$p \approx 0.0685, \tag{8.88c}$$

which means that the amplitude

$$|b_{\max}|^2 \propto |D - D_L| \exp(p/\varepsilon), \tag{8.89}$$

increases rapidly for only a small change in $D - D_L$. Furthermore, since $q \neq 0$ also, $D - D_L$ changes sign when q/ε changes by an amount π. This happens with a very small change of ε of order $O(\varepsilon^2)$. So whether the bifurcation is sub-critical or super-critical has very sensitive dependence on the size of ε. Similar sensitivity has been noted in other contexts (see, e.g. (Le Dizès *et al.*, 1993).

8.3.3.2. The Worledge–Knobloch–Tobias–Proctor model

Here, as above, the function adjoint to (8.83a) is $\mathcal{B}_A(\theta) = \mathcal{B}(\theta_A)$ (see (8.86c)), with a similar identity for $\mathcal{A}_A(\theta)$. In place of (3.87) we have

$$\int_0^{\pi/2} [-(D - D_L)\mathcal{B}\mathcal{B}_A + i(\omega - \omega_L)(\mathcal{A}\mathcal{B}_A + \mathcal{B}\mathcal{A}_A)] \, d\theta = -\frac{3}{4}\mathcal{D}_T \int_0^{\pi/2} |\mathcal{B}|^2 \mathcal{B}\mathcal{B}_A \, d\theta. \qquad (8.90)$$

The terms on the left-hand side are dominated by the large $O(\varepsilon^{-1})$ mainstream term defined by (8.82a). Using (8.84b), that gives a very large $O(\varepsilon^{-2})$ contribution which is readily evaluated. The integrand on the right hand side is $O(1)$ at $\theta = 0$ but decays exponentially on the $O(\varepsilon)$ boundary layer length scale. It yields

$$\frac{4k_T^2}{3} \frac{D - D_L}{\mathcal{D}_T} - i\frac{2}{3}(\omega - \omega_L) = \cdots \varepsilon^3 |b_{\max}|^2, \qquad (8.91)$$

where $|b_{\max}|$ is the maximum amplitude of $\mathcal{B}$ in the boundary layer and the omitted constant coefficient on the right hand side is $O(1)$.

Tobias *et al.* (1998) undertook a similar analysis in the context of the CGL equation. Since that equation is only second order like (8.82a), there is no boundary layer and the mainstream representation (8.84b) of $\mathcal{B}$ is uniformly valid throughout the region; thus they were able to use it to calculate the value of the coefficient on the right-hand side of (8.91). Though this result also gives the correct order of magnitude for the Worledge–Knobloch–Tobias–Proctor model, the realised value of the coefficient is influenced by the other boundary layer terms in (8.83b). Specifically, its precise value is determined from lower order terms, that we have not calculated and which are particularly sensitive to the precise form of the boundary conditions at $\theta = 0$.

The main point to note is that the amplitude of the field again increases rapidly with only small increments of the dynamo number. Nevertheless, since the dependence on ε is only algebraic, it is not as explosive as the Bassom *et al.* (1999) result (8.89), which exhibits exponential dependence.

8.3.4. Finite amplitude Parker waves and fronts

8.3.4.1. The large amplitude solution for $D \approx \mathcal{D}_T$

They are two key ideas that form the basis of the construction of the full nonlinear solution. First, for a given frequency ω, there is a dynamo number $\mathcal{D}_P$ at which a spatially wave like solution is possible and that will occur at θ_P (say), where $\mathcal{D}(\theta_P) = \mathcal{D}_P$. We noted the existence of such a solution even for the linear case (see (8.78)). Note, however, that there is a range $\mathcal{I}_P = [\theta_P, \theta_{AP}]$, $(\theta_{AP} := \theta_A(\theta_P))$ over which $\mathcal{D}(\theta) \geq \mathcal{D}_P$. It suggests that we construct a nonlinear Parker dynamo wave with $\mathcal{D}_{\text{eff}} = \mathcal{D}_P$ on $\mathcal{I}_P$. There the solution continues to be of the linear WKBJ form (8.73) with $k(\theta) = k_P$ (real), while its amplitude modulation is fixed by (8.72d). Elsewhere, the solution is linear with $\mathcal{D}_{\text{eff}} = \mathcal{D}$.

Though the idea described is attractive, it is not entirely correct. The most obvious way to see its shortcomings is to reconsider the linear kinematic dynamo problem for the case in which the local dynamo number is actually defined by the effective dynamo number just constructed. According to linear theory, the local dynamo number must achieve the value $\mathcal{D}_T$, which it now evidently does not. A more mathematical way to eliminate the proposal is to analyse the transitional boundary layer between the linear and nonlinear spatial domains. In the neighbourhood of θ_P, the solution is governed locally by

$$\frac{d\tilde{b}}{d\eta} = ik'_P\left(\eta - \frac{3\mathcal{D}_P}{4\mathcal{D}'_P}|\tilde{b}|^2\right)\tilde{b} \qquad \left(\mathcal{B} := \tilde{b}(\eta)\exp\left[ik_P\frac{\theta - \theta_P}{\varepsilon}\right]\right) \qquad (8.92)$$

(cf. Pier and Huerre, 1996), where $\eta := (\theta - \theta_P)/\varepsilon^{1/2}$ and is illustrated by the broken curves on Fig. 8.8. Since $\text{Im}\{k'_P\} > 0$, a solution exists which connects the linear solution decaying exponentially, as $\eta \downarrow -\infty$, to the finite amplitude solution $|\tilde{b}|^2 = (4\mathcal{D}'_P/3\mathcal{D}_P)\eta$, appropriate as $\eta \uparrow \infty$. On the other hand, no comparable solution exists at θ_{AP}, where $k'(\theta_{AP}) (= -k'_P$ for our symmetric local dynamo number (8.70b)) has negative imaginary part. In other words, the unwanted exponentially growing linear solution ahead of θ_{AP} cannot be eliminated by the nonlinearity behind.

Evidently, our construction is plausible near θ_P but we must limit the extent of the nonlinear solution ($\mathcal{D}_{\text{eff}} = \mathcal{D}_P$) to the range $\theta_P < \theta < \theta_E$; elsewhere it is linear ($\mathcal{D}_{\text{eff}} = \mathcal{D}$). Accordingly, we ensure that the maximum value of the effective dynamo number is located inside the linear range at $\theta = \theta_E$, where $\mathcal{D}_{\text{eff}} = D \approx \mathcal{D}_T$. By this device, we meet all the criteria that we advocated for the linear problem; that is the second idea.

Significantly, there is jump in the value of $\mathcal{D}_{\text{eff}}$ across $\theta = \theta_E$. It is linked to an abrupt drop in the amplitude $|\mathcal{B}|$ and corresponds to a front. This behaviour is well illustrated by Figs. 8.8a and 8.9a.

8.3.4.2. The extension to $D > \mathcal{D}_T$

With the further increase of the dynamo number D beyond $\mathcal{D}_T$, all the ideas mentioned above continue to apply. In detail, the front location θ_F moves to higher latitudes, where it remains close to the threshold angle θ_T, which solves $\mathcal{D}(\theta_T) = \mathcal{D}_T$. The other end of the nonlinear range θ_P, which solves $\mathcal{D}(\theta_P) = \mathcal{D}_P$ moves to lower latitudes. In short, the zeroth order solution satisfies

$$\mathcal{D}_{\text{eff}}(\theta, |\mathcal{B}|) = \begin{cases} \mathcal{D}_P & \text{on } \theta_P < \theta < \theta_F; \\ \mathcal{D}(\theta) & \text{elsewhere.} \end{cases} \qquad (8.93)$$

Thus, the finite amplitude solution emerges smoothly from the linear solution at θ_P at the equatorial end and collapses at the front θ_F at the polar end. This behaviour is well illustrated by Figs. 8.8b,c and 8.9b,c.

Our description of the solution of (8.71) in terms of an effective dynamo is purely pedagogical. This rather blunt approach hides some of the quantitative (as well as qualitative) details of the true numerical solution illustrated in Figs. 8.8 and 8.9. Whereas the analytic development of Meunier *et al.* (1997), which pivoted on (8.72), relied heavily on the notion

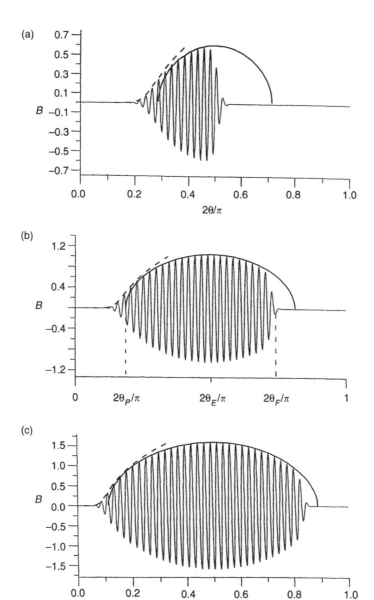

Figure 8.8 The finite amplitude solution B versus $2\theta/\pi$ of (8.71) with local dynamo number (8.70b) and $\varepsilon = \pi/600$. The oscillatory continuous line denotes Meunier *et al.*'s (1997) numerical results. The broken line denotes the solution of (8.92) valid near $2\theta_P/\pi$. The smooth continuous line is the local finite amplitude Parker wave computed for uniform conditions. (a) $D = 9.0, \omega/\omega_T = 1.0043$; (b) $D = 13.0, \omega/\omega_T = 1.0247$; (c) $D = 21.0, \omega/\omega_T = 1.0376$. (Courtesy of Bassom *et al.*, 1999.)

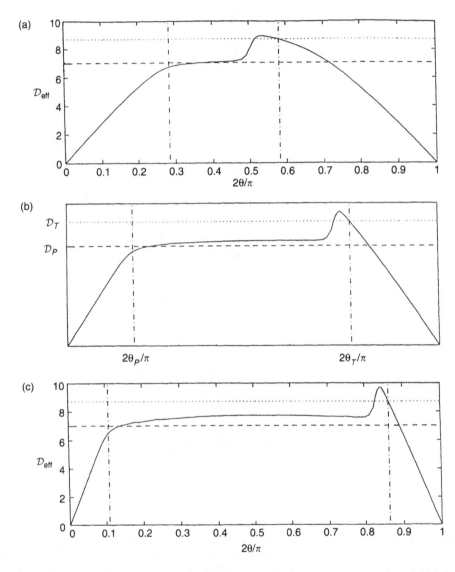

Figure 8.9 The effective dynamo number $\mathcal{D}_{\text{eff}}$ versus $2\theta/\pi$. The solid line plots $\mathcal{D}_{\text{eff}}$ (8.72d) for the numerically computed B of Fig. 8.8. (Courtesy of Bassom *et al.*, 1999.)

of an effective dynamo number, Bassom *et al.* (1999) undertook a proper asymptotic study of (8.71). They considered the finite amplitude spatially periodic Parker wave that occurs under homogeneous conditions and determined its amplitude for the appropriate local conditions. They plotted it by the smooth continuous curve in Fig. 8.8, which provides an excellent approximation to the true envelope. In contrast, the effective dynamo approach has a discrepancy which is quantified by the small difference $\mathcal{D}_{\text{eff}} - \mathcal{D}_P$ evident in Fig. 8.9, which increases with increasing D. A further qualitative feature lost by the effective dynamo number approach is the finite thickness of the front of $O(\varepsilon)$ on the length scale of the Parker wave itself. Fortunately, this is shorter than any other length scale of interest.

In the subsection that follows we examine more closely conditions across the front. The main thrust of the development is the determination of a boundary condition on the modulated linear solution ahead of the front.

8.3.4.3. The front jump conditions

Correct to lowest order the front is located at θ_T, where the complex group velocity vanishes; this frontal condition was proposed by Dee and Langer (1983) (see also Dee, 1985). It means that the wave number for the linear WKJB solution ahead of the front is characterised by the repeated root k_T. As usual, this implies that the associated solution is modulated on a longer length scale that we will call δ. We, therefore, introduce the local coordinate

$$X := (\theta - \theta_0)/\delta \quad (\varepsilon \ll \delta \ll 1) \tag{8.94a}$$

and identify the front location by

$$X_F := (\theta_F - \theta_0)/\delta = O(1), \tag{8.94b}$$

where the origin θ_0 is chosen at our convenience.

On the $O(\delta)$ length scale the value of the effective dynamo number ahead of the front is $\mathcal{D}_T$, while behind it is $\mathcal{D}_P$. The only acceptable local solution in the neighbourhood of the front is of the form

$$\mathcal{B} \sim \begin{cases} \bar{b}_P \exp\left(\mathrm{i}k_P \dfrac{\theta - \theta_F}{\varepsilon}\right) + \bar{b}_{P-} \exp\left(\mathrm{i}k_{P-} \dfrac{\theta - \theta_F}{\varepsilon}\right) & (\theta < \theta_F); \\[2mm] \dfrac{\delta}{\varepsilon}\hat{b}(X) \exp\left(\mathrm{i}k_T \dfrac{\theta - \theta_F}{\varepsilon}\right) + \bar{b}_{T+} \exp\left(\mathrm{i}k_{T+} \dfrac{\theta - \theta_F}{\varepsilon}\right) & (\theta > \theta_F), \end{cases} \tag{8.95}$$

with a similar representation for $\mathcal{A}$. Here k_P is the real Parker wave number (8.78a), k_{P-} is the only root in (8.78b) with negative imaginary part $(\mathrm{Im}\{k_{P-}\} < 0)$, while k_{T+} is the only root in (8.77) (other than $k_A = k_T$) with positive imaginary part $(\mathrm{Im}\{k_{T+}\} > 0)$. Thus the modes, isolated in (8.95) are the only modes with acceptable continuations to the pole and equator.

The amplitude $|\bar{b}_P|$ of the Parker wave is fixed and determined as the solution of

$$\mathcal{D}_{\text{eff}}(\theta_F, |\bar{b}_P|) = \mathcal{D}_P. \tag{8.96}$$

Though the wave is propagating equatorward $k_P > 0$ away from the front, it should be regarded as the given incident wave and, via our representation (8.95), we demand that it only triggers evanescent disturbances. Across the front we require continuity of $\mathcal{A}$, $\mathrm{d}\mathcal{A}/\mathrm{d}\theta$, $\mathcal{B}$ and $\mathrm{d}\mathcal{B}/\mathrm{d}\theta$. This leads us to a transmission problem from which it is necessary to determine the amplitudes $\bar{b}_{P-}$, $\bar{b}_{T+}$, $\hat{b}(X_F)$, $\mathrm{d}\hat{b}/\mathrm{d}X(X_F)$, each of which are linked by algebraic expressions to $\bar{a}_{P-}$, $\bar{a}_{T+}$, $\hat{a}(X_F)$, $\mathrm{d}\hat{a}/\mathrm{d}X(X_F)$. Specifically, the continuity conditions require

$$\bar{b}_P + \bar{b}_{P-} = \hat{b}(X_F) + \bar{b}_{T+},$$

$$\mathrm{i}(k_P\bar{b}_P + k_{P-}\bar{b}_{P-}) = \left[\mathrm{i}k_T \frac{\delta}{\varepsilon}\hat{b}(X_F) + \frac{\mathrm{d}\hat{b}}{\mathrm{d}X}(X_F)\right] + \mathrm{i}k_{T+}\bar{b}_{T+} \tag{8.97}$$

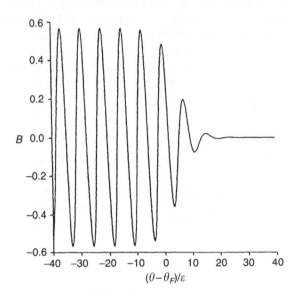

Figure 8.10 The magnetic field B versus frontal boundary layer coordinate $(\theta - \theta_F)/\varepsilon$. (Courtesy of Bassom *et al.*, 1999.)

together with two similar equations for the $\mathcal{A}$-unknowns. Since the $\mathcal{A}$ and the $\mathcal{B}$ unknowns are linked, (8.97) provides us with four equations for four unknowns, whose values are proportional to $\bar{b}_P$.

The problem posed has many similarities with that discussed in Section 8.3.2.2 (see particularly (8.83)). In essence the large parameter δ/ε appearing in (8.97) implies that the system of equations only has a solution when

$$\hat{b}(X_F) = O\left(\frac{\varepsilon}{\delta}\right). \tag{8.98}$$

Within the framework of the effective dynamo number approach, this is as far as it is reasonable to take the analysis. Bassom *et al.* (1999) went to next order and related $d\hat{b}/dX(X_F)$ to $(\delta/\varepsilon)\hat{b}(X_F)$. This, however, involved solving the front boundary layer problem on the short $O(\varepsilon)$ length numerically. The result is illustrated in Fig. 8.10 and is comparable to that obtained from solving the complete equations (see fig. 2 of Tobias *et al.*, 1997).

8.3.5. The frequency and front location

In this section we consider the nature of the modulated WKBJ solution

$$\mathcal{B} := \hat{b}(X) \exp\left(ik_T \frac{\theta - \theta_F}{\varepsilon}\right), \tag{8.99}$$

which dominates the solution (8.95) ahead of the front, when $X = O(1)$. The amplitude solves

$$\frac{d^2\hat{b}}{dX^2} + \Gamma(X)\hat{b} = 0 \qquad \left[\Gamma := \frac{\delta^2}{\varepsilon^2}\left(\frac{4k_T^2}{3}\frac{\mathcal{D} - \mathcal{D}_T}{\mathcal{D}_T} - i\frac{2}{3}(\omega - \omega_T)\right)\right] \tag{8.100a}$$

subject to the boundary conditions

$$\hat{b}(X_F) = 0 \qquad \text{and} \qquad \hat{b} \to 0 \quad \text{as } X \uparrow \infty. \tag{8.100b}$$

The problem posed is of the wall mode variety discussed in Section 8.1.2.3.

Given the dynamo number D, our objective is to determine the small corrections to the frequency $\omega - \omega_T$ and front location $\theta_F - \theta_T$. These are the key quantities that close off the nonlinear dynamo problem at a low order, yet simultaneously establishing the self-consistency of the asymptotic approach.

8.3.5.1. Transition: $|D - \mathcal{D}_T| \ll |\mathcal{D}_T|$

When $\mathcal{D}$ is close to $\mathcal{D}_T$, the front is close to the local dynamo number maximum and so we choose

$$\theta_0 := \theta_E \qquad \text{and set} \qquad \delta := \varepsilon^{1/2}. \tag{8.101a,b}$$

In this way (8.99) reduces to (8.80) with solution (8.81a) provided that

$$U\left(-\tfrac{1}{2} - \Delta, e^{i\pi/12}\sqrt{2|k_E'|}X_F\right) = 0, \tag{8.102a}$$

If we regard the front location X_F as a given real constant, we can solve (8.102a) for complex Δ:

$$\Delta_F := \Delta(X_F). \tag{8.102b}$$

The earlier linear solution corresponds to

$$X_F \downarrow -\infty, \qquad \Delta_F = 0, 1, 2, \ldots \tag{8.103}$$

(see (8.81b)).

Let us restrict attention to the lowest order mode $n = 0$ of the linear system. Then if we express Δ_F in the polar form

$$\Delta_F := \rho \exp(-i\Theta), \tag{8.104a}$$

we may solve (8.80a) to obtain

$$2|k_T|^2 \frac{D - \mathcal{D}_T}{\varepsilon \mathcal{D}_T} \approx 3|k_E'|\left[\frac{\sqrt{3}}{2} + 2\rho \cos\left(\Theta - \frac{\pi}{6}\right)\right],$$

$$\frac{\omega - \omega_T}{\varepsilon} \approx 3|k_E'|\left[\frac{1}{2} + 2\rho \sin\left(\Theta + \frac{\pi}{6}\right)\right]. \tag{8.104b}$$

The linear solution corresponds to $\rho = 0$ and its non-zero value determines the nonlinear corrections. As X_F increases from $-\infty$, Bassom *et al.* (1999) show that the solution of (8.102a) has the parametric representation

$$\rho \sim \sqrt{\frac{2\Theta}{\pi} + \frac{1}{6}} \exp\left[-\sqrt{3}\left(\Theta + \frac{\pi}{12}\right)\right], \qquad X_F \sim -\sqrt{\frac{2}{|k_E'|}}\left(\Theta + \frac{\pi}{12}\right), \tag{8.105a,b}$$

provided that

$$1 \ll \Theta \ll \varepsilon^{-1}. \qquad (8.105c)$$

The intriguing feature of the solution is that $D - D_L \propto \cos(\Theta - (\pi/6))$. It means that D oscillates rapidly about D_L with increasing X_F albeit with small amplitude. Though this solution is only valid for $X_F = O(1)$, it is tempting to move the front back to θ_P, where $-X_F = O(\varepsilon^{-1/2})$ and identify that limit with the weakly nonlinear solution. Then with Θ proportional to ε^{-1}, we find the same sensitivity on ε as to whether the bifurcation is sub or supercritical that we obtained from our earlier weakly nonlinear result (8.88a).

Presumably the weakly nonlinear solution of Section 8.3.3.1 quickly develops a frontal structure at θ_F, which moves rapidly from θ_P up to θ_E in the manner described above. The asymptotic representation (8.105) turns out to be a good approximation up to the final oscillation about D_L. As the front location moves ahead of θ_E, we obtain the new asymptotic solution

$$\rho \sim \frac{1}{2}|k_E'|X_F^2, \qquad \Theta \sim -\frac{\pi}{6}, \qquad (8.106a,b)$$

valid when

$$1 \ll X_F \ll \varepsilon^{-1/2}. \qquad (8.106c)$$

It simply corresponds to the general requirement that θ_F is close to θ_T; that point moves poleward as D increases.

Before continuing we mention briefly, the corresponding solution for the constant $\mathcal{D}$ case. For that, we simply solve (8.82) again. The required solution, which satisfies $\hat{b}(\theta_F) = \hat{b}(\pi/2) = 0$, was obtained by Tobias *et al.* (1998). It is

$$\hat{b} := \bar{b}\sin[\Lambda(\theta - \theta_F)], \qquad \Lambda[(1/2) - (\theta_F/\pi)] = 1, 2, \ldots. \qquad (8.107)$$

For comparison with (8.104b) above, it gives

$$\frac{\omega - \omega_T}{\omega_T} = \frac{2}{3}\frac{D - \mathcal{D}_T}{\mathcal{D}_T} = \frac{3}{2}(\varepsilon\Lambda)^2. \qquad (8.108a)$$

Unlike the variable $\mathcal{D}$ case, D increases monotonically with θ_F. Furthermore, the calculation remains valid provided the length scale $\delta = O(\Lambda^{-1})$ is large compared to ε. Hence the linear increase of ω with D predicted by (8.108a) holds provided that

$$\varepsilon\Lambda \ll 1. \qquad (8.108b)$$

8.3.5.2. *Fully developed nonlinearity:* $D - \mathcal{D}_T = O(\mathcal{D}_T)$

Once D exceeds $\mathcal{D}_T$ by an $O(1)$ amount, the front is located near θ_T and so we choose

$$\theta_0 := \theta_T \qquad \text{and set} \qquad \delta := \varepsilon^{2/3}. \qquad (8.109a,b)$$

In this way the coefficient $\Gamma(X)$ reduces to the expression

$$\Gamma := \frac{4k_T^2\mathcal{D}_T'}{3\mathcal{D}_T}X - i\frac{2}{3}\frac{\omega - \omega_T}{\varepsilon^{2/3}} \qquad (8.110)$$

linear in X. Accordingly, (8.100a) reduces to the Airy equation (8.20c). The appropriate solution satisfying the boundary conditions (8.100b), is the Airy function

$$\hat{b} := \bar{b}\,\mathrm{Ai}\left(\left[\frac{4|k_T|^2|\mathcal{D}'_T|}{3|\mathcal{D}_T|}\right]^{1/3}\,\mathrm{e}^{\mathrm{i}\pi/9}(X - X_F) - s_i\right) \qquad (8.111a)$$

provided that

$$\frac{\mathrm{i}\mathcal{D}_T}{2k_T^2\,\mathcal{D}'_T}\,\frac{\omega - \omega_T}{\varepsilon^{2/3}} - X_F = s_i\left(\frac{3|\mathcal{D}_T|}{4|k_T|^2|\mathcal{D}'_T|}\right)^{1/3}\exp\left(-\frac{\mathrm{i}\pi}{9}\right), \qquad (8.111b)$$

where $-s_i (< 0)$ are the zeros of the Airy function. From this we deduce

$$\omega - \omega_T = 3s_i\left(\varepsilon\frac{4|k_T|^2|\mathcal{D}'_T|}{3|\mathcal{D}_T|}\right)^{2/3}\sin\left(\frac{\pi}{9}\right) + \mathrm{O}(\varepsilon),$$

$$\frac{\theta_F - \theta_T}{\varepsilon} = -2s_i\left(\varepsilon\frac{4|k_T|^2|\mathcal{D}'_T|}{3|\mathcal{D}_T|}\right)^{-1/3}\cos\left(\frac{2\pi}{9}\right) + \mathrm{O}(1). \qquad (8.111c)$$

The result (8.111) was obtained by Meunier *et al.* (1997). Though the lowest order results $\omega \approx \omega_T$ and $\theta_F \approx \theta_T$, derived according to the Dee and Langer (1983) criterion, are clearly confirmed by the numerical results, the $\mathrm{O}(\varepsilon^{2/3})$ corrections only give the right qualitative behaviour; the quantitative comparison, however, is unsatisfactory. In an attempt to rectify the situation, Bassom *et al.* (1999) determined the next order $\mathrm{O}(\varepsilon)$ terms in (8.111c). There is some improvement but the convergence is not convincing (see Fig. 8.11). Since, the expansion parameter is $\varepsilon^{1/3}$, which was only about 0.2 for the numerical problem, it is likely that a smaller parameter value is required for good agreement.

8.3.6. Further developments; the case $\mathcal{D}$ = constant

For the case of constant $\mathcal{D}(\theta) = D$, the analysis of Section 8.3.5.2 is inappropriate. Instead, the transitional regime terminates when $D - \mathcal{D}_T$ attains an $\mathrm{O}(\mathcal{D}_T)$ value. Then the front merges with the wall boundary layer $(\pi/2) - \theta_F = \mathrm{O}(\varepsilon)$. As Tobias *et al.* (1998) point out, a fully nonlinear frontal boundary layer calculation is required which involves applying the wall boundary conditions. In this way, the frequency ω is determined as a function of D.

Tobias *et al.* (1997, 1998) remark that as D is increased their finite amplitude Parker wave, which fills the domain $0 \le \theta \le \pi/2$ eventually suffers a secondary instability at some $\mathcal{D}_{\mathrm{sec}} (> \mathcal{D}_T)$. Apparently, a second frequency is introduced. It characterises a new wave train which rides on the original Parker wave. It builds up from $\theta = 0$ and terminates at a front $\theta_{F\,\mathrm{sec}}$, leaving the original wave ahead of it undisturbed. Tobias *et al.* (1997) argue that the essential mechanism is the same as the finite amplitude evolution of the primary instability, i.e. it is a secondary absolute instability (see Brevdo and Bridges, 1996).

Tobias *et al.* (1998) support their argument by an analysis of the CGL equation. They extend that analysis to the complex version of (8.71) with the ensuing simplification that solutions of the form $B_P(\theta, t) := \bar{b}_P \exp[\mathrm{i}(\varepsilon^{-1}k_P\theta - \omega_P t)]$ having $|\bar{b}_P|^2 = (D - \mathcal{D}_P)/\mathcal{D}_P$ are exact. With this simplification the secondary instability can be sought of the modulated form,

$$b := B_P(\theta, t)\left\{1 + \tilde{b}_+ \exp\left[\mathrm{i}(\varepsilon^{-1}\tilde{k}\theta - \tilde{\omega}t)\right] + \tilde{b}_- \exp\left[-\mathrm{i}(\varepsilon^{-1}\tilde{k}^*\theta - \tilde{\omega}t)\right]\right\}. \qquad (8.112)$$

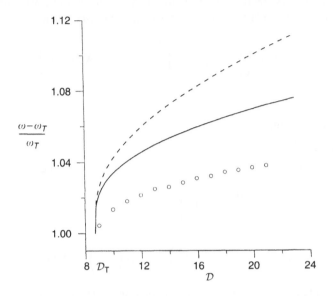

Figure 8.11 The normalised frequency increment $(\omega - \omega_T)/\omega_T$ versus the dynamo number D for $\varepsilon = \pi/600$. The circles denote Meunier *et al.*'s (1997) full numerical solution as exemplified by the oscillatory continuous curves of Fig. 8.8. The broken dashed line gives the lowest order $\varepsilon^{2/3}$ solution (8.111c); the continuous curve gives Bassom *et al.*'s (1999) $O(\varepsilon)$ correction. (Courtesy of Bassom *et al.*, 1999.)

They obtain a dispersion relation for $\tilde{\omega}$ (real) as an analytic function of $\tilde{k}$. They then apply the condition of vanishing group velocity $\tilde{\omega}_{,\tilde{k}} = 0$ to determine the onset of secondary instability.

8.3.7. *Further developments; the case* $\mathcal{D} = D \sin(2\theta)$

Tobias and Proctor (1999, Private communication) have sought secondary instabilities of the complete system (8.71) subject to the spatially varying local dynamo number (8.70b). They found none, consistent with Meunier *et al.*'s (1997) prediction on the basis of a simplified local analysis. Nevertheless, by changing the quenching coefficient in (8.71a) from $(1 + |B|^2)^{-1}$ to $(1 + |\partial A/\partial \theta|^2)^{-1}$, the situation changed dramatically. Under this change, a secondary instability was found at moderate dynamo number, which exhibited a second front as found in the constant $\mathcal{D}$ case with the original quenching. This sensitivity of the secondary instability to the form of quenching is a striking and unexpected feature of the nonlinear system and provides a cautionary note for future investigations.

8.4. Concluding remarks

The nonlinear study of Section 8.3 points to a wide variety of problems that can be approached in a similar way; not least the applications to galactic dynamos and accretion discs as suggested in Section 8.2.1. The $\alpha\Omega$-stellar shell dynamo has recently been extended by Griffiths *et al.* (2001) to the case of $\alpha^2\Omega$-dynamos. The key new feature that emerges from the added complexity is the fact that the saddle point θ_S for the linear problem is not located at the dynamo number maximum θ_E but elsewhere in the complex plane. Though the linear problem is conventional in the sense of Section 8.1, its nonlinear extension has some surprises. For example, in the weakly nonlinear regime, the amplitude does not necessarily grow explosively with small variations of D, as characterised by $p > 0$ in (8.88), but can be strongly suppressed $p < 0$. Such behaviour has recently been isolated in a related problem of spherical Couette flow (Harris *et al.*, 2000). Furthermore application of the Dee and Langer (1983) frontal criteria for the nonlinear solutions yields the intriguing result that the smallest dynamo numbers, which define fully developed finite amplitude states, are subcritical by O(1) amounts. Since the frequency of the nonlinear part of the solution is fixed by the linear part ahead of the front, that linear part controls the entire solution. From this point of view, the linear part concerns a wall mode as envisaged in Section 8.1.2.2, for which the front location x_F provides a boundary. Whereas, in the previous studies (Meunier *et al.*. 1997; Bassom *et al.*, 1999), it is assumed that the turning point x_T is close the wall x_F, Griffiths *et al.* (2001) (see also Bassom and Soward, 2001) consider the possibility that x_T and x_F are well separated. They conclude that such modes are in principle possible, but that they are arguably unstable to other frontal solutions. The issue here is that of uniqueness. So, if we make the plausible hypothesis that the realised solution is the one which maximises the spatial extent of the finite amplitude section, it appears that this hypothesis is generally met when x_T and x_F are close together as determined by the first zero of the Airy function (8.20).

We briefly comment on the reliability of the asymptotic approximations to the proposed astrophysical applications. The crucial feature of an asymptotic series is that it is generally divergent. For a given ε there is an optimum number of terms $N(\varepsilon)$ (say) that gives the best approximation to the true solution. Generally N decreases with increasing ε. For the applications presented in this chapter ε is not that small. So, though the lowest order results based simply on notions like the vanishing of the group velocity and phase mixing look sensible and provide a useful physical picture of what is happening, the higher order approximations, giving for example frequency corrections, do not necessarily improve the results in astrophysical parameter ranges. This aspect is quite clear from Table 8.1 for the oscillatory galactic dynamo and Fig 8.11 for the shell dynamos.

On a related matter, we remark that the value $\varepsilon = \pi/600$ is clearly far too small for the true solar case of roughly $\varepsilon = 1/3$. Our small choice was adopted to stress the frontal aspect of the solutions. Even for other stellar applications, we expect at most only a few spatial oscillations, possibly spanning a width only a little wider than the front itself. In the solar case there is only one oscillation identifiable from the standard butterfly diagram, which is indeed an extreme situation for application of our model. Nevertheless, the results do suggest an underlying mechanism that may lead to an equator-pole asymmetry.

Acknowledgements

I wish to thank Prof. J.B. Taylor for drawing my attention to the related plasma physics literature. I have benefited from discussions with Jean-Marc Chomaz, Chris Jones and Mike Proctor for which I am grateful. Andrew Bassom, Carlo Barenghi, Georgina Griffiths provided helpful comments, which I much appreciate.

References

Baliunas, S.L., Nesme-Ribes, E., Sokoloff, D., and Soon, W., "A dynamo interpretation of stellar activity cycles," *Astrophys. J.* **460**, 848–857 (1996).

Baliunas, S.L., Donahue, R.A., Soon, W.H., Horne, J.H., Frazer, J., Woodard-Eklund, L., Bradford, M., Rao, L.M., Wilson, O.C., Zhang, Q., Bennett, W., Briggs, J., Caroll, S.M., Duncan, D.K., Figueroa, D., Lanning, H.H., Mish, T., Mueller, J., Noyes, R.W., Poppe, D., Porter, A.C., Robinson, C.R., Russel, J., Shelton, J.C., Soyumer, T., Vaughan, A.H. and Whitney, J.H., "Chromospheric variations in main-sequence stars. II," *Astrophys. J.* **438**, 269–287 (1995).

Baryshnikova, Yu., Ruzmaikin, A., Sokoloff, D. and Shukurov, A., "Generation of large-scale magnetic fields in spiral galaxies," *Astron. Astrophys.* **177**, 27–41 (1987).

Bassom, A.P., Kuzanyan, K.M. and Soward, A.M., "A nonlinear dynamo wave riding on a spatially varying background," *Proc. R. Soc. Lond.* A **455**, 1443–1481 (1999).

Bassom, A.P. and Soward, A.M., "On the frontal condition for finite amplitude $\alpha^2\Omega$-dynamo wave trains in stellar shells," *Geophys. Astrophys. Fluid Dynam.* **95**, 285–328 (2001).

Beck, R., Kronberg, P.P. and Wielebinski, R. (Eds.), *Galactic and Intergalactic Magnetic Fields*, Proceedings of the 140th IAU Symposium, Kluwer Academic Publishers (1990).

Beck, R., Brandenburg, A., Moss, D., Shukurov, A. and Sokoloff, D., "Galactic magnetism: recent developments and perspectives," *Annu. Rev. Astron. Astrophys.* **34**, 155–206 (1996).

Braginsky, S.I., "Oscillation spectrum of the hydromagnetic dynamo of the Earth," *Geomag. Aeron.* **10**, 172–181 (1970).

Brandenburg, A., "Solar dynamos: computational background," in: *Lectures on Solar and Planetary Dynamos* (Eds. M.R.E. Proctor and A.D. Gilbert), Cambridge University Press, pp. 117–159 (1994).

Brandenburg, A., "Computational aspects of astrophysical MHD turbulence," in this volume, Ch. 9, pp. 269–344 (2003).

Brandenburg, A. and Campbell, C.G., "Modelling magnetized accretion discs," in: *Accretion Discs – New Aspects* (Eds. E. Meyer-Hoffmeister and H. Spruit), Springer, Vol. 487, pp. 109–124 (1997).

Brandenburg, A., Saar, S.H. and Turpin, C.R., "Time evolution of the magnetic activity cycle period," *Astrophys. J.* **498**, L51–L54 (1998).

Brandenburg, A., Moss, D., Rüdiger, G. and Tuominen, I., "The nonlinear solar dynamo and differential rotation: a Taylor number puzzle?," *Solar Phys.* **128**, 243–251 (1990).

Bräuer, H., "The nonlinear dynamo problem: small oscillatory solutions in a strongly simplified model," *Astron. Nachr.* **300**, 43–49 (1979).

Brevdo, L. and Bridges, T.J., "Absolute and convective instabilities of spatially periodic flows," *Phil. Trans. R. Soc. Lond.* A **354**, 1027–1064 (1996).

Bruns, H., *Abhandl. kgl. säcks. ges. Wiss., Math.-phys. Kl.* **21**, 323 (1895).

Bykov, A., Popov, V., Shukurov, A. and Sokoloff, D., "Anomalous persistence of bisymmetric magnetic structures in spiral galaxies," *Mon. Not. R. Astron. Soc.* **292**, 1–10 (1997).

Campbell, C.G., "The effects of dynamo-generated magnetic fields on accretion disc structure," *Geophys. Astrophys. Fluid Dynam.* **90**, 113–126 (1999).

Chomaz, J.M., "Absolute and convective instabilities in nonlinear systems," *Phys. Rev. Let.* **69**, 1931–1934 (1992).

Connor, J.W., Hastie, R.J. and Taylor, J.B., "High mode number stability of an axisymmetric toroidal plasma," *Proc. R. Soc. Lond.* A **365**, 1–17 (1979).

Couairon, A., "Modes globaux fortement non-linéaires dans les écoulements ouverts," *Thèse École Polytechnique Laboratoire d'Hydrodynamique* (1997).

Couairon, A. and Chomaz, J.M., "Global instability in fully nonlinear systems," *Phys. Rev. Let.* **77**, 4015–4018 (1996).

Couairon, A. and Chomaz, J.M., "Pattern selection in the presence of a cross flow," *Phys. Rev. Let.* **79**, 2666–2669 (1997a).

Couairon, A. and Chomaz, J.M., "Absolute and convective instabilities, front velocities and global modes in nonlinear systems," *Physica D* **108**, 236–276 (1997b).

Couairon, A. and Chomaz, J.M., "Fully nonlinear global modes in slowly varying flows," *Phys. Fluids.* **11**(12), 3688–3708 (1999a).

Couairon, A. and Chomaz, J.M., "Primary and secondary nonlinear global instability," *Physica* D **132**, 428–456 (1999b).

Covas, E., Tavakol, R., Tworkowski, A. and Brandenburg, A., "Axisymmetric mean field dynamos with dynamic and algebraic α-quenchings," *Astron. Astrophys.* **329**, 350–360 (1998).

Dee, G., "Dynamical properties of propagating front solutions of the amplitude equation," *Physica* D **15**, 295–304 (1985).

Dee, G. and Langer, J.S., "Propagating pattern selection," *Phys. Rev. Let.* **50**, 383–386 (1983).

Dicke, R.H., "The phase variations of the solar cycle," *Solar Phys.* **115**, 171–181 (1988).

Dobler, W. and Rädler, K.-H., "An integral equation approach to kinematic dynamo models," *Geophys. Astrophys. Fluid Dynam.* **89**, 45–74 (1998).

Dormy, E., Soward, A.M., Jones, C.A., Jault, D. and Cardin, P., "The onset of thermal convection in rotating spherical shells," *J. Fluid Mech.*, submitted (2002).

Fedoryuk, M.V., *Asymptotic Analysis, Linear Ordinary Differential Equations*, Translation by A. Rodick. Springer-Verlag (1991).

Fioc, M., Nesme-Ribes, E. and Sokoloff, D., "Asymptotic properties of dynamo waves," in: *Physical Processes in Astrophysics* (Eds. I.W. Roxburgh and J.-L. Masnou), Springer, pp. 213–217 (1995).

Galitsky, V. and Sokoloff, D., "Dynamo waves in the theory of cosmic magnetism and probability waves in quantum mechanics," *Acta Astron. et Geophys. Univ. Comenianae* **XIX**, 1–12 (1997).

Galitsky, V. and Sokoloff, D., "The spectrum of the Parker equations," *Astron. Rep.* **42**, 127–133 (1998).

Galitsky, V. and Sokoloff, D., "Kinematic dynamo wave in the vicinity of the solar pole," *Geophys. Astrophys. Fluid Dynam.* **91**, 147–167 (1999).

Griffiths, G., Bassom, A.P., Soward, A.M. and Kuzanyan, K.M., "Nonlinear $\alpha^2\Omega$-waves in stellar shells," *Geophys. Astrophys. Fluid Dynam.* **94**, 85–133 (2001).

Harris, D., Bassom, A.P. and Soward, A.M., "An inhomogeneous Landau equation with application to spherical Couette flow," *Physica* D **137**, 260–276 (2000).

Hastie, R.J., Hesketh, K.W. and Taylor, J.B., "Shear damping of two-dimensional drift waves in a large-aspect-ratio Tokamak," *Nuclear Fusion* **19**, 1223–1233 (1979).

Heading, J., *An Introduction to Phase-Integral Methods*, Methuen, London (1962).

Heyvaerts, J. and Priest, E.R., "Coronal heating by phase-mixed shear Alfvén waves," *Astron. Astrophys.* **117**, 220–234 (1983).

Huerre, P. and Monkewitz, P.A., "Local and global instabilities in spatially developing flows," *Annu. Rev. Fluid Mech.* **22**, 473–537 (1990).

Jennings, R.L., "Symmetry breaking in a nonlinear αω-dynamo," *Geophys. Astrophys. Fluid Dynam.* **57**, 147–189 (1991).

Jennings, R.L. and Weiss, N.O., "Symmetry breaking in stellar dynamos," *Mon. Not. R. Astron. Soc.* **252**, 249–260 (1991).

Jones, C.A., Soward, A.M. and Mussa, A.I., "The onset of thermal convection in a rapidly rotating sphere," *J. Fluid Mech.* **405**, 157–179 (2000).

Killworth, P.D., "Barotropic and baroclinic instability in rotating stratified fluids," *Dynam. Atmos. Oceans* **4**, 143–184 (1980).

Knobloch, E., Tobias, S.M. and Weiss, N.O., "Modulation symmetry in stellar dynamos," *Mon. Not. R. Astron. Soc.* **297**, 1123–1138 (1997).

Krasheninnikova, Yu., Ruzmaikin, A., Sokoloff, D. and Shukurov, A., "Configuration of large-scale magnetic fields in spiral galaxies," *Astron. Astrophys.* **213**, 19–28 (1989).

Krause, F. (Ed.), *Magnetic Fields in Galaxies*, Gordon and Breach, Geophysical and Astrophysical Fluid Dynamics 50 (1990a).

Krause, F., "How well developed is the dynamo theory of flat objects?" *Geophys. Astrophys. Fluid Dynam.* **50**, 67–78 (1990b).

Krause, F. and Rädler, K.-H., *Mean-Field Electrodynamics and Dynamo Theory*, Pergamon Press, Oxford (1980).

Kuzanyan, K. and Sokoloff, D., "A dynamo wave in an inhomogeneous medium," *Geophys. Astrophys. Fluid Dynam.* **81**, 113–129 (1995).

Kuzanyan, K. and Sokoloff, D., "A dynamo wave in a thin shell," *Astron. Rep.* **40**, 425–430 (1996).

Kuzanyan, K. and Sokoloff, D., "Half-width of a solar dynamo wave in Parker's migratory dynamo," *Solar Phys.* **173**, 1–14 (1997).

Lan, S., Kuang, W. and Roberts, P.H., "Ideal instabilities in rapidly rotating MHD systems that have critical layers," *Geophys. Astrophys. Fluid Dynam.* **69**, 133–160 (1993).

Le Dizés, Huerre, P., Chomaz, J.-M. and Monkewitz, P.A., "Nonlinear stability analysis of slowly-diverging flows: limitations of the weakly nonlinear approach," in: *Bluff-Body Wakes, Dynamics and Instabilities* (Eds. H. Eclelmann, J.M.R. Graham, P. Huerre and P.A. Monkewitz), Proceedings of IUTAM Symposium, Springer, Berlin, pp. 147–152 (1993).

Lighthill, J., *Waves in Fluids*, Cambridge University Press (1978).

Mestel, L. and Subramanian, K., "Galactic dynamos and density wave theory," *Mon. Not. R. Astron. Soc.* **248**, 677–687 (1991).

Meunier, N., Nesme-Ribes, E. and Sokoloff, D., "Dynamo wave in a $\alpha^2\omega$ dynamo," *Astron. Rep.* **40**, 415–423 (1996).

Meunier, N., Proctor, M.R.E., Sokoloff, D., Soward, A.M. and Tobias, S.M., "Asymptotic properties of a nonlinear $\alpha\omega$-dynamo wave: period, amplitude and latitude dependence," *Geophys. Astrophys. Fluid Dynam.* **86**, 249–285 (1997).

Moffatt, H.K., *Magnetic Field Generation in Electrically Conducting Fluids*, Cambridge University Press (1987).

Moss, D., Shukurov, A. and Sokoloff, D., "Boundary effects and propagating magnetic fronts in disc dynamos," *Geophys. Astrophys. Fluid Dynam.* **89**, 285–308 (1998).

Moss, D., Shukurov, A. and Sokoloff, D., "Galactic dynamos driven by magnetic buoyancy," *Astron. Astrophys.* **343**, 120–131 (1999).

Moss, D., Shukurov, A., Sokoloff, D., Beck, R. and Fletcher, A., "Magnetic fields in barred galaxies. II. Dynamo models," *Astron. Astrophys.* **380**, 55–71 (2001).

Moss, D., Tuominen, I. and Brandenburg, A., "Buoyancy-limited thin shell dynamos," *Astron. Astrophys.* **240**, 142–149 (1990).

Noyes, R.W., Weiss, N.O. and Vaughan, A.H., "The relationship between stellar rotation rate and activity cycle periods," *Astrophys. J.* **287**, 769–773 (1984).

Parker, E.N., "Hydromagnetic dynamo models," *Astrophys. J.* **122**, 293–314 (1955).

Parker, E.N., "The generation of magnetic fields in astrophysical bodies. II. The galactic field," *Astrophys. J.* **163**, 255–278 (1971).

Parker, E.N., *Cosmical Magnetic Fields: Their Origin and Activity*, Clarendon Press, Oxford (1979).

Phillips, A., Brooke, J. and Moss, D. "The importance of physical structure in solar dynamo models," *Astron. Astrophys.* **392**, 713–727 (2002).

Pier, B. and Huerre, P., "Fully nonlinear global modes in spatially developing media," *Physica D* **97**, 206–222 (1996).

Pier, B., Huerre, P., Chomaz, J.-M. and Couairon, A., "Steep nonlinear global modes in spatially developing media," *Phys. Fluids* **10**, 2433–2435 (1998).

Poezd, A., Shukurov, A. and Sokoloff, D., "Global magnetic patterns in the Milky Way and Andromoda nebula," *Mon. Not. R. Astron. Soc.* **264**, 285–297 (1993).

Priklonsky, V., Shukurov, A., Sokoloff, D. and Soward, A., "Non-local effects in the mean-field disc dynamo," *Geophys. Astrophys. Fluid Dynam.* **93**, 97–114 (2000).

Proctor, M.R.E. and Spiegel, E.A., "Waves of solar Activity," in: *The Sun and Cool Stars: Activity, Magnetism, Dynamos* (Eds. I. Tuominen, D. Moss and G. Rüdiger), Springer, Lecture notes in Physics, Vol. 380, pp. 117–128 (1991).

Pudritz, R.E., "Dynamo action in turbulent accretion discs around black holes-II. The mean magnetic field," *Mon. Not. R. Astron. Soc.* **195**, 897–914 (1981).

Rädler, K.-H. and Bräuer, H.-J., "On the oscillatory behaviour of kinematic mean field dynamos," *Astron. Nachr.* **308**, 101–109 (1987).

Roberts, P.M. and Soward, A.M., "Dynamo theory," *Annu. Rev. Fluid Mech.* **24**, 459–512 (1992).

Rüdiger, G. and Arlt, R., "Physics of the solar cycle," in this volume, Ch. 6, pp. 147–194 (2003).

Rüdiger, G. and Brandenburg, A., "A solar dynamo in the overshoot layer: cycle period and butterfly diagram," *Astron. Astrophys.* **296**, 557–566 (1995).

Ruzmaikin, A. and Starchenko, S., "Kinematic turbulent geodynamo of the mean field," *Geomag. Aeron.* **28**, 402–406 (1988).

Ruzmaikin, A., Shukurov, A. and Sokoloff, D., *Magnetic Fields in Galaxies*, Kluwer (1988a).

Ruzmaikin, A., Shukurov, A., Sokoloff, D. and Starchenko, S., "Maximally-efficient-generation-approach in dynamo theory," *Geophys. Astrophys. Fluid Dynam.* **52**, 125–139 (1990).

Ruzmaikin, A., Sokoloff, D. and Starchenko, S., "Excitation of non-axially symmetric modes of the sun's mean magnetic field," *Solar Physics* **115**, 5–15 (1988b).

Schiff, L.I., *Quantum Mechanics*, McGraw-Hill, New York (1955).

Schüssler, M. and Ferriz-Mas, A., "Magnetic flux tubes and the dynamo problem," in this volume, Ch. 5, pp. 123–146 (2003).

Sokoloff, D., and Soward, A.M., "Asymptotic properties of a nonlinear $\alpha\omega$-dynamo," *Studia Geoph. Geod.* **42**, 309–313 (1998).

Sokoloff, D., Fioc, M. and Nesme-Ribes, E., "Asymptotic properties of dynamo wave," *Magneto-hydrodyn.* **31**, 19–40 (1995a).

Sokoloff, D., Nesme-Ribes, E. and Fioc, M., "Asymptotic properties of dynamo waves," in: *Physical Processes in Astrophysics* (Eds. I.W. Roxburgh and J.-L. Masnou), Springer, Lecture notes in Physics, Vol. 458, pp. 213–217 (1995b).

Sokoloff, D., Shukurov, A. and Ruzmaikin, A., "Asymptotic solution of the α^2 dynamo problem," *Geophys. Astrophys. Fluid Dynam.* **25**, 293–307 (1983).

Soward, A.M., "A thin disc model of the Galactic dynamo," *Astron. Nachr.* **299**, 25–33 (1978).

Soward, A.M., "Thin disc $\alpha\omega$-dynamo models I. Long length scale modes," *Geophys. Astrophys. Fluid Dynam.* **64**, 163–199 (1992a).

Soward, A.M., "Thin disc $\alpha\omega$-dynamo models II. Short length scale modes," *Geophys. Astrophys. Fluid Dynam.* **64**, 201–225 (1992b).

Soward, A.M., Bassom, A.P. and Ponty, Y., "Alpha-quenched $\alpha^2\Omega$-dynamo waves in: stellar shells," in: *Dynamo and Dynamics; a Mathematical Challenge* (Eds. P. Chossat, D. Armbrusto and I. Oprea), Kluwer Academic Publishers, pp. 297–304 (2001).

Soward, A.M. and Jones, C.A., "The linear stability of the flow in the narrow gap between two concentric rotating spheres," *Q. Jl. Mech. appl. Math.* **36**, 19–42 (1983).

Spiegel, E.A., "The chaotic solar cycle," in: *Lectures on Solar and Planetary Dynamos* (Eds. M.R.E. Proctor and A.D. Gilbert), Cambridge University Press, pp. 245–265 (1994).

Starchenko, S., "Dynamo models with strong generation Kinematic solution and axisymmetric $\alpha^2\omega$-dynamo," *Geophys. Astrophys. Fluid Dynam.* **77**, 55–77 (1994).

Starchenko, S. and Kono, M., "A comparison of numerical and asymptotic MEGA solutions of the $\alpha\omega$-dynamo problem," *Geophys. Astrophys. Fluid Dynam.* **82**, 93–123 (1996).

Stepinski, T.F. and Levy, E.H., "Generation of dynamo magnetic fields in thin Keplerian discs," *Astrophys. J.* **362**, 318–332 (1990).

Stix, M., "Nonlinear dynamo waves," *Astron. Astrophys.* **20**, 9–12 (1972).

Stix, M., "The galactic dynamo," *Astron. Astrophys.* **42**, 85–89 (1975).

Stix, M., Erratum "The galactic dynamo," *Astron. Astrophys.* **68**, 459 (1978).

Stix, M., *The Sun. An Introduction*, Springer, Berlin (1989).

Stix, T.H., *The Theory of Plasma Waves*, Advanced Physics Monograph Series, McGraw-Hill, New York (1962).

Taylor, J.B., "Some recent advances in plasma stability theory," *Physikalische Blätter* **35**(12), 611–616 (1979).

Tobias, S.M., "Diffusivity quenching as a mechanism for Parker's surface dynamo wave," *Astrophys. J.* **467**, 870–880 (1996).

Tobias, S.M., "Properties of nonlinear dynamo waves," *Geophys. Astrophys. Fluid Dynam.* **86**, 287–343 (1997a).

Tobias, S.M., "The solar cycle: parity interactions and amplitude modulation," *Astron. Astrophys.* **322**, 1007–1017 (1997b).

Tobias, S.M., "Relating stellar cycle periods to dynamo calculations," *Mon. Not. R. Astron. Soc.* **296**, 653–661 (1998).

Tobias, S.M., Proctor, M.R.E. and Knobloch, E., "The rôle of absolute instability in the solar dynamo," *Astron. Astrophys.* **318**, L55–L58 (1997).

Tobias, S.M., Proctor, M.R.E. and Knobloch, E., "Convective and absolute instabilities of fluid flows in finite geometry," *Physica* D **113**, 43–72 (1998).

Tuominen, I., Rüdiger, G. and Brandenburg, A., "Observational constraints for solar-type dynamos," in: *Activity in Cool Star Envelopes* (Eds. O. Havnes, B.R. Pettersen, J.H.M.M. Schmitt and J.E. Solheim), Kluwer, pp. 13–20 (1988).

Tworkowski, A., Tavakol, R., Brandenburg, A., Brooke, J.M., Moss, D. and Tuominen, I., "Intermittent behaviour in axisymmetric mean-field dynamo models in spherical shells," *Mon. Not. R. Astron. Soc.* **296**, 287–295 (1998).

van Saarloos, W., "Fronts, pulses, sources and sinks in the generalized complex Ginzburg–Landau equations," *Physica* D **54**, 303–367 (1992).

Walker, M.R. and Barenghi, C.F., "High resolution numerical dynamos in the limit of a thin disk galaxy," *Geophys. Astrophys. Fluid Dynam.* **76**, 265–281 (1994).

Walker, M.R. and Barenghi, C.F., "A model of nonlinear dynamo action in lenticular galaxies," *Astrophys. J.* **450**, 540–546 (1995).

Wasow, W.R., *Linear Turning Point Theory*, Springer-Verlag (1985).

Weiss, N.O., "Solar and stellar dynamos," in: *Lectures on Solar and Planetary Dynamos* (Eds. M.R.E. Proctor and A.D. Gilbert), Cambridge University Press, pp. 59–95 (1994).

Weiss, N.O. and Tobias, S.M., "Modulation of solar and stellar activity cycles," in: *Solar and Heliospheric Plasma Physics* (Eds. G.M. Simnett, C.E. Alissandrakis and L. Vlahos), Springer, Lecture Notes in Physics, pp. 25–47 (1997).

Weiss, N.O., Cattaneo, F. and Jones, C.A., "Periodic and aperiodic dynamo waves," *Geophys. Astrophys. Fluid Dynam.* **30**, 305–341 (1984).

White, M.P., PhD dissertation, University of Durham, UK (1977).

Worledge, D., Knobloch, E., Tobias, S. and Proctor, M.R.E., "Dynamo waves in semi-infinite and finite domains," *Proc. R. Soc. Lond.* A **453**, 119–143 (1997).

Yano, J.-L., "Asymptotic theory of thermal convection in rapidly rotating systems," *J. Fluid Mech.* **243**, 103–131 (1992).

Zeldovich, Ya., Ruzmaikin, A. and Sokoloff, D., *Magnetic Fields in Astrophysics*, Gordon and Breach, New York (1980).

9 Computational aspects of astrophysical MHD and turbulence

Axel Brandenburg

Nordic Institute for Theoretical Physics (NORDITA), Blegdamsvej 17, DK-2100
Copenhagen Ø, Denmark, E-mail: brandenb@nordita.dk

The advantages of high-order finite difference scheme for astrophysical MHD and turbulence simulations are highlighted. A number of one-dimensional test cases are presented ranging from various shock tests to Parker-type wind solutions. Applications to magnetized accretion discs and their associated outflows are discussed. Particular emphasis is placed on the possibility of dynamo action in three-dimensional turbulent convection and shear flows, which is relevant to stars and astrophysical discs. The generation of large scale fields is discussed in terms of an inverse magnetic cascade and the consequences imposed by magnetic helicity conservation are reviewed with particular emphasis on the issue of α-quenching.

9.1. Introduction

Over the past 20 years multidimensional astrophysical gas simulations have become a primary tool to understand the formation, evolution, and the final fate of stars, galaxies, and their surrounding medium. The assumption that those processes happen smoothly and in a non-turbulent manner can at best be regarded as a first approximation. This is evidenced by the ever improving quality of direct imaging techniques, e.g. using space telescopes. At the same time not only have computers become large enough to run three-dimensional simulations with relatively little effort, there have also been substantial improvements in the algorithms that are used. In fact, there is now a vast literature on numerical astrophysics. An excellent book was published recently by LeVeque *et al.* (1998) where both numerical methods and astrophysical applications were discussed in great detail. Most of the applications focused however on rather more "violent" processes such as supersonic jets, supernova explosions, core collapse, and on radiative transfer problems, while hydromagnetic phenomena and turbulence problems were only touched upon briefly. Meanwhile, hydromagnetic turbulence simulations have become crucial for understanding viscous dissipation in accretion discs (Hawley *et al.*, 1995), and for understanding magnetic field generation by dynamo action in discs (Brandenburg *et al.*, 1995, 1996a; Hawley *et al.*, 1996; Stone *et al.*, 1996), stars (Nordlund *et al.*, 1992; Brandenburg *et al.*, 1996b), and planets (Glatzmaier and Roberts, 1995, 1996).

Much of the present day astrophysical hydrodynamic work is based on the ZEUS code, which has been documented in great detail and described with a number of test cases in a series of papers by Stone and Norman (1992a,b). The main advantage is its flexibility in dealing with arbitrary orthogonal coordinates which makes the code applicable to a wide variety of astrophysical systems. The code, which is freely available on the net, uses artificial viscosity for stability and shock capturing, and is based on an operator split method with

2nd-order finite differences on a staggered mesh. Another approach used predominantly in turbulence research are spectral methods (e.g. Canuto *et al.*, 1988), which have the advantage of possessing high accuracy. Although these methods are most suitable for incompressible flows (imposing the solenoidality condition is then straightforward), they have also been applied to compressible flows (e.g. Passot and Pouquet, 1987). As a compromise one may resort to high-order finite difference methods, which have the advantage of being easy to implement and yet have high accuracy. Compact methods (e.g. Lele, 1992) are a special variety of high-order finite difference methods, but the truncation error is smaller than for an explicit scheme of the same order. For example, compact schemes have been used by Nordlund and Stein (1990) in simulations of solar convection (Stein and Nordlund, 1989, 1998) and convective dynamos (Nordlund *et al.*, 1992; Brandenburg *et al.*, 1996b).

The use of compact methods involves solving tridiagonal matrix equations, making this method essentially non-local in that all points are now coupled at once. This is problematic for massively parallel computations, which is why Nordlund and Galsgaard (1995, see also Nordlund *et al.*, 1994) began to use explicit high-order schemes for their work on coronal heating by reconnection (Galsgaard and Nordlund, 1996, 1997a,b). In their code the equations are solved in a semi-conservative fashion using a staggered mesh. This code was also used by Padoan *et al.*, (1997) and Padoan and Nordlund (1999) in models of isothermal interstellar turbulence in molecular clouds, and by Rögnvaldsson and Nordlund (2000) in simulations of cooling flows and galaxy formation.

A somewhat different code was used by Brandenburg (1999) and Bigazzi (1999) in simulations of the inverse magnetic cascade, by Kerr and Brandenburg (1999) in a work on the possibility of a singularity of the non-resistive and inviscid MHD equations, and by Sanchez-Salcedo and Brandenburg (1999, 2001) in simulations of dynamical friction. A two-dimensional version of the code modelling outflows from magnetized accretion discs has been described by Brandenburg *et al.* (2000). This code uses 6th-order explicit finite differences in space and 3rd-order Runge–Kutta timestepping. It employs central finite differences, so the extra cost of recentering a large number of variables between staggered meshes each timestep is avoided.

Apart from high numerical accuracy, another important requirement for astrophysical gas simulations is the capability to deal with a large dynamical range in density and temperature. This requirement favors the use of non-conservative schemes, because then logarithmic variables can be used which vary much less than linear density and energy density per unit volume. Solving the non-conservative form of the equations can be more accurate than solving the conservative form. The conservation properties can then be used as an indicator for the overall accuracy.

In this chapter we concentrate on numerical astrophysical turbulence aspects starting with a discussion of different numerical methods and a description of the results of various numerical test problems. This is a good way of assessing the quality of a numerical scheme and of comparing with other methods; see Stone and Norman (1992a,b) for a series of tests using the ZEUS code. After that we discuss particular astrophysical applications including stellar convection, accretion disc turbulence and associated outflows, as well as the generation of magnetic fields (small and large scale) from turbulence in various astrophysical settings.

9.2. The Navier–Stokes equations

The discussion of magnetic fields will be postponed until later, because the inclusion of the Lorentz force in the momentum equation is straightforward. We begin by writing down the Navier–Stokes equations in non-conservative form and rewrite them such that the main thermodynamical variables are entropy and either logarithmic density or potential enthalpy. These variables have the advantage of varying spatially much less than, e.g. linear pressure and density.

The primitive form of the continuity equation is

$$\frac{\partial \rho}{\partial t} = -\nabla \cdot (\rho \mathbf{u}), \tag{9.1}$$

which means that the local change of density is given by the divergence of the mass flux at that point. The Navier–Stokes equation can be written as

$$\rho \frac{D\mathbf{u}}{Dt} = -\nabla p - \rho \nabla \Phi + \mathbf{F} + \nabla \cdot \boldsymbol{\tau}, \tag{9.2}$$

where $D/Dt = \partial/\partial t + \mathbf{u} \cdot \nabla$ is the advective derivative, p is the pressure, Φ is the gravitational potential, $\mathbf{F}$ is a body force (e.g. the Lorentz force), and $\boldsymbol{\tau}$ is the stress tensor.

The Navier–Stokes equation is here written in terms of forces per unit volume. As argued above, if the density contrast is large it is advantageous to write it in terms of forces per unit mass and to divide by ρ. Before we can replace p and ρ by entropy and logarithmic density or potential entropy we first have to define some thermodynamic quantities.

Internal energy, e, and specific enthalpy, h are related to each other by

$$h = e + pv, \tag{9.3}$$

where $v = 1/\rho$ is the specific volume and ρ the density. The specific entropy is defined by

$$T \, ds = de + p \, dv, \tag{9.4}$$

where T is temperature. The specific heats at constant pressure and constant volume are defined as $c_p = dh/dT |_p$ and $c_v = de/dT |_v$, their ratio is $\gamma = c_p/c_v$, and their difference is $\mathcal{R}/\mu = c_p - c_v$, where $\mathcal{R}$ is the universal gas constant and μ the specific molecular weight.

In the following we assume c_p and c_v to be constant for all processes considered. Ionization and recombination processes are therefore ignored here, although this is not a major obstacle; see, e.g. simulations of Nordlund (1982, 1985), Steffen *et al.* (1989), Stein and Nordlund (1989, 1998), Rast *et al.* (1993), and Rast and Toomre (1993a,b) where realistic equations of state have been used.

We now assume that c_p and c_v are constant, so internal energy and specific enthalpy are given by

$$h = c_p T \quad \text{and} \quad e = c_v T. \tag{9.5}$$

This allows us then to write the specific entropy (up to an additive constant) as

$$s = c_v \ln p - c_p \ln \rho. \tag{9.6}$$

The pressure gradient term in the momentum equation can then be written as

$$\frac{1}{\rho} \nabla p = \frac{p}{\rho} \nabla \ln p = \frac{\gamma p}{\rho} (\nabla \ln \rho + \nabla s/c_p) = c_s^2 (\nabla \ln \rho + \nabla s/c_p), \tag{9.7}$$

where we have used

$$c_s^2 = \gamma p/\rho = \frac{\gamma p_0}{\rho_0} \exp[(\gamma - 1)\ln(\rho/\rho_0) + \gamma s/c_p] = c_{s0}^2 \left(\frac{\rho}{\rho_0}\right)^{\gamma-1} \exp(\gamma s/c_p), \quad (9.8)$$

where c_s is the adiabatic sound speed, and $c_{s0}^2 = \gamma p_0/\rho_0$.

With these preparations the evolution of velocity $\mathbf{u}$, logarithmic density $\ln \rho$, and specific entropy s can be expressed as follows:

$$\frac{D\mathbf{u}}{Dt} = -c_s^2(\nabla \ln \rho + \nabla s/c_p) - \nabla \Phi + \mathbf{f} + \frac{1}{\rho}\nabla \cdot (2\nu\rho\mathbf{S}), \quad (9.9)$$

$$\frac{D\ln \rho}{Dt} = -\nabla \cdot \mathbf{u}, \quad (9.10)$$

$$T\frac{Ds}{Dt} = 2\nu\mathbf{S}^2 + \Gamma - \rho\Lambda, \quad (9.11)$$

where $\mathbf{f} = \mathbf{F}/\rho$ is the body force per unit mass, Γ and Λ are heating and cooling functions, ν kinematic viscosity and $\mathbf{S}$ is the (traceless) strain tensor with the components

$$S_{ij} = \tfrac{1}{2}\left(u_{i,j} + u_{j,i} - \tfrac{2}{3}\delta_{ij}u_{k,k}\right). \quad (9.12)$$

In the presence of an additional kinematic bulk viscosity, ζ, the term $2\nu S_{ij}$ under the divergence in (9.9) would need to be replaced by $2\nu S_{ij} + \zeta \delta_{ij}\nabla \cdot \mathbf{u}$, and the viscous heating term, $2\nu\mathbf{S}^2$, in (9.11) would need to be replaced by $2\nu\mathbf{S}^2 + \zeta(\nabla \cdot \mathbf{u})^2$.

Instead of using ρ as a dependent variable one can also use the specific enthalpy h, which allows us to write the pressure gradient as

$$-\frac{1}{\rho}\nabla p = -\nabla h + T\nabla s. \quad (9.13)$$

This formulation is particularly useful if the entropy is nearly constant (or if the gas is barotropic, i.e. $p = p(\rho)$) and if there is a gravitational potential Φ, so that the potential enthalpy $H \equiv h + \Phi$ can be used as dependent variable. In order to express (9.10) in terms of h we write down the total differential of the specific entropy,

$$ds/c_p = \frac{1}{\gamma}d\ln p - d\ln \rho = \frac{1}{\gamma}d\ln h - \left(1 - \frac{1}{\gamma}\right)d\ln \rho, \quad (9.14)$$

so

$$\frac{D\ln h}{Dt} = \gamma\frac{Ds/c_p}{Dt} + (\gamma - 1)\frac{D\ln \rho}{Dt}. \quad (9.15)$$

Furthermore, $T\nabla s = h\nabla s/c_p$, and so the final set of equations is

$$\frac{D\mathbf{u}}{Dt} = -\nabla H + h\nabla s/c_p + \mathbf{f} + \nabla \cdot (2\nu\rho\mathbf{S}), \tag{9.16}$$

$$\frac{Ds/c_p}{Dt} = \frac{1}{h}\left(2\nu\mathbf{S}^2 + \Gamma - \rho\Lambda\right), \tag{9.17}$$

$$\frac{DH}{Dt} = \mathbf{u}\cdot\nabla\Phi + \gamma h\frac{Ds/c_p}{Dt} - c_s^2\nabla\cdot\mathbf{u}, \tag{9.18}$$

where we have absorbed Φ in the potential enthalpy $H = h + \Phi$. In this formulation the density can be recovered as

$$\rho = \rho_0\left[\left(1 - \frac{1}{\gamma}\right)\left(\frac{h}{c_{s0}^2}\right)e^{-\gamma s/c_p}\right]^{1/(\gamma-1)} \tag{9.19}$$

(in dimensional form) or, for $\gamma = 5/3$ and in non-dimensional form (where $\rho_0 = p_0 = c_p = 1$),

$$\rho = (0.4h)^{1.5}e^{-2.5s}. \tag{9.20}$$

We shall use either of the two sets of the equations, (9.9)–(9.11) or (9.16)–(9.18), in some of the following sections, especially in connection with shock tests and stellar wind problems. In these cases the gravity potential Φ is important and it turns out that the potential enthalpy $H \equiv h + \Phi$ varies only very little near the central object even though Φ itself tends to become singular.

The heating and cooling terms (Γ and Λ) are important, e.g. in the case of interstellar turbulence which is driven primarily by supernova explosions which inject a certain amount of thermal energy ($\int\rho\Gamma dV$) with each supernova explosion. MHD turbulence simulations of this type were performed recently by Korpi *et al.* (1999). At the same time there is cooling through various processes (e.g. bremsstrahlung at high temperatures) which transports energy either non-locally via a cooling term $\Lambda(T)$, or locally via thermal conduction or radiative diffusion. In the radiative diffusion approximation we express Λ as $-\rho^2\Lambda = \nabla\cdot K\nabla T$, where K is the radiative conductivity which is in general a function of temperature and density. The radiative diffusivity (which has the same dimensions as the kinematic viscosity ν) is given by $\chi = K/(\rho c_p)$, so

$$-\rho\Lambda/(c_pT) = \frac{1}{\rho c_pT}\nabla\cdot\rho\chi c_p\nabla T. \tag{9.21}$$

Since we shall use a non-conservative scheme with centered finite differences it is important to isolate second derivative terms, so

$$-\rho\Lambda/h = \chi(\nabla^2\ln T + \nabla\ln p\cdot\nabla\ln T), \tag{9.22}$$

where we have assumed for simplicity that χ is constant. In terms of s/c_p and $\ln\rho$ we have

$$-\rho\Lambda/h = \chi\gamma\left[\nabla^2 s/c_p + \nabla_{ad}\nabla^2\ln\rho + \gamma(\nabla s/c_p + \nabla\ln\rho)\cdot(\nabla s/c_p + \nabla_{ad}\nabla\ln\rho)\right], \tag{9.23}$$

where $\nabla_{ad} = 1 - 1/\gamma$ is a commonly used abbreviation in stellar astrophysics. For $\gamma = 5/3$ we have $\nabla_{ad} = 2/5 = 0.4$. We shall use (9.23) later in connection with shock and wind calculations. However, we begin by discussing first a suitable numerical scheme which will be used in most of the cases presented below.

9.3. The advantage of higher-order derivative schemes

Spectral methods are commonly used in almost all studies of ordinary (usually incompressible) turbulence. The use of this method is justified mainly by the high numerical accuracy of spectral schemes. Alternatively, one may use high-order finite differences that are faster to compute and that can possess almost spectral accuracy. Nordlund and Stein (1990) and Brandenburg *et al.* (1995) use high-order finite difference methods, e.g. 4th- and 6th-order compact schemes (Lele 1992).[1]

In this section we demonstrate, using simple test problems, some of the advantages of high-order schemes. We begin by defining various schemes including their truncation errors and their high wavenumber characteristics. We consider centered finite differences of 2nd-, 4th-, 6th-, 8th-, and 10th-order, which are given respectively by the formulae

$$f_i' = (-f_{i-1} + f_{i+1})/(2\delta x), \tag{9.24}$$

$$f_i' = (f_{i-2} - 8f_{i-1} + 8f_{i+1} - f_{i+2})/(12\delta x), \tag{9.25}$$

$$f_i' = (-f_{i-3} + 9f_{i-2} - 45f_{i-1} + 45f_{i+1} - 9f_{i+2} + f_{i+3})/(60\delta x), \tag{9.26}$$

$$f_i' = (3f_{i-4} - 32f_{i-3} + 168f_{i-2} - 672f_{i-1} + 672f_{i+1}$$
$$- 168f_{i+2} + 32f_{i+3} - 3f_{i+4})/(840\delta x), \tag{9.27}$$

$$f_i' = (-2f_{i-5} + 25f_{i-4} - 150f_{i-3} + 600f_{i-2} - 2100f_{i-1}$$
$$+ 2100f_{i+1} - \cdots)/(2520\delta x), \tag{9.28}$$

for the first derivative, and

$$f_i'' = (f_{i-1} - 2f_i + f_{i+1})/(\delta x^2), \tag{9.29}$$

$$f_i'' = (-f_{i-2} + 16f_{i-1} - 30f_i + 16f_{i+1} - f_{i+2})/(12\delta x^2), \tag{9.30}$$

$$f_i'' = (2f_{i-3} - 27f_{i-2} + 270f_{i-1} - 490f_i$$
$$+ 270f_{i+1} - 27f_{i+2} + 2f_{i+3})/(180\delta x^2), \tag{9.31}$$

$$f_i'' = (-9f_{i-4} + 128f_{i-3} - 1008f_{i-2} + 8064f_{i-1} - 14350f_i$$
$$+ 8064f_{i+1} - 1008f_{i+2} + \cdots)/(5040\delta x^2), \tag{9.32}$$

$$f_i'' = (8f_{i-5} - 125f_{i-4} + 1000f_{i-3} - 6000f_{i-2} + 42000f_{i-1}$$
$$- 73766f_i + 42000f_{i+1} - \cdots)/(25200\delta x^2), \tag{9.33}$$

for the second derivative. The expressions for one-sided and semi-one-sided finite difference formulae are given in Appendix A.

9.3.1. High wavenumber characteristics

The chief advantage of high-order schemes is their high fidelity at high wavenumber. Suppose we differentiate the function $\sin kx$, we are supposed to get $k \cos kx$, but when k is close to the Nyquist frequency, $k_{Ny} \equiv \pi/\delta x$, where δx is the mesh spacing, numerical schemes yield *effective* wavenumbers, k_{eff}, that can be significantly less than the actual wavenumber k. Here we calculate k_{eff} from

$$(\cos kx)'_{num} = -k_{eff} \sin kx. \tag{9.34}$$

When $k = k_{Ny}$, every centered difference scheme will give $k_{eff} = 0$, because then the function values of $\cos kx$ are just $-1, +1, -1, \ldots$, so the function values on the left and the right are the same, and the difference that enters the scheme gives therefore zero.

It is useful to mention at this point that for a staggered mesh, where the first derivative is evaluated *between* mesh points, the value of the first derivative remains finite at the Nyquist frequency, provided one does not need to remesh back to the original mesh. Especially in the context with magnetic fields, however, remeshing needs to be done quite frequently, which therefore diminishes the advantage of a staggered mesh.

In Fig. 9.1 we plot effective wavenumbers for different schemes. Apart from the different *explicit* finite difference schemes given above, we also consider a *compact* scheme of sixth order, which can be written in the form

$$\tfrac{1}{3}f'_{i-1} + f'_i + \tfrac{1}{3}f'_{i+1} = (f_{i-2} - 28f_{i-1} + 28f_{i+1} - f_{i+2})/(36\delta x), \tag{9.35}$$

for the first derivative, and

$$\tfrac{2}{11}f''_{i-1} + f''_i + \tfrac{2}{11}f''_{i+1} = (3f_{i-2} + 48f_{i-1} - 102f_i + 48f_{i+1} + 3f_{i+2})/(44\delta x^2), \tag{9.36}$$

for the second derivative. As we have already mentioned in the introduction, this scheme involves obviously solving tridiagonal matrix equations and is therefore effectively non-local.

In the second panel of Fig. 9.1 we have plotted effective wavenumbers for second derivatives, which were calculated as

$$(\cos kx)''_{num} = -k^2_{eff} \cos kx. \tag{9.37}$$

Of particular interest is the behavior of the second derivative at the Nyquist frequency, because that is relevant for damping zig-zag modes. For a 2nd-order finite difference scheme k^2_{eff} is only 4, which is less than half the theoretical value of $\pi^2 = 9.87$. For 4th-, 6th-, and 10th-order schemes this value is respectively 5.33, 6.04, and 6.83. The last value is almost the same as for the 6th-order compact scheme, which is 6.86. Significantly stronger damping at the Nyquist frequency can be obtained by using hyperviscosity, which Nordlund and Galsgaard (1995) treat as a quenching factor that diminishes the value of the second derivative for wavenumbers that are small compared with the Nyquist frequency. Accurate high-order second derivatives (with no quenching factors) are important when calculating the current $\mathbf{J}$ in the Lorentz force $\mathbf{J} \times \mathbf{B}$ from a vector potential $\mathbf{A}$ using $-\mu_0\mathbf{J} = \nabla^2\mathbf{A} - \nabla(\nabla \cdot \mathbf{A})$. This will be important in the MHD calculations presented below.

9.3.2. The truncation error

One can express f_{i-1}, f_{i+1}, etc., in terms of the derivatives of f at point i, so

$$f_{i-1} = f_i - \delta x f'_i + \tfrac{1}{2}\delta x^2 f''_i - \tfrac{1}{6}\delta x^3 f'''_i + \cdots, \tag{9.38}$$

$$f_{i+1} = f_i + \delta x f'_i + \tfrac{1}{2}\delta x^2 f''_i + \tfrac{1}{6}\delta x^3 f'''_i + \cdots. \tag{9.39}$$

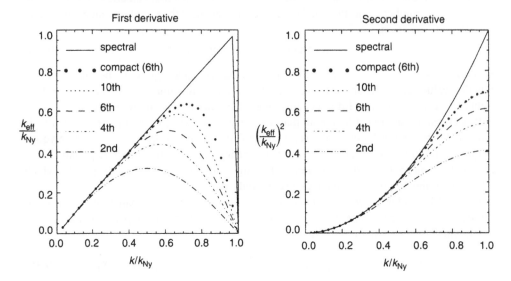

Figure 9.1 Effective wavenumbers for first and second derivatives using different schemes. Note that for the second derivatives the 6th-order compact scheme is almost equivalent to the 10th-order explicit scheme. For the first derivative the 6th-order compact scheme is still superior to the 10th-order explicit scheme.

Inserting this into the finite difference expressions yields for the 2nd-order formula

$$(f_i')_{2\text{nd}} \equiv (f_{i+1} - f_{i-1})/(2\delta x) = f_i' + \tfrac{1}{6}\delta x^2 f_i'''. \tag{9.40}$$

The error scales quadratically with the mesh size, which is why the method is called second order. The truncation error is proportional to the third derivative of the function. Because this is an odd derivative it corresponds to a dispersive (as opposed to diffusive) error. Schemes that are only first order (or of any odd order) have diffusive errors, and it is this what is sometimes referred to as *numerical diffusivity*, which is not to be confused with artificial diffusivity that is sometimes used for stability and shock capturing. For the other schemes given in (9.25)–(9.28) the truncation errors are

$$(f_i')_{4\text{th}} = f_i' + 3 \times 10^{-2}\,\delta x^4 f_i^{(\text{v})}, \tag{9.41}$$

$$(f_i')_{6\text{th}} = f_i' + 7 \times 10^{-3}\,\delta x^6 f_i^{(\text{vii})}, \tag{9.42}$$

$$(f_i')_{10\text{th}} = f_i' + 3 \times 10^{-4}\,\delta x^{10} f_i^{(\text{xi})}. \tag{9.43}$$

For the 6th-order compact scheme the error scales like for the 6th-order explicit scheme, but the coefficient in front of the truncation error is about 10 times smaller, so

$$(f_i')_{6\text{th}}^{\text{compact}} = f_i' + 5.1 \times 10^{-4}\,\delta x^6 f_i^{(\text{vii})}. \tag{9.44}$$

For the second derivatives we have

$$(f_i'')_{2nd} = f_i'' + 8 \times 10^{-2} \delta x^2 f_i^{(iv)}, \tag{9.45}$$

$$(f_i'')_{4th} = f_i'' - 1 \times 10^{-2} \delta x^4 f_i^{(vi)}, \tag{9.46}$$

$$(f_i'')_{6th} = f_i'' + 2 \times 10^{-3} \delta x^6 f_i^{(viii)}, \tag{9.47}$$

$$(f_i'')_{10th} = f_i'' - 5 \times 10^{-5} \delta x^{10} f_i^{(xii)}. \tag{9.48}$$

Again, for the 6th-order compact scheme the scaling is the same as for the 6th-order explicit scheme, but the coefficient in front of the truncation error is about five times less, so

$$(f_i'')_{6th}^{compact} = f_i'' + 3.2 \times 10^{-4} \delta x^6 f_i^{(viii)}. \tag{9.49}$$

This information about the accuracy of schemes would obviously be of little use if the various schemes did not perform well when applied to real problems. For this reason we now begin by carrying out various tests, including advection and shock tests.

9.3.3. Advection tests

As a first test we compare the various schemes by performing inviscid advection tests and solve the equation $Df/Dt = 0$, i.e.

$$\dot{f} = -uf', \tag{9.50}$$

on a periodic mesh. It is advantageous to use a relatively small number of meshpoints (here we use $N_x = 8$ meshpoints), because that way we see deficiencies most clearly. This case is actually also relevant to real applications, because in practice one will always have small scale structures that are just barely resolved.

After some time an initially sinusoidal signal will suffer a change in amplitude and phase. We have calculated the amplitude and phase errors for schemes of different spatial order. For the time integration we use high-order Runge–Kutta methods of third- or fourth-order, RK3 and RK4, respectively. In most cases considered below we use the RK3 scheme that allows reasonable use of storage. It can be written in three steps (Rogallo, 1981)

$$
\begin{array}{lll}
\text{1st step:} & f = f + \gamma_1 \delta t \, \dot{f}, & g = f + \zeta_1 \delta t \, \dot{f}, \\
\text{2nd step:} & f = g + \gamma_2 \delta t \, \dot{f}, & g = f + \zeta_2 \delta t \, \dot{f}, \\
\text{3rd step:} & f = g + \gamma_3 \delta t \, \dot{f},
\end{array} \tag{9.51}
$$

where

$$\gamma_1 = \frac{8}{15}, \quad \gamma_2 = \frac{5}{12}, \quad \gamma_3 = \frac{3}{4}, \quad \zeta_1 = -\frac{17}{60}, \quad \zeta_2 = -\frac{5}{12}, \tag{9.52}$$

where, f and g always refer to the current value (so the same space in memory can be used), but $\dot{f}$ is evaluated only once at the beginning of each of the three steps at $t = t_0$, $t_{1/3} = t_0 + \gamma_1 \delta t \approx t_0 + 0.5333\delta t$, and at $t_{2/3} = t_0 + (\gamma_1 + \zeta_1 + \gamma_2)\delta t = t_0 + (2/3)\delta t$. Even more memory-effective are the so-called $2N$-schemes that require one set of variables less to be hold in memory. Such schemes work for arbitrarily high order, although not all Runge–Kutta schemes can be written as $2N$-schemes (Williamson, 1980; Stanescu and Habashi, 1998). These schemes work iteratively according to the formula

$$w_i = \alpha_i w_{i-1} + \delta t \, F(t_{i-1}, u_{i-1}), \quad u_i = u_{i-1} + \beta_i w_i. \tag{9.53}$$

Table 9.1 Possible coefficients for different $2N$-RK3 schemes

Label	α_2	α_3	β_1	β_2	β_3	$(t_1 - t_0)/\delta t$	$(t_2 - t_0)/\delta t$
Symmetric (i)	$-2/3$	-1	$1/3$	1	$1/2$	$1/3$	$2/3$
Symmetric (ii)	$-1/3$	-1	$1/3$	$1/2$	1	$1/3$	$2/3$
Predictor–corrector	$-1/4$	$-4/3$	$1/2$	$2/3$	$1/2$	$1/2$	1
Inhomogeneous	$-17/32$	$-32/27$	$1/4$	$8/9$	$3/4$	$1/4$	$2/3$
Quadratic	-0.367	-1.028	0.308	0.540	1	0.308	0.650
Williamson (1980)	$-5/9$	$-153/128$	$1/3$	$15/16$	$8/15$	$1/3$	$3/4$

For a three-step scheme we have $i = 1, \ldots, 3$. In order to advance the variable u from $u^{(n)}$ at time $t^{(n)}$ to $u^{(n+1)}$ at time $t^{(n+1)} = t^{(n)} + \delta t$ we set in (9.53)

$$u_0 = u^{(n)} \quad \text{and} \quad u^{(n+1)} = u_3, \tag{9.54}$$

with u_1 and u_2 being intermediate steps. In order to be able to calculate the first step, $i = 1$, for which no $w_{i-1} \equiv w_0$ exists, we have to require $\alpha_1 = 0$. Thus, we are left with five unknowns, $\alpha_2, \alpha_3, \beta_1, \beta_2$, and β_3. Three conditions follow from the fact that the scheme be third order, so we have to have two more conditions. One possibility is to choose the fractional times at which the right-hand side is evaluated, e.g. $(0, 1/3, 2/3)$ or even $(0, 1/2, 1)$. In the latter case the right-hand side is evaluated twice at the same time. It is therefore some sort of "predictor–corrector" scheme. In the following, these two schemes are therefore referred to as "symmetric" and "predictor–corrector" schemes. Yet another possibility is to require that inhomogeneous equations of the form $\dot{u} = t^n$ with $n = 1$ and 2 are solved exactly. Such schemes are abbreviated as "inhomogeneous" schemes. The detailed method of calculating the coefficients for such 3rd-order Runge–Kutta schemes with $2N$-storage is discussed in detail in Appendix B. Several possible sets of coefficients are listed in Table 9.1 and compared with the favorite scheme of Williamson (1980). Note that the 1st-order Euler scheme corresponds to $\beta_1 = 1$ and the classic second-order to $\alpha_2 = -1/2, \beta_1 = 1/2$, and $\beta_2 = 1$.

We estimate the accuracy of these schemes by solving the homogeneous differential equation

$$\dot{u} = nu^{1-1/n}, \quad u(1) = 1. \tag{9.55}$$

The exact solution is $u = t^n$. In Table 9.2 we list the rms error with respect to the exact solution, for the range $1 < t \le 4$ and fixed timestep $\delta t = 0.1$ using $n = -1, 2$, or 3.

The length of the timestep must always be a certain fraction of the Courant–Friedrich–Levy condition, i.e. $\delta t = k_{CFL}\delta x / U_{max}$, where $k_{CFL} = \mathcal{O}(1)$ and U_{max} is the maximum transport speed in the system (taking into account advection, sound waves, viscous transport, etc.). Too long a timestep can not only lead to instability, but it also increases the error.

In Table 9.3 we give amplitude and phase errors for the various schemes. The most important conclusion to be drawn from this is the fact that low-order spatial schemes result in large *phase errors*. In the case of a 2nd-order scheme the phase error is $36°$ after a single passage of a barely resolved wave through a periodic mesh. Higher-order schemes have easily a hundred times smaller phase errors. The amplitude error, on the other hand, is virtually not affected by the spatial order of the scheme. The amplitude error is mainly affected both by the temporal order of the scheme and by the length of the timestep; see also Table 9.4. Therefore,

Table 9.2 Errors (in units of 10^{-6}) for different 2N-RK3 schemes, obtained by solving (9.55) in the range $1 < t \leq 4$ with $\delta t = 0.1$ and different values of n. Minimum values in each row are indicated in bold

Label	$n = -1$	$n = 2$	$n = 3$
Symmetric (i)	69	103	193
Symmetric (ii)	226	119	411
Predictor–corrector	469	346	1068
Inhomogeneous	84	**6**	**97**
Quadratic	197	94	339
Williamson (1980)	**68**	10	123
For comparison: RK3	66	13	134

Table 9.3 Amplitude and phase errors for inviscid advection of the function $f = \cos kx$ with $k = 2\pi$ and $N = 8$ meshpoints after $t = 20$, corresponding to 20 revolutions in a periodic mesh. The amplitude error is counted positive when the amplitude decreases. A positive phase error means that the solution lags behind the theoretical one

δt scheme	2nd-order	4th-order	6th-order	10th-order	Spectral
RK4	1%	2%	2%	2%	2%
$c_{\delta t} = 0.6$	720°	87°	13°	2.9°	2.8°
RK3	10%	14%	14%	15%	15%
$c_{\delta t} = 0.4$	716°	83°	8°	−2.1°	−2.3°
RK3	4%	6.3%	6.6%	6.6%	6.6%
$c_{\delta t} = 0.3$	717°	84°	10°	−0.6°	−0.8°

Table 9.4 Dependence of the amplitude and phase errors on the length of the timestep and the scheme used for the timestep. In all cases spectral x-derivatives are used

δt scheme $c_{\delta t}$	RK3 0.3	RK3 0.4	RK4 0.4	RK4 0.6	RK4 1.0
Amplitude error	6.6%	15%	0.3%	2%	21%
Phase error	−0.8°	−2.3°	0.6°	2.8°	18°

high-order schemes with low dissipation and dispersion are particularly important in computational acoustics (Stanescu and Habashi, 1998). However, in applications to turbulence a certain amount of viscosity is always necessary. This would decrease the amplitude of the wave further and would eventually be even more important. (This additional viscosity could be the real one, an explicit artificial, or an implicit numerical viscosity that would result from the discretisation error or the numerical scheme; see Section 9.3.2.)

A common criticism of high-order schemes is their tendency to produce Gibbs phenomena (ripples) near discontinuities. Consequently one needs a small amount of diffusion to damp out the modes near the Nyquist frequency. Thus, one needs to replace (9.50) by the equation

$$\dot{f} = -uf' + vf''. \tag{9.56}$$

The question is now how much diffusion is necessary, and how this depends on the spatial order of the scheme.

A perfect step function would produce large start-up errors; it is better to use a smoothed profile, e.g. one of the form

$$f(x) = \tanh\left(\frac{\cos x}{\delta x}\right),$$ (9.57)

where δx is the mesh width. For a periodic mesh of length L one would obviously use $f = f(kx)$, where $k = 2\pi/L$. In that case the step width would be $k\delta x$. In the following we consider a periodic domain of size $L_x = 1$ with $N_x = 60$ meshpoints, so we use $k = 2\pi/L_x = 2\pi$.

In Fig. 9.2 we plot the result of advecting the periodic step-like function, $f(kx)$, over five wavelengths, corresponding to a time $T = L/u$. The goal is to find the minimum diffusion coefficient ν necessary to avoid wiggles in the solution. In the first two panels one sees that for a 6th-order scheme the diffusion coefficient has to be approximately $\nu = 0.01\, u\delta x$. For $\nu = 0.005\, u\delta x$ there are still wiggles. For a 10th-order scheme one can still use $\nu = 0.005\, u\delta x$ without producing wiggles, while for a spectral scheme of nearly infinite order one can go down to $\nu = 0.002\, u\delta x$ without any problems.

We may thus conclude that all these schemes need some diffusion, but that the diffusion coefficient can be much reduced when the spatial order of the scheme is high. In that sense it is therefore not true that high-order schemes are particularly vulnerable to Gibbs phenomena, but rather the contrary!

In Fig. 9.3 we compare the corresponding results of advection tests for 2nd- and 4th-order schemes with the 6th-order scheme. It is evident that a 2nd-order scheme requires a relatively high diffusion coefficient, typically around $\nu = 0.05\, u\delta x$, but this leads to rather unacceptable distortions of the original profile. (It may be noted that, if one uses at the same time a 1st-order temporal scheme, which has antidiffusive properties, and a timestep which is not too short, then the antidiffusive error of the timestep scheme would partially compensate the actual diffusion and one could reduce the value of ν, but this would be a matter of tuning and hence not generally useful for arbitrary profiles.)

9.3.4. Burgers equation

In the special case where the velocity itself is being advected, i.e. $f = u$, (9.56) turns into the Burgers equation,

$$\dot{u} = -uu' + \nu u''.$$ (9.58)

In one dimension there is an analytic solution for a kink,

$$u = U\left(1 - \tanh\frac{x - Ut}{2\delta}\right),$$ (9.59)

where $\delta = \nu/U$ is the shock thickness (e.g. Dodd *et al.*, 1982). Note that the amplitude of the kink is twice its propagation speed. Expressed in terms of the Reynolds number, $Re = UL/\nu$, we have $\delta = L Re^{-1}$. (We note in passing that the dissipative cutoff scale in ordinary turbulence is somewhat larger; $\delta = L Re^{-3/4}$.)

In order to have a stationary shock we use the initial condition

$$u = -U \tanh(x/2\delta).$$ (9.60)

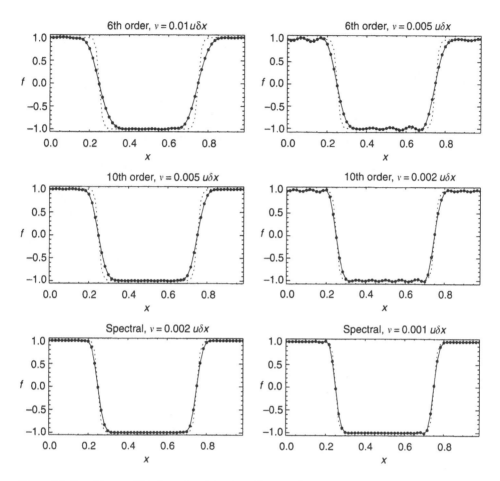

Figure 9.2 Resulting profile after advecting a step-like function five times through the periodic mesh. The dots on the solid line give the location of the function values at the computed meshpoints and the dotted line gives the original profile. For the panels on the right-hand side the diffusion coefficient is too small and the profile shows noticeable wiggles. $\delta x = 1/60$.

In Fig. 9.4 we present numerical solutions using the 6th-order explicit scheme with different values of the mesh Reynolds number, $\delta x\, U/v$, which was varied by changing the value of v. Here we used $N_x = 100$ meshpoints in the range $-1 \leq x \leq 1$. Note that the overall error, defined here as $\max(|f - f_{\text{exact}}|)$, decreases with decreasing mesh width like δx^5.

The test cases considered so far were not directly related to the Navier–Stokes equation, e.g. which permits sound waves that can pile up to form shocks. This will be considered in the next section.

9.3.5. Shock tube tests

A popular test problem for compressible codes is the shock tube problem of Sod (1951). On the one hand, one can assess the sharpness of the various fronts. On the other hand, and

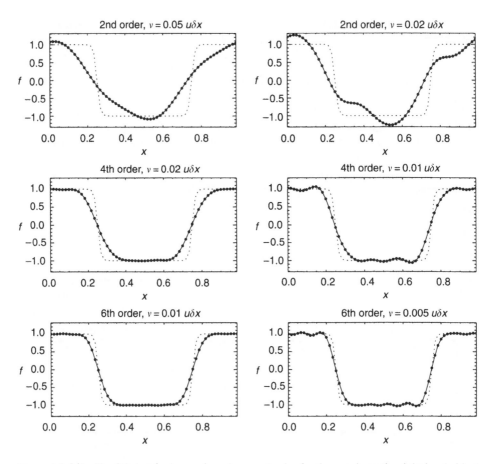

Figure 9.3 Like Fig. 9.2, but for low-order schemes. Again, for the panels on the right-hand side the diffusion coefficient is too small and the profile shows noticeable wiggles. For the 2nd-order scheme one needs a viscosity of $\nu = 0.05\,u\delta x$ to prevent wiggles, but then the resulting distortion of the original profile becomes rather unacceptable. $\delta x = 1/60$.

perhaps most importantly, it allows one to test the conversion of kinetic energy to thermal energy via viscous heating.

In the following we use the formulation of the compressible Navier–Stokes equations in terms of entropy and enthalpy (9.16)–(9.18). We use units where $p_0 = \rho_0 = c_p = 1$ and adopt the abbreviations $\Lambda = \ln\rho$ (not to be confused with the cooling function used in Section 9.2). In one dimension (with $\nu = $ const) these equations reduce to

$$\dot{u} = -uu' - c_s^2(\Lambda' + s') + \tilde{\nu}(u'' + \Lambda' u'), \tag{9.61}$$

$$\dot{s} = -us' + (\gamma - 1)\tilde{\nu}u'^2/c_s^2 + Q_s, \tag{9.62}$$

$$\dot{\Lambda} = -u\Lambda' - u', \tag{9.63}$$

where dots and primes refer respectively to time and space derivatives, Q_s describes the change of entropy due to radiative diffusion, and $\Lambda = \ln\rho$ is the logarithmic density. In

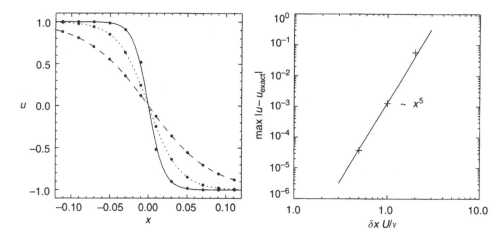

Figure 9.4 Numerical solution of the Burgers equation using the 6th-order explicit scheme with (9.60) as initial condition. In the left-hand panel the lines give the exact solution (9.60) and the dots give the numerical solution for the corresponding value of the mesh Reynolds number: $\delta x\, U/\nu = 0.5$ (solid line), 1.0 (dotted line), and 2.0 (dashed line). In the right-hand panel the scaling of the error with δx is shown.

(9.61)–(9.63) we have used the abbreviation

$$c_s^2 = \gamma p/\rho = \gamma \exp[(\gamma - 1)\Lambda + \gamma s] \qquad (9.64)$$

for the adiabatic sound speed squared, and $\tilde{\nu} = \frac{4}{3}\nu$ is the effective viscosity for compressive motions. This 4/3 factor comes from the fact that in one-dimensional

$$\mathbf{S} = \text{diag}\left(\tfrac{2}{3}, -\tfrac{1}{3}, -\tfrac{1}{3}\right) u_{x,x}, \qquad (9.65)$$

and therefore

$$\frac{1}{\rho}\nabla\cdot(2\nu\rho\mathbf{S}) = \frac{4}{3}\nu[u_{x,xx} + (\ln\rho)_{,x}u_{x,x}], \qquad (9.66)$$

so $\mathbf{S}^2 = \frac{2}{3}u_{x,x}^2$, or $2\nu\mathbf{S}^2 = \tilde{\nu}u_{x,x}^2$. In the radiative diffusion approximation we have $Q_s = -\Lambda_{\text{cond}}/(c_p T)$, and so (9.23) gives in one dimension

$$Q_s = \chi\gamma[s'' + \nabla_{\text{ad}}\Lambda'' + \gamma(\Lambda' + s')(s' + \nabla_{\text{ad}}\Lambda')] \qquad (9.67)$$

In Fig. 9.5 we show the solution for an initial density and pressure jump of $1:10$ and the viscosity is now $\nu = 0.6\delta x\, c_s$. In this case a small amount of thermal diffusion (with Prandtl number $\chi/\nu = 0.05$) has been adopted to remove wiggles in the entropy.

For stronger shocks velocity and entropy excess increase; see Figs. 9.6 and 9.7, where the initial pressure jumps are $1:100$ and $1:1000$, respectively, and the viscosities are chosen to be $\nu = 1.6\delta x\, c_s$ and $\nu = 2.4\delta x\, c_s$. (For the stationary shock problem considered below we also find that the viscosity must increase with the Mach number and, moreover, that the two should be proportional to each other.) In the cases shown in Figs. 9.6 and 9.7 we were able to put $\chi = 0$ without getting any wiggles in s. However, in the case of strong shocks

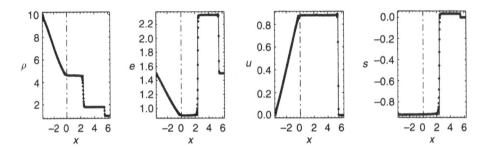

Figure 9.5 Standard shock tube test with an initial pressure jump of $1:10$ and $\nu = 0.6\delta x\, c_s$ and $\chi/\nu = 0.05$. The solid line indicates the analytic solution (in the limit $\tilde{\nu} \to 0$) and the dots the numerical solution. Note the small entropy excess on the right of the initial entropy discontinuity. $t = 2.7$.

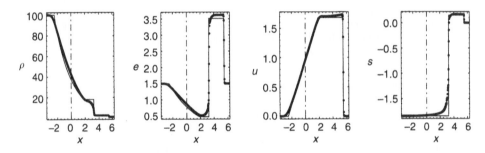

Figure 9.6 Same as Fig. 9.5, but for an initial pressure jump of $1:100$ and $\nu = 1.6\delta x\, c_s$. $t = 1.9$.

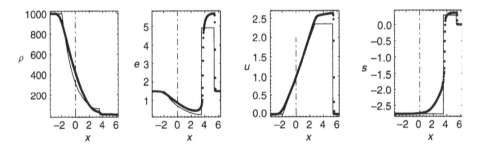

Figure 9.7 Same as Fig. 9.6, but for an initial pressure jump of $1:1000$ and $\nu = 2.4\delta x\, c_s$. $t = 1.5$.

(pressure ratio $1:1000$) the discrepancy between numerical and analytical solutions becomes quite noticeable.

In many practical applications shocks occur only in a small portion of space. One can therefore reduce the viscosity outside shocks or, conversely, use a small viscosity everywhere except in the locations of shock, i.e. where the flow is convergent (negative divergent). This leads to the concept of an artificial (Neumann–Richtmyer) shock viscosity,

$$\nu_{shock} = c_{shock}\delta x^2 \, \langle(-\nabla \cdot \mathbf{u})_+\rangle_{n.n.},\qquad(9.68)$$

where $\langle\ldots\rangle_{n.n.}$ indicates averaging over nearest neighbors and the subscript $+$ means that only the positive part is taken.

The last panel in Figs. 9.6 and 9.7 shows quite clearly how the entropy increases behind the shock. This entropy increase is just a consequence of the viscous heating term, $\tilde{\nu}u'^2/h$. Without this term the solution would obviously be wrong everywhere behind the shock, especially when the shock is strong.

A somewhat simpler situation is encountered with standing shocks. In Fig. 9.8 we give an example of a numerically determined solution at Ma $= 100$. The agreement in the jump for the numerically determined solution (dotted line and dots) and the theoretical solution (solid lines) is very good, although the position of the jump has moved away somewhat from the initial location ($x = 0$), but this is merely a consequence of having used more-or-less arbitrarily a tanh profile to smooth the initial jumps. After some initial adjustment phase the profiles do indeed remain stationary. Note also, however, that the entropy profile is slightly shifted relative to the profiles of u and Λ.

It is interesting to note that when solving the Rankine–Hugoniot jump conditions for shocks one is allowed to use the inviscid equations provided they are written in conservative form. Sometimes one finds in the literature the inviscid Navier–Stokes equations written in non-conservative form. This is not strictly correct, because without viscosity there would be no viscous heating and hence no entropy increase behind the shock. Moreover, it is quite common to consider a polytropic equation of state, $p = K\rho^\Gamma$. Again, in this case the entropy is constant, and so energy conservation is violated. Nevertheless, given that polytropic equations of state are often considered in astrophysics we consider this case in more detail in the next subsection.

9.3.6. *Polytropic and isothermal shocks*

For polytropes with $p = K\rho^\Gamma$, but $\Gamma \neq \gamma$ in general, we can write

$$-\nabla h + h\nabla s = -\frac{1}{\rho}\nabla p = -\nabla\left(\frac{\Gamma K}{\Gamma - 1}\rho^{\Gamma-1}\right) \equiv -\nabla\tilde{h},\qquad(9.69)$$

so we can introduce a pseudo enthalpy $\tilde{h}$ as

$$\tilde{h} = \frac{\Gamma K}{\Gamma - 1}\rho^{\Gamma-1} = \left[\left(1 - \frac{1}{\gamma}\right)\Big/\left(1 - \frac{1}{\Gamma}\right)\right]h.\qquad(9.70)$$

This is consistent with a fixed entropy dependence, where s only depends on ρ like

$$s = s(\rho) = \frac{1}{\gamma}\ln\left(p/\rho^\gamma\right) = \frac{1}{\gamma}\ln\left(K\,\rho^{\Gamma-\gamma}\right) = \frac{1}{\gamma}\ln K + \left(\frac{\Gamma}{\gamma} - 1\right)\ln\rho,\qquad(9.71)$$

which implies that in the polytropic case (9.62) is discarded. In the adiabatic case, $\Gamma = \gamma$, entropy is constant. In the isothermal case, $p = c_s^2\rho$, we have $\Gamma \to 1$, so entropy is not constant, but it varies only in direct relation to $-\ln\rho$ and not as a consequence of viscous heating behind the shock.

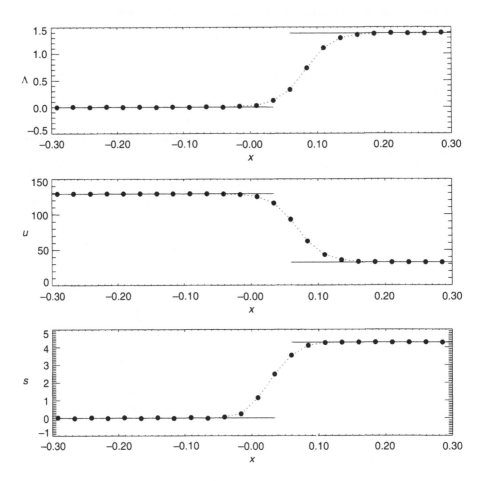

Figure 9.8 Example of a very strong standing shock with Ma $= 100$. Note the relative shift of the position where s increases relative to where $\Lambda = \ln \rho$ increases. The viscosity is chosen to be $\nu = \text{Ma} \times \delta x$.

In deriving the Rankine–Hugionot jump conditions one uses the conservation of mass, momentum, and energy in a comoving frame, where the following three quantities are constants of motion:

$$J = \rho u, \quad I = \rho u^2 + p, \quad E = \frac{1}{2} f u^2 + \frac{\gamma}{\gamma - 1} \frac{p}{\rho}. \tag{9.72}$$

The values of these three constants can be calculated when all three variables, u, p, and ρ, are known on one side of the shock. For polytropic equations of state, with $p = K\rho^\gamma$, the energy equation is no longer used, so there are only the following two conserved quantities,

$$J = \rho u, \quad I = \rho u^2 + K\rho^\gamma. \tag{9.73}$$

The dependence of the velocity, density, pressure, and entropy jumps on the upstream Mach number is plotted in Fig. 9.9 for the case $\gamma = 5/3$ and compared with the polytropic case using $\Gamma = \gamma$ (Fig. 9.10).

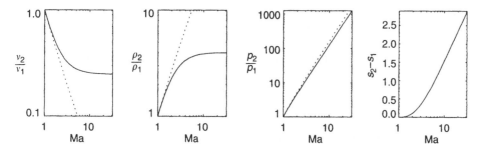

Figure 9.9 The dependence of the velocity, density, pressure, and entropy jumps on the upstream Mach number for the case $\gamma = 5/3$ (solid line) and comparison with the polytropic case using $\Gamma = \gamma$ (dotted line). Note that the velocity and density jumps saturate at 1/4 and 4, respectively, while there is no such saturation for the polytropic shock.

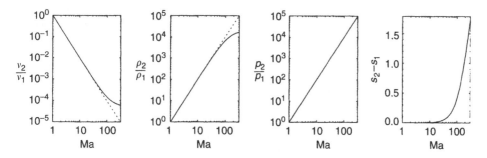

Figure 9.10 Like Fig. 9.9, but for $\gamma = \Gamma = 1.0001$. Note that the velocity and density jumps saturate at much more extreme values than for $\gamma = 5/3$. Thus, the results using polytropic and ideal gas equations agree up to much larger Mach numbers than for $\gamma = 5/3$.

Note that the pressure jump, p_2/p_1, is almost independent of the value of γ and does also not significantly depend on the polytropic assumption.

9.4. Non-uniform and lagrangian meshes

In many cases it is useful to consider non-uniform meshes, either by adding more points in places where large gradients are expected, or by letting the points move with the flow (lagrangian mesh). The lagrangian mesh is particularly useful in one-dimensional cases, because then the mesh topology (i.e. the ordering of mesh points) remains unchanged. This approach is related to the method of Smooth Particle Hydrodynamics; see Maron and Howes (2001) for references and an improvement of this method. Another method that gains constantly in popularity is adaptive mesh refinement (e.g. Grauer *et al.*, 1998), which will not be discussed here.

9.4.1. Non-uniform topologically cartesian meshes

The implementation of non-uniform meshes can be relatively easy when each of the new coordinates depend on only one variable, e.g. when $x = x(\tilde{x})$, $y = y(\tilde{y})$, and $z = z(\tilde{z})$. Here, $\tilde{x}$, $\tilde{y}$, and $\tilde{z}$ are cartesian coordinates on a uniform mesh. In the more general case, however, we have

$$x = x(\tilde{x}, \tilde{y}, \tilde{z}), \quad y = y(\tilde{x}, \tilde{y}, \tilde{z}), \quad z = z(\tilde{x}, \tilde{y}, \tilde{z}), \tag{9.74}$$

so that x, y, and z derivatives of a function f can be calculated using the chain rule,

$$\frac{\partial f}{\partial x} = \frac{\partial f}{\partial \tilde{x}}\frac{\partial \tilde{x}}{\partial x} + \frac{\partial f}{\partial \tilde{y}}\frac{\partial \tilde{y}}{\partial x} + \frac{\partial f}{\partial \tilde{z}}\frac{\partial \tilde{z}}{\partial x} \equiv \frac{\partial f}{\partial \tilde{x}_i}\frac{\partial \tilde{x}_i}{\partial x}. \tag{9.75}$$

Corresponding formulae apply obviously for the other two directions, so in general we can write

$$\frac{\partial f}{\partial x_j} = \frac{\partial f}{\partial \tilde{x}_i}J_{ij}, \quad \text{where} \quad J_{ij} = \frac{\partial \tilde{x}_i}{\partial x_j} \tag{9.76}$$

is the jacobian of this coordinate transformation. This method allows one to have high resolution, e.g. near a central object, without however having high resolution anywhere else far away from the central object. This is useful in connection with outflows from jets.

We discuss here one particular application that is relevant for simulating flows in a sphere. It is possible to transform a cartesian mesh to cover a sphere without a coordinate singularity. It will turn out, however, that there is a discontinuity in the jacobian. We discuss this here in two-dimension. We denote the coordinate mesh by a tilde, so $(\tilde{x}, \tilde{y})$ are the coordinates in a uniform cartesian mesh. We want to stretch the mesh such that points on the $\tilde{x}$ and $\tilde{y}$ axes are not affected, and that the distance of points on the diagonal is reduced by a factor $1/\sqrt{2}$ (or by $1/\sqrt{3}$ in three-dimension). This can be accomplished by introducing new coordinates (x, y) as

$$\begin{pmatrix} x \\ y \end{pmatrix} = \begin{pmatrix} \tilde{x} \\ \tilde{y} \end{pmatrix} \frac{(\tilde{x}^n + \tilde{y}^n)^{1/n}}{(\tilde{x}^2 + \tilde{y}^2)^{1/2}}, \tag{9.77}$$

where n is a large even number. In the limit $n \to \infty$ we may substitute

$$(\tilde{x}^n + \tilde{y}^n)^{1/n} \to \max(|\tilde{x}|, |\tilde{y}|). \tag{9.78}$$

Examples of the resulting meshes for two different values of n are given in Fig. 9.11.

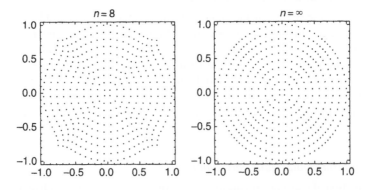

Figure 9.11 Examples of the resulting meshes for $n = 8$ and $n \to \infty$ in which case (9.78) is used.

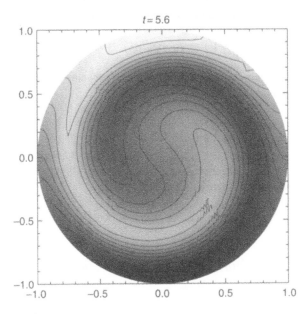

Figure 9.12 Example of an advection experiment on a $n = 8$ mesh.

In order to obtain the jacobian of this transformation, $\partial \tilde{x}_i / \partial x_j$, we have to consider separately the cases $\tilde{x} \geq \tilde{y}$ and $\tilde{x} \leq \tilde{y}$. The derivation is given in Appendix C, and is most
concisely expressed in terms of the logarithmic derivative, so

$$
\begin{pmatrix}
\dfrac{\partial \ln \tilde{x}}{\partial \ln x} & \dfrac{\partial \ln \tilde{x}}{\partial \ln y} \\[2mm]
\dfrac{\partial \ln \tilde{y}}{\partial \ln x} & \dfrac{\partial \ln \tilde{y}}{\partial \ln y}
\end{pmatrix}
=
\begin{pmatrix}
+1 - \left(\dfrac{\tilde{y}}{\tilde{r}}\right)^2 & +1 - \left(\dfrac{\tilde{x}}{\tilde{r}}\right)^2 \\[3mm]
-1 + \left(\dfrac{\tilde{x}}{\tilde{r}}\right)^2 & +1 + \left(\dfrac{\tilde{y}}{\tilde{r}}\right)^2
\end{pmatrix}
\qquad \text{if } |\tilde{x}| \geq |\tilde{y}|,
\tag{9.79}
$$

$$
\begin{pmatrix}
\dfrac{\partial \ln \tilde{x}}{\partial \ln x} & \dfrac{\partial \ln \tilde{x}}{\partial \ln y} \\[2mm]
\dfrac{\partial \ln \tilde{y}}{\partial \ln x} & \dfrac{\partial \ln \tilde{y}}{\partial \ln y}
\end{pmatrix}
=
\begin{pmatrix}
+1 + \left(\dfrac{\tilde{x}}{\tilde{r}}\right)^2 & -1 + \left(\dfrac{\tilde{y}}{\tilde{r}}\right)^2 \\[3mm]
+1 - \left(\dfrac{\tilde{y}}{\tilde{r}}\right)^2 & +1 - \left(\dfrac{\tilde{x}}{\tilde{r}}\right)^2
\end{pmatrix}
\qquad \text{if } |\tilde{x}| \leq |\tilde{y}|,
\tag{9.80}
$$

where $\tilde{r}^2 = \tilde{x}^2 + \tilde{y}^2$. Note that the jacobian is discontinuous on the diagonals. This is
a somewhat unfortunate feature of this transformation. It is not too surprising however that
something like this happens, because the diagonals are the locations where a rotating flow
must turn direction by 90° in the coordinate mesh. Nevertheless, it is possible to obtain reasonably well behaved solutions; see Fig. 9.12 for an advection experiment using a prescribed
differentially rotating flow.

The fluid equations are still solved in rectangular cartesian coordinates, so, e.g. the equation
$Ds/Dt = 0$ is solved in the form

$$
\frac{\partial s}{\partial t} = -u_x \frac{\partial s}{\partial x} - u_y \frac{\partial s}{\partial y},
\tag{9.81}
$$

where the spatial derivatives are evaluated according to (9.75). For the velocity field, stress-free boundary conditions, e.g. would be written in the form

$$\hat{r}_j u_j = 0, \quad \hat{\phi}_i S_{ij} \hat{r}_j = 0, \tag{9.82}$$

where S_{ij} is the rate of strain tensor, $\hat{r}_j$ and $\hat{\phi}_i$ are the cartesian components ($i = x, y, z$) of the radial and azimuthal unit vectors, i.e.

$$\hat{\mathbf{r}} = \frac{1}{r}\begin{pmatrix} x \\ r \end{pmatrix} y \quad \text{and} \quad \hat{\boldsymbol{\phi}} = \frac{1}{r}\begin{pmatrix} -y \\ x \end{pmatrix} \tag{9.83}$$

are unit vectors in the r and ϕ directions and $r = \sqrt{x^2 + y^2}$ is the distance from the rotation axis. The stress-free boundary conditions are then

$$x u_x + y u_y = 0 \tag{9.84}$$

and

$$(x^2 - y^2)(u_{x,y} + u_{y,x}) - 2xy(u_{x,x} - u_{y,y}) = 0. \tag{9.85}$$

9.4.2. Lagrangian meshes

We now consider a simple one-dimensional lagrangian mesh problem. Assume that ℓ labels the particle, then the lagrangian derivative is

$$\frac{Ds}{Dt} \equiv \left(\frac{\partial s}{\partial t}\right)_{\ell=\text{const}} = \left(\frac{\partial s}{\partial t}\right)_{x=\text{const}} + \left(\frac{\partial x}{\partial t}\right)_{\ell=\text{const}} \frac{\partial s}{\partial x}. \tag{9.86}$$

Now, because

$$\left(\frac{\partial x}{\partial t}\right)_{\ell=\text{const}} = \frac{Dx}{Dt} = u, \tag{9.87}$$

we have the well-known equation

$$\frac{Ds}{Dt} = \frac{\partial s}{\partial t} + u\frac{\partial s}{\partial x}. \tag{9.88}$$

As an example we now consider the Burgers equation,

$$\frac{Du}{Dt} = \tilde{\nu}\frac{\partial^2 u}{\partial x^2}. \tag{9.89}$$

We now take $u(x, t) = u(\ell(x), t)$ to be a function of the coordinate variable ℓ which, in turn, is a function of x. The x-derivatives are obtained using the chain rule, i.e.

$$\frac{\partial u}{\partial x} = \frac{\partial \ell}{\partial x}\frac{\partial u}{\partial \ell} = \frac{u'}{x'}, \tag{9.90}$$

and likewise for the second derivative

$$\frac{\partial^2 u}{\partial x^2} = \frac{u''x' - u'x''}{x'^3}. \tag{9.91}$$

Thus, the Burgers equation can then be written as

$$\frac{\partial u}{\partial t} = \tilde{\nu}\,\frac{u''x' - u'x''}{x'^3},$$

(9.92)

where the x variable is given by

$$\frac{\partial x}{\partial t} = u.$$

(9.93)

A solution of these two equations is given in Fig. 9.13.

In the test problem above the initial meshpoint distribution was uniform. Although this is not quite suitable for this problem, it shows that subsequently the mesh spacing became narrower still, which means that the timestep in now governed by viscosity, $\delta t \leq 0.06\delta x_{\min}^2/\tilde{\nu}$, where the numerical factor is empirical. However, the mesh spacing does not need to be governed by (9.93), so it is quite possible to come up with other prescriptions for the mesh spacing.

Consider as another example the isothermal eulerian equations

$$\frac{Du}{Dt} = -c_s^2\frac{\partial \ln \rho}{\partial x} + \tilde{\nu}\left(\frac{\partial^2 u}{\partial x^2} + \frac{\partial \ln \rho}{\partial x}\frac{\partial u}{\partial x}\right),$$

(9.94)

$$\frac{D \ln \rho}{Dt} = -\frac{\partial u}{\partial x}.$$

(9.95)

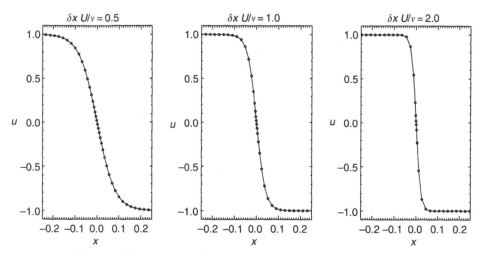

Figure 9.13 Solution of the Burgers equation using a lagrangian mesh combined with a 6th-order explicit scheme. The values of the mesh Reynolds number vary between $\delta x\, U/\nu = 0.5$ and 2.0, where δx refers to the initially uniform mesh spacing.

In lagrangian form they take the form

$$\frac{\partial u}{\partial t} = -c_s^2 \frac{(\ln \rho)'}{x'} + \tilde{\nu} \left(\frac{u''x' - u'x''}{x'^3} + \frac{(\ln \rho)'}{x'} \frac{u'}{x'} \right), \qquad (9.96)$$

$$\frac{\partial \ln \rho}{\partial t} = -\frac{u'}{x'}, \qquad (9.97)$$

$$\frac{\partial x}{\partial t} = u. \qquad (9.98)$$

The example above demonstrates clearly the problem that lagrangian mesh points can continue to pile up near convergence points of the flow. This is a general problem with fully lagrangian schemes. One possible alternative is to use *lagrangian-eulerian* schemes (e.g. Benson, 1992; Peterkin *et al.*, 1998; Arber *et al.*, 2001), which combine the advantages of lagrangian and eulerian codes, but involve obviously some kind of interpolation. Another alternative is to use a semi-lagrangian code which advects the mesh points not with the actual gas velocity **u**, but with a more independent mesh velocity **U**. Clearly, we want to avoid too small distances between neighboring points, so one could artificially lower the effective mesh velocity by involving, e.g. the modulus of the jacobian, $|\mathbf{J}|$, which becomes large when the concentration of mesh points is high. Thus, one could choose, e.g. $\mathbf{U} = \mathbf{u}/(1 + |\mathbf{J}|)$. In the present case, $|\mathbf{J}| = |x'|^{-1}$. In the following we discuss the formalism that needs to be invoked in order to calculate first and second derivatives on an advected mesh.

9.4.3. Non-lagrangian mesh advection

The main advantage of a lagrangian mesh is that it allows higher resolution locally. Another advantage, which is however less crucial, is that the nonlinear advection term drops out. The main disadvantage is however that a lagrangian mesh may become too distorted and overconcentrated, as seen in the previous section. In this subsection we address the possibility of advecting the mesh with a velocity **U** that can be different from the fluid velocity. This way one can remove the swirl of the mesh by taking a velocity that is the gradient of some other quantity, i.e.

$$\mathbf{U} = -\nabla \Phi_{\text{mesh}}, \qquad (9.99)$$

where Φ_{mesh} should be large in those regions where many points are needed. One possible criterion would be to require that the number of scale heights per meshpoint, $|\delta \mathbf{x} \cdot \nabla \ln \rho|$, does not exceed an empirical value of $1/3$, say. Thus, $3|\delta \mathbf{x} \cdot \nabla \ln \rho| < 1$ would be a necessary condition. Another possibility would be to let Φ_{mesh} evolve itself according to some suitable advection–diffusion equation. However, no generally satisfactory method seems to be available as yet. In order to calculate the jacobian for the coordinate transformation one can make use of the fact that the mesh evolves only gradually from one timestep to the next. For a more extended discussion of mesh advection schemes we refer to the article by Dorfi in the book by LeVeque *et al.* (1998).

9.4.3.1. Calculating the jacobian

Initially, at $t = 0$, we have $\mathbf{x} = \tilde{\mathbf{x}}$. After the nth timestep, at $t = n\delta t$, we calculate the new $\mathbf{x}$-mesh, $\mathbf{x}^{(n+1)}$, from the previous one, $\mathbf{x}^{(n)}$, i.e.

$$\mathbf{x}^{(n+1)} = \mathbf{x}^{(n)} + \mathbf{U}\left(\mathbf{x}^{(n)}, t\right) \delta t. \tag{9.100}$$

Here, $\mathbf{x}^{(0)} = \tilde{\mathbf{x}}$ is just the original coordinate mesh. Differentiating the ith component with respect to the jth component, as we have done in Section 9.4.1, we obtain

$$\delta_{ij} = \frac{\partial x_i^{(n+1)}}{\partial x_j^{(n+1)}} = \frac{\partial x_i^{(n)}}{\partial x_j^{(n+1)}} + \frac{\partial U_i^{(n)}}{\partial x_j^{(n+1)}} \delta t, \tag{9.101}$$

where $U_i^{(n)} = U_i\left(\mathbf{x}^{(n)}, t\right)$. In the expression above we have U_i on the mesh $\mathbf{x}^{(n)}$, but we need to differentiate with respect to the new mesh $\mathbf{x}^{(n+1)}$. This can be fixed by another factor $\partial x_i^{(n)}/\partial x_j^{(n+1)}$. Thus, we have

$$\delta_{ij} = \frac{\partial x_i^{(n)}}{\partial x_j^{(n+1)}} + \frac{\partial x_k^{(n)}}{\partial x_j^{(n+1)}} \frac{\partial U_i\left(\mathbf{x}^{(n)}, t\right)}{\partial x_k^{(n)}} \delta t = \left[\delta_{ik} + \frac{\partial U_i^{(n)}}{\partial x_k^{(n)}} \delta t\right] \frac{\partial x_k^{(n)}}{\partial x_j^{(n+1)}}. \tag{9.102}$$

This can be written in matrix form,

$$\delta_{ij} = M_{ik} J_{kj}^{(n)}, \tag{9.103}$$

where

$$M_{ik} = \delta_{ik} + \frac{\partial U_i\left(\mathbf{x}^{(n)}, t\right)}{\partial x_k^{(n)}} \delta t \tag{9.104}$$

is a transformation matrix and

$$J_{ij}^{(n)} = \frac{\partial x_i^{(n)}}{\partial x_j^{(n+1)}} \tag{9.105}$$

is the incremental jacobian, so $\mathbf{J}^{(n)} = \mathbf{M}^{-1}$. To obtain the jacobian at $t = 2\delta t$, e.g. we calculate

$$\frac{\partial x_i^{(0)}}{\partial x_j^{(2)}} = \frac{\partial x_i^{(0)}}{\partial x_k^{(1)}} \frac{\partial x_k^{(1)}}{\partial x_j^{(2)}} = J_{ik}^{(0)} J_{kj}^{(1)} \equiv \left(\mathbf{J}^{(0)} \mathbf{J}^{(1)}\right)_{ij}. \tag{9.106}$$

The jacobian at $t = n\delta t$ is then obtained by successive matrix multiplication from the right, so

$$J_{ij}^{(0 \to n+1)} = J_{ik}^{(0 \to n)} J_{kj}^{(n)}, \tag{9.107}$$

where $J_{ij}^{(0 \to n+1)}$ and $J_{ij}^{(0 \to n)}$ are the full (as opposed to incremental) jacobians at the new and previous timesteps, respectively.

9.4.3.2. Calculating the 2nd-order jacobian

A corresponding calculation (see Appendix D) for the second derivatives of a function f shows that

$$\frac{\partial^2 f}{\partial x_i \partial x_j} = \frac{\partial^2 f}{\partial \tilde{x}_p \partial \tilde{x}_q} J_{pi} J_{qj} + \frac{\partial f}{\partial \tilde{x}_k} K_{kij}, \tag{9.108}$$

where

$$K_{kij} = \frac{\partial^2 \tilde{x}_k}{\partial x_i \partial x_j} \tag{9.109}$$

is the 2nd-order jacobian. Like for the first derivative the 2nd-order jacobian can be obtained by successive tensor multiplication,

$$K_{kij}^{(0 \to n+1)} = K_{kpq}^{(0 \to n)} J_{pi}^{(n)} J_{qj}^{(n)} + J_{kl}^{(0 \to n)} K_{lij}^{(n)}, \tag{9.110}$$

where $K_{kij}^{(0 \to n+1)}$ and $K_{kij}^{(0 \to n)}$ are the 2nd-order jacobians at the new and previous timesteps, respectively, and

$$K_{kij}^{(n)} = \frac{\partial^2 x_k^{(n)}}{\partial x_i^{(n+1)} \partial x_j^{(n+1)}} \tag{9.111}$$

is the incremental 2nd-order jacobian, which is calculated at each timestep as

$$K_{kij}^{(n)} = -\delta t \left(\mathbf{M}^{-1} \right)_{kl} U_{l,pq} J_{pi}^{(n)} J_{qj}^{(n)}, \tag{9.112}$$

where $\mathbf{M}$ was defined in (9.104) and

$$U_{l,pq} = \frac{\partial^2 U_l \left(\mathbf{x}^{(n)}, t \right)}{\partial x_p^{(n)} \partial x_q^{(n)}} \tag{9.113}$$

is the 2nd-order velocity gradient matrix on the physical mesh. Since $\mathbf{M}^{-1} = \mathbf{J}^{(n)}$ is just the incremental jacobian, we can write (9.112) as

$$K_{kij}^{(n)} = -J_{kl}^{(n)} \delta t \, U_{l,pq} J_{pi}^{(n)} J_{qj}^{(n)}. \tag{9.114}$$

Since the expressions (9.110) and (9.114) involve both a multiplication with $J_{pi}^{(n)} J_{qj}^{(n)}$, we can simplify (9.110) to give

$$K_{kij}^{(0 \to n+1)} = \left[K_{kpq}^{(0 \to n)} - J_{kl}^{(0 \to n)} J_{lm}^{(n)} \delta t \, U_{m,pq} \right] J_{pi}^{(n)} J_{qj}^{(n)}. \tag{9.115}$$

Here the expression $J_{kl}^{(0 \to n)} J_{lm}^{(n)}$ is of course the new jacobian, $J_{kl}^{(0 \to n+1)}$.

So in summary, the new 1st- and 2nd-order jacobians are obtained from the previous ones via the formulae

$$J_{ij}^{(0\to n+1)} = J_{ik}^{(0\to n)} J_{kj}^{(n)}, \tag{9.116}$$

$$K_{kij}^{(0\to n+1)} = \left[K_{kpq}^{(0\to n)} - J_{kl}^{(0\to n+1)} \delta t \, U_{l,pq} \right] J_{pi}^{(n)} J_{qj}^{(n)}, \tag{9.117}$$

$$\frac{\partial f}{\partial x_i} = \frac{\partial f}{\partial \tilde{x}_p} J_{pi}, \tag{9.118}$$

$$\frac{\partial^2 f}{\partial x_i \partial x_j} = \frac{\partial^2 f}{\partial \tilde{x}_p \partial \tilde{x}_q} J_{pi} J_{qj} + \frac{\partial f}{\partial \tilde{x}_k} K_{kij}, \tag{9.119}$$

where $\mathbf{J} \equiv \mathbf{J}^{(0\to n+1)}$ and $\mathbf{K} \equiv \mathbf{K}^{(0\to n+1)}$ has been assumed.

Since now the mesh is moving in time with the local speed $\mathbf{U}$ which is different from the gas velocity $\mathbf{u}$, the lagrangian derivative is

$$\frac{D}{Dt} = \frac{\partial}{\partial t} + (\mathbf{u} - \mathbf{U}) \cdot \nabla. \tag{9.120}$$

In all other respects the basic equations, written in cartesian form, are still unchanged, provided all x, y, and z derivatives (first and second) are evaluated, as in (9.75) and (9.108), with the components of the jacobian. As an example we show in Fig. 9.14 the result of a kinematic collapse calculation where $Du/Dt = -\nabla\phi$ and $Ds/Dt = 0$ with a smoothed but localized gravitational potential ϕ. In Fig. 9.15 we compare the results of an eulerian and a lagrangian calculation using the same number of meshpoints. Already after some short time the eulerian calculation begins to become underresolved and develops wiggles while the lagrangian calculation proceeds without problems.

9.4.4. Unstructured meshes

We now discuss how we can calculate spatial derivatives of our variables from a non-uniformly spaced ensemble of points. Consider the function $f(x, y, z)$, which stand for one of the components of a vector (velocity or magnetic vector potential) or a scalar, such as $\ln \rho$. We approximate the function $f(x, y, z)$ in the neighborhood of the point $\mathbf{x}_i = (x_i, y_i, z_i)$ by a multidimensional polynomial of degree N,

$$f(x, y, z) = f(x_i, y_i, z_i) + \sum_{l+m+n \leq N} c_{lmn} (x - x_i)^l (y - y_i)^m (z - z_i)^n, \tag{9.121}$$

where l, m, and n are non-negative integers and c_{lmn} are coefficients that are to be determined separately for each point by applying (9.121) to all neighboring points $\mathbf{x}_j$. Note that $c_{000} = 0$ and does not need to be considered. Thus, for each point j we have a system of equations

$$f_{ij} = \sum_{l+m+n \leq N} \frac{1}{l! \, m! \, n!} c_{lmn} x_{ij}^l \, y_{ij}^m \, z_{ij}^n, \tag{9.122}$$

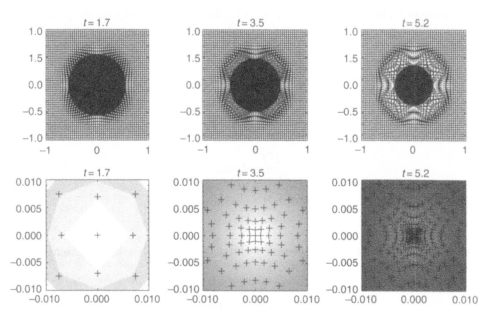

Figure 9.14 Example of a collapse calculation. The second row shows only the inner parts with $|x|, |y| \leq 0.01$ at the same times as in the upper row.

where $f_{ij} = f(x_i, y_i, z_i) - f(x_j, y_j, z_j)$ and $\mathbf{x}_{ij} = \mathbf{x}_i - \mathbf{x}_j$. This system of equations can be written in matrix form

$$f_\alpha = \mathsf{M}_{\alpha\beta} C_\beta, \tag{9.123}$$

where $1 \leq (\alpha, \beta) \leq M$ and M is the spatial dimension of the matrix, which is related to N and the dimension as follows:

$$M = \begin{cases} N & \text{in one-dimension,} \\ (N+1)(N+2)/2 - 1 & \text{in two-dimension,} \\ (N+1)(3N/2) & \text{in three-dimension.} \end{cases} \tag{9.124}$$

When $N = 2$ the matrix M is given by

$$\mathsf{M} = \begin{pmatrix} x_{ij_1} & y_{ij_1} & z_{ij_1} & \frac{1}{2}x_{ij_1}^2 & x_{ij_1}y_{ij_1} & \frac{1}{2}y_{ij_1}^2 & y_{ij_1}z_{ij_1} & \frac{1}{2}z_{ij_1}^2 & z_{ij_1}x_{ij_1} \\ x_{ij_2} & y_{ij_2} & z_{ij_2} & \frac{1}{2}x_{ij_2}^2 & x_{ij_2}y_{ij_2} & \frac{1}{2}y_{ij_2}^2 & y_{ij_2}z_{ij_2} & \frac{1}{2}z_{ij_2}^2 & z_{ij_2}x_{ij_2} \\ \cdots & \cdots & \cdots & \cdots & \cdots & \cdots & \cdots & \cdots & \cdots \\ x_{ij_M} & y_{ij_M} & z_{ij_M} & \frac{1}{2}x_{ij_M}^2 & x_{ij_M}y_{ij_M} & \frac{1}{2}y_{ij_M}^2 & y_{ij_M}z_{ij_M} & \frac{1}{2}z_{ij_M}^2 & z_{ij_M}x_{ij_M} \end{pmatrix} \tag{9.125}$$

and

$$\mathbf{C} = (c_{100}, c_{010}, c_{001}, c_{200}, c_{110}, c_{020}, c_{011}, c_{002}, c_{101})^{\mathsf{T}}. \tag{9.126}$$

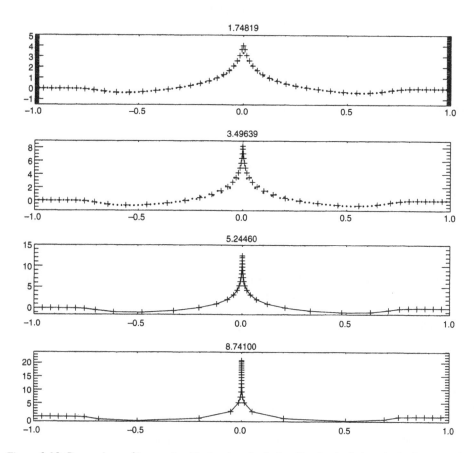

Figure 9.15 Comparison of lagrangian (+ signs) and eulerian (dots) calculations in the first two plots, and later development (last two plots) where the eulerian no longer works. Note that already in the second plot the eulerian calculation has developed noticeable wiggles which the lagrangian proceeds without problems.

Here, j_n ($n = 1, 2, \ldots, M$) are the M nearest neighbors of the point i. In general the matrix can be written in the form

$$M_{\alpha\beta}^{(J)} = x_{\alpha J}^{l(\beta)} \, y_{\alpha J}^{m(\beta)} \, z_{\alpha J}^{n(\beta)}, \tag{9.127}$$

where J is the index of the point at which the derivative is to be calculated. The set of exponents $l(\beta)$, $m(\beta)$, and $n(\beta)$ is given here for the case $N = 4$ in three-dimensional

$$l(\beta) = (1, 0, 0|2, 1, 0, 0, 0, 1|3, 2, 1, 0, 0, 0, 0, 1, 2|4, 3, 2, 1, 0, 0, 0, 0, 0, 1, 2, 3), \tag{9.128}$$

$$m(\beta) = (0, 1, 0|0, 1, 2, 1, 0, 0|0, 1, 2, 3, 2, 1, 0, 0, 0|0, 1, 2, 3, 4, 3, 2, 1, 0, 0, 0, 0), \tag{9.129}$$

$$n(\beta) = (0, 0, 1|0, 0, 0, 1, 2, 1|0, 0, 0, 0, 1, 2, 3, 2, 1|0, 0, 0, 0, 0, 1, 2, 3, 4, 3, 2, 1), \tag{9.130}$$

where the vertical bars separate the sets of exponents that correspond to increasing orders. Once the C vector has been obtained, the first derivatives of f are simply given by

$$\frac{\partial f}{\partial x} = c_{100}, \quad \frac{\partial f}{\partial y} = c_{010}, \quad \frac{\partial f}{\partial z} = c_{001}. \tag{9.131}$$

Likewise, the second derivatives are given by

$$\frac{\partial^2 f}{\partial x^2} = c_{200}, \quad \frac{\partial^2 f}{\partial y^2} = c_{020}, \quad \frac{\partial^2 f}{\partial z^2} = c_{002}, \tag{9.132}$$

and the mixed second derivatives are given by

$$\frac{\partial^2 f}{\partial x \partial y} = c_{110}, \quad \frac{\partial^2 f}{\partial y \partial z} = c_{011}, \quad \frac{\partial^2 f}{\partial z \partial x} = c_{101}. \tag{9.133}$$

Although this method can be used for meshes that are static in time, it can also be used in connection with multidimensional lagrangian schemes. In that case there may arise the problem that neighboring points get very close together, and so small errors strongly affect the coefficients. A good way out of this is to use a few more points and to solve the linear matrix equation using singular value decomposition. An example of such a calculation is shown in Fig. 9.16, where a passive scalar, with the initial distribution $A(\mathbf{x}, 0) = x$, is advected by the velocity, $\mathbf{u} = \dot{\mathbf{r}}$, which in turn is obtained by solving Kepler's equation, $\ddot{\mathbf{r}} = -GM\mathbf{r}/r^3$, using the normalization $GM = 1$. This windup problem corresponds to the windup of initially horizontal magnetic field lines.

In diffusivity used in Fig. 9.16 was $\eta = 0.02$, but due to the coarse resolution and the implicit smoothing resulting from the singular value decomposition technique the effective diffusivity is somewhat larger.

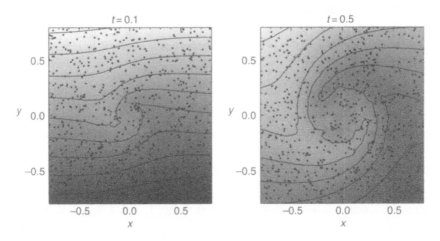

Figure 9.16 Two-dimensional advection problem on an unstructured lagrangian mesh. The dots indicate the 1000 lagrangian particles which constitute the unstructured mesh.

9.5. Implementing magnetic fields

As mentioned in Section 9.2, implementing magnetic fields is relatively straightforward. On the one hand, the magnetic field causes a Lorentz force, $\mathbf{J} \times \mathbf{B}$, where $\mathbf{B}$ is the flux density, $\mathbf{J} = \nabla \times \mathbf{B}/\mu_0$ is the current density, and μ_0 is the vacuum permeability. Note, however, that $\mathbf{J} \times \mathbf{B}$ is the force per unit volume, so in (9.9) we need to add the term $\mathbf{J} \times \mathbf{B}/\rho$ on the right-hand side. On the other hand, $\mathbf{B}$ itself evolves according to the Faraday equation,

$$\frac{\partial \mathbf{B}}{\partial t} = -\nabla \times \mathbf{E}, \tag{9.134}$$

where the electric field $\mathbf{E}$ can be expressed in terms of $\mathbf{J}$ using Ohm's law in the laboratory frame, $\mathbf{E} = -\mathbf{u} \times \mathbf{B} + \mathbf{J}/\sigma$, where $\sigma = (\eta\mu_0)^{-1}$ is the electric conductivity and η is the magnetic diffusivity.

In addition we have to satisfy the condition $\nabla \cdot \mathbf{B} = 0$. This is most easily done by solving not for $\mathbf{B}$, but instead for the magnetic vector potential $\mathbf{A}$, where $\mathbf{B} = \nabla \times \mathbf{A}$. The evolution of $\mathbf{A}$ is governed by the uncurled form of (9.134),

$$\frac{\partial \mathbf{A}}{\partial t} = -\mathbf{E} - \nabla\phi, \tag{9.135}$$

where ϕ is the electrostatic potential, which takes the role of an integration constant which does not affect the evolution of $\mathbf{B}$. The choice $\phi = 0$ is most advantageous on numerical grounds. (By contrast, the Coulomb gauge $\nabla \cdot \mathbf{A} = 0$, which is very popular in analytic considerations, would obviously be of no advantages, since one still has the problem of solving a the solenoidality condition.)

Solving for $\mathbf{A}$ instead of $\mathbf{B}$ has significant advantages, even though this involves taking another derivative. However, the total number of derivatives taken in the code is essentially the same. In fact, when centered finite differences are employed, Alfvén waves are better resolved when $\mathbf{A}$ is used, because then the system of equations for one-dimensional Alfvén waves in the presence of a uniform B_{x0} field in a medium of constant density ρ_0 reduces to

$$\dot{u}_z = \frac{1}{\mu_0\rho_0} B_{x0} A_y'', \quad \dot{A}_y = B_{x0} u_z, \tag{9.136}$$

where a second derivative is taken only once (primes denote x-derivatives). If, instead, one solves for the B_z field, one has

$$\dot{u}_z = \frac{1}{\mu_0\rho_0} B_{x0} B_z', \quad \dot{B}_z = B_{x0} u_z', \tag{9.137}$$

where a first derivative is applied twice, which is far less accurate at small scales if a centered finite difference scheme is used. At the Nyquist frequency, e.g. the first derivative is zero and applying an additional first derivative gives still zero. By contrast, taking a second derivative once gives of course not zero. The use of a staggered mesh circumvents this difficulty. However, such an approach introduces additional complications which hamper the ease with which the code can be adapted to other problems.

Another advantage of using $\mathbf{A}$ is that it is straightforward to evaluate the magnetic helicity $\langle \mathbf{A} \cdot \mathbf{B} \rangle$, which is a particularly important quantity to monitor in connection with dynamo and reconnection problems.

The main advantage of solving for $\mathbf{A}$ is of course that one does not need to worry about the solenoidality of the $\mathbf{B}$-field, even though one may want to employ irregular meshes or complicated boundary conditions.

As we have emphasized before, when centered meshes are used, it is usually a good idea to avoid taking first derivatives of the same variable twice, because it is more accurate to take instead a second derivative only once. For this reason we calculate the current not as $\mathbf{J} = \mu_0^{-1}\nabla \times (\nabla \times \mathbf{A})$, but as

$$\mathbf{J} = \mu_0^{-1}\left[-\nabla^2\mathbf{A} + \nabla(\nabla \cdot \mathbf{A})\right]. \tag{9.138}$$

Taking the gradient of $\nabla \cdot \mathbf{A}$ involves of course also taking first derivatives of the same variable twice, but these contributions are cancelled by corresponding components of the $\nabla^2\mathbf{A}$ term. There are some advantages relying here on the numerical cancellation, which is of course not exact. The reason is that the full $\nabla^2\mathbf{A}$ term is important when used in the magnetic diffusion term. If the diagonal terms, $\partial^2 A_x/\partial x^2$, $\partial^2 A_y/\partial y^2$, and $\partial^2 A_z/\partial z^2$, which would all drop out analytically, were taken out there would be no diffusion of $\mathbf{A}$ in the direction of $\mathbf{A}$.

There is one more aspect that is often useful keeping in mind. There is a particular gauge that allows one to rewrite the uncurled induction equation in such a form that the evolution of $\mathbf{A}$ is controlled by the advective derivative of $\mathbf{A}$. The calculation is easy. Write the induction term $\mathbf{u} \times \mathbf{B}$ in component form and express $\mathbf{B}$ in terms of $\mathbf{A}$, so

$$(\mathbf{u} \times \nabla \times \mathbf{A})_i = u_j(\partial_i A_j - \partial_j A_i) = \partial_i(u_j A_j) - A_j\partial_i u_j - u_j\partial_j A_i. \tag{9.139}$$

Here the last term contributes to the advective derivative, the first term can be removed by a gauge transformation and the middle term is a modified stretching term, so the induction equation takes the form

$$\frac{DA_i}{Dt} = -A_j\partial_i u_j - \eta\mu_0 J_i. \tag{9.140}$$

This gauge was used by Brandenburg *et al.* (1995) in order to treat a linear velocity shear using pseudo-periodic (shearing box) boundary conditions. The formulation (9.140) can also be useful when solving the induction equation using lagrangian methods. Note, however, that the non-resistive evolution of $\mathbf{A}$ differs from that of $\mathbf{B}$ in that the indices of the matrix $\mathsf{U}_{ij} \equiv \partial u_i/\partial x_j$ are interchanged and that the sign is different; positive for the $\mathbf{B}$-equation,

$$\frac{DB_i}{Dt} = +\mathsf{U}_{ij} B_j + \text{other terms}, \tag{9.141}$$

and negative for the $\mathbf{A}$-equation,

$$\frac{DA_i}{Dt} = -A_j\mathsf{U}_{ji} + \text{other terms}. \tag{9.142}$$

These two formulations are particularly advantageous when the velocity has a constant gradient, as in the case of linear shear. In local simulations of accretion discs, e.g. the shear component is $u_y(x) = -\frac{3}{2}\Omega x$, so $\mathsf{U}_{yx} = -\frac{3}{2}\Omega$, and all other U_{ij} vanish. Hence

$$\frac{DA_x}{Dt} = +\frac{3}{2}\Omega A_y + \text{other terms} \tag{9.143}$$

for the **A**-formulation, or

$$\frac{DB_y}{Dt} = -\frac{3}{2}\Omega B_x + \text{other terms} \tag{9.144}$$

for the **B**-formulation. In these two formulations all dependent variables are clearly periodic (or rather pseudo-periodic), so there is no term that is explicitly non-periodic such as $u_y(x) = -\frac{3}{2}\Omega x$. In the following, whenever magnetic fields are present, we use the **A**-formulation, mainly because it guarantees the solenoidality of **B** everywhere (including the boundaries), and also because it is easy to use.

9.5.1. Cache-efficient coding

Unlike the CRAY computers that dominated supercomputing in the eighties and early nineties, all modern computers have a cache that constitutes a significant bottleneck for many codes. This is the case if large three-dimensional arrays are constantly used within each timestep. The advantage of this way of coding is clearly the conceptual simplicity of the code. A more cache-efficient way of coding is to calculate an entire timestep (or a corresponding substep in a three-stage $2N$-Runge–Kutta scheme) only along a one-dimensional pencil of data within the box. On Linux and Irix architectures, e.g. this leads to a speed-up by 60%. An additional advantage is a drastic reduction in temporary storage that is needed for auxiliary variables within each timestep.

9.6. Application to astrophysical outflows

9.6.1. The isothermal Parker wind

Before discussing outflows from accretion discs it is illuminating to consider first the one-dimensional example of pressure-driven outflows in spherical geometry. A particularly simple case is the *isothermal* wind problem, which is governed by the equations

$$\frac{\partial u}{\partial t} + u\frac{\partial u}{\partial r} = -c_s^2\frac{\partial \ln \rho}{\partial r} - \frac{\partial \Phi}{\partial r}, \tag{9.145}$$

$$\frac{\partial \ln \rho}{\partial t} + u\frac{\partial \ln \rho}{\partial r} = -\frac{1}{r^2}\frac{\partial}{\partial r}(r^2 u) + \frac{\dot{M}\xi(r)}{\rho}, \tag{9.146}$$

where c_s is the isothermal sound speed (assumed constant), $\dot{M}$ is the mass loss rate, and $\xi(r)$ is a prescribed function of position, normalized such that $\int 4\pi r^2 \xi(r)\, dr = 1$, and non-vanishing only near $r = 0$. For a point mass the gravity potential Φ would be $-GM/r$, but this becomes singular at the origin. Therefore we use the expression $\Phi = -GM/(r^n + r_0^n)^{1/n}$ instead, where we choose $n = 5$ in all cases, and $1/r_0$ gives the depth of the potential well. In Fig. 9.17 we show radial velocity and density profiles for different values of $\dot{M}$. Note that the velocity profile is independent of the value of $\dot{M}$, but the density profile changes by a constant factor. In the steady case the equations can be combined to

$$\left(u^2 - c_s^2\right)\frac{d\ln|u|}{dr} = \frac{2c_s^2}{r} - \frac{GM}{r^2}, \tag{9.147}$$

so the sonic point, $|u| = c_s$, is at $r = r^* = GM/2c_s^2$. In Fig. 9.17 we have chosen $GM = 2$ and $c_s = 1$, so $r^* = 1$, which is consistent with the graph of u.

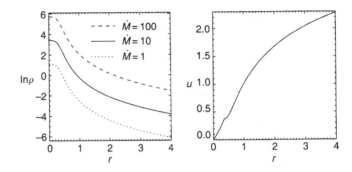

Figure 9.17 Isothermal Parker wind solutions for different values of $\dot{M}$. Note that the u profile is independent of the value of $\dot{M}$. $GM = 2$, $c_s = 1$, $r_0 = 0.4$.

9.6.2. The polytropic or adiabatic wind

In the following we make the assumption that the entropy is constant. In that case it is particularly useful to solve for the *potential enthalpy*, $H \equiv h + \Phi$, which varies much less than either h or Φ. Using H as dependent variable is particularly useful if one solves the equations all the way to the origin, $r = 0$, where Φ tends to become singular (or at least strongly negative if a smoothed potential is used). In terms of H the governing equations are

$$\frac{\partial u}{\partial t} + u \frac{\partial u}{\partial r} = -\frac{\partial H}{\partial r} - u \dot{M} \xi, \tag{9.148}$$

$$\frac{\partial H}{\partial t} + u \frac{\partial H}{\partial r} = u \frac{\partial \Phi}{\partial r} + c_s^2 \left[-\frac{1}{r^2} \frac{\partial}{\partial r}(r^2 u) + \frac{\dot{M} \xi(r)}{\rho} \right], \tag{9.149}$$

where $H = h + \Phi$ is the potential enthalpy, $h = p/\rho + e$ is the enthalpy, and $c_s^2 = (\gamma - 1)h = (\gamma - 1)(H - \Phi)$ for a ideal gas, where c_s is the adiabatic sound speed and $h = c_p T$ is the enthalpy. These equations are also valid in the nonisothermal case ($\gamma \neq 1$). The isothermal case may be recovered by putting $\gamma = 1$ and replacing h by $c_s^2 \ln \rho$. In Fig. 9.18 we show solutions for different values of $\dot{M}$ and $\gamma = 5/3$. Again we put $GM = 2$ and $c_{s0} = 1$.

We note that, depending on the strength of the mass source, the polytropic wind problem allows a variety of different velocity and Mach number profiles, whereas for the isothermal wind problem there was only one solution possible, independent of the strength of the mass source. The velocity profile was always the same and also the density was the same up to some scaling factor that changes with $\dot{M}$. This is connected with the additional degree of freedom introduced through the polytropic constant $K = p/\rho^\gamma$. Since c_s is no longer constant, the position of the sonic point is no longer fixed and different solutions are possible.

In Fig. 9.19 we show solutions where $\dot{M}$ is kept constant, but the depth of the potential well, GM/r_0, is changed by varying the value of r_0. Note that the deeper the potential well, the higher the wind speed. The density far away from the source is then correspondingly smaller, so as to maintain the same mass flux.

As we have seen in Section 9.3.6, a polytropic equation of state is unphysical. Therefore we now consider the case where the energy equation is included. To be somewhat more general we consider first the basic equations in conservative form with mass, momentum, and energy

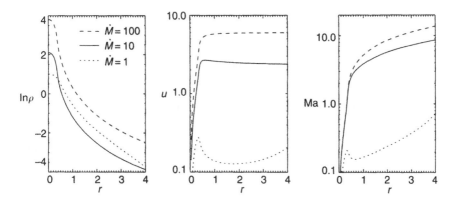

Figure 9.18 Polytropic Parker wind solutions for different values of $\dot{M}$. $GM = 2$, $c_{s0} = 1$, $r_0 = 0.4$.

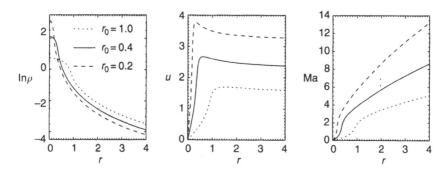

Figure 9.19 Density ρ, velocity u, and Mach number Ma $= u/c_s$ for the polytropic Parker wind solutions for different values of r_0. $GM = 2$, $c_{s0} = 1$, $\dot{M} = 10$.

sources included, i.e.

$$\frac{\partial \rho}{\partial t} + \frac{\partial}{\partial x_j}\left(\rho u_j\right) = \dot{M}\xi, \tag{9.150}$$

$$\frac{\partial}{\partial t}\left(\rho u_i\right) + \frac{\partial}{\partial x_j}\left(\rho u_i u_j + p\delta_{ij} - \tau_{ij}\right) = \dot{I}_i\xi, \tag{9.151}$$

$$\frac{\partial}{\partial t}\left(\tfrac{1}{2}\rho \mathbf{u}^2 + \rho e\right) + \frac{\partial}{\partial x_j}\left[u_j\left(\tfrac{1}{2}\rho \mathbf{u}^2 + \rho h\right) - u_i \tau_{ij}\right] = \dot{E}\xi, \tag{9.152}$$

where $\dot{M}$, $\dot{\mathbf{I}}$, and $\dot{E}$ are the rates of mass, momentum, and energy injection into the system, $\tau_{ij} = 2\nu\rho S_{ij}$ is the viscous stress tensor, and S_{ij} is the (traceless) rate of strain tensor; see (9.12). Rewriting the energy equation in non-conservative form we have

$$\frac{De}{Dt} + \frac{p}{\rho}\nabla \cdot \mathbf{u} \equiv T\frac{Ds}{Dt} = 2\nu \mathbf{S}^2 + \left[\dot{E} - \mathbf{u} \cdot \left(\dot{\mathbf{I}} - \mathbf{u}\dot{M}\right) - \left(\tfrac{1}{2}\mathbf{u}^2 + e\right)\dot{M}\right]\frac{\xi(r)}{\rho}, \tag{9.153}$$

which can also be rewritten in terms of entropy, so the final system of non-conservative equations with source terms is

$$\frac{D \ln \rho}{Dt} + \nabla \cdot \mathbf{u} = \dot{M}\frac{\xi(r)}{\rho}, \tag{9.154}$$

$$\frac{D\mathbf{u}}{Dt} + c_s^2 (\nabla \ln \rho + \nabla s) = \frac{1}{\rho}\nabla \cdot (2\nu\rho\mathbf{S}) + \left(\dot{\mathbf{I}} - \mathbf{u}\dot{M}\right)\frac{\xi(r)}{\rho}, \tag{9.155}$$

$$T\frac{Ds}{Dt} = 2\nu\mathbf{S}^2 + \left[\dot{E} - \mathbf{u}\cdot\dot{\mathbf{I}} + \left(\tfrac{1}{2}\mathbf{u}^2 - e\right)\dot{M}\right]\frac{\xi(r)}{\rho}, \tag{9.156}$$

where T can be replaced by $c_s^2/(\gamma - 1)$ (remember that $c_p = 1$), and c_s^2 is given by (9.8).

In Fig. 9.20 we present solutions of (9.154)–(9.156) for different values of $\dot{E}$ and $\dot{M}$. The main effect of varying the value of $\dot{E}$ is to change the value of the entropy in the wind. Outside the acceleration region, however, the value of the entropy is fairly constant, so the polytropic assumption appears to be reasonably good here.

While outflows of some very early-type stars are driven mostly by the $\dot{\mathbf{I}}$ term (resulting from the radiation pressure in lines), the winds of cool stars are driven mostly by the $\dot{E}$ term (resulting from the hot coronae). Similar differences may also explain why some jets are massive (e.g. stellar jets), whilst others are not (jets from active galactic nuclei, e.g. or those anticipated in gamma-ray bursters).

9.6.3. Relevance to outflows and jets

The pressure-driven outflows discussed in the previous section may take the form of more collimated outflows once a magnetic field is involved. This applies to the case of magnetized accretion discs. These discs are generally magnetized both because of dynamo action within

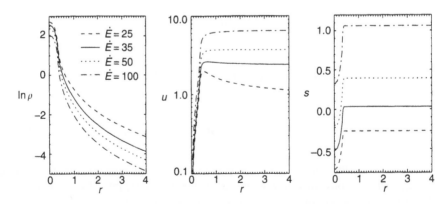

Figure 9.20 Wind solutions for different values of $\dot{E}$ and $\dot{M} = 10$. Note that the solution with $\dot{E} = 35$ is quantitatively very similar to the polytropic solution with the same value of $\dot{M}$. $GM = 2$, $c_{s0} = 1$, $r_0 = 0.4$.

the disc and because of external fields that were dragged into the disc from outside due to the accretion flow.

At least in some types of jets the outflows may be driven by hot coronae. Other possibilities for driving outflows involve the magneto-centrifugal effect. It is well known that outflows can be driven from a magnetized disc if the angle between the field and the disc is less than 60° (Blandford and Payne, 1982). Recent work in this field was directed to the question whether this angle is the result of some self-regulating process (Ouyed *et al.*, 1997; Ouyed and Pudritz, 1997a,b, 1999) and whether it can be obtained automatically from a dynamo operating within the disc (Campbell, 1999, 2000; Dobler *et al.*, 1999; Rekowski *et al.*, 2000). This latter question is particularly interesting in view of the fact that jets in star-forming regions are not really pointing in a similar direction (e.g. Hodapp and Ladd, 1995), as one might expect from jet models that start off with a prescribed large scale field.

In Fig. 9.21 we present a particular model of Dobler *et al.* (1999) and Brandenburg (2000); see Brandenburg *et al.* (2000) for a full account of this work. In these models the outflow is driven by mass sources whose strength is proportional to the local density deficit relative to that in the original equilibrium solution of the disc. Such a density deficit was initially caused by slow gas motions that resulted from an instability of the initial equilibrium solution, because a cool disc embedded in a hot corona is non-rotating outside the disc, and it is the resulting vertical shear profile that causes the instability (cf. Urpin and Brandenburg, 1998). At later times, of course, the outflow makes the corona corotating, but by that time the outflow is driven by a persistent density deficit in the disc relative to the initial references solution.

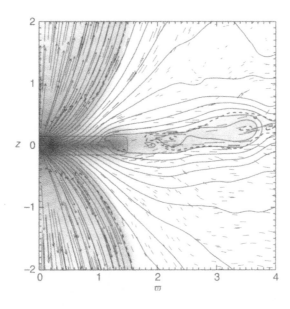

Figure 9.21 Poloidal velocity vectors and poloidal magnetic field lines superimposed on a gray-scale image of the logarithmic density. Dark means high density. The thick dashed line denotes the location where the poloidal flow speed equals the fast magnetosonic speed. The thin solid line gives the location of the disc surface. The slight asymmetry in the field is a relic from the mixed-parity initial condition. (Adapted from Brandenburg, 2000.)

In this model the magnetic field was generated by an α–Ω dynamo operating within the disc. However α is negative in the upper disc plane (see Brandenburg *et al.*, 1995), and then the most preferred field geometry is dipolar (Campbell, 1999; Rekowski *et al.*, 2000). The field parity is sensitive to details in the disc physics assumed in the particular model (aspect ratio, disc thickness, the presence of outflows, and the conductivity in the disc and the exterior). Nevertheless, both dipolar and quadrupolar fields are equally well able to contribute to wind launching, at least in the outer parts of the disc where the angle between the field and the disc plane is less than 60°, the critical angle for magneto-centrifugal wind launching (Blandford and Payne, 1982). We note, however, that the more detailed analysis of Campbell (1999) suggests that the critical angle can be significantly smaller.

In our models the outflow is only weakly collimated (if at all). This is probably connected with the fact that here the fast magnetosonic surface is rather close to the disc surface, making it difficult for the field to become strong enough to channel the magnetic field. Instead, the field lines themselves are still being controlled too strongly by the outflow. However, outflows with rather large opening angles are actually seen in some star-forming regions; see Greenhill *et al.* (1998).

While most of the disc mass is ejected in a cone of half-opening angle around 25°, most of the disc angular momentum is ejected at rather low latitudes, almost in the direction of the disc plane away from the central object. The timescales for these various processes are comparable. In Fig. 9.22 we show the azimuthally integrated mass flux, angular momentum flux, and magnetic (Poynting) flux as a function of polar angle, and compare with a non-magnetic run. We find that in the magnetic run the outflow is more strongly concentrated towards the axis. Also, the amount of angular momentum loss (dash-dotted line) is larger when the disc is magnetized. We emphasize in particular that in the magnetic run significant amounts of magnetic field are eject from the system. In the following section we discuss the significance of such magnetic flux ejection for magnetizing the interstellar medium into which the outflow is streaming. This discussion is similar to a corresponding discussion for the contamination of the intergalactic medium via outflows from active galactic nuclei (Brandenburg, 2000).

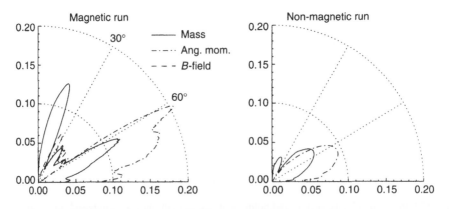

Figure 9.22 Comparison of the angular dependence of azimuthally integrated fluxes for magnetic and non-magnetic outflows. The solid line refers to mass flux, the dashed line to angular momentum flux, and the dash-dotted line (in the second panel) corresponds to the Poynting flux. The units of all quantities are thus $1/[t]$.

9.6.4. Magnetic contamination from outflows

It may at first appear somewhat unrealistic to expect significant magnetization of the interstellar medium from outflows. However, the following calculation shows that the effect may be quite significant. Assume that every star did undergo a phase of strong accretion with associated outflows, so $N = 10^{11}$ for the whole galaxy. The duration of intense outflow activity is 10^5 years, say, but it could even be 10^6 years. The magnetic luminosity is $L_{mag} = 0.05\dot{M}_w c_s^2$ (Brandenburg *et al.*, 2000), where $c_s \approx 10$ km/s is the average sound speed of the interstellar medium, and $\dot{M}_w = 0.1\dot{M}_d$ (see Pelletier and Pudritz, 1992), where $\dot{M}_d \approx 10^{-7} M_\odot$/year is a conservative estimate for the disc accretion rate. Again this value may be larger. With the above numbers the magnetic luminosity from all $N = 10^{11}$ sources is then $NL_{mag} = 7 \times 10^{39}$ erg/s and the total energy output delivered from all stars at some early point in the life time is therefore $E_{mag} = \tau NL_{mag} = 2 \times 10^{52}$ erg. Diluting this over a volume of a galaxy of 300 kpc^3 (radius 10 kpc, height 1 kpc) gives 2×10^{-15} erg/cm^3. Multiplying this by 8π and taking the square root gives $0.2\,\mu$G. Expressed more concisely in a formula we have for the rms magnetic field strength

$$\langle \mathbf{B}^2 \rangle^{1/2} \approx \left(8\pi \, \frac{F_{Poy}}{F_{kin}} \, \frac{N\dot{M}_w c_s^2}{V} \, \delta t \right)^{1/2}, \tag{9.157}$$

where the efficiency factor F_{Poy}/F_{kin} (=0.05 in our model) may be lower in systems where the disc dynamo is less strong.

The parameters for a corresponding estimate for outflows from young galactic discs (active galactic nuclei) are as follows. Assuming $N \sim 10^4$ galaxies per cluster, each with $\dot{M}_w \approx 0.1 M_\odot$/year $= 10^{25}$ g/s, and $c_s = 1000$ km/s for the sound speed in the intracluster gas, the rate of magnetic energy injection for all galaxies together is $L_{mag} = 10^{44}$ erg/s. Distributing this over the volume of the cluster of $V \sim 1$ Mpc^3, and integrating over a duration of $\delta t = 1$ Gyr, this corresponds to a mean magnetic energy density of $\langle \mathbf{B}^2/8\pi \rangle \approx 10^{-13}$ erg/cm^3, so $\langle \mathbf{B}^2 \rangle^{1/2} \approx 10^{-6}$ G, which is indeed of the order of the field strength observed in galaxy clusters. We note that our estimate has been rather optimistic in places (e.g. $\dot{M}_w$ could be lower, or the relevant δt could be shorter), but it does show that outflows are bound to produce significant magnetization of the intracluster gas and the interstellar medium (see also Völk and Atoyan, 1999). In the latter case it will provide a good seed field for the galactic dynamo. A dynamo is still necessary to shape the magnetic field and to prevent if from decaying in the galactic turbulence. Similarly, many galaxy clusters undergo merging and this too can enhance and reorganize the magnetic field. The necessity for a recent merger event would also be consistent with the fact that not all halos are observed to have strong magnetic fields. Recent simulations by Roettiger *et al.* (1999) suggest that after a merger the field strength may increase by a factor of at least 20 (and this value increases with improving observational resolution).

As an alternative consideration for causing the magnetization in clusters of galaxies, *primordial* magnetic fields are sometimes discussed. There are numerous mechanisms that could generate relatively strong fields at an early time, e.g. during inflation (age $\sim 10^{-36}$ s) or during the electroweak phase transition (age $\sim 10^{-10}$ s). Such fields would now still be at a very small scale if one considers only the cosmological expansion. However, depending on the degree of magnetic helicity in this primordial field, the magnetic energy can be transferred to larger scales that are now on the scale of galaxies. For a recent discussion of these results see Brandenburg (2001a).

9.7. Hydromagnetic turbulence and dynamos

As mentioned in the beginning, accurate high-order schemes are essential in all applications to turbulent flows. Nevertheless, we should mention that one often attempts solutions of the inviscid and nonresistive equations using low-order finite differences combined with monotonicity schemes that result in some kind of effective diffusion. The piece-wise parabolic method (PPM) of Colella and Woodward (1984) is an example. However, unlike the Smagorinsky scheme (see Chan and Sofia, 1986, 1989; Steffen *et al.*, 1989; Fox *et al.*, 1991 for applications to convection simulations), PPM and similar methods cannot be proven to converge to the original Navier–Stokes equation in the limit of infinite resolution. Nevertheless, they are rather popular in astrophysical gas simulations. These schemes are rather robust and have also been applied to high-resolution simulations of compressible turbulence (Porter *et al.*, 1992, 1994). While the results from those simulations are generally quite plausible, the power spectrum shows a k^{-1} subrange at large wavenumbers, which is still not fully understood. This was sometimes regarded as an artifact of PPM, and should therefore only occur at small scales. However, as the resolution was increased further (up to 1024^3), the k^{-1} subrange just became more extended.

A similar feature was found in cascade models of turbulence when the ordinary ∇^2 diffusion operator was replaced by a $-\nabla^4$ "hyperdiffusion" operator (Lohse and Müller-Groeling, 1995). Whatever the outcome of this puzzle is, it is clear that with schemes that cannot be proven to converge to the actual Navier–Stokes equations in the limit of infinite resolution, there would always remain some uncertainty and debate. On the other hand, especially in the incompressible case the use of hyperviscosity does generally allow the exploration of larger Reynolds numbers and broader inertial ranges.

MHD simulations with the highest resolution to date have been performed by Biskamp and Müller (1999), who considered decaying turbulence with and without magnetic helicity. They found that in the presence of magnetic helicity the magnetic energy decay is significantly slower. In particular, they found the magnetic energy decays like $t^{-1/2}$, as opposed to t^{-1} found earlier by Mac Low *et al.* (1998) for compressible turbulence.

Before we start discussing dynamo action in turbulence simulations representative of more astrophysical settings, such as accretion discs and the solar convection zone, let us first illustrate the mechanism of the inverse cascade that is believed to be an important ingredient of large scale magnetic field generation.

9.7.1. Isotropic MHD turbulence

Most developments in the theory of turbulence have been carried out under the assumptions of homogeneity and isotropy. This is certainly true of the work on the inverse cascade (or turbulent cascades in general), but it is also true of much of the work on the α-effect which – like the inverse cascade – describes the generation of large scale fields. However, unlike the inverse cascade process, the energy comes here directly from the velocity field at the scales of the energy-carrying eddies and not from the velocity and magnetic field at successively smaller scales, which are usually larger than the scale of the energy-carrying eddies.

It is not easy to see whether any of these effects is actually responsible for the large scale field generation in astrophysical bodies or even the simulations. In simulations of accretion disc turbulence there is certainly some evidence for the presence of an α-effect, but it is extremely noisy (Brandenburg *et al.*, 1995; Brandenburg and Donner, 1997; Ziegler and Rüdiger, 2000). Evidence for the inverse magnetic cascade comes mostly from the magnetic energy spectra (Balsara and Pouquet, 1999; Brandenburg, 2001b), which show a marked peak at large scales, but this is convincing only in cases where the flow is driven at a wavenumber that is clearly larger than the smallest wavenumber in the box. In practice, e.g. in convectively driven turbulence, the flow is driven at all scales including the large scale making it difficult to see a marked peak at the smallest wavenumber (see a corresponding discussion in Meneguzzi and Pouquet, 1989).

From the seminal papers of Frisch *et al.* (1975) and Pouquet *et al.* (1976) it is clear that amplification of large scale fields can also be explained by an inverse cascade of magnetic helicity. In those papers the authors also showed that the inverse cascade is a consequence of the fact that the magnetic helicity $\langle \mathbf{A} \cdot \mathbf{B} \rangle$, is conserved by the nonresistive equations. ($\mathbf{A}$ is the magnetic vector potential giving the magnetic field as $\mathbf{B} = \nabla \times \mathbf{A}$.) The inverse magnetic cascade effect too is rather difficult to isolate in simulations of astrophysical turbulence. However, under somewhat more idealized conditions, e.g. when magnetic energy is injected at high wavenumbers, one clearly sees how the magnetic energy increases at large scales; see Fig. 9.23. Further details of this model have been published in the proceedings of the helicity meeting in Boulder (Brandenburg, 1999).

In the model considered above the flow was forced magnetically. This may be motivated by the recent realization that strong magnetic field generation in accretion discs can be facilitated by *magnetic instabilities*, such as the Balbus–Hawley instability. Other examples of magnetic instabilities include the magnetic buoyancy instability, which can lead to an α-effect (e.g. Brandenburg and Schmitt, 1998; Thelen, 2000), and the reversed field pinch which also leads to a dynamo effect (e.g. Ji *et al.*, 1996). Before returning to the accretion disc dynamo in Section 9.7.9 we should emphasize that strong large scale field generation is also possible with purely hydrodynamic forcing. Simulations in this type were considered recently by Brandenburg (2001b). There are many similarities compared with the case of

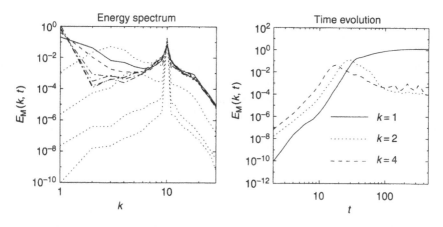

Figure 9.23 Spectral magnetic energy, $E_M(k, t)$, as a function of wavenumber k for different times: dotted lines are for early times ($t = 2, 4, 10, 20$), the solid and dashed lines are for $t = 40$ and 60, respectively, and the dotted-dashed lines are for later times ($t = 80, 100, 200, 400$). Here magnetic energy is injected at wavenumber 10. Note the occurrence of a sharp secondary peak of spectral magnetic energy at $k = 10$. By the time the energy at $k = 1$ has reached equipartition the energies in $k = 2$ and $k = 4$ become suppressed.

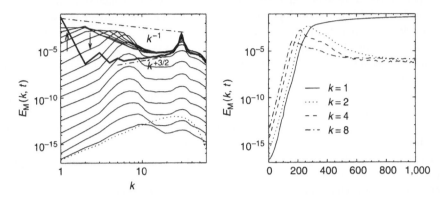

Figure 9.24 Left: Magnetic energy spectra for a run with forcing at $k = 30$. The times range from 0 (dotted line) to 10, 30, ..., 290 (solid lines). The thick solid line gives the final state at $t = 1000$. Note that at early times the spectra peaks at $k_{max} \approx 7$. The k^{-1} and $k^{+3/2}$ slopes are given for orientation as dash-dotted lines. Right: Evolution of spectral magnetic energy for selected wavenumbers in a simulation with hydrodynamical forcing at $k = 30$.

magnetic forcing. The evolution of magnetic energy spectra in the presence of hydrodynamic forcing is shown in Fig. 9.24. Like in the case of magnetic forcing (Fig. 9.23) there are marked peaks both at the forcing scale and at the largest scale of the box. Furthermore, the evolution of spectral energy at the largest scales shows similar behavior: the magnetic energy with wavenumber $k = 8$ increases, reaches a maximum, and begins to decrease when the magnetic energy at $k = 4$ reaches a maximum. The same happens for the next larger scales (wavenumbers $k = 4$ and 2, until the scale of the box (with $k = 1$) is reached.

The *suppression* of magnetic energy at intermediate scales, $2 \le k \le 8$, is quite essential for the development of a well-defined large scale field. In a recent letter Brandenburg and Subramanian (2000) showed that this type of *self-cleaning* effect can also be simulated by using ambipolar diffusion as nonlinearity and ignoring the Lorentz force altogether. Without any nonlinearity, however, there would be no interaction between different scales and the magnetic energy would increase at all scales, especially at small scales, which would soon swamp the large scale field structure with small scale fields.

The model presented in Fig. 9.24 has large scale separation in the sense that there is a large gap between the forcing wavenumber ($k = k_f = 30$) and the wavenumber of the box ($k = k_1 = 1$). One sees that during the growth phase there is a clear secondary maximum at $k = 7$. This is indeed expected for an α^2 dynamo, whose maximum growth rate is at $k_{max} = \frac{1}{2}\alpha/\eta_T$, where η_T is the total (turbulent plus microscopic) magnetic diffusion coefficient.

The disadvantage of a high forcing wavenumber is that for modest resolution (here we used 120^3 meshpoints) no inertial range can develop. This is different if once forces at $k_f = 5$, keeping otherwise the same resolution. In Fig. 9.25 we show spectra for different cases with $k_f = 5$ where we compare the results for different values of the magnetic Reynolds and magnetic Prandtl number. In Fig. 9.26 we show cross-sections of one field component at different times. In this model (Run 3 of Brandenburg, 2001b) the forcing is at $k_f = 5$, so there is now a clear tendency for the build-up of an inertial range in $8 \le k \le 25$.

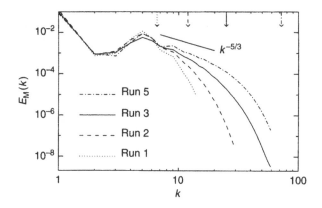

Figure 9.25 Comparison of time averaged magnetic energy spectra for Runs 1–3 ($t = 600 - 1000$) with a non-averaged spectrum for Run 5 (large magnetic Prandtl number) taken at $t = 1600$. To compensate for different field strengths and to make the spectra overlap at large scales, two of the three spectra have been multiplied by a scaling factor. There are clear signs of the gradual development of an inertial subrange for wavenumbers larger than the forcing scale. The $k^{-5/3}$ slope is shown for orientation. The dissipative magnetic cutoff wavenumbers, $\langle \mathbf{J}^2/\eta^2 \rangle^{1/4}$, are indicated by arrows at the top.

9.7.2. The inverse cascade in decaying turbulence

We now turn to the case of *decaying* turbulence, which is driven only by an initial kick to the system. There are several circumstances in astrophysics where this could be relevant: early universe, neutron stars, and mergers of galaxy clusters. In all those cases one is interested in the development of large scale fields. In the context of the early universe the possibility of energy conversion from small to large scale fields was pointed out by Brandenburg *et al.* (1996) who found that fields generated at the horizon scale of 3 cm after the electroweak phase transition would now have a scale on the order of kiloparsecs, even though the cosmological expansion alone would only lead to scales on the order of 1 AU. These results were only based on either two-dimensional simulations or three-dimensional cascade model calculation (e.g. Biskamp, 1994). Therefore we now turn to fully three-dimensional simulations.

In the absence of any forcing and with no kinetic energy initially an initial magnetic field can only decay. However, if initially most of the magnetic energy is in the small scales, there is the possibility that magnetic helicity and thereby also magnetic energy is transferred to large scales. This is exactly what happens (Fig. 9.27), provided there is initially some net helicity. The inset of Fig. 9.27 shows that in the absence of initial net helicity the field at large scales remains unchanged, until diffusion kicks in and destroys the remaining field at very late times.

If the magnetic field has the possibility to tap energy also from the large scale velocity the situation is somewhat different again and there is the possibility that a large scale magnetic field can also be driven without net helicity. In that case the large scale field can increase due to dynamo action from the incoherent α–Ω-effect (Vishniac and Brandenburg, 1997). In astrophysical settings there is usually large scale shear from which energy can be tapped.

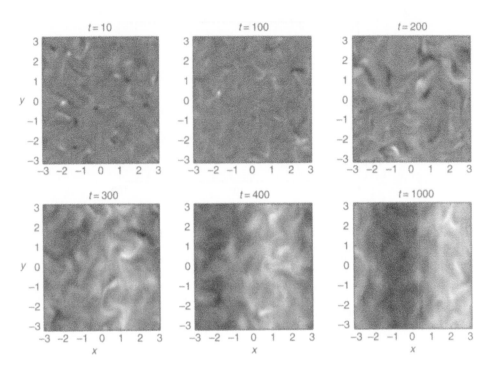

Figure 9.26 Gray-scale images of cross-sections of $B_x(x, y, 0)$ for Run 3 of Brandenburg (2001b) at different times showing the gradual build-up of the large scale magnetic field after $t = 300$. Dark (light) corresponds to negative (positive) values. Each image is scaled with respect to its min and max values.

Before we discuss simulations with imposed shear in more detail we first present a simple argument that makes the link between the inverse cascade and helicity conservation.

9.7.3. The connection with magnetic helicity conservation

In the following we give a simple argument due to Frisch *et al.* (1975) that helps to understand why the magnetic helicity conservation property leads to the occurrence of an inverse cascade. We define in the following magnetic energy and helicity spectra, $M(k)$ and $H(k)$, respectively. Now, because of Schwartz inequality, we have

$$|\hat{\mathbf{B}}(k)|^2 = |(i\mathbf{k} \times \hat{\mathbf{A}}) \cdot \hat{\mathbf{B}}| \geq |\mathbf{k}||\hat{\mathbf{A}} \cdot \hat{\mathbf{B}}| \qquad (9.158)$$

we have a lower bound on the spectral magnetic energy at each wavenumber $k = |\mathbf{k}|$. In terms of shell integrated magnetic energy and helicity spectra this corresponds to

$$M(k) \geq \tfrac{1}{2}k|H(k)|, \qquad (9.159)$$

where the 1/2-factor comes simply from the 1/2-factor in the definition of the magnetic energy. Assuming that two wavenumbers q and p interact such that they produce power at a new wavenumber k, then

$$M(p) + M(q) = M(k), \quad H(p) + H(q) = H(k). \qquad (9.160)$$

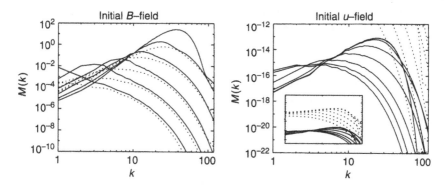

Figure 9.27 Power spectra of magnetic energy (solid lines) and kinetic energy (dotted lines) in a decay run with helicity. The left-hand panel is for a case where the flow is only driven by an initial helical magnetic field. In the right-hand panel the field is weak and governed by strong decaying fluid turbulence. The inset shows both velocity and magnetic spectra in the same plot. The Prandtl number ν/η is equal to one in both cases, but the mesh Reynolds number, which is kept constant at all times, is different in the two cases: 20 in the left-hand panel and 50 in the right-hand panel. The times are 0, 0.01, 0.1, etc., till $t = 10^2$ in the left-hand panel and $t = 10^3$ in the right-hand panel.

For simplicity we consider the case $p = q$, so

$$2M(p) = M(k), \quad 2H(p) = H(k). \tag{9.161}$$

Assume also that initially the constraint was sharp (maximum helicity), then

$$M(p) = \tfrac{1}{2} p H(p). \tag{9.162}$$

Now, from the constrain again we have

$$\tfrac{1}{2} k H(k) \le M(k) = 2M(p) = pH(p) = \tfrac{1}{2} p H(k), \tag{9.163}$$

so

$$k \le p, \tag{9.164}$$

that is the wavenumber of the target result must be larger or equal to the wavenumbers of the initial field.

The argument given above is of course quite rough, because it ignores, e.g. the detailed angular dependence of the wave vectors. This was taken into account properly already in the early paper by Pouquet *et al.* (1976), but this approach was based on closure assumptions for the higher moments, which is in principle open to criticism. Thus, numerical simulations, like those presented above, are necessary for an independent confirmation that the inverse cascade really works. In this connection one should mention that there are some parallels with the inverse cascade of enstrophy in two-dimensional hydrodynamic (non-magnetic) turbulence. In that case the enstrophy (i.e. the mean squared vorticity) is conserved because of the absence of vortex stretching in two dimensions. The inverse hydrodynamic cascade has some significance in meteorology and perhaps in low aspect ratio convection experiments, where one finds a peculiar energy and entropy spectrum that is referred to as Bolgiano scaling; see Brandenburg (1992) and Suzuki and Toh (1995) for corresponding shell model calculations and Toh and Iima (2000) for direct simulations.

9.7.4. Inverse cascade or α-effect?

In Section 9.7.1 we made a distinction between inverse cascade and α-effect in the sense that, although both lead to large scale field generation, in the inverse cascade there is a gradual transfer of magnetic helicity and energy to ever larger scales, whereas the α-effect produces large scale magnetic fields directly from small scale fields. Thus, the distinction is really one between local and non-local inverse cascades.

In Fig. 9.28 we show the normalized spectral energy transfer function $T(k, p, t)$ for $k = 1$ and 2 as a function of p, and at different times t. The index k signifies the gain or losses of the field at wavenumber k, and the index p indicates the wavenumber of the velocity from which the energy comes from. This function shows that most of the energy of the large scale field at $k = 1$ comes from velocity and magnetic field fluctuations at the forcing scale, which is here $k = k_f = 5$. At early times this is also true of the energy of the magnetic field at $k = 2$, but at late times, $t = 1000$, the gain from the forcing scale, $k = 5$, has diminished, and instead there is now a net loss of energy into the next larger scale, $k = 3$, suggestive of a direct cascade operating at $k = 2$, and similarly at $k = 3$.

Based on these results we may conclude that in the saturated state the magnetic energy at $k = 1$ is sustained by a *non-local* inverse cascade from the forcing scale directly to the largest scale of the box. This is characteristic of the α-effect of mean-field electrodynamics, except that here nonlinearity plays an essential role in isolating the large scale from the small scale "magnetic trash", as Parker used to say.

A closer look at Fig. 9.24, where $k = k_f = 30$, suggests that once the scale separation is large enough the energy is at first transferred not to the scale of the box, but instead to a somewhat smaller scale (here at wavenumber $k = 7$). Following the corresponding discussion in Brandenburg (2001b), this wavenumber is close to the wavenumber, $k_{max} = \frac{1}{2}|\alpha|/\eta_T$, where the α^2 dynamo grows fastest.

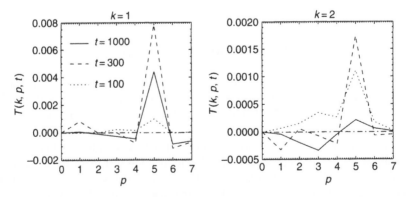

Figure 9.28 Spectral energy transfer function $T(k, p, t)$, normalized by $\langle \mathbf{B}^2 \rangle$ for three different times, for $k = 1$ and 2. Run 3 of Brandenburg (2001b).

In the following section we address the issue of magnetic helicity conservation which has important consequences for the timescale after which the large scale field begins to develop. This has also a bearing on the widely discussed controversy of the so-called "catastrophic α-quenching" of Vainshtein and Cattaneo (1992).

9.7.5. *Approximate helicity conservation*

The magnetic helicity, $H = \langle \mathbf{A} \cdot \mathbf{B} \rangle$, is conserved by the non-resistive MHD equations. For a closed or periodic box $\langle \mathbf{A} \cdot \mathbf{B} \rangle$ satisfies the equation

$$\frac{d}{dt} \langle \mathbf{A} \cdot \mathbf{B} \rangle = -2\eta\mu_0 \langle \mathbf{J} \cdot \mathbf{B} \rangle, \qquad (9.165)$$

where $\langle \mathbf{J} \cdot \mathbf{B} \rangle$ is the current helicity, and angular brackets denote volume averages. Note that for a periodic box $\langle \mathbf{A} \cdot \mathbf{B} \rangle$ is gauge invariant, i.e. $\langle \mathbf{A} \cdot \mathbf{B} \rangle$ does not change after a gauge transformation, $\mathbf{A} \rightarrow \mathbf{A} + \nabla\varphi$. This is a direct consequence of the solenoidality of the magnetic field, because $\langle \nabla\varphi \cdot \mathbf{B} \rangle = -\langle \varphi \nabla \cdot \mathbf{B} \rangle = 0$ owing to $\nabla \cdot \mathbf{B} = 0$.

In order to judge whether $\langle \mathbf{A} \cdot \mathbf{B} \rangle$ is small or large we calculate the length scale

$$\ell_H = |\langle \mathbf{A} \cdot \mathbf{B} \rangle| / \langle \mathbf{B}^2 \rangle. \qquad (9.166)$$

In Fig. 9.29 we see that the evolution of ℓ_H proceeds in three distinct phases: (i) a very short period ($t < 1$) where ℓ_H is very small and comparable to the numerical noise level, so magnetic helicity almost perfectly conserved, (ii) an intermediate interval ($2 < t < 200$) where ℓ_H is much larger, but still only roughly equal to the mesh size of the calculation, and then (iii) a regime where ℓ_H is of order unity. The latter is only possible because of the presence of helicity in the system, which leads to a large scale magnetic field configuration that is nearly force-free.

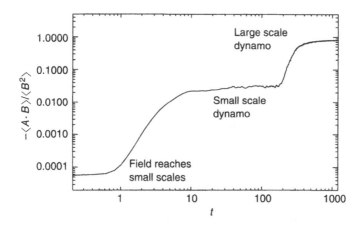

Figure 9.29 Evolution of the (negative) magnetic helicity length scale in a double-logarithmic plot. Note the presence of three distinct phases: very approximate helicity conservation near zero, followed by a phase of larger magnetic helicity scale (three orders of magnitude), and finally a phase where the magnetic helicity scale has reached the scale of the box.

9.7.6. Resistively limited growth of the large scale field

The approximate conservation of magnetic helicity has an important consequence for the generation of large scale fields: in order to build up a large scale field with magnetic helicity one has to *change* the value of $\langle \mathbf{A} \cdot \mathbf{B} \rangle$ from its initial value of zero to its certain final value. This final value of $\langle \mathbf{A} \cdot \mathbf{B} \rangle$ is such that the length scale ℓ_H is close to the maximum value possible for a certain geometry. It also implies that when interpreting the results in terms of mean-field electrodynamics (α-effect and turbulent diffusion), the α-effect must be quenched early on, well before the field has reached its final value. Nevertheless, this strong quenching, which was first anticipated by Vainshtein and Cattaneo (1992) and later confirmed numerically by Cattaneo and Hughes (1996), does not prevent the field from reaching its final super-equipartition field strength.

Before we come to the details we mention already now that the helicity constraint is probably too severe to be acceptable for astrophysical conditions, so one must look for possible escape routes. The most plausible way of relaxing the helicity constraint is in allowing for open boundary conditions (Blackman and Field, 2000; Kleeorin *et al.*, 2000), but the situation can still be regarded as inconclusive. Given that much of the work on large scale dynamos so far assumes periodic boundaries, we shall now consider this particular case in more detail. In the periodic case the final field geometry can be, e.g. of the form $\bar{\mathbf{B}} = B_0(\cos k_1 z, \sin k_1 z, 0)$, where $k_1 = 1$ is the smallest possible wavenumber in the box, B_0 is the field amplitude, and $\langle \bar{\mathbf{B}}^2 \rangle = B_0^2$. Alternatively, the field may vary in the x or y direction, and there may be an arbitrary phase shift; examples of these possibilities have been reported in Brandenburg (2001b). Anyway, for $\bar{\mathbf{B}} = B_0(\cos k_1 z, \sin k_1 z, 0)$ the corresponding vector potential is $\bar{\mathbf{A}} = -(B_0/k_1)(\cos k_1 z, \sin k_1 z, 0) + \nabla \phi$, where ϕ is an arbitrary gauge which does not affect the value of $\langle \bar{\mathbf{A}} \cdot \bar{\mathbf{B}} \rangle$. In this example we have

$$\langle \bar{\mathbf{A}} \cdot \bar{\mathbf{B}} \rangle = -\langle \bar{\mathbf{B}}^2 \rangle / k_1, \tag{9.167}$$

where we have included the k_1 factor, even though in the present case $k_1 = 1$. (The minus sign in (9.167) would turn into a plus if the forcing had negative helicity.) The mean current density is given by $\bar{\mathbf{J}} = -B_0 k_1 (\cos k_1 z, \sin k_1 z, 0)$, so the current helicity of the mean field is given by

$$\mu_0 \langle \bar{\mathbf{J}} \cdot \bar{\mathbf{B}} \rangle = -k_1 \langle \bar{\mathbf{B}}^2 \rangle. \tag{9.168}$$

Before we can use (9.167) and (9.168) in (9.165) we need to relate the magnetic and current helicities of the mean field to those of the actual field. We can generally split up the two helicities into contributions from large and small scales, i.e.

$$\langle \mathbf{A} \cdot \mathbf{B} \rangle = \langle \bar{\mathbf{A}} \cdot \bar{\mathbf{B}} \rangle + \langle \mathbf{a} \cdot \mathbf{b} \rangle, \tag{9.169}$$

$$\langle \mathbf{J} \cdot \mathbf{B} \rangle = \langle \bar{\mathbf{J}} \cdot \bar{\mathbf{B}} \rangle + \langle \mathbf{j} \cdot \mathbf{b} \rangle. \tag{9.170}$$

As the large scale magnetic field begins to saturate, the magnetic helicity has to become constant and so (9.165) dictates that $\langle \mathbf{J} \cdot \mathbf{B} \rangle$ must go to zero in the steady state. Consequently, the contribution from $\langle \mathbf{j} \cdot \mathbf{b} \rangle$ must be as large as that of $\langle \bar{\mathbf{J}} \cdot \bar{\mathbf{B}} \rangle$, and of opposite sign, so that the two cancel. This, together with (9.168), allows us immediately to write down an expression for the equilibrium strength of the mean field;

$$\langle \bar{\mathbf{B}}^2 \rangle = \mu_0 |\langle \mathbf{j} \cdot \mathbf{b} \rangle| / k_1, \tag{9.171}$$

which is now valid for both signs of the helicity of the forcing. The *residual* helicity (Pouquet et al., 1976),

$$H_{\text{res}} = \langle \boldsymbol{\omega} \cdot \mathbf{u} \rangle - \langle \mathbf{j} \cdot \mathbf{b} \rangle / \rho_0, \tag{9.172}$$

is small in the nonlinear saturated state and nearly vanishing. [We mention that this is also the case in the models of Brandenburg and Subramanian (2000).] Furthermore, for forced turbulence with a well defined forcing wavenumber the kinetic helicity may be estimated as $\langle \boldsymbol{\omega} \cdot \mathbf{u} \rangle \approx k_f \langle \mathbf{u}^2 \rangle$. Together with (9.171) we have

$$\langle \bar{\mathbf{B}}^2 \rangle = \frac{k_f}{k_1} B_{\text{eq}}^2, \tag{9.173}$$

where $B_{\text{eq}}^2 = \mu_0 \langle \rho \mathbf{u}^2 \rangle$ and B_{eq} is the equipartition field strength. Thus, the mean field can exceed (!) the equipartition field by the factor $(k_f / k_1)^{1/2}$. This estimate agrees well with the results of the simulations; see Brandenburg (2001b).

Using (9.169) and (9.170) together with (9.167) and (9.168) we can rewrite (9.165) in the form

$$\frac{\mathrm{d}}{\mathrm{d}t} \langle \bar{\mathbf{B}}^2 \rangle = -2\eta k_1^2 \langle \bar{\mathbf{B}}^2 \rangle + 2\eta k_1 \mu_0 |\langle \mathbf{j} \cdot \mathbf{b} \rangle|, \tag{9.174}$$

where we have taken into account the contribution of the small scale current helicity which is of similar magnitude as the large scale current helicity. For the magnetic helicity, on the other hand, the small scale contribution is negligible, because

$$|\langle \mathbf{a} \cdot \mathbf{b} \rangle| \approx \mu_0 |\langle \mathbf{j} \cdot \mathbf{b} \rangle| / k_f^2 \approx \mu_0 |\langle \bar{\mathbf{J}} \cdot \bar{\mathbf{B}} \rangle| / k_f^2 \approx |\langle \bar{\mathbf{A}} \cdot \bar{\mathbf{B}} \rangle| \left(\frac{k_1}{k_f} \right)^2 \ll |\langle \bar{\mathbf{A}} \cdot \bar{\mathbf{B}} \rangle|. \tag{9.175}$$

After the saturation at small and intermediate scales the small scale current helicity is approximately constant and can be estimated as

$$|\langle \mathbf{j} \cdot \mathbf{b} \rangle| \approx \rho_0 |\langle \boldsymbol{\omega} \cdot \mathbf{u} \rangle| \approx k_f \langle \rho \mathbf{u}^2 \rangle = k_f B_{\text{eq}}^2 / \mu_0. \tag{9.176}$$

The solution of (9.174) is given by

$$\langle \bar{\mathbf{B}}^2 \rangle = \epsilon_0 B_{\text{eq}}^2 \left[1 - e^{-2\eta k_1^2 (t - t_{\text{sat}})} \right], \tag{9.177}$$

where ϵ_0 is a coefficient which, in the present model with a well-defined forcing wavenumber, can be approximated by $\epsilon_0 \approx k_f / k_1$.

This is indeed also the limiting behavior found for α^2-dynamos with simultaneous α and η quenching of the form

$$\alpha = \frac{\alpha_0}{1 + \alpha_B \bar{\mathbf{B}}^2 / B_{\text{eq}}^2}, \quad \eta_t = \frac{\eta_{t0}}{1 + \eta_B \bar{\mathbf{B}}^2 / B_{\text{eq}}^2}, \tag{9.178}$$

where $\alpha_B = \eta_B$ is assumed. Assuming that the magnetic energy density of the mean field, $\bar{\mathbf{B}}^2$, is approximately uniform (which is well satisfied in the simulations) we can obtain the solution $\bar{\mathbf{B}} = \bar{\mathbf{B}}(t)$ of (9.301) in the form

$$\bar{\mathbf{B}}^2 / (1 - \bar{\mathbf{B}}^2 / B_{\text{fin}}^2)^{1 + \lambda / \eta k_1^2} = B_{\text{ini}}^2 \, e^{2\lambda t}, \tag{9.179}$$

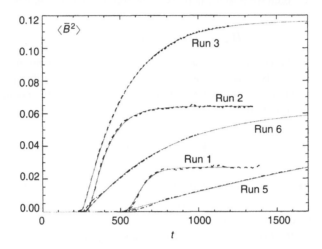

Figure 9.30 Evolution of $\langle \bar{\mathbf{B}}^2 \rangle$ for Runs 1–3, 5, and 6, compared with the solution (9.179) of the dynamo equations using (9.179).

where

$$\lambda = |\alpha_0| k_1 - \eta_{T0} k_1^2 \tag{9.180}$$

is the kinematic growth rate of the dynamo, B_{ini} is the initial field strength, and B_{fin} is the final field strength of the large scale field, which is related to α_B and η_B via

$$\alpha_B = \eta_B = \frac{\lambda}{\eta k_1^2} \left(\frac{B_{\text{eq}}}{B_{\text{fin}}} \right)^2. \tag{9.181}$$

The full derivation is given in Appendix E. The significance of this result is that it provides an excellent fit to the numerical simulations; see Fig. 9.30 where we present the evolution of $\langle \bar{\mathbf{B}}^2 \rangle$ for the different runs of Brandenburg (2001b). Equation (9.179) can therefore be used to extrapolate to astrophysical conditions. The time it takes to convert the small scale field generated by the small scale dynamo to a large scale field, τ_{eq}, increases linear with the magnetic Reynolds number, R_{m}. Apart from some coefficients of order unity the ratio of τ_{eq} to the turnover time is therefore just R_{m}. For the sun this ratio would be 10^8–10^{10}. However, before interpreting this result further one really has to know whether or not the presence of open boundary conditions could alleviate the issue of very long timescales for the mean magnetic field. Furthermore, it is not clear whether the long timescales discussed above have any bearing on the cycle period in the case of oscillatory solutions. The reason this is not so clear is because for a cyclic dynamo the magnetic helicity in each hemisphere stays always of the same sign and is only slightly modulated. It is likely that this modulation pattern is advected precisely with the meridional circulation, in which case the helicity could be nearly perfectly conserved in a lagrangian frame. This could provide an interesting clue for why the solar dynamo is migrating. The relation between meridional circulation and dynamo wave propagation has been advocated by Durney (1995) and Choudhuri *et al.* (1995), but helicity conservation would strongly lock the two aspects.

It is clear that virtually all astrophysical bodies are open, allowing for constant loss of magnetic helicity. In the case of the sun significant amounts of magnetic helicity are indeed observed at the solar surface (Berger and Ruzmaikin, 2000). Significant losses of magnetic helicity are particularly obvious in the case of accretion discs which are almost always accompanied by strong outflows that can sometimes be collimated into jets. Thus, dynamo action from accretion disc turbulence would be a good candidate for clarifying the significance of open boundaries on the nature of the dynamo. Another reason why accretion disc turbulence is a fruitful topic for understanding dynamo action is because the shear is extremely strong. In the following we discuss some recent progress that has been made in this field.

9.7.7. Joule dissipation from mean and fluctuating fields

In an MHD flow the mean magnetic Joule dissipation per unit volume is given by

$$Q_{\text{Joule}} = \eta \mu_0 \langle \mathbf{J}^2 \rangle. \tag{9.182}$$

Whilst in may astrophysical flows η may be very small, $|\mathbf{J}|$ can be large so that Q_{Joule} remains finite even in the limit $\eta \to 0$. One example where this is very important is accretion discs, where Joule dissipation (together with viscous dissipation) are important in heating the disc. These viscous and resistive processes are indeed the only significant sources of energy supply in discs, and yet the luminosities of discs that result from the conversion of magnetic and kinetic energies into heat and radiation can be enormous. Much of the work on discs involves mean-field modelling, so it would be interesting to see how the Joule dissipation, $Q_{\text{Joule}}^{(\text{mf})}$, predicted from a mean-field model,

$$Q_{\text{Joule}}^{(\text{mf})} = \eta_t \mu_0 \langle \bar{\mathbf{J}}^2 \rangle, \tag{9.183}$$

relates to the actual Joule dissipation. In Fig. 9.31 we show the evolution of actual and mean-field Joule dissipation and compare with an estimate for the rate of total energy dissipation, B_{eq}^2 / τ, where τ is the turnover time. Here we have taken into account that η_t is "catastrophically" quenched using the formulae of Brandenburg (2001a) with the parameters for Run 3.

There is no reason *a priori* that the magnetic energy dissipations from the mean-field model should agree with the actual one. It turns out that the mean-field dissipation is a fourth of the actual one, so it is definitely significant. It would therefore be interesting so see how those two dissipations compare with each other in other models.

9.7.8. Possible pitfalls in connection with hyperresistivity

In many astrophysical applications hyperresistivity and hyperviscosity are sometimes used in order to concentrate the effects of magnetic diffusion and viscosity to the smallest possible scale. The purpose of this section is to highlight possible spurious artifacts associated with this procedure. As we have seen above, large scale dynamos can depend on the microscopic magnetic diffusivity and must therefore be affected when it is replaced by hyperresistivity. The resulting modifications that are to be expected are easily understood: on the right hand side of (9.165) the term $\langle \mathbf{J} \cdot \mathbf{B} \rangle$ needs to be replaced by $\langle (\nabla^4 \mathbf{A}) \cdot \mathbf{B} \rangle$. This leads to a change of the relative importance of small and large scale contributions, which therefore changes (9.173) to

$$\langle \bar{\mathbf{B}}^2 \rangle = \left(\frac{k_f}{k_1} \right)^3 B_{\text{eq}}^2. \tag{9.184}$$

Thus, the final field strength will be even larger than before: instead of a factor of 5 super-equipartition (for $k_f = 5$) one now expects a factor of 125. Recent simulations by Brandenburg and Sarson (2002) have indeed confirmed this tendency. The main conclusion is that hyper-resistivity can therefore be used to address certain issues regarding large magnetic Reynolds numbers that are otherwise still inaccessible. On the other hand, the results are in some ways distorted and need therefore be interpreted carefully.

9.7.9. Remarks on accretion disc turbulence

We have already mentioned the possibility of dynamo action in accretion discs. Accretion discs have been postulated some 30 years ago in order to explain the incredibly high luminosities of quasars. Only in the past few years has direct imaging of accretion discs become possible, mostly due to the Hubble Space Telescope. Accretion discs form in virtually all collapse processes, such as galaxy and star formation. In the latter case the central mass is of the order of one solar mass, while in the former it is around 10^8 solar masses and is concentrated in such a small volume that that it must be a black hole. If the surrounding matter was nonrotating, it would fall radially towards the center. But this is unrealistic and even the slightest rotation relative to the central object would become important eventually as matter falls closer to the center.

If there was no effective diffusive process in discs, the angular momentum of the matter would stay with the gas parcels, and since the gravity force is balanced by the corresponding centrifugal force, the gas would never accrete. However, the angular velocity of the gas follows a $r^{-3/2}$ Kepler law, so the gas is differentially rotating and one may expect shear instabilities to occur that would drive turbulence and hence turbulent dissipation. Unfortunately, however, the story is not so simple. Discs are both linearly stable (Stewart, 1975) and probably also nonlinearly stable (Hawley *et al.*, 1996). Nevertheless, in the presence of a magnetic field there is a powerful *linear* instability (Balbus and Hawley, 1991), and subsequent work has shown that this instability is indeed capable of driving the instability and hence turbulence.

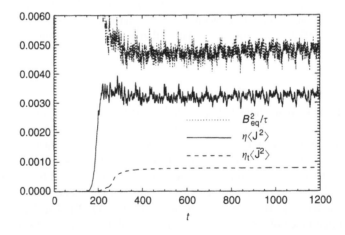

Figure 9.31 Joule dissipation for Run 3 (solid line), compared with the Joule dissipation estimated for a corresponding mean-field model (dashed line). An estimate for the rate of total energy dissipation, B_{eq}^2/τ, is also given.

One of the key outcomes of such simulations is the rate of turbulent dissipation, which determines the rate of angular momentum transport and correspondingly the rate at which orbital kinetic and potential energy is liberated in the form of heat. This is normally expressed in terms of a turbulent viscosity (e.g. Frank *et al.*, 1992), but it may equally well be expressed in terms of the horizontal components of the Reynolds and Maxwell stress tensors. The stress may then be normalized by $\langle \rho c_s^2 \rangle$ to give a non-dimensional measure (called α_{SS}) of the ability of the turbulence to transport angular momentum outward (if $\alpha_{SS} > 0$). This α_{SS} is indeed always positive, see Fig. 9.32, but it fluctuates significantly about a certain mean value. These fluctuations are in fact correlated with the energy in the mean magnetic field, $\langle \mathbf{B} \rangle^2$, as is shown in the right hand panel of Fig. 9.32. This mean magnetic field shows regular reversals combined with a migration away from the midplane, as can be seen in Fig. 9.33.

The evolution of the mean magnetic field found in the simulations is reminiscent of the behavior known from mean-field α–Ω dynamos. Further details regarding this correspondence (relation between the value of α and cycle period, field parity for different boundary conditions, etc.) can be found in recent reviews of the subject (e.g. Brandenburg, 1998, 2000).

9.7.10. Connection with the solar dynamo problem

The disc simulations have shown that a global large scale field can be obtained even in cartesian geometry. The detailed behavior of this large scale field depends of course on the boundary conditions adopted (Brandenburg, 1998), and will therefore be different in different geometries. Nevertheless, the very fact that large scale dynamo action is possible already in simple cartesian geometry is interesting.

In Fig. 9.34 we show the evolution of magnetic and kinetic energies as well as the magnitudes of the large scale field for a simulation of a convectively driven dynamo in the presence of large scale shear (Brandenburg *et al.*, 2001). It turns out that the ratio of the magnetic energies in large scale fields relative to the total field, $\langle \mathbf{B} \rangle^2 / \langle \mathbf{B}^2 \rangle$, which is a measure of the filling factor of the magnetic field, is around 15% when the field has reached saturation, i.e. when the field growth has stopped. This is similar to the case of isotropic nonmirror-symmetric turbulence considered in Section 9.7.1. On the average, however, the magnetic field is then directed

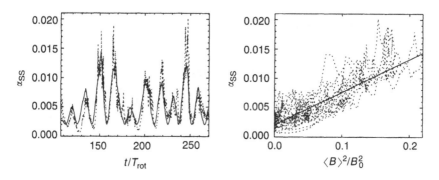

Figure 9.32 Dependence of α_{SS} on time and the mean magnetic field strength for a local accretion disc model (Run B of Brandenburg *et al.*, 1996a). Here $B_0 = \langle \mu_0 \rho c_s^2 \rangle^{1/2}$ is the thermal equipartition field strength and $T_{rot} = 2\pi / \Omega$ the local rotation period. In the left hand panel the dotted line represents the actual data and the solid line gives the fit obtained by correlating α_{SS} with the mean magnetic field (right-hand panel).

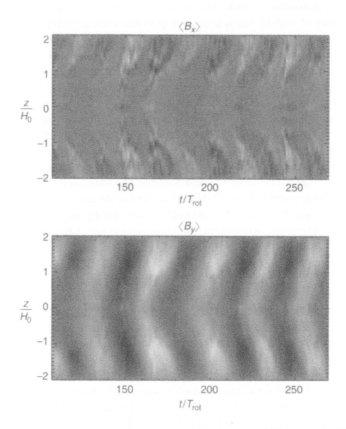

Figure 9.33 Butterfly (space-time) diagram of the poloidal and toroidal magnetic field components averaged over the two horizontal (x and y) directions for the local accretion disc model of (Brandenburg *et al.*, 1996a, Run B). Note that the poloidal field is much more noisy than the toroidal field, and that there is a clear outward migration of magnetic field.

into the negative y-direction (corresponding to the negative azimuthal direction in spherical geometry), but there is a weak and more noisy cross-stream field component directed in the positive x-direction (pointing north).

When the field has reached saturation, the mean field direction is approximately constant. Although the magnitude of this mean field fluctuates somewhat, the sign is always the same. Thus, this simulation shows no cycles, which are so characteristic of the solar dynamo. However, since those features, including the field geometry depend strongly on boundary condition and on the location of the boundary conditions, this disagreement is to be expected, and one would really need to resort to global simulations in spherical geometry.

9.7.11. Dynamics of the overshoot layer

Late-type stars with outer convection zones have an interface between the convection zone proper and the radiative interior. This leads to some additional dynamics that is important to include, especially in connection with the dynamo problem. This interface is the layer where magnetic flux can accumulate, i.e. *not* necessarily the layer where the dynamo operates;

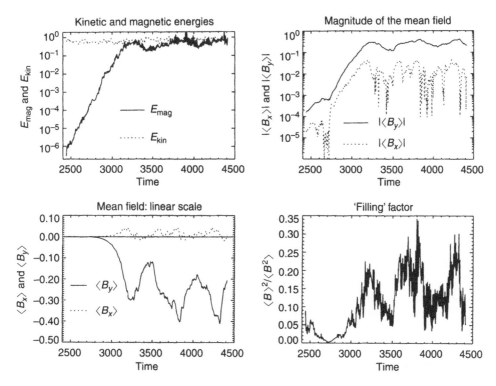

Figure 9.34 Evolution of several quantities for a convective dynamo model with shear: kinetic and magnetic energies (dotted and solid lines; first panel), mean latitudinal and toroidal fields (dotted and solid lines; second panel), mean magnetic field in a linear scale (third panel), and the filling factor (fourth panel). Energies and mean magnetic fields are given in units of the equipartition value, $B_{eq} = 4\pi \langle \mathbf{u}^2 \rangle$.

see the discussion in Brandenburg (1994). The accumulation is a consequence of turbulent pumping down the turbulence intensity gradient and the effect was seen clearly in video animations reported by Brandenburg and Tuominen (1991) and was analyzed in detail by Nordlund *et al.* (1992). Tobias *et al.* (1998) have studied the effect in isolation starting with an initial magnetic field distribution as opposed to a dynamo-generated field.

The flow dynamics changes drastically as one enters this overshoot layer. The stabilizing buoyancy effect provides a restoring force on a downward moving element, which can give rise to gravity waves that could be driven by individual plumes. This leads to a marked wavy pattern that can extend deep into the lower overshoot layer, as seen in Fig. 9.35 where we have plotted the vertical rms velocity as a function of depth and time. These waves extend a major fraction into the stably stratified layer beneath the convection zone, but are damped eventually. The typical period of such events is seen to be around 20 (in units of $\sqrt{d/g}$), where d is the depth of the unstable (convective) layer. This is comparable with the mean Brunt–Väisälä frequency,

$$N^2_{BV} = -\mathbf{g} \cdot \mathbf{\nabla}(s/c_p), \qquad (9.185)$$

which is around 0.3 in the overshoot layer; see Fig. 9.36.

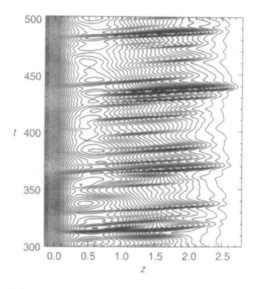

Figure 9.35 Space-time diagram of the vertical rms velocities for a nonmagnetic convection zone model of Brandenburg *et al.* (2001). Note the propagation of isolated plumes in more or less regular time intervals. Note also that the wavy pattern extends well into the convection zone proper ($0.5 \leq z \leq 1$), and that the plumes appear to propagate at an approximately constant speed towards the bottom. This speed is around 0.1, which is comparable to the rms velocity in the runs.

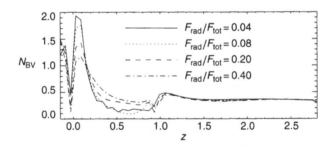

Figure 9.36 Modulus of the Brunt–Väisälä frequency for a run with polytropic index $m = -0.9$ and a nondimensional input flux $\mathcal{F}_{conv} = 0.01$. The various curves are for different values of the radiative flux, but fixed nominal convective flux. Small values of $\mathcal{F}_{rad}/\mathcal{F}_{tot}$ are typical for the upper layers of the solar convection zone.

Technically the presence of a lower overshoot layer provides a formidable challenge, because there the dynamics is governed by the very slow thermal timescale. This can lead to problems if the properties of the upper surface layers change which could affect the entropy in deeper parts of the convection zone. This in turn will also affect the stratification in the rest of the radiative interior. Since this can only happen on a thermal timescale (there is no turbulence in these layers to speed things up) it takes very long before one arrives at a new statistically stable state. This problem is of course also encountered when starting with initial conditions that are derived under too unrealistic assumptions. There are essentially two different approaches to this: one either considers a toy model where dynamical and thermal timescales are artificially brought closer together or one adopts an implicit code which allows the use of somewhat longer timesteps. The former approach implies that one adopts input fluxes that exceed those in real stars, but the good news is that turbulent velocities and temperature fluctuations vary with changing flux in exactly the way that is expected from mixing length theory; see Brandenburg *et al.* (2001) for details.

9.8. Conclusions

Many phenomena in astrophysics show direct manifestations of turbulence. As in the case of accretion discs, without turbulence there would be no enhanced dissipation, no heating of the disc, and hence no emission. Magnetic fields are of major importance as is evidenced again by the example of accretion discs, where the turbulence is a direct consequence of the presence of magnetic fields (Balbus–Hawley instability). Although magnetic fields in discs could have primordial origin and could just have been compressed during their formation, it is also clear that discs are actually a favorable candidate for producing strong large scale magnetic fields, as shown by the local simulations discussed above.

Other bodies where strong dynamo action is possible are stars. Simulations of convection have shown that strong small scale magnetic fields are possible (Meneguzzi and Pouquet, 1989; Nordlund *et al.*, 1992; Brandenburg *et al.*, 1996b; Cattaneo, 1999), but if there is strong shear an intense large scale field can also emerge.

The use of non-conservative high-order schemes has proved useful in all those applications. They are easy to implement and to modify, but they are also reasonably accurate. In this chapter we have illustrated the behavior of such schemes using various test problems. Using potential enthalpy and entropy as the main thermodynamic variables has a number of advantages, especially in connection with strongly stratified flows near a central object with a deep potential well, which is relevant to studying outflow phenomena. Contrary to common belief, high-order schemes are not more vulnerable to Gibbs phenomena near discontinuities. Instead, in simple advection tests high-order methods are able to produce smoother solutions with less viscosity, which is important for accurate modeling of turbulence.

In the last part of this chapter we have briefly mentioned some astrophysical applications of simulations using high-order schemes where hydromagnetic turbulence played an important role. In the next few years we may expect a dramatic increase in the quality and predictive power of such simulations, as larger computers become available. Already now a number of very promising simulations are emerging. There is important work addressing the stability of astrophysical jets in three dimensions (Ouyed *et al.*, 2000). Also worthwhile mentioning are recent high resolution simulations by Hawley (2000) of three-dimensional accretion tori in global geometry. What remains to be done in this field is a proper connection between disc physics and the launching mechanism of jets. This would require incorporating proper thermodynamics allowing for radiative cooling and magnetic heating in particular. Global simulations would also be highly desirable to address the global stellar dynamo problem. For example, it would be interesting to see how the dynamo works in fully convective stars. This problem is in some ways simpler than the solar dynamo problem, because one does not need to worry about the lower overshoot layer where the relevant timescales are much longer than in the convection zone proper.

Appendix A. Centered, one-sided and semi-one-sided derivatives

In Section 9.3 we gave the centered finite difference formulae for schemes of different order. Here we first describe the method for determining the finite difference formulae for second order, but the generalization to higher order is straightforward. We also give the corresponding expressions for one-sided and semi-one-sided finite difference formulae.

We want to write the derivative $f'(x)$ at the point x_i as

$$f_i' = a_{-1} f_{i-1} + a_0 f_0 + a_1 f_1, \tag{9.186}$$

where $f_i' = f'(x_i)$, $f_{i-1} = f(x_i - \delta x)$, and $f_{i+1} = f(x_i + \delta x)$. To determine the coefficients $a_{-1}, a_0,$ and a_1 we expand $f(x)$ up to second order

$$f(x) = c_0 + c_1 x + c_2 x^2. \tag{9.187}$$

The first derivative is then

$$f'(x) = c_1 + 2c_2 x. \tag{9.188}$$

In particular, the value at $x = 0$ is just $f'(0) = c_1$. Likewise, we have $f''(0) = 2c_2$. To determine all coefficients we make use of our knowledge at the neighboring points around x_i, i.e. we use the function values $f(x_i - \delta x) \equiv f_{i-1}$, $f(x_i) \equiv f_i$, and $f(x_i + \delta x) \equiv f_{i+1}$, so we have

$$f_{i-1} = c_0 + c_1(-\delta x) + c_2(-\delta x)^2, \tag{9.189}$$

$$f_i = c_0, \tag{9.190}$$

$$f_{i+1} = c_0 + c_1(+\delta x) + c_2(+\delta x)^2. \tag{9.191}$$

This can be written in matrix form

$$\begin{pmatrix} f_{i-1} \\ f_i \\ f_{i+1} \end{pmatrix} = \begin{pmatrix} (-1)^0 & (-1)^1 & (-1)^2 \\ 0^0 & 0^1 & 0^2 \\ 1^0 & 1^1 & 1^2 \end{pmatrix} \begin{pmatrix} c_0 \\ c_1 \delta x \\ c_2 \delta x^2 \end{pmatrix} \tag{9.192}$$

(where $(-1)^0 = 0^0 = 1^0 = 1$), or

$$\mathbf{f} = \mathbf{Mc}, \tag{9.193}$$

and so we obtain the coefficients as

$$\mathbf{c} = \mathbf{M}^{-1}\mathbf{f}. \tag{9.194}$$

To calculate f' we need the value of c_1, see (9.188), and so the coefficients a_n needed to express the derivative are $a_{-1} = (\mathbf{M}^{-1})_{10}$, $a_0 = (\mathbf{M}^{-1})_{11}$, and $a_1 = (\mathbf{M}^{-1})_{12}$, i.e. all points of the inverted matrix in the second row. The resulting formula for f_i' is well known,

$$f_i' = (-f_{i-1} + f_{i+1})/(2\delta x). \tag{9.195}$$

The corresponding result for the second derivative is

$$f_i'' = (f_{i-1} - 2f_i + f_{i+1})/(\delta x^2). \tag{9.196}$$

On the boundaries we have to calculate for derivative using only points inside the domain, which is explained in the next subsection for second order accuracy, but again the generalization to higher order is straightforward and only the results will be listed.

A.1. One-sided 2nd-order derivatives

Again, we want to write the derivative $f'(x)$ as

$$f'_i = a_0 f_0 + a_1 f_1 + a_2 f_2, \tag{9.197}$$

but now

$$\begin{pmatrix} f_i \\ f_{i+1} \\ f_{i+2} \end{pmatrix} = \begin{pmatrix} 0^0 & 0^1 & 0^2 \\ 1^0 & 1^1 & 1^2 \\ 2^0 & 2^1 & 2^2 \end{pmatrix} \begin{pmatrix} c_0 \\ c_1 \delta x \\ c_2 \delta x^2 \end{pmatrix}. \tag{9.198}$$

Thus, one arrives at

$$f'_i = (-3 f_i + 4 f_{i+1} - f_{i+2})/(2 \delta x). \tag{9.199}$$

Correspondingly, for the second derivative we have

$$f''_i = (2 f_i - 5 f_{i+1} + 4 f_{i+2} - f_{i+3})/\delta x^2. \tag{9.200}$$

A.2. 4th-order derivatives

First derivatives

$$f'_i = (f_{i-2} - 8 f_{i-1} + 8 f_{i+1} - f_{i+2})/(12 \delta x), \tag{9.201}$$

$$f'_i = (-3 f_{i-1} - 10 f_i + 18 f_{i+1} - 6 f_{i+2} + f_{i+3})/(12 \delta x), \tag{9.202}$$

$$f'_i = (-25 f_i + 48 f_{i+1} - 36 f_{i+2} + 16 f_{i+3} - 3 f_{i+4})/(12 \delta x). \tag{9.203}$$

Second derivatives

$$f''_i = (-f_{i-2} + 16 f_{i-1} - 30 f_i + 16 f_{i+1} - f_{i+2})/(12 \delta x^2), \tag{9.204}$$

$$f''_i = (11 f_{i-1} - 20 f_i + 6 f_{i+1} + 4 f_{i+2} - f_{i+3})/(12 \delta x^2), \tag{9.205}$$

$$f''_i = (35 f_i - 104 f_{i+1} + 114 f_{i+2} - 56 f_{i+3} + 11 f_{i+4})/(12 \delta x^2). \tag{9.206}$$

A.3. 6th-order derivatives

First derivatives

$$f'_i = (-f_{i-3} + 9 f_{i-2} - 45 f_{i-1} + 45 f_{i+1} - 9 f_{i+2} + f_{i+3})/(60 \delta x), \tag{9.207}$$

$$f'_i = (2 f_{i-2} - 24 f_{i-1} - 35 f_i + 80 f_{i+1} - 30 f_{i+2} + 8 f_{i+3} - f_{i+4})/(60 \delta x), \tag{9.208}$$

$$f'_i = (-10 f_{i-1} - 77 f_i + 150 f_{i+1} - 100 f_{i+2} + 50 f_{i+3} \\ - 15 f_{i+4} + 2 f_{i+5})/(60 \delta x), \tag{9.209}$$

$$f'_i = (-147 f_i + 360 f_{i+1} - 450 f_{i+2} + 400 f_{i+3} - 225 f_{i+4} \\ + 72 f_{i+5} - 10 f_{i+6})/(60 \delta x). \tag{9.210}$$

Second derivatives

$$f_i'' = (2f_{i-3} - 27f_{i-2} + 270f_{i-1} - 490f_i + 270f_{i+1}$$
$$- 27f_{i+2} + 2f_{i+3})/(180\delta x^2), \tag{9.211}$$

$$f_i'' = (-13f_{i-2} + 228f_{i-1} - 420f_i + 200f_{i+1} + 15f_{i+2}$$
$$- 12f_{i+3} + 2f_{i+4})/(180\delta x^2), \tag{9.212}$$

$$f_i'' = (137f_{i-1} - 147f_i - 255f_{i+1} + 470f_{i+2} - 285f_{i+3}$$
$$+ 93f_{i+4} - 13f_{i+5})/(180\delta x^2), \tag{9.213}$$

$$f_i'' = (812f_i - 3132f_{i+1} + 5265f_{i+2} - 5080f_{i+3}$$
$$+ 2970f_{i+4} - 972f_{i+5} + 137f_{i+6})/(180\delta x^2). \tag{9.214}$$

Appendix B. The 2*N*-RK3 scheme

If N is the number of variables to be updated from one timestep to the next, the $2N$-schemes require only $2 \times N$ variables to be stored in memory at any time. This is better than for the standard Runge–Kutta schemes. The general iteration formula is

$$w_i = \alpha_i w_{i-1} + hF(t_{i-1}, u_{i-1}), \quad u_i = u_{i-1} + \beta_i w_i. \tag{9.215}$$

For a 3-step scheme we have $i = 1, \ldots, 3$. In order to advance the variable u from $u^{(n)}$ at time $t^{(n)}$ to $u^{(n+1)}$ at time $t^{(n+1)} = t^{(n)} + h$ we set in (9.215)

$$u_0 = u^{(n)} \quad \text{and} \quad u^{(n+1)} = u_3, \tag{9.216}$$

with u_1 and u_2 being intermediate steps. In order to be able to calculate the first step, $i = 1$, for which no $w_{i-1} \equiv w_0$ exists, we have to require $\alpha_1 = 0$. Thus, we are left with five unknowns, $\alpha_2, \alpha_3, \beta_1, \beta_2$, and β_3. We write down (9.215) in explicit form for $i = 1, \ldots, 3$:

$$w_1 = hF(t_0, u_0), \qquad u_1 = u_0 + \beta_1 w_1, \tag{9.217}$$

$$w_2 = \alpha_2 w_1 + hF(t_1, u_1), \quad u_2 = u_1 + \beta_2 w_2, \tag{9.218}$$

$$w_3 = \alpha_3 w_2 + hF(t_2, u_2), \quad u_3 = u_2 + \beta_3 w_3. \tag{9.219}$$

Written in explicit form, we have, for $i = 1$,

$$u_1 = u_0 + \beta_1 hF(t_0, u_0). \tag{9.220}$$

The $i = 2$ step yields

$$w_2 = \alpha_2 hF(t_0, u_0) + hF(t_1, u_1), \tag{9.221}$$

$$u_2 = u_0 + h[(\beta_1 + \beta_2 \alpha_2)F(t_0, u_0) + \beta_2 F(t_1, u_1)], \tag{9.222}$$

and the $i = 3$ step gives

$$w_3 = h[\alpha_2 \alpha_3 F(t_0, u_0) + \alpha_3 F(t_1, u_1) + F(t_2, u_2)], \tag{9.223}$$

$$u_3 = u_0 + [\beta_1 + \alpha_2(\beta_2 + \alpha_3\beta_3)]h F(t_0, u_0)$$
$$+ (\beta_2 + \beta_3\alpha_3)h F(t_1, u_1) + \beta_3 h F(t_2, u_2). \tag{9.224}$$

The corresponding times can be calculated by putting $F = 1$. This yields

$$t_1 = t_0 + \beta_1 h, \tag{9.225}$$
$$t_2 = t_0 + h[\beta_1 + \beta_2(1 + \alpha_2)], \tag{9.226}$$
$$t_3 = t_0 + h[\beta_1 + \beta_3 + (1 + \alpha_2)(\beta_2 + \alpha_3\beta_3)]. \tag{9.227}$$

The last expression can also be written in the form

$$t_3 = t_0 + h\{\beta_1 + \beta_2(1 + \alpha_2) + \beta_3[1 + (1 + \alpha_2)\alpha_3)]\}. \tag{9.228}$$

Next we need to determine the conditions that the scheme is indeed of third order. This can be done by considering the differential equation

$$du/dt = u \quad \text{with} \quad u(0) = u_0 \tag{9.229}$$

for $u = u(t)$, where u_0 is the initial value of u. The exact solution of (9.229) is $u_0 e^t$. Its Taylor expansion for $t = t_0 + h$ is

$$u(t_0 + h) = u_0\left[1 + h + \tfrac{1}{2}h^2 + \tfrac{1}{6}h^3 + \cdots\right]. \tag{9.230}$$

The solution based on (9.224) is

$$u_3 = u_0 + [\beta_1 + \alpha_2(\beta_2 + \alpha_3\beta_3)]\, h u_0, +(\beta_2 + \alpha_3\beta_3)h u_1 + \beta_3 h u_2. \tag{9.231}$$

In order to compare with (9.230) we need the explicit expressions for u_1 and u_2, which are

$$u_1 = (1 + h\beta_1)u_0, \tag{9.232}$$

$$u_2 = \left\{1 + h[\beta_1 + (1 + \alpha_2)\beta_2] + h^2\beta_1\beta_2\right\} u_0. \tag{9.233}$$

Hence we can write

$$u_3 = u_0 + \gamma_1 h + \gamma_2 h^2 + \gamma_3 h^3 \tag{9.234}$$

with

$$\gamma_1 = \beta_1 + \beta_3 + (1 + \alpha_2)(\beta_2 + \alpha_3\beta_3), \tag{9.235}$$
$$\gamma_2 = \beta_1\beta_2 + \beta_3[\beta_2(1 + \alpha_2) + \beta_1(1 + \alpha_3)], \tag{9.236}$$
$$\gamma_3 = \beta_1\beta_2\beta_3. \tag{9.237}$$

In order for the scheme to be third order we have to require $\gamma_1 = 1$, $\gamma_2 = 1/2$, and $\gamma_3 = 1/6$; see (9.230). Thus, we have now *three* equations for *five* unknowns. We now have to come up with two more equations to solve for the five unknowns.

If we assume that the intermediate timesteps are evaluated in equidistant time intervals, we have to require that the time increments in (9.225) and (9.226) are $1/3$ and $2/3$, respectively. This yields

$$\beta_1 = \beta_2(1 + \alpha_2) = 1/3 \tag{9.238}$$

with two particular solutions[2]

$$\begin{array}{llll}
\alpha_2 = -2/3, & \alpha_3 = -1, & \beta_1 = 1/3, & \beta_2 = 1, & \beta_3 = 1/2, \\
\alpha_2 = -1/3, & \alpha_3 = -1, & \beta_1 = 1/3, & \beta_2 = 1/2, & \beta_3 = 1.
\end{array} \tag{9.239}$$

These are in fact the simplest $2N$-RK3 schemes that also lead to comparatively small residual errors.

Alternatively, one can move the times closer to the end time of the timestep and evaluate the right-hand side at times $t_1 - t_0 = \frac{1}{2}h$ and $t_2 - t_0 = h$. This gives the particular solution

$$\alpha_2 = -1/4, \quad \alpha_3 = -4/3, \quad \beta_1 = 1/2, \quad \beta_2 = 2/3, \quad \beta_3 = 1/2. \tag{9.240}$$

Again, there could be other solutions.

Another possibility is to require that the inhomogeneous equation

$$du/dt = t^n \qquad \text{with} \qquad u(0) = 0 \tag{9.241}$$

is solved exactly up to some n. The exact solutions for $t = h$ are $u = \frac{1}{2}h^2$ for $n = 1$ and $u = \frac{1}{3}h^3$ for $n = 2$.

The case $n = 0$ was already considered in (9.225)–(9.227). For $n = 1$ we have $F(t_1, u_1) = \frac{1}{3}h$ and $F(t_2, u_2) = \frac{2}{3}h$, so (9.224) gives

$$u_3 = u_0 + (\beta_2 + \beta_3\alpha_3)\frac{1}{3}h^2 + \beta_3\frac{2}{3}h^2, \tag{9.242}$$

or

$$u_3 = u_0 + \frac{1}{3}[\beta_2 + \beta_3(\alpha_3 + 2)]h^2. \tag{9.243}$$

Comparing with the exact solution this yields the additional equation

$$\beta_2 + \beta_3(\alpha_3 + 2) = \frac{3}{2}. \tag{9.244}$$

For $n = 2$ we have $F(t_1, u_1) = \frac{1}{9}h^2$ and $F(t_2, u_2) = \frac{4}{9}h^2$. So (9.224) gives

$$u_3 = u_0 + (\beta_2 + \beta_3\alpha_3)\frac{1}{9}h^3 + \beta_3\frac{4}{9}h^3, \tag{9.245}$$

or

$$u_3 = u_0 + \frac{1}{9}[\beta_2 + \beta_3(\alpha_3 + 4)]h^3. \tag{9.246}$$

Again, comparing with the exact solution one obtains

$$\beta_2 + \beta_3(\alpha_3 + 4) = 3. \tag{9.247}$$

This gives the solution

$$\alpha_2 = -17/32, \quad \alpha_3 = -32/27, \quad \beta_1 = 1/4, \quad \beta_2 = 8/9, \quad \beta_3 = 3/4, \tag{9.248}$$

which implies that the right hand sides are evaluated at the times $t_1 - t_0 = \frac{1}{4}h$ and $t_2 - t_0 = \frac{2}{3}h$. In tables 9.1 and 9.2 this scheme is referred to as "inhomogeneous."

Yet another idea (W. Dobler, private communication) is to obtain the additional two equations by requiring that the quadratic differential equation $du/dt = u^2$ with $u_0 = 1$ is solved exactly. The solution is $u = (1 - t)^{-1}$, of which we only need the expansion up to h^2, so we have $u_3 \approx 1 + h + h^2$. Again, we use (9.224), but now with $F = u^2$:

$$u_3 = 1 + [\beta_1 + \alpha_2(\beta_2 + \alpha_3\beta_3)]\,h + (\beta_2 + \alpha_3\beta_3)hu_1^2 + \beta_3 hu_2^2. \tag{9.249}$$

We need u_1^2 and u_2^2 only up to the term linear in h. Using (9.232) and (9.233) we have

$$u_1^2 = 1 + 2\beta_1 h + O(h^2), \quad u_2^2 = 1 + 2[\beta_1 + (1 + \alpha_2)\beta_2]\,h + O(h^2). \tag{9.250}$$

Inserting this in (9.249) yields

$$u_3 = 1 + \delta_1 h + \delta_2 h^2 + O(h^3) \tag{9.251}$$

with

$$\delta_1 = \beta_1 + \alpha_2(\beta_2 + \alpha_3\beta_3) + (\beta_2 + \alpha_3\beta_3) + \beta_3 \tag{9.252}$$

and

$$\delta_2 = 2\beta_1(\beta_2 + \alpha_3\beta_3) + 2\beta_3[\beta_1 + (1 + \alpha_2)\beta_2]. \tag{9.253}$$

Thus, the two additional equations are

$$\beta_1 + (1 + \alpha_2)(\beta_2 + \alpha_3\beta_3) + \beta_3 = 1, \tag{9.254}$$

$$\beta_1[\beta_2 + (1 + \alpha_3)\beta_3] + (1 + \alpha_2)\beta_2 = \tfrac{1}{2}. \tag{9.255}$$

The numerical solution is

$$\alpha_2 = -0.36726297, \quad \alpha_3 = -1.0278248,$$
$$\beta_1 = 0.30842796, \quad \beta_2 = 0.54037434, \quad \beta_3 = 1 \tag{9.256}$$

which implies that the right-hand sides are evaluated at the times $t_1 - t_0 = 0.308h$ and $t_2 - t_0 = 0.650h$. In tables 9.1 and 9.2 this scheme is referred to as "quadratic."

Appendix C. Derivation of the jacobian for transformation on a sphere

Here we give the explicit derivation of (9.79) and (9.80). We first use the transformation in the form

$$x = \tilde{x}^2/\tilde{r}, \quad y = \tilde{x}\tilde{y}/\tilde{r}, \quad \text{if } \tilde{x} \geq \tilde{y}, \tag{9.257}$$

$$x = \tilde{x}\tilde{y}/\tilde{r}, \quad y = \tilde{y}^2/\tilde{r}, \quad \text{if } \tilde{x} \leq \tilde{y}. \tag{9.258}$$

To obtain the jacobian we differentiate with respect to x and, so we have

$$1 = 2\frac{\partial \ln \tilde{x}}{\partial \ln x} - \frac{\partial \ln \tilde{r}}{\partial \ln x}, \qquad 0 = \frac{\partial \ln \tilde{x}}{\partial \ln x} + \frac{\partial \ln \tilde{y}}{\partial \ln x} - \frac{\partial \ln \tilde{r}}{\partial \ln x}, \qquad \text{if } \tilde{x} \geq \tilde{y}, \tag{9.259}$$

$$1 = \frac{\partial \ln \tilde{x}}{\partial \ln x} + \frac{\partial \ln \tilde{y}}{\partial \ln x} - \frac{\partial \ln \tilde{r}}{\partial \ln x}, \qquad 0 = 2\frac{\partial \ln \tilde{y}}{\partial \ln x} - \frac{\partial \ln \tilde{r}}{\partial \ln x}, \qquad \text{if } \tilde{x} \leq \tilde{y}. \tag{9.260}$$

We now differentiate with respect to y:

$$0 = 2\frac{\partial \ln \tilde{x}}{\partial \ln y} - \frac{\partial \ln \tilde{r}}{\partial \ln y}, \qquad 1 = \frac{\partial \ln \tilde{x}}{\partial \ln y} + \frac{\partial \ln \tilde{y}}{\partial \ln y} - \frac{\partial \ln \tilde{r}}{\partial \ln y}, \qquad \text{if } \tilde{x} \geq \tilde{y}, \tag{9.261}$$

$$0 = \frac{\partial \ln \tilde{x}}{\partial \ln y} + \frac{\partial \ln \tilde{y}}{\partial \ln y} - \frac{\partial \ln \tilde{r}}{\partial \ln y}, \qquad 1 = 2\frac{\partial \ln \tilde{y}}{\partial \ln y} - \frac{\partial \ln \tilde{r}}{\partial \ln y}, \qquad \text{if } \tilde{x} \leq \tilde{y}. \tag{9.262}$$

The derivatives of $\tilde{r}$ can be written as

$$\frac{\partial \ln \tilde{r}}{\partial \ln x} = \left(\frac{\tilde{x}}{\tilde{r}}\right)^2 \frac{\partial \ln \tilde{x}}{\partial \ln x} + \left(\frac{\tilde{y}}{\tilde{r}}\right)^2 \frac{\partial \ln \tilde{y}}{\partial \ln x}, \tag{9.263}$$

$$\frac{\partial \ln \tilde{r}}{\partial \ln y} = \left(\frac{\tilde{x}}{\tilde{r}}\right)^2 \frac{\partial \ln \tilde{x}}{\partial \ln y} + \left(\frac{\tilde{y}}{\tilde{r}}\right)^2 \frac{\partial \ln \tilde{y}}{\partial \ln y}. \tag{9.264}$$

In all cases we have

$$\frac{x}{y} = \frac{\tilde{x}}{\tilde{y}}, \tag{9.265}$$

so

$$+1 = \frac{\partial \ln \tilde{x}}{\partial \ln x} - \frac{\partial \ln \tilde{y}}{\partial \ln x}, \tag{9.266}$$

$$-1 = \frac{\partial \ln \tilde{x}}{\partial \ln y} - \frac{\partial \ln \tilde{y}}{\partial \ln y}, \tag{9.267}$$

and so

$$1 = 2\frac{\partial \ln \tilde{x}}{\partial \ln x} - \left(\frac{\tilde{x}}{\tilde{r}}\right)^2 \frac{\partial \ln \tilde{x}}{\partial \ln x} - \left(\frac{\tilde{y}}{\tilde{r}}\right)^2 \left(\frac{\partial \ln \tilde{x}}{\partial \ln x} - 1\right), \qquad \text{if } \tilde{x} \geq \tilde{y}, \tag{9.268}$$

$$0 = \frac{\partial \ln \tilde{x}}{\partial \ln x} + \frac{\partial \ln \tilde{y}}{\partial \ln x} - \left(\frac{\tilde{x}}{\tilde{r}}\right)^2 \frac{\partial \ln \tilde{x}}{\partial \ln x} - \left(\frac{\tilde{y}}{\tilde{r}}\right)^2 \left(\frac{\partial \ln \tilde{x}}{\partial \ln x} - 1\right), \qquad \text{if } \tilde{x} \geq \tilde{y}, \tag{9.269}$$

so

$$1 = \frac{\partial \ln \tilde{x}}{\partial \ln x} + \left(\frac{\tilde{y}}{\tilde{r}}\right)^2, \qquad \text{if } \tilde{x} \geq \tilde{y}, \tag{9.270}$$

and so

$$\frac{\partial \ln \tilde{x}}{\partial \ln x} = 1 - \left(\frac{\tilde{y}}{\tilde{r}}\right)^2, \qquad \frac{\partial \ln \tilde{y}}{\partial \ln x} = -\left(\frac{\tilde{y}}{\tilde{r}}\right)^2, \qquad \text{if } \tilde{x} \geq \tilde{y}, \tag{9.271}$$

and correspondingly

$$1 = \frac{\partial \ln \tilde{x}}{\partial \ln x} + \frac{\partial \ln \tilde{y}}{\partial \ln x} - \left(\frac{\tilde{x}}{\tilde{r}}\right)^2 \frac{\partial \ln \tilde{x}}{\partial \ln x} - \left(\frac{\tilde{y}}{\tilde{r}}\right)^2 \left(\frac{\partial \ln \tilde{x}}{\partial \ln x} - 1\right), \quad \text{if } \tilde{x} \le \tilde{y}, \tag{9.272}$$

$$0 = 2\frac{\partial \ln \tilde{x}}{\partial \ln x} - \left(\frac{\tilde{x}}{\tilde{r}}\right)^2 \frac{\partial \ln \tilde{x}}{\partial \ln x} - \left(\frac{\tilde{y}}{\tilde{r}}\right)^2 \left(\frac{\partial \ln \tilde{x}}{\partial \ln x} - 1\right), \quad \text{if } \tilde{x} \le \tilde{y}, \tag{9.273}$$

so

$$1 = \frac{\partial \ln \tilde{y}}{\partial \ln x} + \left(\frac{\tilde{y}}{\tilde{r}}\right)^2, \quad \text{if } \tilde{x} \le \tilde{y}, \tag{9.274}$$

and so

$$\frac{\partial \ln \tilde{x}}{\partial \ln x} = 2 - \left(\frac{\tilde{y}}{\tilde{r}}\right)^2, \quad \frac{\partial \ln \tilde{y}}{\partial \ln x} = 1 - \left(\frac{\tilde{y}}{\tilde{r}}\right)^2, \quad \text{if } \tilde{x} \le \tilde{y}. \tag{9.275}$$

Hence note that there is a discontinuity of the jacobian along the diagonals. Now for the y-derivatives we have

$$0 = 2\frac{\partial \ln \tilde{x}}{\partial \ln y} - \left(\frac{\tilde{x}}{\tilde{r}}\right)^2 \frac{\partial \ln \tilde{x}}{\partial \ln y} - \left(\frac{\tilde{y}}{\tilde{r}}\right)^2 \left(\frac{\partial \ln \tilde{x}}{\partial \ln y} + 1\right), \quad \text{if } \tilde{x} \ge \tilde{y}, \tag{9.276}$$

$$1 = \frac{\partial \ln \tilde{x}}{\partial \ln y} + \frac{\partial \ln \tilde{y}}{\partial \ln y} - \left(\frac{\tilde{x}}{\tilde{r}}\right)^2 \frac{\partial \ln \tilde{x}}{\partial \ln y} - \left(\frac{\tilde{y}}{\tilde{r}}\right)^2 \left(\frac{\partial \ln \tilde{x}}{\partial \ln y} + 1\right), \quad \text{if } \tilde{x} \ge \tilde{y}, \tag{9.277}$$

so

$$0 = \frac{\partial \ln \tilde{x}}{\partial \ln y} - \left(\frac{\tilde{y}}{\tilde{r}}\right)^2, \quad \text{if } \tilde{x} \ge \tilde{y}, \tag{9.278}$$

and so

$$\frac{\partial \ln \tilde{x}}{\partial \ln y} = +\left(\frac{\tilde{y}}{\tilde{r}}\right)^2, \quad \frac{\partial \ln \tilde{y}}{\partial \ln y} = 1 + \left(\frac{\tilde{y}}{\tilde{r}}\right)^2, \quad \text{if } \tilde{x} \ge \tilde{y}, \tag{9.279}$$

and correspondingly

$$0 = \frac{\partial \ln \tilde{x}}{\partial \ln y} + \frac{\partial \ln \tilde{y}}{\partial \ln y} - \left(\frac{\tilde{x}}{\tilde{r}}\right)^2 \frac{\partial \ln \tilde{x}}{\partial \ln y} - \left(\frac{\tilde{y}}{\tilde{r}}\right)^2 \left(\frac{\partial \ln \tilde{x}}{\partial \ln y} + 1\right), \quad \text{if } \tilde{x} \le \tilde{y}, \tag{9.280}$$

$$-1 = 2\frac{\partial \ln \tilde{x}}{\partial \ln y} - \left(\frac{\tilde{x}}{\tilde{r}}\right)^2 \frac{\partial \ln \tilde{x}}{\partial \ln y} - \left(\frac{\tilde{y}}{\tilde{r}}\right)^2 \left(\frac{\partial \ln \tilde{x}}{\partial \ln y} + 1\right), \quad \text{if } \tilde{x} \le \tilde{y}, \tag{9.281}$$

so

$$0 = \frac{\partial \ln \tilde{x}}{\partial \ln y} + 1 - \left(\frac{\tilde{y}}{\tilde{r}}\right)^2, \quad \text{if } \tilde{x} \le \tilde{y}, \tag{9.282}$$

and so

$$\frac{\partial \ln \tilde{x}}{\partial \ln y} = -1 + \left(\frac{\tilde{y}}{\tilde{r}}\right)^2, \quad \frac{\partial \ln \tilde{y}}{\partial \ln y} = +\left(\frac{\tilde{y}}{\tilde{r}}\right)^2, \quad \text{if } \tilde{x} \le \tilde{y}. \tag{9.283}$$

So, in summary, we have

$$
\begin{pmatrix}
\dfrac{\partial \ln \tilde{x}}{\partial \ln x} & \dfrac{\partial \ln \tilde{x}}{\partial \ln y} \\[2mm]
\dfrac{\partial \ln \tilde{y}}{\partial \ln x} & \dfrac{\partial \ln \tilde{y}}{\partial \ln y}
\end{pmatrix}
=
\begin{pmatrix}
+1 - \left(\dfrac{\tilde{y}}{\tilde{r}}\right)^2 & +1 - \left(\dfrac{\tilde{x}}{\tilde{r}}\right)^2 \\[3mm]
-1 + \left(\dfrac{\tilde{x}}{\tilde{r}}\right)^2 & +1 + \left(\dfrac{\tilde{y}}{\tilde{r}}\right)^2
\end{pmatrix}
\qquad \text{if } \tilde{x} \geq \tilde{y},
\tag{9.284}
$$

$$
\begin{pmatrix}
\dfrac{\partial \ln \tilde{x}}{\partial \ln x} & \dfrac{\partial \ln \tilde{x}}{\partial \ln y} \\[2mm]
\dfrac{\partial \ln \tilde{y}}{\partial \ln x} & \dfrac{\partial \ln \tilde{y}}{\partial \ln y}
\end{pmatrix}
=
\begin{pmatrix}
+1 + \left(\dfrac{\tilde{x}}{\tilde{r}}\right)^2 & -1 + \left(\dfrac{\tilde{y}}{\tilde{r}}\right)^2 \\[3mm]
+1 - \left(\dfrac{\tilde{y}}{\tilde{r}}\right)^2 & +1 - \left(\dfrac{\tilde{x}}{\tilde{r}}\right)^2
\end{pmatrix}
\qquad \text{if } \tilde{x} \leq \tilde{y}.
\tag{9.285}
$$

Appendix D. Derivation of the incremental jacobian for second derivatives

Here we present the explicit derivation of (9.110). To calculate the second derivative of a function f that is represented on a coordinate mesh $\tilde{\mathbf{x}}$, is given by

$$
\frac{\partial^2 f}{\partial x_i \partial x_j} = \frac{\partial}{\partial x_i}\left(\frac{\partial f}{\partial x_j}\right) = \frac{\partial}{\partial x_i}\left(\frac{\partial f}{\partial \tilde{x}_k}\frac{\partial \tilde{x}_k}{\partial x_j}\right),
\tag{9.286}
$$

so

$$
\frac{\partial^2 f}{\partial x_i \partial x_j} = \frac{\partial^2 f}{\partial \tilde{x}_l \partial \tilde{x}_k}\frac{\partial \tilde{x}_l}{\partial x_i}\frac{\partial \tilde{x}_k}{\partial x_j} + \frac{\partial f}{\partial \tilde{x}_k}\frac{\partial^2 \tilde{x}_k}{\partial x_i \partial x_j},
\tag{9.287}
$$

or

$$
\frac{\partial^2 f}{\partial x_i \partial x_j} = \frac{\partial^2 f}{\partial \tilde{x}_p \partial \tilde{x}_q}\mathsf{J}_{pi}\mathsf{J}_{qj} + \frac{\partial f}{\partial \tilde{x}_k}\mathsf{K}_{kij},
\tag{9.288}
$$

which is just (9.108), using (9.109) for the definition of K_{kij} of the 2nd-order jacobian.

To obtain the 2nd-order jacobian by successive tensor multiplication we differentiate twice the evolution equation for $\mathbf{x}$:

$$
x_k^{(n+1)} = x_k^{(n)} + u_k^{(n)}\,\delta t,
\tag{9.289}
$$

so

$$
\frac{\partial^2 x_k^{(n+1)}}{\partial x_i^{(n+1)}\partial x_j^{(n+1)}} = \frac{\partial^2 x_k^{(n)}}{\partial x_i^{(n+1)}\partial x_j^{(n+1)}} + \frac{\partial^2 u_k^{(n)}}{\partial x_i^{(n+1)}\partial x_j^{(n+1)}}\delta t.
\tag{9.290}
$$

The expression on the left-hand side is just the derivative of a Kronecker delta, see (9.102), so it is zero. Thus we have

$$
0 = \frac{\partial^2 x_k^{(n)}}{\partial x_i^{(n+1)}\partial x_j^{(n+1)}} + \frac{\partial}{\partial x_i^{(n+1)}}\left(\frac{\partial u_k^{(n)}}{\partial x_q^{(n)}}\frac{\partial x_q^{(n)}}{\partial x_j^{(n+1)}}\right)\delta t,
\tag{9.291}
$$

or

$$
0 = \frac{\partial^2 x_q^{(n)}}{\partial x_i^{(n+1)}\partial x_j^{(n+1)}}\left(\delta_{kq} + \frac{\partial u_k^{(n)}}{\partial x_q^{(n)}}\delta t\right) + \frac{\partial^2 u_k^{(n)}}{\partial x_p^{(n)}\partial x_q^{(n)}}\mathsf{J}_{pi}^{(n)}\mathsf{J}_{qj}^{(n)}\delta t,
\tag{9.292}
$$

which can be written as

$$0 = M_{kq}K_{qij}^{(n)} + u_{k.pq}J_{pi}^{(n)}J_{qj}^{(n)}\delta t, \tag{9.293}$$

so

$$K_{kij}^{(n)} = -\delta t \left(M^{-1}\right)_{kl} u_{lpq}J_{pi}^{(n)}J_{qj}^{(n)}, \tag{9.294}$$

which is just (9.112).

We now need to derive the equation that relates the incremental 2nd-order jacobians to the 2nd-order jacobian of the previous timestep. To this end we begin with the 2nd-order jacobian at time $2\delta t$, so

$$K_{kij}^{(0\to2)} \equiv \frac{\partial^2 x_k^{(0)}}{\partial x_i^{(2)}\partial x_j^{(2)}} = \frac{\partial}{\partial x_i^{(2)}}\left(\frac{\partial x_k^{(0)}}{\partial x_q^{(1)}}\frac{\partial x_q^{(1)}}{\partial x_j^{(2)}}\right), \tag{9.295}$$

or

$$K_{kij}^{(0\to2)} \equiv \frac{\partial^2 x_k^{(0)}}{\partial x_p^{(1)}\partial x_q^{(1)}}\frac{\partial x_p^{(1)}}{\partial x_i^{(2)}}\frac{\partial x_q^{(1)}}{\partial x_j^{(2)}} + \frac{\partial^2 x_q^{(1)}}{\partial x_i^{(2)}\partial x_j^{(2)}}\frac{\partial x_k^{(0)}}{\partial x_q^{(1)}}, \tag{9.296}$$

or

$$K_{kij}^{(0\to2)} = K_{kpq}^{(0\to1)}J_{pi}^{(1)}J_{qj}^{(1)} + J_{kl}^{(0\to1)}K_{lij}^{(1)}. \tag{9.297}$$

For the next step we have

$$K_{kij}^{(0\to3)} \equiv \frac{\partial^2 x_k^{(0)}}{\partial x_i^{(3)}\partial x_j^{(3)}} = \frac{\partial}{\partial x_i^{(3)}}\left(\frac{\partial x_k^{(0)}}{\partial x_q^{(1)}}\frac{\partial x_q^{(1)}}{\partial x_p^{(2)}}\frac{\partial x_p^{(2)}}{\partial x_j^{(3)}}\right), \tag{9.298}$$

so

$$K_{kij}^{(0\to3)} = K_{kpq}^{(0\to1)}J_{pr}^{(1)}J_{qs}^{(1)}J_{ri}^{(2)}J_{sj}^{(2)} + J_{kl}^{(0\to1)}K_{lrs}^{(1)}J_{ri}^{(2)}J_{sj}^{(2)} + J_{kr}^{(0)}J_{rs}^{(1)}K_{sij}^{(2)}. \tag{9.299}$$

This can be written as

$$K_{kij}^{(0\to3)} = K_{kpq}^{(0\to2)}J_{pi}^{(2)}J_{qj}^{(2)} + J_{kl}^{(0\to2)}K_{lij}^{(2)}, \tag{9.300}$$

which, for the general step from 0 to n, becomes (9.110).

Appendix E. Solution for α and η_t quenched α^2-dynamo

Here we present the explicit derivation of (9.181). According to mean-field theory for non-mirror symmetric isotropic homogeneous turbulence with no mean flow the mean magnetic field is governed by the equation

$$\frac{\partial}{\partial t}\bar{B} = \nabla \times \left(\alpha\bar{B} - \eta_T\eta_0\bar{J}\right), \tag{9.301}$$

where bars denote the mean fields and $\eta_T = \eta + \eta_t$ is the total (microscopic plus turbulent) magnetic diffusion. Both α-effect and turbulent diffusion are assumed to be quenched in the same way, so

$$\alpha = \frac{\alpha_0}{1 + \alpha_B \, \bar{\mathbf{B}}^2 / B_{eq}^2}, \qquad \eta_t = \frac{\eta_{t0}}{1 + \eta_B \, \bar{\mathbf{B}}^2 / B_{eq}^2}. \tag{9.302}$$

In the following we assume $\alpha_B = \eta_B$ and denote

$$\alpha_B / B_{eq}^2 = 1 / B_0^2 = \eta_B / B_{eq}^2. \tag{9.303}$$

We emphasize that only the *turbulent* magnetic diffusivity is quenched, not of course the total one. It is only because of the presence of microscopic diffusion that saturation is possible.

In the simulations $\bar{\mathbf{B}}^2$ is to a good approximation spatially uniform. Defining the magnetic energy as $M(t) = \frac{1}{2}\langle \bar{\mathbf{B}}^2 \rangle$ we have

$$\bar{\mathbf{B}}^2 = 2M, \tag{9.304}$$

which is only a function of time.

Consider the particular example where the large scale field varies only in the z direction (9.301) becomes

$$\dot{\bar{\mathbf{B}}}_x = -\alpha \bar{\mathbf{B}}'_y + \eta_T \bar{\mathbf{B}}''_x, \tag{9.305}$$

$$\dot{\bar{\mathbf{B}}}_y = +\alpha \bar{\mathbf{B}}'_x + \eta_T \bar{\mathbf{B}}''_y, \tag{9.306}$$

where dots and primes denote differentiation with respect to t and z, respectively. Since $\alpha < 0$, the solution can be written in the form

$$\bar{\mathbf{B}}_x(y, t) = b_x(t) \cos(z + \phi), \tag{9.307}$$

$$\bar{\mathbf{B}}_y(y, t) = b_y(t) \sin(z + \phi), \tag{9.308}$$

where $b_x(t)$ and $b_z(t)$ are positive functions of time that satisfy

$$\dot{b}_x = |\alpha| b_y - \eta_T b_x, \tag{9.309}$$

$$\dot{b}_y = |\alpha| b_x - \eta_T b_y. \tag{9.310}$$

We now choose the special initial condition, $b_x = b_y \equiv b$, so we have only one equation for the variable b. Note also that in the quenching factor $\bar{\mathbf{B}}^2 = b_x^2 + b_y^2 = 2b^2$. Thus, we have

$$\frac{db}{dt} = \frac{\alpha_0 k_1 - \eta_{t0} k_1^2}{1 + 2b^2 / B_0^2} b - \eta k_1^2 b. \tag{9.311}$$

Multiplying with b yields

$$\frac{1}{2} \frac{db^2}{dt} = \frac{\alpha_0 k_1 - \eta_{t0} k_1^2}{1 + 2b^2 / B_0^2} b^2 - \eta k_1^2 b^2. \tag{9.312}$$

Using the definition $M = \frac{1}{2}b^2$ we have

$$\frac{1}{2}\frac{1}{M}\frac{dM}{dt} = \frac{\alpha_0 k_1 - \eta_{t0} k_1^2}{1 + 2M/M_0} - \eta k_1^2, \tag{9.313}$$

where $M_0 = \frac{1}{2}B_0^2$. Thus, we have

$$\left(1 + 2\frac{M}{M_0}\right)\frac{1}{2}\frac{1}{M}\frac{dM}{dt} = \alpha_0 k_1 - \eta_{T0} k_1^2 - 2\eta k_1^2\frac{M}{M_0}, \tag{9.314}$$

where we have defined $\eta_{T0} = \eta_{t0} + \eta$. We define the abbreviations $\lambda = \alpha_0 k_1 - \eta_{T0} k_1^2$ for the kinematic growth rate of the dynamo and $\lambda_{t0} = \eta_{t0} k_1^2$ for the turbulent decay rate if there were no dynamo action, and arrive thus at the integral

$$\int_{M_{ini}}^{M}\frac{1 + 2M'/M_0}{1 - (2\eta k_1^2/\lambda)\,M'/M_0}\frac{dM'}{M'} = 2\lambda t. \tag{9.315}$$

We now also define the abbreviation

$$\mathcal{M} = 2M'/M_0 \tag{9.316}$$

and have

$$\int_{M_{ini}}^{M}\frac{1 + \mathcal{M}}{1 - (\eta k_1^2/\lambda)\,\mathcal{M}}\frac{d\mathcal{M}}{\mathcal{M}} = 2\lambda t, \tag{9.317}$$

which can be split into two integrals,

$$\int_{M_{ini}}^{M}\frac{d\mathcal{M}}{[1 - (\eta k_1^2/\lambda)\,\mathcal{M}]\mathcal{M}} + \int_{M_{ini}}^{M}\frac{d\mathcal{M}}{1 - (\eta k_1^2/\lambda)\,\mathcal{M}} = 2\lambda t. \tag{9.318}$$

To solve these integrals we note that

$$\int\frac{dx}{1 - x/x_0} = -x_0 \ln(x_0 - x), \tag{9.319}$$

$$\int\frac{dx}{(1 - x/x_0)x} = \int\frac{dx}{x_0 - x} + \int\frac{dx}{x} = \ln x - \ln(1 - x/x_0) = \ln\left(\frac{x}{x_0 - x}\right). \tag{9.320}$$

So (9.318) becomes

$$\ln\left(\frac{\mathcal{M}}{\lambda/\eta k_1^2 - \mathcal{M}}\right) + \frac{\lambda}{\eta k_1^2}\ln\left(\frac{1}{\lambda/\eta k_1^2 - \mathcal{M}}\right) = 2\lambda(t - t_0), \tag{9.321}$$

where t_0 is an integration constant. Exponentiation yields

$$\frac{\mathcal{M}}{(\lambda/\eta k_1^2 - \mathcal{M})^{1+\lambda/\eta k_1^2}} = e^{2\lambda(t-t_0)}. \tag{9.322}$$

In terms of the original variables, $\bar{\mathbf{B}}^2/B_0^2 = 2b^2/B_0^2 = 2M/M_0 = \mathcal{M}$, this becomes

$$\frac{\bar{\mathbf{B}}^2}{B_0^2} \bigg/ \left(\frac{\lambda}{\eta k_1^2} - \frac{\bar{\mathbf{B}}^2}{B_0^2}\right)^{1+\lambda/\eta k_1^2} = e^{2\lambda(t-t_0)}. \tag{9.323}$$

The final field strength,

$$B_{\text{fin}} = \lim_{t \to \infty} |\bar{\mathbf{B}}(t)|, \tag{9.324}$$

is obtained by requiring the denominator to vanish, which yields

$$\frac{B_{\text{fin}}^2}{B_0^2} = \frac{\lambda}{\eta k_1^2}. \tag{9.325}$$

Rewriting (9.323) in terms of B_{fin} we have

$$\frac{\bar{\mathbf{B}}^2}{B_{\text{fin}}^2} \bigg/ \left(1 - \frac{\bar{\mathbf{B}}^2}{B_{\text{fin}}^2}\right)^{1+\lambda/\eta k_1^2} = \left(\frac{\lambda}{\eta k_1^2}\right)^{1+\lambda/\eta k_1^2} e^{2\lambda(t-t_0)}. \tag{9.326}$$

We can express t_0 in terms of the initial field strength, $B_{\text{ini}} = |\bar{\mathbf{B}}(0)|$, and if the initial field strength is much weaker than the final field strength, i.e. $B_{\text{ini}} \ll B_{\text{fin}}$, then we can rewrite (9.326), in the form

$$\bar{\mathbf{B}}^2 \bigg/ \left(1 - \frac{\bar{\mathbf{B}}^2}{B_{\text{fin}}^2}\right)^{1+\lambda/\eta k_1^2} = B_{\text{ini}}^2\, e^{2\lambda t}. \tag{9.327}$$

Thus, for early times we have the familiar relation

$$|\bar{\mathbf{B}}| \approx B_{\text{ini}}\, e^{\lambda t} \quad (|\bar{\mathbf{B}}| \ll B_{\text{fin}}), \tag{9.328}$$

whereas for late times near the final field strength we have

$$\left(1 - \frac{\bar{\mathbf{B}}^2}{B_{\text{fin}}^2}\right)^{-(1+\lambda/\eta k_1^2)} \approx \left(\frac{B_{\text{ini}}}{B_{\text{fin}}}\right)^2 e^{2\lambda t} \quad (|\bar{\mathbf{B}}| \approx B_{\text{fin}}), \tag{9.329}$$

or

$$\bar{\mathbf{B}}^2 \approx B_{\text{fin}}^2 \left[1 - e^{-2\lambda(t-t_{\text{sat}})}\right]^{\eta k_1^2/(\lambda+\eta k_1^2)} \quad (|\bar{\mathbf{B}}| \approx B_{\text{fin}}), \tag{9.330}$$

where $t_{\text{sat}} = \lambda^{-1} \ln(B_{\text{fin}}/B_{\text{ini}})$ is the time it takes to reach saturation. If the Reynolds number is large we have $\lambda \gg \eta k_1^2$, and so

$$\bar{\mathbf{B}}^2 \approx B_{\text{fin}}^2 \left[1 - e^{-2\eta k_1^2(t-t_{\text{sat}})}\right] \quad (|\bar{\mathbf{B}}| \approx B_{\text{fin}}), \tag{9.331}$$

which is identical to the result obtained from helicity conservation.

Note that the solution (9.327) is governed by four parameters: B_{ini}, B_{fin}, λ, and ηk_1^2. The latter is known from the input data to the simulation, B_{ini} and λ can be determined from the linear growth phase of the dynamo (characterized by properties of the small scale dynamo!) and so B_{fin} is the only parameter that is determined by the nonlinearity of the dynamo and can easily be determined from the simulations. Once B_{fin} is measured from numerical experiments we know immediately the quenching parameters

$$\alpha_B = \eta_B = \frac{\lambda}{\eta k_1^2} \left(\frac{B_{\text{eq}}}{B_{\text{fin}}}\right)^2, \tag{9.332}$$

and since $B_{\text{fin}}^2 / B_{\text{eq}}^2 \approx k_{\text{f}} / k_1$ we have

$$\alpha_B = \eta_B \approx \frac{\lambda}{\eta k_1 k_{\text{f}}}, \tag{9.333}$$

which shows that α_B and η_B are proportional to the magnetic Reynolds number.

Acknowledgments

I am indebted to Åke Nordlund for teaching me many of the methods and techniques that are reflected to some extend in the present work. I thank Wolfgang Dobler and Petri Käpylä for their careful reading of the manuscript and for pointing out a number of mistakes in the manuscript.

Notes

1 The 4th-order compact scheme is really identical to calculating derivatives from a cubic spline, as was done in Nordlund and Stein (1990). In the book by Collatz (1966) the compact methods are also referred to as *Hermitian methods* or as *Mehrstellen-Verfahren*, because the derivative in one point is calculated using the derivatives in neighboring points.

2 I thank Petri Käpylä from Oulu for pointing out the second of these solutions.

References

Arber, T.D., Longbottom, A.W., Gerrard, C.L. and Milne, A.M., "A staggered grid, Lagrangian-Eulerian remap code for 3D MHD simulations," *J. Comp. Phys.* **171**, 151–181 (2001).

Balbus, S.A. and Hawley, J.F., "A powerful local shear instability in weakly magnetized disks. I. Linear analysis," *Astrophys. J.* **376**, 214–222 (1991).

Balsara, D. and Pouquet, A., "The formation of large-scale structures in supersonic magnetohydrodynamic flows," *Phys. Plasmas* **6**, 89–99 (1999).

Benson, D.J., "Computational methods in Lagrangian and Eulerian hydrocodes," *Comp. Meth. Appl. Mech. Eng.* **99**, 235–394 (1992).

Berger, M.A. and Ruzmaikin, A., "Rate of helicity production by solar rotation," *J. Geophys. Res.* **105**, 10481–10490 (2000).

Biskamp, D., "Cascade models for magnetohydrodynamic turbulence," *Phys. Rev. E* **50**, 2702–2711 (1994).

Biskamp, D. and Müller, W.-C., "Decay laws for three-dimensional magnetohydrodynamic turbulence," *Phys. Rev. Lett.* **83**, 2195–2198 (1999).

Bigazzi, A., "*Models of small-scale and large-scale dynamo action*," PhD thesis, Università Degli Studi Dell'Aquila (1999).

Blackman, E.G. and Field, G.F., "Constraints on the magnitude of α in dynamo theory," *Astrophys. J.* **534**, 984–988 (2000).

Blandford, R.D. and Payne, D.G., "Hydromagnetic flows from accretion discs and the production of radio jets," *Mon. Not. R. Astr. Soc.* **199**, 883–903 (1982).

Brandenburg, A., "Energy spectra in a model for convective turbulence," *Phys. Rev. Lett.* **69**, 605–608 (1992).

Brandenburg, A., "Solar dynamos: computational background," in: *Lectures on Solar and Planetary Dynamos* (Eds. M.R.E. Proctor and A.D. Gilbert), Cambridge University Press, pp. 117–159 (1994).

Brandenburg, A., "Disc Turbulence and Viscosity," in: *Theory of Black Hole Accretion Discs* (Eds. M.A. Abramowicz, G. Björnsson and J.E. Pringle), Cambridge University Press, pp. 61–86 (1998).

Brandenburg, A., "Helicity in large-scale dynamo simulations," in: *Magnetic Helicity in Space and Laboratory Plasmas* (Eds. M.R. Brown, R.C. Canfield and A.A. Pevtsov), Geophys. Monograph **111**, American Geophysical Union, Florida, pp. 65–73 (1999).

Brandenburg, A., "Dynamo-generated turbulence and outflows from accretion discs," *Phil. Trans. Roy. Soc. Lond.* A **358**, 759–776 (2000).

Brandenburg, A., "Magnetic mysteries," *Science* **292**, 2440–2441 (2001a).

Brandenburg, A., "The inverse cascade and nonlinear alpha-effect in simulations of isotropic helical hydromagnetic turbulence," *Astrophys. J.* **550**, 824–840 (2001b).

Brandenburg, A. and Donner, K.J., "The dependence of the dynamo alpha on vorticity," *Mon. Not. R. Astr. Soc.* **288**, L29–L33 (1997).

Brandenburg, A. and Schmitt, D., "Simulations of an alpha-effect due to magnetic buoyancy," *Astron. Astrophys.* **338**, L55–L58 (1998).

Brandenburg, A. and Subramanian, K., "Large scale dynamos with ambipolar diffusion nonlinearity," *Astron. Astrophys.* **361**, L33–L36 (2000).

Brandenburg, A. and Tuominen, I., "The solar dynamo," in: *The Sun and Cool Stars: Activity, Magnetism, Dynamos, IAU Coll. 130* (Eds. I. Tuominen, D. Moss and G. Rüdiger), Lecture Notes in Physics **380**, Springer-Verlag, pp. 223–233 (1991).

Brandenburg, A., Enqvist, K. and Olesen, P., "Large-scale magnetic fields from hydromagnetic turbulence in the very early universe," *Phys. Rev.* D **54**, 1291–1300 (1996).

Brandenburg, A., Nordlund, Å. and Stein, R.F., "Astrophysical convection and dynamos," in: *Geophysical and Astrophysical Convection* (Eds. P.A. Fox and R.M. Kerr), Gordon and Breach Science Publishers, pp. 85–105 (2000).

Brandenburg, A., Nordlund, Å. and Stein, R.F., "Simulation of a convective dynamo with imposed shear," Astron. Astrophys. (to be submitted) (2001).

Brandenburg, A., Nordlund, Å., Stein, R.F. and Torkelsson, U., "Dynamo generated turbulence and large scale magnetic fields in a Keplerian shear flow," *Astrophys. J.* **446**, 741–754 (1995).

Brandenburg, A., Nordlund, Å., Stein, R.F. and Torkelsson, U., "The disk accretion rate for dynamo generated turbulence," *Astrophys. J. Lett.* **458**, L45–L48 (1996a).

Brandenburg, A., Jennings, R.L., Nordlund, Å., Rieutord, M., Stein, R.F. and Tuominen, I., "Magnetic structures in a dynamo simulation," *J. Fluid Mech.* **306**, 325–352 (1996b).

Brandenburg, A., Dobler, W., Shukurov, A. and von Rekowski, B. "Pressure-driven outflow from a dynamo active disc, *Astron. Astrophys.* (submitted). astro-ph/0003174 (2000).

Brandenburg, A., Chan, K.L., Nordlund, Å. and Stein, R.F., "The effect of the radiative background flux in convection simulations," *Astron. Astrophys.* (to be submitted) (2001).

Brandenburg, A. and Sarson, G.R.,"The effect of hyperdiffusivity on turbulent dynamos with helicity," *Phys. Rev. Lett.* **88**, 055003, 1–4 (2002).

Campbell, C.G., "Launching of accretion disc winds along dynamo-generated magnetic fields," *Mon. Not. R. Astr. Soc.* **310**, 1175–1184 (1999).

Campbell, C.G., "An accretion disc model with a magnetic wind and turbulent viscosity," *Mon. Not. R. Astr. Soc.* **317**, 501–527 (2000).

Canuto, C., Hussaini, M.Y., Quarteroni, A. and Zang, T.A., *Spectral Methods in Fluid Dynamics,* Springer, Berlin (1988).

Cattaneo, F., "On the origin of magnetic fields in the quiet photosphere," *Astrophys. J.* **515**, L39–L42. (1999).

Cattaneo, F. and Hughes, D.W., "Nonlinear saturation of the turbulent alpha effect," *Phys. Rev.* E **54**, R4532–R4535 (1996).

Chan, K.L. and Sofia, S., "Turbulent compressible convection in a deep atmosphere. II. Tests on the validity and limitation of the numerical approach," *Astrophys. J.* **307**, 222–241 (1986).

Chan, K.L. and Sofia, S., "Turbulent compressible convection in a deep atmosphere. III. Results of three-dimensional computations," *Astrophys. J.* **336**, 1022–1040 (1989).

Choudhuri, A.R., Schüssler, M. and Dikpati, M., "The solar dynamo with meridional circulation," *Astron. Astrophys.* **303**, L29–L32 (1995).

Collatz, L., *The Numerical Treatment of Differential Equations.* Springer-Verlag, New York, p. 164.

Colella, P. and Woodward, P.R. "The piecewise parabolic method (PPM) for gas-dynamical simulations," *J. Comp. Phys.* **54**, 174–201 (1984).

Dobler, W., Brandenburg, A. and Shukurov, A., "Pressure-driven outflow and magneto-centrifugal wind from a dynamo active disc," in: *Plasma Turbulence and Energetic Particles in Astrophysics* (Eds. M. Ostrowski and R. Schlickeiser), Publ. Astron. Obs. Jagiellonian Univ., Cracow, pp. 347–352 (1999).

Dodd, R.K., Eilbeck, J.C., Gibbon, J.D. and Morris, H.C., *Solitons and Nonlinear Wave Equations.* Academic Press, London (1982).

Durney, B.R., "On a Babcock-Leighton dynamo model with a deep-seated generating layer for the toroidal magnetic field. II," *Solar Phys.* **166**, 231–260 (1995).

Fox, P.A., Theobald, M.L. and Sofia, S., "Compressible magnetic convection: formulation and two-dimensional models," *Astrophys. J.* **383**, 860–881 (1991).

Frank, J., King, A.R. and Raine, D.J., *Accretion Power in Astrophysics.* Cambridge Univ. Press, Cambridge (1992).

Frisch, U., Pouquet, A., Léorat, J. and Mazure, A., "Possibility of an inverse cascade of magnetic helicity in hydrodynamic turbulence," *J. Fluid Mech.* **68**, 769–778 (1975).

Galsgaard, K. and Nordlund, Å., "Heating and activity of the solar corona: I. boundary shearing of an initially homogeneous magnetic-field," *J. Geophys. Res.* **101**, 13445–13460 (1996).

Galsgaard, K. and Nordlund, Å., "Heating and activity of the solar corona: II. Kink instability in a flux tube," *J. Geophys. Res.* **102**, 219–230 (1997a).

Galsgaard, K. and Nordlund, Å., "Heating and activity of the solar corona: III. Dynamics of a low beta plasma with three-dimensional null points," *J. Geophys. Res.* **102**, 231–248 (1997b).

Glatzmaier, G.A. and Roberts, P.H., "A three-dimensional self-consistent computer simulation of a geomagnetic field reversal," *Nature* **377**, 203–209 (1995).

Glatzmaier, G.A. and Roberts, P.H., "Rotation and magnetism of Earth's inner core," *Science* **274**, 1887–1891 (1996).

Grauer, R., Marliani, C. and Germaschewski, K., "Adaptive mesh refinement for singular solutions of the incompressible Euler equations," *Phys. Rev. Lett.* **80**, 4177–4180 (1998).

Greenhill, L.J., Gwinn, C.R., Schwartz, C., Moran, J.M. and Diamond, P.J., "Coexisting conical bipolar and equatorial outflows from a high-mass protostar," *Nature* **396**, 650–653 (1998).

Hawley, J.F., "Global magnetohydrodynamical simulations of accretion tori," *Astrophys. J.* **528**, 462–479 (2000).

Hawley, J.F., Gammie, C.F. and Balbus, S.A., "Local three-dimensional magnetohydrodynamic simulations of accretion discs," *Astrophys. J.* **440**, 742–763 (1995).

Hawley, J.F., Gammie, C.F. and Balbus, S.A., "Local three dimensional simulations of an accretion disk hydromagnetic dynamo," *Astrophys. J.* **464**, 690–703 (1996).

Hodapp, K.-W. and Ladd, E.F., "Bipolar jets from extremely young stars observed in molecular hydrogen emission," *Astrophys. J.* **453**, 715–720 (1995).

Ji, H., Prager, S.C., Almagri, A.F., Sarff, J.S. and Toyama, H., "Measurement of the dynamo effect in a plasma," *Phys. Plasmas* **3**, 1935–1942 (1996).

Kerr, R.M. and Brandenburg, A., "Evidence for a singularity in ideal magnetohydrodynamics: implications for fast reconnection," *Phys. Rev. Lett.* **83**, 1155–1158 (1999).

Kleeorin, N.I., Moss, D., Rogachevskii, I. and Sokoloff, D., "Helicity balance and steady-state strength of the dynamo generated galactic magnetic field," *Astron. Astrophys.* **361**, L5–L8 (2000).

Korpi, M.J., Brandenburg, A., Shukurov, A., Tuominen, I. and Nordlund, Å., "A supernova regulated interstellar medium: simulations of the turbulent multiphase medium," *Astrophys. J. Lett.* **514**, L99–L102 (1999).

Lele, S.K., "Compact finite difference schemes with spectral-like resolution," *J. Comp. Phys.* **103**, 16–42 (1992).

LeVeque, R.J., Mihalas, D., Dorfi, E.A. and Müller, E., *Computational Methods for Astrophysical Fluid Flow.* Springer, Berlin (1998).

Lohse, D. and Müller-Groeling, A., "Bottleneck effects in turbulence: scaling phenomena in r versus p space," *Phys. Rev. Lett.* **74**, 1747–1750 (1995).

Mac Low, M.-M., Klessen, R.S. and Burkert, A., "Kinetic energy decay rates of supersonic and super-Alfvénic turbulence in star-forming clouds," *Phys. Rev. Lett.* **80**, 2754–2757 (1998).

Maron, J.L. and Howes, G.G., *Gradient particle magnetohydrodynamics,* astro-ph/0107454 (2001).

Meneguzzi, M. and Pouquet, A., "Turbulent dynamos driven by convection," *J. Fluid Mech.* **205**, 297–312.

Nordlund, Å., "Numerical simulations of the solar granulation I. Basic equations and methods," *Astron. Astrophys.* **107**, 1–10 (1982).

Nordlund, Å., "Solar convection," *Solar Phys.* **100**, 209–235 (1985).

Nordlund, Å. and Galsgaard, K., *A 3D MHD code for Parallel Computers.* http://www.astro.ku.dk/aake/NumericalAstro/papers/kg/mhd.ps.gz.

Nordlund, Å. and Stein, R.F., "3-D simulations of solar and stellar convection and magnetoconvection," *Comput. Phys. Commun.* **59**, 119–125 (1990).

Nordlund, Å., Galsgaard, K. and Stein, R.F., "Magnetoconvection and Magnetoturbulence," in: *Solar surface magnetic fields,* vol. 433 (Eds. R.J. Rutten and C.J. Schrijver), NATO ASI Series, pp. 471–498 (1994).

Nordlund, Å., Brandenburg, A., Jennings, R.L., Rieutord, M., Ruokolainen, J., Stein, R.F. and Tuominen, I., "Dynamo action in stratified convection with overshoot," *Astrophys. J.* **392**, 647–652 (1992).

Ouyed, R., Pudritz, R.E. and Stone, J.M., "Episodic jets from black holes and protostars," *Nature* **385**, 409–414 (1997).

Ouyed, R. and Pudritz, R.E., "Numerical simulations of astrophysical jets from keplerian discs. I. Stationary models," *Astrophys. J.* **482**, 712–732 (1997a).

Ouyed, R. and Pudritz, R.E., "Numerical simulations of astrophysical jets from keplerian discs. II. Episodic outflows," *Astrophys. J.* **484**, 794–809 (1997b).

Ouyed, R. and Pudritz, R.E., "Numerical simulations of astrophysical jets from keplerian discs. III. The effects of mass loading," *Mon. Not. R. Astr. Soc.* **309**, 233–244 (1999).

Ouyed, R., Clarke, D.A. and Pudritz, R.E., "3-Dimensional simulations of jets from keplerian disks: regulatory stability," *Astrophys. J.* **582**, 292–319 (2003).

Padoan, P., Nordlund, Å. and Jones, B.J.T., "The universality of the stellar mass function," *Astrophys. J.* **288**, 145–152 (1997).

Padoan, P. and Nordlund, Å., "A super-Alfvnic model of dark clouds," *Astrophys. J.* **526**, 279–294 (1999).

Passot, T. and Pouquet, A., "Numerical simulation of compressible homogeneous flows in the turbulent regime," *J. Fluid Mech.* **181**, 441–466 (1987).

Pelletier, G. and Pudritz, R.E., "Hydromagnetic disk winds in young stellar objects and active galactic nuclei," *Astrophys. J.* **394**, 117–138 (1992).

Peterkin, R.E., Frese, M.H. and Sovinec, C.R., "Transport of magnetic flux in an arbitrary coordinate ALE code," *J. Comp. Phys.* **140**, 148–171 (1998).

Porter, D.H., Pouquet, A. and Woodward, P.R., "Three-dimensional supersonic homogeneous turbulence: a numerical study," *Phys. Rev. Lett.* **68**, 3156–3159 (1992).

Porter, D.H., Pouquet, A. and Woodward, P.R., "Kolmogorov-like spectra in decaying 2-dimensional supersonic flows," *Phys. Fluids* **6**, 2133–2142 (1994).

Pouquet, A., Frisch, U. and Léorat, J., "Strong MHD helical turbulence and the nonlinear dynamo effect," *J. Fluid Mech.* **77**, 321–354 (1976).

Rast, M.P., Nordlund, Å., Stein, R.F. and Toomre, J., "Ionization effects in three-dimensional granulation simulations," *Astrophys. J. Lett.* **408**, L53–L56 (1993).

Rast, M.P. and Toomre, J., "Compressible convection with ionization. I. Stability, flow asymmetries, and energy transport," *Astrophys. J.* **419**, 224–239 (1993a).

Rast, M.P. and Toomre, J., "Compressible convection with ionization. II. Thermal boundary-layer instability," *Astrophys. J.* **419**, 240–254 (1993b).

Rekowski, M.v., Rüdiger, G. and Elstner, D., "Structure and magnetic configurations of accretion disk-dynamo models," *Astron. Astrophys.* **353**, 813–822 (2000).

Roettiger, K., Stone, J.M. and Burns, J.O., "Magnetic field evolution in merging clusters of galaxies," *Astrophys. J.* **518**, 594–602 (1999).

Rögnvaldsson, Ö.E. and Nordlund, Å. *Magnetic fields in young galaxies.* astro-ph/0010499 (2000).

Rogallo, R.S., *Numerical experiments in homogeneous turbulence.* NASA Tech. Memo. 81315 (1981).

Sánchez-Salcedo, F.J. and Brandenburg, A., "Deceleration by dynamical friction in a gaseous medium," *Astrophys. J. Lett.* **522**, L35–L38 (1999).

Sánchez-Salcedo, F.J. and Brandenburg, A., "Dynamical friction of bodies orbiting in a gaseous sphere," *Mon. Not. R. Astr. Soc.* **322**, 67–78 (2001).

Sod, G.A., "A survey of several finite difference methods for systems of nonlinear hyperbolic conservation laws," *J. Comp. Phys.* **27**, 1–31 (1978).

Stanescu, D. and Habashi, W.G., "$2N$-storage low dissipation and dispersion Runge-Kutta schemes for computational acoustics," *J. Comp. Phys.* **143**, 674–681 (1998).

Steffen, M., Ludwig, H.-G. and Krüß, A., "A numerical study of solar granular convection in cells of different horizontal dimensions," *Astron. Astrophys.* **123**, 371–382 (1989).

Stein, R.F. and Nordlund, Å., "Topology of convection beneath the solar surface," *Astrophys. J. Lett.* **342**, L95–L98 (1989).

Stein, R.F. and Nordlund, Å., "Simulations of solar granulation. I. General properties," *Astrophys. J.* **499**, 914–933 (1998).

Stone, J.M. and Norman, M., "ZEUS-2D: A radiation magnetohydrodynamics code for astrophysical flows in two space dimensions: I. The hydrodynamic algorithms and tests," *Astrophys. J.* **80**, 753–790 (1992a).

Stone, J.M. and Norman, M., "ZEUS-2D: A radiation magnetohydrodynamics code for astrophysical flows in two space dimensions: II. The magnetohydrodynamic algorithms and tests," *Astrophys. J.* **80**, 791–818 (1992b).

Stone, J.M., Hawley, J.F., Gammie, C.F. and Balbus, S.A., "Three dimensional magnetohydrodynamical simulations of vertically stratified accretion disks," *Astrophys. J.* **463**, 656–673 (1996).

Stewart, J.M., "The hydrodynamics of accretions discs I: Stability," *Astron. Astrophys.* **42**, 95–101 (1975).

Suzuki, E. and Toh, S., "Entropy cascade and temporal intermittency in a shell model for convective turbulence," *Phys. Rev.* **E 51**, 5628–5635 (1995).

Thelen, J.-C., "A mean electromotive force induced by magnetic buoyancy instabilities," *Mon. Not. R. Astr. Soc.* **315**, 155–164 (2000).

Tobias, S.M., Brummell, N.H., Clune, T.L. and Toomre, J., "Pumping of magnetic fields by turbulent penetrative convection," *Astrophys. J. Lett.* **502**, L177–L177 (1998).

Toh, S. and Iima, M., "Dynamical aspect of entropy transfer in free convection turbulence," *Phys. Rev.* **E 61**, 2626–2639 (2000).

Urpin, V. and Brandenburg, A., "Magnetic and vertical shear instabilities in accretion discs," *Mon. Not. R. Astr. Soc.* **294**, 399–406 (1998).

Vainshtein, S.I. and Cattaneo, F., "Nonlinear restrictions on dynamo action," *Astrophys. J.* **393**, 165–171 (1992).

Vishniac, E.T. and Brandenburg, A., "An incoherent α–Ω dynamo in accretion disks," *Astrophys. J.* **475**, 263–274 (1997).

Völk, H.J. and Atoyan, A.M., "Clusters of galaxies: magnetic fields and nonthermal emission," *Astroparticle Phys.* **11**, 73–82 (1999).

Williamson, J.H., "Low-storage Runge-Kutta schemes," *J. Comp. Phys.* **35**, 48–56 (1980).

Ziegler, U. and Rüdiger, G., "Angular momentum transport and dynamo-effect in stratified, weakly magnetic disks," *Astron. Astrophys.* **356**, 1141–1148 (2000).

10 Topological quantities in magnetohydrodynamics

Mitchell A. Berger

Department of Mathematics, University College London,
Gower Street, London WC1E 6BT, UK, E-mail: m.berger@ucl.ac.uk

Increasingly, observations, experiments, and numerical simulations reveal a rich nonlinear world full of complex structures. Geometric and topological techniques can provide valuable assistance in making sense of these structures. This chapter describes crossing numbers and helicity integrals, which measure different aspects of topological complexity in vector fields. These measures can be applied to vortex structures in fluid mechanics and to magnetic field line structures in magnetized fluids. We also discuss in detail astrophysical applications concerning magnetic fields in the sun and solar atmosphere.

Helicity integrals measure the linking of field lines, and are pseudo-scalar under spatial reflection. They are conserved for vector fields dragged by fluid motions. Crossing numbers, on the other hand, do not measure the sign of linkings. They are not conserved, but have a positive lower bound for a given field topology which provides a good measure of complexity. Simple relations exist between crossing numbers and magnetic energy. If crossing numbers can be predicted for a field, then the energy stored in that field can also be predicted. Crossing numbers will be described for both magnetic fields inside a sphere, and braided fields connecting two parallel planes.

To illustrate these ideas, we apply crossing numbers and braid theory to the problem of solar coronal heating and the source of coronal microflares. Observations of structures such as X-ray loops in the solar atmosphere have gained in resolution over the past few decades, revealing more and more small-scale structure. This small-scale structure originates in the turbulent motions below the surface. Simple models predict how much structure (as measured by crossing numbers) is generated by the turbulence, and hence how much energy is available for heating and flaring. These models provide sufficient energy to heat the corona.

Helicity integrals quantify vector field features such as shear, twist, and braiding. In ideal magnetized plasmas, helicity is absolutely conserved. Even with resistivity the decay of helicity obeys rather severe bounds. No other knot-like invariants of magnetic field lines possess similar bounds. Approximate helicity conservation implies that reconnecting flux tubes become twisted. This result has relevance to the observation of twisted field structures in solar and space physics. Recently space scientists have become highly interested in helicity. Observations show an abundance of helical structure in the solar corona and solar wind; furthermore, the sign of the helicity depends on hemisphere. The solar dynamo is probably responsible for this asymmetry. We discuss the contributions to helicity generation from both the α-effect and from differential rotation.

10.1. The invariants of knots, links, and braids

This section describes in general terms knots, links, and braids. These curves or sets of curves possess two kinds of invariants. The first kind (isotopic invariants) consists of numbers or other objects (e.g. polynomials) which do not change under deformation of the curves. The meaning of these invariants, particularly the knot polynomials, can be obscure. However, linking numbers and winding numbers do have simple geometric interpretations. We will cover these latter invariants in some detail.

A second kind of invariant involves quantities which do change during deformation of the curves. The quantity must be bounded in some way so that well-defined extrema can be found. For example, a trefoil knot when seen in projection always has at least three crossings, so the number 3 is an invariant for the trefoil (its minimum crossing number). This second kind of invariant often provides a measure of complexity.

10.1.1. Knots

Knots have many uses in technology, from securing ropes on a sailboat to tying shoelaces. The description of a knot in these applications differs from the description preferred by mathematicians. In most uses, a finite length of a cord is knotted somewhere in the middle, sufficiently far from the ends so that friction prevents the knot from unraveling. The existence of the two ends allows us to create the knot to begin with, but also complicates the topological description. Thus mathematicians usually join the ends together.

The mathematical definition of a knot starts with a closed curve in $\mathcal{R}^3$. Let S_1 be the unit circle parametrized by $\theta, 0 \leq \theta < 2\pi$. Consider a curve

$$\kappa : S^1 \to \mathcal{R}^3 \qquad (10.1)$$

which is one-to-one, i.e. if $\theta_1 \neq \theta_2$ then $\kappa(\theta_1) \neq \kappa(\theta_2)$. There are infinitely many such closed curves, and many will look very similar to each other. We first say that two curves κ_0 and κ_1 are topologically equivalent if κ_0 can be smoothly deformed into κ_1 without letting the curve pass through itself. Next we chop up the set of all closed curves into equivalence classes, each one representing a particular *knot type* – the *unknot* (e.g. a circle), the trefoil knot, the square knot, etc. Then we can proceed to look at the properties and invariants of the different knot types.

More formally, let $\mathcal{K}$ be the set of all curves κ. A homotopy between κ_0 and κ_1 is a path in this function space. We parametrize this path by the letter t and visualize the homotopy as a deformation from κ_0 to κ_1 proceeding from time $t = 0$ to time $t = 1$. Define a continuous mapping $K : S^1 \times [0, 1] \to \mathcal{R}^3$ where $K(\theta, t) = \kappa_t(\theta)$ and in particular

$$K(\theta, 0) = \kappa_0(\theta); \qquad (10.2)$$

$$K(\theta, 1) = \kappa_1(\theta). \qquad (10.3)$$

In the mathematics literature K is called an isotopic deformation. Since each $\kappa_t(\theta)$ is a one-to-one function of θ the curve cannot pass through itself during the deformation. By definition, κ_0 and κ_1 belong to the same knot type if and only if an isotopic deformation exists between them. A fluid dynamicist may wish to think of the isotopic deformation in terms of fluid motions with velocity $\mathbf{v}(\mathbf{x}, t)$. In particular, suppose that

$$\frac{\mathrm{d}K(\theta, t)}{\mathrm{d}t} = \mathbf{v}(K(\theta, t), t). \qquad (10.4)$$

Then the knot moves with the fluid without slipping; it is *frozen into the fluid*.

Traditionally, most knot invariants have been found by looking at the knot in projection. The knot will possess a set of crossings; the sequence of overcrossings and undercrossings contains a rich amount of topological information. Isotopic invariants built from the crossing sequence include the famous knot polynomials (Gilbert and Porter, 1994). The minimum crossing number, as mentioned above, is not an isotopic invariant but provides the standard measure of knot complexity. Other measures of complexity exist. For example, the unknotting number equals the minimum number of crossings which must be removed in order to obtain the unknot.

Recently, knot theorists have searched for invariants which do not involve projections. Three dimensional flows and fields can help in this search (Ricca and Berger, 1996). For example, the minimum energy of a knotted magnetic tube (with some constraints, such as constant total volume) provides a measure of complexity (Moffatt, 1985; Moffatt and Ricca, 1992). Some isotopic invariants can also be defined without projections. For example, Witten employed functional integral techniques from quantum field theory to derive the Jones polynomial (Witten, 1989).

10.1.2. Links and linking numbers

A collection of several closed curves κ_i, $i = 1, \ldots, n$ is called a *link*. The study of links brings us closer to fluid mechanics and magnetism, where one may wish to study a collection of vortex lines or magnetic filed lines.

The linking number introduced by Gauss plays a special role in knot theory because the calculation can be done using line integrals. Thus linking numbers provide a connection between topology and vector calculus. As soon as we express a topological idea with vector calculus, the hope arises of extending that idea to vector fields.

Let γ_1 and γ_2 be two closed curves (i.e. two knots or unknots). Parametrize γ_1 by arclength σ (measured from some arbitrary point on the curve). Positions along γ_1 are given by $\mathbf{x}(\sigma)$. Assign similar symbols τ and $\mathbf{y}(\tau)$ to γ_2. Also let $\mathbf{r} = \mathbf{y} - \mathbf{x}$ be the relative position vector. The Gauss linking number between γ_1 and γ_2 is

$$L(\gamma_1, \gamma_2) = \frac{1}{4\pi} \oint_{\gamma_1} \oint_{\gamma_2} \frac{d\mathbf{x}}{d\sigma} \cdot \frac{\mathbf{r}}{r^3} \times \frac{d\mathbf{y}}{d\tau} \, d\tau \, d\sigma. \tag{10.5}$$

This double integral results in an integer measuring the linking of the two curves (see Fig. 10.1). This number is invariant to deformations of the two curves which do not involve one curve passing through the other (each curve is allowed to pass through itself).

10.1.3. Braids

A geometrical braid is a set of N curves

$$(x_i(z), y_i(z)), \quad 0 \le z \le L, \quad i = 1, \ldots, N \tag{10.6}$$

stretching between two planes. Note the following connection to the theory of dynamical systems: If we replace z by t, we can consider a geometrical braid as the time history of a set of N particles moving in a plane.

A topological braid is an equivalence class of geometrical braids: two geometrical braids are defined to be equivalent if one can be deformed into the other without the curves passing through each other. During the deformation, the endpoints on the two boundary planes must be fixed.

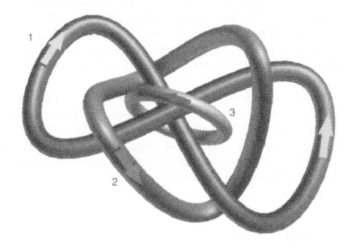

Figure 10.1 The Gauss linking number. For this figure, $L(\gamma_1, \gamma_2) = L(\gamma_2, \gamma_3) = 0$ while $L(\gamma_3, \gamma_1) = 2$.

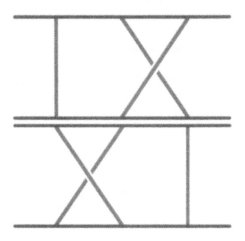

Figure 10.2 Combining two braids.

The set of topological braids of N strings has a group structure. For simplicity one usually assumes that the endpoints in both planes are identical:

$$\{(x_i(0),\, y_i(0))\} = \{(x_i(L),\, y_i(L))\}. \tag{10.7}$$

The group operation which combines a braid B_1 and B_2 into the braid $B_1 B_2$ (see Fig. 10.2) then consists of placing B_2 on top of B_1 (and rescaling z by a factor of $1/2$).

Fig. 10.3 shows two topologically equivalent braids; the braid on the right has the smallest possible number of crossings. How do we find minimal crossing configurations for braids? This problem is analogous to the problem of finding minimum energy equilibrium magnetic fields. Here we sketch the solution for the case $N = 3$; the general problem is still unsolved.

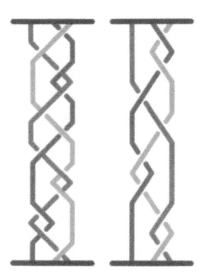

Figure 10.3 Two topologically equivalent braids. The left braid has $C = 12$; the right braid has $C = C_{\min} = 8$.

Consider a 3-braid and let t denote its height. The braid can be described by three curves in the complex plane $a(t)$, $b(t)$, and $c(t)$ with $0 \le t \le 1$. The curves intersect a plane $t = \text{const.}$ in three points, which form the vertices of a triangle. We would like to keep track of the interior angles of this triangle as a function of t. These interior angles will change as t increases and the points $a(t)$, $b(t)$, and $c(t)$ move about each other. Rather crudely, one might use a quantity like $\text{Im}[\log(b - a)/(c - a)]$ to calculate the interior angle at vertex a. The difficulty lies in defining the log function. As the points move about each other, an angle may go all the way through 2π, but we do not desire any discontinuities at branch cuts. Instead we define

$$\lambda_a(t) \equiv \frac{1}{2\pi} \int_0^t \left(\frac{b - a}{c - a}\right)^{-1} \frac{\mathrm{d}}{\mathrm{d}t'} \left(\frac{b - a}{c - a}\right) \mathrm{d}t'. \tag{10.8}$$

The three functions $\text{Im}\,\lambda_a(t)$, $\text{Im}\,\lambda_b(t)$, and $\text{Im}\,\lambda_c(t)$ can be used to find minimum crossing states as well as topological invariants (Berger, 1994a). Let $\gamma(t) = \text{Im}\,(\lambda_a, \lambda_b, \lambda_c)(t)$ be the curve described by the angles as t goes from 0 to 1. This curve travels in a two dimensional plane: the sum of interior angles (minus their sum at $t = 0$) always vanishes. Furthermore there are forbidden regions in this plane, corresponding to mathematically impossible combinations of interior angles (Fig. 10.4). The method of minimizing crossing numbers consists of distorting $\gamma(t)$ so that it passes through the minimum possible number of vertices in this plane (Berger, 1990, 1994a).

The net winding angle of any two curves provides a set of isotopic invariants. There are in fact a hierarchy of winding number invariants. There is one third order invariant (Berger, 1994a).

$$\int_0^1 \left(\lambda_a \frac{\mathrm{d}\lambda_b}{\mathrm{d}t} + \lambda_b \frac{\mathrm{d}\lambda_c}{\mathrm{d}t} + \lambda_c \frac{\mathrm{d}\lambda_a}{\mathrm{d}t}\right) \mathrm{d}t. \tag{10.9}$$

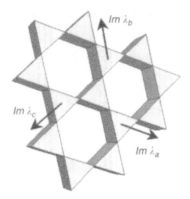

Figure 10.4 The plane $\text{Im}\,\lambda_a(t) + \text{Im}\,\lambda_b(t) + \text{Im}\,\lambda_c(t) = 0$. The hexagonal regions are forbidden. Vertices correspond to a, b, and c colinear. The direction perpendicular to the plane corresponds to uniform rotation of all three vertices.

This quantity measures the number of holes the curve $\gamma(t)$ encircles in the plane of Fig. 10.4, and equals 1 for a standard pigtail braid.

10.2. Magnetic energy and crossing numbers

Equilibrium states of a magnetized plasma are important in several areas of magnetohydrodynamics (MHD). Fusion technologists would like to confine their 100 million degree plasma in a magnetic field; this field should be as quiescent as possible so that the plasma does not hit the walls and cool down. In the solar atmosphere many structures such as X-ray loops and prominences last for days or weeks, much longer than dynamical timescales (e.g. the Alfvén wave travel time from one end of a loop to the other end). These structures must be near some stable equilibrium state. Furthermore, equilibrium states provide a background field for studying wave phenomena.

When an equilibrium field has a translational or rotational symmetry, the field depends on only two coordinates. This makes the problem of finding the equilibria much simpler. However, a highly structured field may not have any symmetries, and so we must consider fully three-dimensional fields. A first possibility could be the use of three-dimensional numerical simulations; however, the configurations of interest often have intense field gradients, making the numerics quite difficult. Instead, we will see how much we can learn about magnetic equilibria purely from topological considerations.

10.2.1. Upper and lower bounds for the minimum energy in a set of topological equivalent magnetic fields

We consider a perfectly conducting (ideal) fluid in a given region $\mathcal{V}$ of three-dimensional space. The magnetic field lines in this region are transported (frozen-in) by the flow of the fluid. To a given velocity field there belongs a flow map $\varphi(t_1, t_2)$, mapping the points of $\mathcal{V}$ at time t_1 to the points in $\mathcal{V}$ at a later time t_2; the fluid motion sends a fluid element at $\mathbf{x}(t_1)$ to the point $\mathbf{x}(t_2) = \varphi(\mathbf{x}(t_1))$. The flow map is bijective and continuous. As a consequence, if a magnetic field $\mathbf{B}_1$ is given at time t_1, the magnetic field $\mathbf{B}_2$ at *any* given time t_2 will be *topologically equivalent* to $\mathbf{B}_1$.

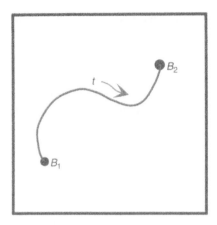

Figure 10.5 The function space [$\mathbf{B}_1$]. Each point corresponds to a magnetic field.

Consider the set [$\mathbf{B}_1$] of all magnetic fields defined on $\mathcal{V}$ topological equivalent to $\mathbf{B}_1$. (see Fig. 10.5). Ideal motions generate a path in this space parametrized by time. Any quantity which is the same for all fields in [$\mathbf{B}_1$] is an ideal invariant. A prominent example is the magnetic helicity $K_{\mathcal{V}}$ described in the next section.

As mentioned earlier (Section 10.1), a second type of knot invariant involve extrema on the set of all curves corresponding to a given knot type. Here we look at invariants which involve extrema for functions on the set [$\mathbf{B}_1$]. We are especially interested in the energy of the magnetic field

$$E = \frac{1}{8\pi} \int_{\mathcal{V}} \mathbf{B} \cdot \mathbf{B} \, d^3x. \tag{10.10}$$

The minimum energy $E_{\min}([\mathbf{B}_1])$ for all the magnetic fields in [$\mathbf{B}_1$] is an example of this second type of invariant. If the magnetic field configuration of minimum energy lies inside of [$\mathbf{B}_1$], then this field will be *stable* to ideal MHD-perturbations. Warning: in some cases the energy $E_{\min}([\mathbf{B}_1])$ may be an asymptotic limit, not realized by any field in [$\mathbf{B}_1$] (Berger, 1985; Aly, 1992).

The problem of finding magnetic equilibria corresponding to a particular topology (Aly, 1992) is particularly difficult. One must find the minimum energy $E_{\min}([\mathbf{B}_1])$ and the corresponding field $\mathbf{B}_{\min}$. As mentioned earlier great difficulties must be overcome:

1 The equilibrium equation $\mathbf{J} \times \mathbf{B} = \nabla p$ is nonlinear. (Here p = pressure, $\mathbf{J}$ = electric current = $(1/\mu)\nabla \times \mathbf{B}$).
2 Most interesting topologies are intrinsically three-dimensional; i.e. no member of [$\mathbf{B}_1$] has an ignorable coordinate.
3 The equilibrium probably has current sheets (discontinuities in $\mathbf{B}$) or at least very narrow current layers (see Parker, 1983).

Let us be less ambitious: can we at least find upper and lower bounds for $E_{\min}([\mathbf{B}_1])$?

10.2.1.1. Upper bounds

If one is able to construct *any* field with the correct topology, then by definition its energy is an upper bound for $E_{\min}$.

10.2.1.2. Lower bounds

A crude lower bound is provided by the helicity K_V: The simplest case is a simply connected volume bounded by a magnetic surface ($\mathbf{B} \cdot \hat{\mathbf{n}} = 0$ on the boundary of V). For fields in such a volume

$$K_V = \int_V \mathbf{A} \cdot \mathbf{B} \, d^3 x, \tag{10.11}$$

$$E_{\min} \geq \frac{\alpha K_V}{8\pi}, \tag{10.12}$$

where $\mathbf{A}$ is the vector potential and α is given by the minimum eigenvalue of the equation $\nabla \times \mathbf{B} = \alpha \mathbf{B}$. A field satisfying this equation is called a linear force-free field. By 'uncurling' this equation, one finds $\mathbf{B} = \alpha \mathbf{A} + \nabla \psi$ where ψ is a gauge function. One can show that linear force free fields minimize the energy over all fields with the same K_V. Unfortunately many different topologies have the same helicity. This is why (10.12) gives only a lower bound.

Freedman and He (1991) provided another lower bound for $E_{\min}$: the crossing number. First, consider how crossing numbers are defined for knotted and linked curves. Fix a plane with normal $\hat{\mathbf{p}}$ in three-dimensional space. Project the curves onto this plane and count the crossings $c(\hat{\mathbf{p}})$ of the projected curves. Freedman and He employ a special way to remove the dependence on the plane of projection: take the average over all possible projection planes. For example, let γ_1 and γ_2 be two curves and $\hat{p}$ the projection angle, then the crossing number c will be in general a function of $\hat{p}$, $c(\gamma_1, \gamma_2, \hat{p})$. However, this $\hat{p}$-dependence can be removed by averaging over all possible $\hat{p}$ (where $\hat{p}$ ranges over the unit sphere):

$$c(\gamma_1, \gamma_2) = \langle c(\gamma_1, \gamma_2, \hat{p}) \rangle = \frac{1}{4\pi} \oint c(\gamma_1, \gamma_2, \hat{p}) d^2 S. \tag{10.13}$$

The quantity can be computed directly from the equations for the two curves. Let $\mathbf{x}(\sigma)$ be a point on γ_1, and $\mathbf{y}(\tau)$ be a point on γ_2. Then (Freedman and He, 1991)

$$c(\gamma_1, \gamma_2) = \frac{1}{2\pi} \oint_{\gamma_1} \oint_{\gamma_2} \left| \frac{d\mathbf{x}}{d\sigma} \cdot \frac{\mathbf{r}}{r^3} \times \frac{d\mathbf{y}}{d\tau} \right| d\tau \, d\sigma. \tag{10.14}$$

The modulus is necessary, because otherwise crossings with a positive sign would cancel with crossings of negative sign. If we removed the modulus we would obtain twice the linking number of the two curves, (10.5). Equation (10.14) gives the total number of crossings independent of sign. Note that while the linking number does not change if the curves are distorted (as long as they do not pass through each other), the crossing number does vary. However, it does have a minimum $c_{\min}$. The above integral can also be done for a single knotted curve κ, by setting $\gamma_2 = \gamma_1 = \kappa$.

Consider, for example, a magnetic flux tube whose axis is the knotted curve κ. Let $\Phi(\kappa)$ be the magnetic flux through the tubular knot and $V(\kappa)$ be its volume. Freedman and He get the following lower bound for the minimum energy:

$$E_{\min}(\kappa) \geq c(\kappa) \left(\frac{16}{\pi V(\kappa)} \right)^{1/3} \Phi^2(\kappa). \tag{10.15}$$

The field lines inside the knot can be twisted. The effect of internal twist was examined by Chui and Moffatt (1995). Let h parametrize the internal twist inside the tube. By dimensional analysis, one expects that

$$E_{\min}(\kappa) = m(\kappa, h)V^{-1/3}(\kappa)\Phi^2(\kappa). \qquad (10.16)$$

Here the function $m(\kappa, h)$ is a dimensionless number which depends on the knot type κ and the twist h.

A lower energy bound can also be found for the case of concentric cylindrical flux surfaces. Let Φ be the toroidal flux contained within a surface, and let Ψ be the poloidal flux between the surface and the outer boundary. Also let $q \equiv d\Phi/d\Psi$. The helicity equals the average twist over all the flux (Berger and Field, 1984)

$$K_V = 2\int_0^{\Psi_0}\int_0^{\Psi_2} q_1\, d\Psi_1\, d\Psi_2. \qquad (10.17)$$

By taking the modulus of q we in effect make all the twist positive. Define the quantity

$$C = 2\int_0^{\Psi_0}\int_0^{\Psi_2} |q_1|d\Psi_1\, d\Psi_2. \qquad (10.18)$$

Then, in a manner similar to the Freedman and He crossing number analysis one can show that

$$E_{\min} \geq \left(\frac{16}{\pi V}\right)^{1/3} C. \qquad (10.19)$$

10.2.2. Crossing numbers for braided fields

Crossing numbers can also be calculated for braided fields, with application to loops in the solar atmosphere. Magnetic fields inside a solar coronal loop have a strong axial component B_z. To a first approximation, this component can be regarded as uniform (van Ballegooijen, 1985). We will model a coronal loop as a cylinder of radius R and length L, with field

$$\mathbf{B} = B_z(b_x, b_y, 1) = B_z(\mathbf{b} + \hat{\mathbf{z}}). \qquad (10.20)$$

Consider two field lines between two planes $z = 0$ and $z = L$. Fig. 10.6 shows a pair of field lines from two different viewpoints. In each viewpoint we may count the apparent number of crossings. Call ϕ the viewing angle (polar angle in the x–y plane) and $c(\phi)$ the crossing number.

A quantity independent of viewing angle is

$$\bar{c} = \frac{1}{\pi}\int_0^{\pi} c(\phi)\, d\phi. \qquad (10.21)$$

The crossing number bears a simple relation to the field line geometry. Let $\mathbf{x}_1(z) = (x_1, y_1)$ and $\mathbf{x}_2(z)$ be vectors along two curves. The relative position vector $\mathbf{r}_{12}(z) = \mathbf{x}_2 - \mathbf{x}_1$ makes an angle $\theta_{12}(z)$ with respect to the x-axis. At each level z, there is crossing as seen from viewing angle ϕ if $\phi = \theta_{12}(z)$. Between z and $z + dz$ there will be crossings seen by viewing angles $\phi \in \{\theta_{12}(z), \theta_{12}(z) + d\theta_{12}(z)\}$. Thus,

$$d\bar{c}_{12} = \frac{1}{\pi}|d\theta_{12}(z)| \qquad (10.22)$$

Figure 10.6 Crossing of two field lines.

and (10.21) becomes

$$\bar{c}_{12} = \frac{1}{\pi} \int_0^L \left| \frac{d\theta_{12}}{dz} \right| dz. \tag{10.23}$$

We generalize to a magnetic field by summing over all pairs of lines, removing double-counting, and weighting by flux:

$$C = \frac{1}{2\pi} \int_0^L \iint B_{z1} B_{z2} \left| \frac{d\theta_{12}}{dz} \right| d^2x_1\, d^2x_2\, dz. \tag{10.24}$$

Now,

$$\frac{d\theta_{12}}{dz} = \frac{(\mathbf{b}_2 - \mathbf{b}_1) \cdot \hat{\boldsymbol{\theta}}_{12}}{r_{12}}, \tag{10.25}$$

so we have

$$\frac{dC}{dz} = \frac{B_z^2}{2\pi} \iint \frac{\left| (\mathbf{b}_2 - \mathbf{b}_1) \cdot \hat{\boldsymbol{\theta}}_{12} \right|}{r_{12}} d^2x_1\, d^2x_2. \tag{10.26}$$

We are looking for a lower bound for the energy, so apply the triangle inequality to the integrand of (10.26):

$$\frac{dC}{dz} \leq \frac{B_z^2}{2\pi} \iint (|\mathbf{b}_1 \cdot \hat{\boldsymbol{\theta}}_{12}| + |\mathbf{b}_2 \cdot \hat{\boldsymbol{\theta}}_{12}|) \frac{1}{r_{12}} d^2x_1 d^2x_2, \tag{10.27}$$

$$= \frac{B_z^2}{\pi} \iint |\mathbf{b}_1 \cdot \hat{\boldsymbol{\theta}}_{12}| \frac{1}{r_{12}} d^2x_1\, d^2x_2. \tag{10.28}$$

After a few more inequalities we find (Berger, 1993)

$$C^2 \leq \frac{13.6}{\pi} L R^4 B_{\underset{\sim}{z}}^4 \int b^2 \, \mathrm{d}^3 x. \tag{10.29}$$

The free magnetic energy is

$$E_f \equiv \frac{B_{\underset{\sim}{z}}^2}{8\pi} \int b^2 \, \mathrm{d}^3 x. \tag{10.30}$$

Equation (10.29) thus gives us an estimate

$$E_f \geq 9.18 \times 10^{-3} \frac{C^2}{L R^2 B_{\underset{\sim}{z}}^2}. \tag{10.31}$$

In practice, this may underestimate E_f by a factor of $\sim$3. For example, a random vector field, with a Gaussian distribution for b_x and b_y and small correlation lengths, has a true energy 3.01 times the lower bound.

Note that C is not a topological invariant. However the minimum crossing number C_{min} for a given topology does provide an invariant:

$$E_f \geq 9.18 \times 10^{-3} \frac{C_{min}^2}{L R^2 B_{\underset{\sim}{z}}^2}. \tag{10.32}$$

Section 10.5.3 discusses an application to solar physics.

10.3. Helicity integrals

Can we define a number for the linking of two vector fields?

The definition should follow naturally from the form of the Gauss linking number for closed curves, (10.5). Some of the individual field lines within a vector field may close upon themselves, and we can certainly measure the Gauss linking of pairs of closed lines. In the unlikely case that all field lines close upon themselves, we could sum linking numbers over all pairs of lines. There are an infinite number of field lines, of course, so the sum must be done as an integral, weighted by flux, which will be called a *helicity integral*. Unfortunately most field lines do not follow simple closed curves with a finite length – some may wander ergodically inside some region of space, and others may twist around a toroidal surface with an irrational twist number. Remarkably, the Gauss linking number can still be integrated over a vector field even when the field lines do not behave themselves (Moffatt, 1969; Arnold and Khesin, 1998).

We will consider two divergence-free vector fields $\mathbf{V}$ and $\mathbf{W}$ inside a domain $\mathcal{V}$ with boundary $\partial \mathcal{V}$ (some or all of the boundary could be at infinity). The helicity integral $H(\mathbf{V}, \mathbf{W})$ measures how much the flux of $\mathbf{V}$ links the flux of $\mathbf{W}$. Common examples are $H(\boldsymbol{\omega}, \boldsymbol{\omega})$, the kinetic helicity, which measures the self linking of vorticity; $K = H(\mathbf{B}, \mathbf{B})$, the magnetic helicity, which measures the self linking of magnetic flux; and $H(\boldsymbol{\omega}, \mathbf{B})$, the cross helicity, which measures the mutual linking between vorticity and magnetic flux. Other quadratic integrals can be written as helicity integrals; e.g. the magnetic energy $M = \int B^2 \, \mathrm{d}^3 x$ measures linking of magnetic flux and electric current, $M = H(\mathbf{B}, \mathbf{J})$.

10.3.1. Helicity in closed simply connected volumes

The general definition of $H(\mathbf{V}, \mathbf{W})$ will be given below. For now we give a restricted definition valid for special boundary conditions. Any $\mathbf{V}$ or $\mathbf{W}$ lines that cross $\partial \mathcal{V}$ at some point can be pictured as forming an arc outside $\mathcal{V}$, only to plunge back through $\partial \mathcal{V}$ at some other point. Of course, we do not really know the shape of these arcs, as we are only concerned with the field structure inside $\mathcal{V}$. For this reason, the Gauss double line integral no longer works when field lines cross the boundary. For now we simply assume all field lines close within $\mathcal{V}$:

$$\mathbf{V} \cdot \hat{\mathbf{n}}|_{\partial \mathcal{V}} = \mathbf{W} \cdot \hat{\mathbf{n}}|_{\partial \mathcal{V}} = 0. \tag{10.33}$$

We call $\mathcal{V}$ a *closed field region* for $\mathbf{V}$ and $\mathbf{W}$.

We now let γ_v be a closed field line (integral curve) of $\mathbf{V}$, and let γ_w be a closed field line of $\mathbf{W}$. Parametrize γ_v by arclength σ, so that a point on γ_v is $\mathbf{x}(\sigma)$. Then

$$\frac{d\mathbf{x}(\sigma)}{d\sigma} = \frac{\mathbf{V}(\mathbf{x}(\sigma))}{|\mathbf{V}(\mathbf{x}(\sigma))|} \equiv \widehat{\mathbf{V}}(\sigma). \tag{10.34}$$

Parametrize γ_w in a similar manner using arclength τ. Using these definitions, the expression for linking number, (10.5) is now beginning to involve the fields $\mathbf{V}$ and $\mathbf{W}$:

$$L(\gamma_v, \gamma_w) = +\frac{1}{4\pi} \oint_{\gamma_v} \oint_{\gamma_w} \widehat{\mathbf{V}} \cdot \frac{\mathbf{r}}{r^3} \times \widehat{\mathbf{W}} \, d\sigma \, d\tau. \tag{10.35}$$

Let us sum linking numbers over all field lines. First, weight the linking numbers by flux. For $\mathbf{V}$ in the neighbourhood of the field line γ_v the differential of flux is $|\mathbf{V}|d^2x$ where d^2x is an area element perpendicular to γ_v. After doing the same for $\mathbf{W}$ in (10.35), and summing over all field lines, the result is

$$H(\mathbf{V}, \mathbf{W}) = +\frac{1}{4\pi} \int_{\mathcal{V}} \int_{\mathcal{V}} \mathbf{V}(\mathbf{x}) \cdot \frac{\mathbf{r}}{r^3} \times \mathbf{W}(\mathbf{y}) \, d^3x \, d^3y. \tag{10.36}$$

The forms of (10.35) and (10.36) are similar; note that $H(\mathbf{V}, \mathbf{W})$ has units of V-flux $\times$ W-flux whereas $L(\gamma_v, \gamma_w)$ is dimensionless.

10.3.2. Using vector potentials

The form of the double integral defining $H(\mathbf{V}, \mathbf{W})$, (10.36), is of particular interest. The first integration performed, either over $\mathbf{x}$ or $\mathbf{y}$, inverts the curl operator. For a divergence free field $\mathbf{V}$ inside a closed field region $\mathcal{V}$ and $\mathbf{r} = \mathbf{y} - \mathbf{x}$, we define (Cantarella *et al.*, 2001)

$$\text{curl}^{-1}\mathbf{W}(\mathbf{x}) \equiv +\frac{1}{4\pi} \int_{\mathcal{V}} \frac{\mathbf{r}}{r^3} \times \mathbf{W}(\mathbf{y}) \, d^3y. \tag{10.37}$$

Then

$$H(\mathbf{V}, \mathbf{W}) = \int_{\mathcal{V}} \mathbf{V} \cdot \text{curl}^{-1}\mathbf{W} \, d^3x = \int_{\mathcal{V}} \text{curl}^{-1}\mathbf{V} \cdot \mathbf{W} \, d^3y. \tag{10.38}$$

In this form we can demonstrate that the Gauss integral really does give an integer measuring linking, as in Fig. 10.1.

To illustrate, we consider two toroidal flux tubes $\mathcal{V}$ and $\mathcal{W}$ with central axes γ_v, γ_w. Here $\mathcal{V}$ contains net $\mathbf{V}$ flux Φ_V and $\mathcal{W}$ contains net $\mathbf{W}$ flux Φ_W. The linking number between the axes of $\mathcal{V}$ and $\mathcal{W}$ is $L(\gamma_v, \gamma_w)$. We first assume that $\mathcal{V}$ is a thin tube, with $\mathbf{V}$ always parallel to its axis. No corresponding assumption will be needed for $\mathcal{W}$. The only information about $\mathcal{W}$ we need is the net flux linking $\mathcal{V}$, but this information is encoded into the vector potential $\text{curl}^{-1}\mathbf{W}$. Let S_v be a surface bounded by γ_v; the $\mathbf{W}$ flux pierces this surface $L(\gamma_v, \gamma_w)$ times. By Stokes' theorem

$$\oint_{\gamma_v} \text{curl}^{-1}\mathbf{W} \cdot d\mathbf{l} = \int_{S_v} \mathbf{W} \cdot \hat{\mathbf{n}}\, d^2x = L(\gamma_v, \gamma_w)\Phi_W. \tag{10.39}$$

Now sum over the $\mathbf{V}$ field. For simplicity we assume for the moment that the field lines of $\mathcal{V}$ are all parallel to γ_v and all close upon themselves. Then

$$\int_{\mathcal{V}} \mathbf{V} \cdot \text{curl}^{-1}\mathbf{W}\, d^3x = \int |\mathbf{V}|\, d^2x \oint_{\gamma_v} \text{curl}^{-1}\mathbf{W} \cdot d\mathbf{l} \tag{10.40}$$

$$= L(\gamma_v, \gamma_w)\Phi_V\Phi_W. \tag{10.41}$$

In this calculation the shape of the field lines of $\mathcal{W}$ were arbitrary (so long as they stayed within $\mathcal{W}$). By symmetry the same holds true for $\mathcal{V}$: the result still holds if the $\mathbf{V}$ lines twist or tangle within $\mathcal{V}$. They do not need to close upon themselves.

10.3.3. Helicity in open and multiply connected volumes

Gauge transformations of the form $\mathbf{A} \rightarrow \mathbf{A} + \nabla\psi$ have no effect on the magnetic helicity in a simply connected volume (e.g. a sphere rather than a torus) bounded by a magnetic surface. Multiply connected volumes, on the other hand, have one or more holes like a torus. Through each hole a magnetic flux must be specified. This leads to the existence of a wider class of gauge transformations, i.e. $\mathbf{A} \rightarrow \mathbf{A} + \mathbf{G}$ where $\nabla \times \mathbf{G} = 0$ but $\mathbf{G}$ is not a gradient. Changing $\mathbf{G}$ changes the net magnetic flux through the hole or holes without changing the magnetic field inside the volume (Taylor, 1986). A second gauge difficulty arises if the volume of interest is not bounded by a magnetic surface; then field lines will have endpoints on the boundary, and linking numbers will no longer be defined.

The general form of the magnetic helicity integral resolves these problems (Berger and Field, 1984; Finn and Antonsen, 1985). Let $\mathcal{V}$ be a volume with boundary $\delta\mathcal{V}$ and magnetic field $\mathbf{B}$. Within $\mathcal{V}$ there exists a unique vacuum or potential field $\mathbf{P}$ which satisfies $\nabla \times \mathbf{P} = 0$ and $\mathbf{P} \cdot \hat{\mathbf{n}}|_{\delta\mathcal{V}} = \mathbf{B} \cdot \hat{\mathbf{n}}|_{\delta\mathcal{V}}$. In a multiply connected volume $\mathbf{P}$ is uniquely specified if the interior fluxes of $\mathbf{P}$ are the same as $\mathbf{B}$ (Cantarella *et al.*, 2001). The gauge-invariant general helicity integral is

$$K_V = \int_{\mathcal{V}} (\mathbf{A} + \mathbf{A_P}) \cdot (\mathbf{B} - \mathbf{P})\, d^3x. \tag{10.42}$$

There are two geometrical interpretations of this expression. The first is illustrated in Fig. 10.7. In the right part of the figure, the potential field has replaced the true field in $\mathcal{V}$, but the extension of the field in the complement of $\mathcal{V}$ remains the same. The difference of total helicities (integrated over all space) can be shown to be independent of the extension field and equal to the expression in (10.42). The vacuum field $\mathbf{P}$ is the optimal reference field, as it has the minimum energy of all possible fields in $\mathcal{V}$ with the same boundary data $\mathbf{B} \cdot \hat{\mathbf{n}}$.

The second interpretation is illustrated in Fig. 10.8. Let $\mathbf{B_{cl}}$ be a closed field with an associated vector potential $\mathbf{A_{cl}}$ defined by (Kusano *et al.*, 1995)

$$\mathbf{B_{cl}} = \mathbf{B} - \mathbf{P}; \quad \mathbf{B_{cl}} = \nabla \times \mathbf{A_{cl}}. \tag{10.43}$$

As $\mathbf{B} = \mathbf{B}_{cl} + \mathbf{P}$, by the bilinearity of $H(\mathbf{B}, \mathbf{B})$ its helicity should be the sum of the self-helicities of $\mathbf{B}_{cl}$ and $\mathbf{P}$, plus their mutual helicity:

$$K_\mathcal{V} = H_\mathcal{V}(\mathbf{B}, \mathbf{B}) = H_\mathcal{V}(\mathbf{B}_{cl}, \mathbf{B}_{cl}) + 2H_\mathcal{V}(\mathbf{P}, \mathbf{B}_{cl}) + H_\mathcal{V}(\mathbf{P}, \mathbf{P}). \qquad (10.44)$$

The self helicity of the closed field $\mathbf{B}_{cl}$ is simply

$$H_\mathcal{V}(\mathbf{B}_{cl}, \mathbf{B}_{cl}) = \int_\mathcal{V} \mathbf{A}_{cl} \cdot \mathbf{B}_{cl} \, d^3x. \qquad (10.45)$$

Meanwhile, the mutual helicity of an open and closed field should also be well defined. A closed structure linking an open structure (e.g. an arch or loop as shown in Fig. 10.8)

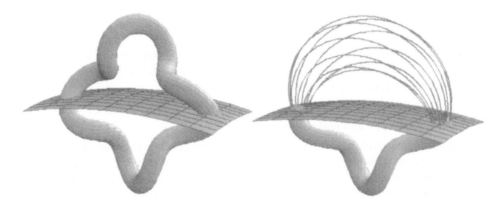

Figure 10.7 Helicity in an open volume is measured relative to that of the potential field.

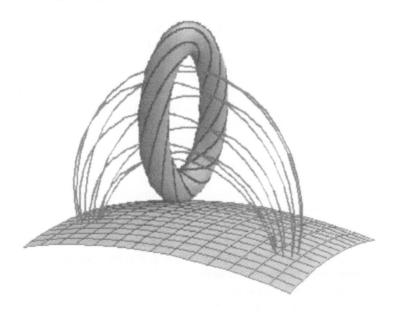

Figure 10.8 An open field decomposed into a closed part and a potential part. In general, these parts will occupy the same space.

cannot be freed. The mutual helicity is

$$2H_{\mathcal{V}}(\mathbf{B}_{cl}, \mathbf{P}) = 2 \int_{\mathcal{V}} \mathbf{A_P} \cdot \mathbf{B}_{cl} \, d^3 x. \tag{10.46}$$

Finally, the self helicity of the vacuum field $H_{\mathcal{V}}(\mathbf{P}, \mathbf{P})$ will simply be set to zero. Equation (10.44) then sums to (10.42). This generalized helicity is gauge invariant in any physical situation (unless magnetic monopoles are shown to be physical)!

Periodic geometries are often used in numerical simulations. In Section 10.4 we show that these can sometimes hide monopoles; this can lead to a breakdown of the helicity invariant.

10.3.4. *Magnetic helicity transport and dissipation*

Choosing a convenient gauge greatly simplifies finding the transport and dissipation equations for helicity (even though $dK_{\mathcal{V}}/dt$ is gauge invariant). We follow (Barnes, 1988a,b; Berger, 1998; Cantarella *et al.*, 2001) and let the vector potential for the vacuum field $\mathbf{A_P}$ satisfy

$$\nabla \cdot \mathbf{A_P} = 0; \quad \mathbf{A_P} \cdot \hat{\mathbf{n}}|_{\partial \mathcal{V}} = 0. \tag{10.47}$$

For multiply connected volumes we further require $\mathbf{A_P}$ to be flux-free (e.g. if regarded as a magnetic field, $\mathbf{A_P}$ would have no net toroidal flux; the surface integral $\int \mathbf{A_P} \cdot \hat{\mathbf{n}} \, d^2 x = 0$ across any cross-section). These requirements uniquely specify $\mathbf{A_P}$ within $\mathcal{V}$. One can then calculate the time derivative

$$\dot{K}_{\mathcal{V}} = -2 \int_{\mathcal{V}} \mathbf{E} \cdot \mathbf{B} \, d^3 x + 2 \oint_{\partial \mathcal{V}} \mathbf{A_P} \times \mathbf{E} \cdot \hat{\mathbf{n}} \, d^2 x, \tag{10.48}$$

where $\mathbf{E}$ is the electric field. This equation is analogous to Poynting's theorem for the dissipation and transport of electromagnetic energy.

Given a simple Ohm's law, $\mathbf{E} = \mathbf{B} \times \mathbf{V} + \eta \mathbf{J}$, the time derivative becomes

$$\dot{K}_{\mathcal{V}} = -2 \int_{\mathcal{V}} \eta \mathbf{J} \cdot \mathbf{B} \, d^3 x + 2 \oint_{\partial \mathcal{V}} ((\mathbf{A_P} \cdot \mathbf{V})\mathbf{B} - (\mathbf{A_P} \cdot \mathbf{B})\mathbf{V}) \cdot \hat{\mathbf{n}} \, d^2 x. \tag{10.49}$$

The first term represents dissipation, the second measures the effect of twisting motions or rotation on the boundary, while the third represents the bulk transport of helical field across the boundary. Helicity transport across boundaries is of crucial importance for the overall magnetic helicity balance of the sun. Section 10.6 discusses how keeping track of magnetic helicity in the sun gives us clues about the nature of the solar dynamo.

The dissipation term is small for reconnection events. Let $M = \int B^2 \, d^3 x$ measure magnetic energy, with Ohmic dissipation $\dot{M}_{\text{Ohm}} = -2 \int \eta J^2 \, d^3 x$. A Schwarz inequality gives

$$|\dot{K}_{\mathcal{V}}|_{\text{Ohm}} \leq \sqrt{2\eta M \left| \frac{dM}{dt} \right|_{\text{Ohm}}}. \tag{10.50}$$

A natural length scale can be derived from the ratio of $H_{\mathcal{V}}$ and M: $L \equiv |H_{\mathcal{V}}|/M$. For example, $L \approx 0.31R$ for a linear force free field inside a spherical magnetic surface of radius R. We then obtain a dissipation time scale for helical structure $\tau_d = L^2/\eta$. In an isolated plasma (i.e. no input of magnetic energy) the helicity cannot be dissipated on faster timescales. Integrating (10.50) over a time Δt gives a bound on helicity change

$$\left| \frac{\Delta K_V}{K_V} \right| \le \sqrt{\frac{\Delta t}{\tau_d}}. \tag{10.51}$$

This inequality shows that ΔK_V is small for any fast reconnection event, where $\Delta t \ll \tau_d$. Note that the maximum helicity dissipation scales as $\eta^{1/2}$ (and hence magnetic Reynolds number $R_m^{-1/2}$). The actual helicity dissipation may be even smaller at high R_m. Geometric considerations (Freedman and Berger, 1993) suggest that helicity dissipation during reconnection may actually scale as R_m^{-2}.

10.4. The topology of periodic domains

Numerical simulations often employ a periodic geometry to avoid physical boundaries. This may be dangerous – the peculiar topology of a periodic domain can have strange effects on physical phenomena inside. Periodic boundary conditions (or any other kind of boundary conditions) should primarily affect structure on large scales, scales similar in size to the domain itself. One phenomenon where large scales are of interest involves the inverse cascade of magnetic helicity. The strange nature of periodic domains shows itself most clearly when considering this inverse cascade (Berger, 1997).

First, consider an infinite domain and let **B** be expressed in terms of a vector potential $\mathbf{B} = \nabla \times \mathbf{A}$. We divide **B** into a mean field $\mathbf{B_0} = B_0\hat{\mathbf{z}}$ and a fluctuating field **b**. These have corresponding vector potentials $\mathbf{A} = \mathbf{A_0} + \mathbf{a}$. A possible choice for $\mathbf{A_0}$ could be

$$\mathbf{A_0} = \frac{B_0}{2}(-y, x). \tag{10.52}$$

The helicity is

$$K = \int \left(\mathbf{A_0} \cdot \mathbf{B_0} + \mathbf{A_0} \cdot \mathbf{b} + \mathbf{a} \cdot \mathbf{B_0} + \mathbf{a} \cdot \mathbf{b} \right) d^3x. \tag{10.53}$$

The first term vanishes while the second and the third are equal. Thus

$$K = K_0 + h; \tag{10.54}$$

$$K_0 = 2 \int \mathbf{A_0} \cdot \mathbf{b} \, d^3x; \tag{10.55}$$

$$h = \int \mathbf{a} \cdot \mathbf{b} \, d^3x. \tag{10.56}$$

The quantity K_0 measures the linking of b with the mean field, while h measures the self-linking of b. The ideal-MHD invariance of helicity implies

$$\frac{dK}{dt} = 0 \ \Rightarrow \ \frac{dK_0}{dt} = -\frac{dh}{dt}. \tag{10.57}$$

Now suppose we wish to extend these definitions to a periodic domain, e.g. a cube with sides $(0, 2\pi)$. A grave difficulty soon appears: $\mathbf{A_0}$ is not periodic! In fact, one can show that no periodic vector potential exists for a mean field in a periodic box. This result can be proved most simply using cohomology theory, but we will not follow this path. Instead we will motivate this result by showing that the mean magnetic field must have as its source not currents but magnetic monopoles. There are no currents because $\nabla \times \mathbf{B}_0 = \mathbf{0}$. But no monopoles seem to be around either. Where is the source of the mean field?

Topologists use the term 3-*torus* to describe a three-dimensional-periodic domain. A 2-*torus* can be visualized as the surface of a donut. However, 3-*torii* are impossible to draw, because we cannot embed them in three dimensions. Instead we will compromise, preserving the strangeness of our situation, by having only two dimensions be periodic. Thus let x and y be periodic, while z ranges from 1 to 2. This can be visualized as a toroidal annulus (see Fig. 10.9), where x and y become angles on the surface, while z becomes distance from the toroidal axis. We choose $[1, 2]$ for z rather than $[0, 1]$ because $z = 0$ is anomalous: it is a curve rather than a surface. Note that the mean field $B_0\hat{\mathbf{z}}$ points uniformly outward from the central $z = 0$ toroidal axis. The mean field is thus the field arising from a ring of monopoles lurking inside $z < 1$. The monopoles are hidden outside our domain! Nevertheless they prevent us from using vector potentials.

In a numerical simulation of three-dimensional-MHD turbulence with a mean magnetic field, Stribling *et al.* (1994) found small scale helicity inverse cascading to larger scales. When the helicity reached the size of the box, it simply disappeared. In particular, $h(t) \to 0$ as the turbulence proceeded. For an infinite domain, or a domain with fixed boundaries, one would simply expect the helicity to be transfered to K_0 according to (10.57). Unfortunately for a periodic box K_0 cannot be defined, because of the lack of any meaningful vector potential which can be placed into (10.55). (Note that the integrand $\mathbf{a} \cdot \mathbf{B}_0$ cannot be substituted into (10.55) since it is not gauge-invariant.)

Defining helicity through linking of flux does not work either. For example, one computes the magnetic flux Φ through a closed loop using the formula

$$\Phi = \oint (\mathbf{A}_0 + \mathbf{a}) \cdot d\mathbf{l}.$$

This integral becomes undefined when the loop goes all the way across the periodic domain.

The old magnetic helicity simply does not work in a periodic domain with mean fields. This is not merely a problem with definitions: something unphysical is happening. To illustrate, we will show that it is possible to turn a twisted flux tube inside out in a periodic universe (this

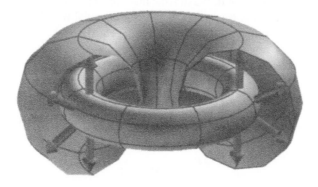

Figure 10.9 Toroidal representation of periodic mean field. The arrows point in the direction of the mean field.

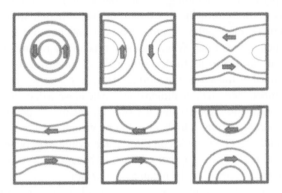

Figure 10.10 Turning a flux tube inside out.

cannot happen in ordinary space). Fig. 10.10 shows the x and y components of the magnetic field of a flux tube invariant in the z direction. The second drawing is the same as the first – the tube has merely been displaced horizontally. A series of reconnections and deformations leads to the outer field lines becoming inner field lines and vice-versa. Furthermore, the sense of twists flips – a right-handed tube (of positive helicity) becomes a left-handed tube (of negative helicity). This change in helicity (supposing we could find a way of defining it) cannot be blamed on dissipation: dissipation of helicity can be made arbitrarily small by assuming the reconnection proceeds sufficiently rapidly (see Section 10.3.4). Instead the helicity change results from the peculiar topology of periodic domains.

   Is there a replacement for helicity which does work in periodic domains? Here we present a well-defined quantity which does measure a form of linking. Unfortunately, it is only an ideal invariant for two-dimensional motions, i.e. with no component parallel to the mean field.

   We consider a single plane $z = $ constant, and look at the field lines in that plane. A uniform z field crosses this plane, so that the flux through any area $\mathcal{A}$ is $\Phi = B_0 \mathcal{A}$. Suppose that γ is a set of two oriented curves which divides the periodic box in two (see Fig. 10.11). We can say that γ links one of the areas $\mathcal{A}_+$ in a positive sense, and the other area $\mathcal{A}_-$ in a negative sense. Thus the net flux linked is $B_0(\mathcal{A}_+ - \mathcal{A}_-)$.

   Now replace γ by a field line of **b**. We sum the flux linked by all the **b** lines to obtain a new helicity measure $\mathcal{H}$. First write **b** in terms of a vector potential $\mathbf{b} = \nabla \times \mathbf{T}(\mathbf{x}, \mathbf{y})\hat{\mathbf{z}}$. A short calculation (Berger, 1997) shows that the new measure equals

$$\mathcal{H} = \iint T \, \mathrm{d}x \, \mathrm{d}y - \frac{1}{2}(2\pi)^2(T_{\max} + T_{\min}). \tag{10.58}$$

The function $T(x, y)$ moves with the fluid in two-dimensional ideal MHD,

$$\frac{\partial T}{\partial t} + \mathbf{v} \cdot \nabla T = 0 \tag{10.59}$$

which implies that $\mathcal{H}$ is an ideal two-dimensional MHD invariant.

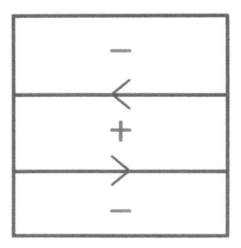

Figure 10.11 Oriented areas.

10.5. Coronal magnetic fields

The solar atmosphere can be divided into three different layers governed by different physical processes. The photosphere, about 500 km thick, is the optical surface of the sun: photons fighting their way out from the solar interior into space usually last scatter inside the photosphere. Strong line emission originates in the next layer, the chromosphere, of about 2000 km depth. Above lies the outer atmosphere, the corona, where density sharply drops to very low values while temperature suddenly rises to 10^6 K. The origin of this extraordinary temperature rise is not fully understood. Several different mechanisms have been proposed, including Alfvén wave heating and current sheet dissipation.

While in the dense photosphere the thermal pressure dominates the average magnetic pressure, the opposite is true in the dilute corona. Hence in the photosphere plasma motions move the field around, while in the upper chromosphere and in the corona the field determines the plasma dynamics. We can assume that the field is almost everywhere frozen-in due to the very high Reynolds number; as a consequence field lines do not break up or slip through the plasma except during reconnection. Because of the dominance of the Lorentz magnetic force over other forces (pressure and gravitational) equilibrium configurations should generally be close to solutions of the force-free condition $\mathbf{J} \times \mathbf{B} = \mathbf{0}$. Thermal and gravitational effects must not be neglected, however, in models of the large-scale corona and in models of how the coronal plasma radiates.

The magnetic distribution in the corona differs substantially from that of the dense photosphere. In the photosphere the high plasma pressure compresses the magnetic flux into thin tubes of radius typically $R \sim 150–200$ km, and field strength $B \sim 2000$ G (Solanki, 1993). In addition there are distributions of flux on smaller scales. This flux does not reach high into the atmosphere; it produces fields resembling moss or a magnetic carpet (Berger *et al.*, 1999). In the corona, where the plasma pressure is much smaller, the magnetic flux expands filling the entire volume. The field intensity varies only slightly, forming bright X-ray loops observed as the typical structures in the corona.

10.5.1. Solar flares

Coronal loops are line-tied to the photosphere at their footpoints: due to the vast differences in gas pressure, coronal magnetic field has little influence on photospheric motion; on the other hand, the photospheric flux elements move around with the dense plasma. Random motions (especially vortical motions) continuously shear and twist the field lines and flux tubes around each other. This increases the complexity of the field structure, and thus increases the magnetic energy. In general, highly structured fields will be intermittent: they contain narrow regions of strong magnetic field gradients (Parker, 1983; Berger, 1991; Bhattacharjee and Wang, 1991). Another process which adds to the field structure happens when new flux emerges through the photosphere and bumps into old flux. Strong field gradients develop at the boundaries between new and old.

Strong field gradients correspond to intense current layers. Rapid reconnection of field lines can be triggered in these layers, leading to a rearrangement of the field. The field loses energy and relaxes to a state with less structure and simpler topology. Solar *flares* result from these sudden releases of magnetic free energy. The maximum energy available is the difference between that stored in the configuration before reconnection and that of a force-free state with the same photospheric flux distribution, i.e. the lowest energy configuration of given helicity (Berger, 1984; Heyvaerts and Priest, 1984).

Solar flares have energies ranging from $< 10^{26}$ ergs (observational limit) to 10^{33} ergs. Nano and microflares are occurring at a continuous rate and may be a major coronal heating source, while big flares are sporadic events. Flares obey a remarkable power law in energy release E (Shimizu, 1995):

$$P(E) \sim E^{-a} \quad a \sim 1.4 - 1.6. \tag{10.60}$$

The distribution of small flare energies is of crucial importance to coronal heating theory. In particular, if microflares and nanoflares are the primary souce of heat for coronal loops then the slope of the flare distribution must steepen somewhere below the present observable limit to about $a \sim 2$. Recent observations provide some evidence for this strengthening (Parnell and Jupp, 2000).

Lu and Hamilton (1991) suggest that solar magnetic fields provide an example of self-organized criticality. They give a statistical 'avalanche model' to explain the power law. In this model, there is an elementary small event, and larger events are composed of many of these small events occuring simultaneously. Small events occasionally trigger a medium event; medium events occasionally trigger large events, and so on. The statistics of this process yield a power law distribution of event sizes.

Note that a power law distribution has no intrinsic scale. But there may be intrinsic scales on the sun, set by the size of elementary flux tubes in the photosphere: reconnection of two tubes of radius 150 km and strength 2000 G releases $\sim 10^{26} - 10^{27}$ ergs (Berger, 1994b). In the simple avalanche model, this may set the energy of the 'elementary' event. The power law may not be valid (at least not with the same exponent) below this level. Events could in fact occur with smaller energy release if flux tubes fragment in the corona (i.e. only part of the flux from a photospheric tube becomes involved in a reconnection event). A wide distribution of photospheric tube sizes could introduce its own power laws.

10.5.2. Magnetic structure in coronal loops

The plasma in active region loops radiates at temperatures of millions of degrees. If the heat source were switched off, the cooling time would be about one day. Heat must be replenished at an average rate per unit area $P = 10^7 \, \text{ergs cm}^{-2} \, \text{s}^{-1}$. As mentioned above, microflares which remove fine-scale structure may release sufficient energy to supply this heat. An alternative mechanism involves the dissipation of upward propagating Alfvén waves. A highly conducting plasma like the corona does not dissipate waves efficiently. However, MHD wave dissipation would be strongly enhanced by interaction with pre-existing fine-scale magnetic structures. In both wave and microflare models, then, the existence of fine-scale structure is important. This idea has led to the construction of hybrid models of coronal heating (Lou and Rosner, 1986; Similon and Sudan, 1989).

Footpoint motions add to the complexity of the coronal field in a manner which is readily treatable by statistical and topological methods (see Fig. 10.12). Processes include *braiding* of flux tubes, *twisting* of individual tubes, and *fragmentation*. Sturrock and Uchida (1981) calculated the energy stored in a randomly twisted tube: they found that the energy stored is proportional to twist T^2. Since the root-mean-square twist T_{rms} increases as time $t^{1/2}$ it follows that the energy $E \sim t$ and the power $P = dE/dt \sim t^0$. The heating rate P is therefore independent of dissipation time (this analysis does require the dissipation time to be longer than the correlation time). Unfortunately, given typical velocities and correlation times in the photosphere, P is too small by a factor of 10–20 for active region heating (Berger, 1991); however enhanced vorticity at granular boundaries may increase P to reasonable levels (Karpen *et al.*, 1993).

Parker (1983) suggested that besides twisting, random motions of many photospheric points lead to a tangled coronal field (Fig. 10.13). The energetics of braiding and fragmentation differ from twisting. As shown below, braiding and fragmentation generate energy at a rate which increases quadratically in time (in the absence of dissipation). This is similar to a spring whose energy increases quadratically with displacement. Thus these processes do depend on dissipation rate, or more precisely on the saturation level, where loss of structure to dissipation or reconnection balances input of structure. To calculate the heating rate, we need to be able to estimate the energy of a highly tangled field, and how fast this energy increases as the endpoints of field lines random walk in the photosphere. This problem will be addressed in the next section.

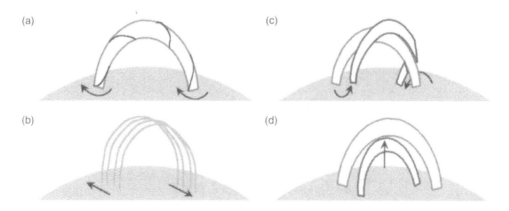

Figure 10.12 Adding structure to coronal fields. (a) Rotational motions twist flux tubes. (b) Shearing motions stress magnetic arcades. (c) Random motions braid tubes about each other. (d) Rising flux bumping into old flux.

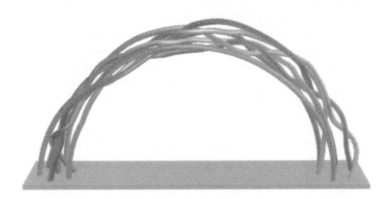

Figure 10.13 A highly tangled field.

10.5.3. *Energy estimates*

The energy of a highly tangled equilibrium coronal loop can be estimated using the crossing number techniques of Section 10.2.2. The basic idea is that any collection of field lines within a loop determines a braid. We apply these ideas to three coronal flux tubes within an X-ray loop. The rate of change of the E_f estimate gives a value for the power input at the photosphere. As mentioned in the previous section, the power requirement is approximately 10^7 ergs cm^{-2} s^{-1}. Assume that the energy input saturates when reconnection liberates energy at the same rate as input. This happens when b reaches some critical value μ. For example, $\mu = 0.28$ if neighboring tubes are sheared by $30°$.

Consider N photospheric tubes of flux Φ random walking about each other. Let V be a typical photospheric velocity (e.g. 1 km s^{-1}), d the typical distance between footpoints (e.g. 10^3 km) and D the diameter of coronal tubes. If the photospheric tubes are uniformly distributed then $d \approx D$; however if the tubes are clumped (at the edges of convection cells) then we may expect $d < D$. Dimensional arguments (Berger, 1993) suggest that $C_{min}(t)$ grows as

$$C_{min} = \epsilon N^{3/2} \frac{Vt}{d} \Phi^2. \qquad (10.61)$$

A random walk simulation with step size λ inside a circular reflecting wall (to keep the typical footpoint spacing d constant) gives $\epsilon = \epsilon_0 \lambda/d$ where $\epsilon_0 \approx 0.125$.

The energy at saturation is

$$E_f = LN\pi \frac{D^2}{4} B_{\tilde{z}}^2 \mu^2. \qquad (10.62)$$

Putting together the last four equations gives power input

$$P \geq 5 \times 10^6 \frac{D\lambda}{d^2} \left(\frac{V}{1 \text{ km s}^{-1}} \right) \left(\frac{B_{\tilde{z}}}{100G} \right)^2 \frac{\mu}{0.28} \text{ergs cm}^{-2} \text{ s}^{-1}. \qquad (10.63)$$

This result seems to be in good agreement with expected values.

Let us now examine the case of fragmented flux tubes. Suppose that individual flux tubes are twisted by rotational motions at the photosphere. They may sometimes twist clockwise and sometimes anti-clockwise, so that the average twist is zero. Of course, there will be a mean square twist which grows linearly with time as in the Sturrock–Uchida model.

Now suppose that occasionally the photospheric tubes break apart. New flux tubes will later form, gathering and recycling the stray flux from the old tubes. Flux from several dead tubes may be mixed together in each new tube. All this photospheric activity complicates the connectivity between photospheric tubes and the coronal field. If a flux tube acquires a random clockwise twist before breaking apart, this twist will no longer be cancelled by an anti-clockwise twist on one of the new tubes, because of the increased topological complexity.

This leads to a heating rate (Berger, 1994b)

$$P = \frac{\Phi}{8} \left(\frac{B_z}{\pi} \right)^{3/2} \frac{\mu V}{\lambda} \left(\frac{3\tau_c}{\tau_f} \right)^{1/2}, \tag{10.64}$$

where Φ is the flux of a coronal tube, λ is the granule size (velocity correlation length), τ_c is the velocity correlation time and τ_f the fragmentation time. Taking $\mu = 0.28$, $B_z = 100\,\mathrm{G}$, $V = 1\,\mathrm{km\,s^{-1}}$, and $\tau_f = 3\tau_c = 2400\,\mathrm{s}$, one obtains $P = 10^7\,\mathrm{ergs\,cm^{-2}\,s^{-1}}$, sufficient for active regions. Furthermore, if throughout the sun all elementary photospheric flux elements have the same flux Φ then heating will scale as $B_z^{3/2}$. For the quiet sun $B_z \approx 10\,\mathrm{G}$, which implies $P = 10^{5.5}\,\mathrm{ergs\,cm^{-2}\,s^{-1}}$, consistent with observed values. Note that fragmentation can give microflares and nanoflares smaller than $10^{26}\,\mathrm{ergs}$, perhaps with a different power law slope than that given by avalanche models above $10^{26}\,\mathrm{ergs}$.

10.6. The solar dynamo and solar chirality

The classical observational evidence for the solar dynamo comes from the 22-year cycle in sunspot numbers and distribution. However, new observations are always welcome, especially if they tie the solar dynamo to other manifestations of solar magnetism, such as the structure of the corona. Recently, observers have begun to examine in detail the sign and magnitude of magnetic helicity of structures in both the corona and solar wind. One clear result stands out: the sign of the magnetic helicity reverses going from northern to southern hemispheres. Since magnetic helicity changes sign under mirror reflection, this says that the handedness of magnetic structures correlates with position: northern structure is predominantly left-handed ($K < 0$), while southern structure is predominantly right-handed ($K > 0$).

The magnitude of the coronal helicity must be studied as well as its sign, as concentrations of helicity can increase activity. When magnetic helicity generated in the solar dynamo emerges in the corona, free magnetic energy builds up. Eventually this energy may trigger flares and coronal mass ejections (Low, 1994; Rust, 1994; Rust and Kumar, 1996).

10.6.1. Observations

Observations of north–south chirality extend from the photosphere to the solar wind (Brown *et al.*, 1999). Vector magnetograms of photospheric fields can be differentiated to give the line-of-sight electric current J_z. The ratio J_z/B_z is a pseudoscalar; for a linear force free field ($\mathbf{J} = \alpha \mathbf{B}$) it yields α, which has the same sign as the magnetic helicity. Observations show negative J_z/B_z in the north and positive in the south (Seehafer, 1990; Pevtsov *et al.*, 1995). Prominence structures also depend on hemisphere: northern prominences tend to have negative magnetic helicity (Martin *et al.*, 1992; Rust and Kumar, 1994).

Further away, the solar wind possesses helical structure on many scales. On the largest scales, helicity flows outwards in the shape of the Parker spiral (Bieber *et al.*, 1987). This helicity arises because solar rotation twists the open field lines. Again one finds negative magnetic helicity in the north and positive helicity in the south. The net helicity flow into the southern wind during an 11-year cycle is about $1 \times 10^{47} \, \mathrm{Mx}^2$ in an 11-year cycle (Berger and Ruzmaikin, 2000). Smaller scale structure also carries magnetic helicity away from the sun. This smaller structure can be observed in magnetic clouds, and may originate in coronal mass ejections (Kumar and Rust, 1996). Estimates give a helicity flow due to CMEs on the order of $10^{45} \, \mathrm{Mx}^2$ each 11-year cycle.

10.6.2. Origin of the helicity flow

The large scale helicity flow into the Parker spiral has a simple explanation in terms of the mean rotation of the sun. Think of the net flux Φ extending into the wind from the Northern hemisphere. This flux acquires one twist each mean solar rotation. The magnetic helicity of a uniformly twisted tube with T twists is $K = T\Phi^2$. Thus the helicity flow dK/dt is roughly equal to Φ^2/τ, where τ is the rotation period (Bieber *et al.*, 1987; Berger and Ruzmaikin, 2000).

The origin of helical structure on less global scales, e.g. in prominences and CMEs, is not so clear. Several possible mechanisms have been considered.

10.6.2.1. Coriolis forces on rising flux tubes

Magnetic flux generated at the base of the convection zone by the solar dynamo can become buoyant and rise to the surface. As it rises coriolis forces distort the axis of the tube. This is an ideal process, so total magnetic helicity is conserved. However, the helicity of a tube can be decomposed into *twist* T_w, measuring the twist of field lines about the axis, and *writhe* W_r, measuring kinking and knotting of the axis itself (Berger and Field, 1984; Moffatt and Ricca, 1992). These two quantities may change as a result of the coriolis forces, with the constraint that $dT_w/dt = -dW_r/dt$. In the northern hemisphere the effect should give positive writhe and negative twist.

Unfortunately, when the tube reaches the corona it forgets its shape. The part of the tube above the photosphere lies in a magnetically dominated region (see Section 10.5), while below the surface the dense plasma dominates. Thus the coronal tube will take a shape consistent with its total helicity, forgetting the decomposition into twist and writhe it had below the surface. Since the coriolis effect produces no net helicity, it cannot contribute to the helical structures found in prominences and CMEs. It may be, however, that the ratio J_z/B_z at the photosphere is influenced more by subsurface dynamics rather than coronal dynamics. In this case, the vector magnetogram observations may reflect the influence of the coriolis forces on subsurface structure (Longcope *et al.*, 1998).

10.6.2.2. Motion of surface fields

Large-scale surface motions (such as differential rotation) can shear pre-existing coronal fields (Van Ballegooijen and Martens, 1990; Priest *et al.*, 1996). Unfortunately, this process generates helicity of the wrong sign more often than the correct sign (Van Ballegooijen *et al.*, 1998). For example, if the polarity inversion line of a polar crown prominence is oriented north–south, differential rotation yields the correct sign, but an east–west orientation yields the wrong sign. These negative results suggest most of the hemispheric asymmetry must originate below the photosphere. Thus, we must look to the solar dynamo to understand the source of magnetic helicity in the corona.

10.6.3. Magnetic helicity balance in the solar dynamo

Here we discuss the influences of the α-effect and the Ω-effect (differential rotation) on magnetic helicity distribution in the sun. Analysis of the Ω-effect is straightforward: in fact, the net helicity generation in each hemisphere can be turned into a surface integral. This integral can be readily evaluated using existing magnetogram and helioseismology data.

On the other hand, several uncertainties accompany the α discussion. First, we do not have a certain idea of the value of α, or even its sign; in fact α may vary considerably through the convection and subconvection layers (Brandenburg *et al.*, 1990; Brummell *et al.*, 1998). Second, the α-effect generates equal and opposite amounts of magnetic helicity in the mean and fluctuating fields (Ji, 1999). But we cannot simply trust the fluctuating helicity to dissipate due to Ohmic resistivity. The problem lies in the direction of the turbulent cascade: magnetic helicity tends to inverse cascade to lower wavenumbers where dissipation is inefficient.

10.6.3.1. The Ω-effect

The Ω-effect involves the twisting up of the interior magnetic field by differential rotation (see Fig. 10.14). This effect produces helicity of the correct sign in each hemisphere (see Fig. 10.15), independent of cycle. The net inflow to a hemisphere is large – observations (Berger and Ruzmaikin, 2000) find -2×10^{46} Mx^2 for an 11-year cycle to the northern hemisphere (2.5×10^{46} Mx^2 for the south). Thus one may expect that subsurface fields store helicity of the correct sign. We only need some 10% of this magnetic helicity input to account for estimates of the helicity shed in mass ejections. The magnetic helicity stored at the base of the convection zone can be carried to the surface in rising flux tubes.

We briefly outline how the observations are analysed. Let $K_N(t)$ be the magnetic helicity in the northern interior of the sun. As this volume (a half sphere) is not bounded by magnetic surfaces (10.42) must be used for the helicity, with time derivative equation (10.49). The first term will be discussed in the next subsection. Here we evaluate the surface term. In particular, we consider the helicity transfer into the northern hemisphere due to the differential rotation velocity $\mathbf{v} = \Omega(r, \theta)r \sin\theta\hat{\varphi}$. There are two contributions to (10.49). Transfer through the northern photosphere gives

$$\dot{K}_N = 2R^3 \int_0^{2\pi} \int_0^{\pi/2} B_r(\theta, \phi) A_{P\phi}(\theta, \phi)\Omega(R, \theta) \sin^2\theta \, d\theta \, d\phi. \tag{10.65}$$

The flux distribution $B_r(\theta, \phi)$ can be determined by time-series of magnetogram data (Zhao and Hoeksema, 1993). Use of spherical harmonics simplifies the evaluation of $\mathbf{A_P}$.

To compute transfer through the equator, we assume the unknown flux distribution is a function of r alone. As the equatorial interior rotates almost uniformly, this can only be a small source of error. We know the total flux through the equator, as it equals the total flux through the northern photosphere. For axial symmetry (Berger and Ruzmaikin, 2000), $\mathbf{A_P} = \Phi(r)\hat{\varphi}/(2\pi r)$. The equatorial contribution is thus

$$\dot{K}_N = 2 \int_0^{R_\odot} \int_0^{2\pi} B_\theta(r)\Phi(r)\Omega(r, \pi/2)r \, dr. \tag{10.66}$$

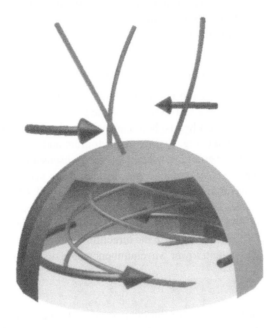

Figure 10.14 Differential rotation twists up magnetic flux within the northern hemisphere of the sun.

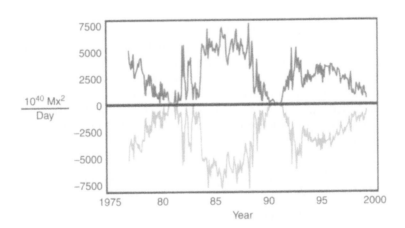

Figure 10.15 Helicity transfer into the northern interior (predominantly negative curve) and into the southern interior (predominantly positive curve) since 1976.

The net transfer of helicity into the northern interior can be evaluated by summing these two integrals. Fig. 10.15 displays the results. Net helicity injection into the northern hemisphere reaches a maximum at solar minima (near 1985 and 1995) when the poloidal flux is strongest.

10.6.3.2. *The α-effect*

In solar dynamo theory, turbulent flows regenerate poloidal magnetic flux from the toroidal field (Moffatt, 1978). The effect of the turbulence is highly nonlinear; but to a first approximation the flows create an effective electromotive force

$$\boldsymbol{\xi} = \alpha\mathbf{B} - \beta\nabla \times \mathbf{B}. \tag{10.67}$$

Here $\mathbf{B}$ is the mean magnetic field. In classical dynamo theory α arises from kinetic helicity (and is opposite in sign); while β represents a turbulent diffusion. The electric field appearing in (10.48) becomes $\mathbf{E} = -\boldsymbol{\xi}$. Thus (Ruzmaikin, 1996; Seehafer, 1996) the α-effect can change the helicity of the mean field. From (10.48) we have

$$\dot{K}_V = 2\int_V (\alpha B^2 - \beta\nabla \times \mathbf{B} \cdot \mathbf{B})\,\mathrm{d}^3x - 2\oint_{\partial V} \mathbf{A_P} \times (\alpha\mathbf{B} - \beta\nabla \times \mathbf{B}) \cdot \hat{\mathbf{n}}\,\mathrm{d}^2x. \tag{10.68}$$

The first (volume) term represents nonlinear transfer of helicity between the mean field and the fluctuating field; no net helicity is created. Both the sign and value of α are subjects of intense debate (Brandenburg *et al.*, 1990; Brummell *et al.*, 1998; Field and Blackman, 1999). Consider the northern hemisphere. Recent simulations give α negative at the convective overshoot layer. This is good: it implies that the mean field receives the correct sign of helicity to match observations (as does differential rotation).

The magnitude of this term can be quite high. As a crude estimate, we consider a dynamo operating in the convective overshoot layer of the sun between the equator and $\theta = \pi/4$. Suppose the layer has a thickness 10^4 km, with $B = 10^4$ G and $\alpha = 0.1\,\mathrm{cm\,s^{-1}}$. Then for the northern hemisphere $K_{\mathrm{N}} = 10^{44}\,\mathrm{Mx^2/day}$. This is an order of magnitude above the differential rotation rate, and perhaps two orders of magnitude higher than the estimated helicity loss in CMEs (Bieber and Rust, 1995). On the other hand, α-effect generation may be inhibited by several factors, e.g. the β term, variations in the sign of α from place to place, and inhibition of fluid turbulence by the magnetic field (Field and Blackman, 1999).

What is the fate of the magnetic helicity transferred to the fluctuating magnetic field? In MHD turbulence, magnetic energy can cascade to small scales due to turbulent processes. However, magnetic helicity does not easily fit into small scales. In fact, theory and numerical simulations (Pouquet *et al.*, 1976; Matthaeus and Montgomery, 1980) suggest that magnetic helicity inverse cascades to larger length scales. Thus while the α and β terms represent transfer of helicity between large (e.g. 10^4–10^5 km for the Sun), and intermediate scales (e.g. 10^2–10^3 km), further transfer to dissipation length scales may be inhibited. At the base of the convection zone the hot plasma ($T = 2 \times 10^6$ K) has a low Spitzer resistive diffusion coefficient, $\eta = 3 \times 10^3\,\mathrm{cm^2\,s^{-1}}$. Magnetic helicity stored in structures with a length scale λ will dissipate in a time $\tau_{\mathrm{d}} = \lambda^2/\eta$. For $\tau_{\mathrm{d}} = 11$ years this gives $\lambda = 10$ km. If magnetic helicity cannot cascade down to this length scale, then dissipation of fluctuating magnetic helicity will be relatively inefficient. Some approximate steady state must be reached, of course. One possibility is a great buildup of positive helicity at small scales, so that dissipation does indeed balance the α effect input. Another possibility is that much of the fluctuating helicity inverse cascades back to large scales. This would weaken the effect: i.e. the naive calculation of mean field helicity generation by α must be decreased.

The second integral in (10.68) represents turbulent diffusion of magnetic helicity across boundaries. If we apply the equation to the northern interior of the sun, the boundaries are the photosphere and the equator. Presumably buoyancy is more important than turbulent diffusion in bringing helical fields through the photosphere. Also, we know the magnitude of helicity flow upwards carried by emerging flux; it can be estimated from the CME rate. However, diffusion of helicity across the equator due to turbulent motions may be quite important for the global helicity balance. Here negative helicity from the north cancels with positive helicity from the south. The diffusion across the equator is

$$\dot{K}_N = 2 \int_{\text{equator}} \mathbf{A_P} \times (\alpha \mathbf{B} - \beta \nabla \times \mathbf{B}) \cdot \hat{\boldsymbol{\theta}} \, d^2 x. \tag{10.69}$$

For a symmetric mean field $\mathbf{A_P}$ only has a ϕ component, while the radial component of $\mathbf{B}$ vanishes at the equator. Thus the α term does not contribute. Meanwhile the radial component of $\nabla \times \mathbf{B}$ is $(1/r)\partial B_\phi/\partial\theta \approx B/R_\odot$. For a thin ($10^4$ km) layer dynamo with a modest $B = 10^4$ G, one obtains $\dot{K}_N = 10^{30}\beta$Mx^2/day (with β in units of cm^2 s^{-1}). A fairly large turbulent diffusivity of $\beta \approx 10^{13}$ cm^2 s^{-1} will make the equatorial diffusion term comparable in magnitude to the helicity flow due to differential rotation.

Acknowledgements

This work has been influenced by numerous people. I thank F. Bacciotti, S. Champeaux, R. Prandi, R. Verzicco, and E. Zienicke for help in writing up lecture notes which have been incorporated in some of the sections.

References

Aly, J.J., "Some properties of finite-energy constant-alpha force-free magnetic fields in a half-space," *Solar Phys.* **138**, 133 (1992).

Arnold, V.I. and Khesin, B.A. *Topological Methods in Hydrodynamics*, Springer, New York (1998).

Barnes, D.C., "Mechanical injection of magnetic helicity," *Phys. Fluids* **31**, 2214 (1988).

Berger, M.A., "Rigorous new limits on magnetic helicity dissipation in the solar corona," *Geophys. and Astrophys. Fluid Dynam.* **30**, 79 (1984).

Berger, M.A., "Structure and stability of constant α force-free fields," *Astrophys. J. Suppl.* **59**, 433 (1985).

Berger, M.A., "An energy formula for nonlinear force-free magnetic fields," *Astronomy Astrophys.* **201**, 355 (1988).

Berger, M.A., "Third-order invariants of randomly braided curves," in *Topological Fluid Mechanics* (Eds. H.K. Moffatt and A. Tsinober), Cambridge University Press, p. 440 (1990).

Berger, M.A., "The role of helicity and other magnetic invariants," in *Mechanisms of Chromospheric and Coronal Heating* (Eds. P. Ulmschneider, E.R. Priest and R. Rosner), Springer-Verlag, Berlin, p. 570 (1991).

Berger, M.A., "Energy-crossing number relations for braided magnetic fields," *Phys. Rev. Lett.* **70**, 705 (1993).

Berger, M.A., "Minimum crossing numbers for three–braids," *J. Phys. A* **27**, 6205 (1994a).

Berger, M.A., "Coronal heating by dissipation of magnetic structure," *Space Sci. Rev.* **68**, 3 (1994b).

Berger, M.A., "Magnetic helicity in a periodic domain," *J. Geophys. Res.* **102**, 2637 (1997).

Berger, M.A. and Field, G.B., "The topological properties of magnetic helicity," *J. Fluid Mech.* **147**, 133 (1984).

Berger, M.A. and Ruzmaikin, A., "Helicity production by differential rotation," *J. Geophys. Res.* **105**, 10481 (2000).

Berger, T.E., DePontieu, B., Schrijver, C.J. and Title, A.M., "High-resolution imaging of the solar chromosphere corona transition region," *Astrophys. J.* **5119**, L97 (1999).

Bhattacharjee, A. and Wang, X., "Current sheet formation and rapid reconnection in the solar corona," *Astrophys. J.* **372**, 321 (1991).

Bieber, J.W., Evenson, P.A. and Matthaeus, W.H., "Magnetic helicity of the Parker field," *Astrophys. J.* **315**, 700 (1987).

Bieber, J.W. and Rust, D.M., "The escape of magnetic flux from the sun," *Astrophys. J.* **453**, 911 (1995).

Brandenburg, A., Nordlund, A., Pulkinnen, P., Stein, R.F. and Tuominen, I., "3-D simulation of turbulent cyclonic magnetoconvection," *Astron. Astrophys.* **232**, 277 (1990).

Brown, M.R., Canfield, R.C. and Pevtsov, A.A. (Eds.), *Magnetic Helicity in Space and Laboratory Plasmas*, AGU Geophysical Monograph Series 111.

Brummell, N.H., Hurlbert, N.E. and Toomre, J., "Turbulent compressible convection with rotation. II. Mean flows and differential rotation," *Astrophys. J.* **493**, 955 (1998).

Cantarella, J., DeTurck, D. and Gluck, H., "The Biot-Savart operator," *J. Math. Phys.* **42**, 876 (2001).

Chui, A.Y.K. and Moffatt, H.K., "The energy and helicity of knotted magnetic flux tubes," *Phil. Trans. R. Soc. London A* **451**, 609 (1995).

Field, G.B. and Blackman, E.G., "Nonlinear alpha-effect in dynamo theory," *Astrophys. J.* **513**, 638 (1999).

Finn, J. and Antonsen, T.M., "Magnetic helicity: what is it, and what is it good for?," *Comm. Plasma Phys. Contr. Fusion.* **9**, 111 (1985).

Freedman, M.H. and Berger, M.A., "Combinatorial Relaxation," *Geophys. Astrophys. Fluid Dynam.* **73**, 91 (1993).

Freedman, M.H. and He, Z.X., "Divergence-free fields – energy and asymptotic crossing number," *Ann. Math.* **134**, 189 (1991).

Gilbert, N.D. and Porter, T., *Knots and Surfaces*, Oxford University Press (1994).

Heyvaerts, J. and Priest, E.R., Astro. Astrophys. "Coronal heating by reconnection in DC current systems – a theory based on Taylor hypothesis," **137**, 63 (1984).

Ji, H.T., "Turbulent dynamos and magnetic helicity," *Phys. Rev. Lett.* **83**, 3198 (1999).

Karpen, J.T., Antiochos, S.K., Dahlburg, R.B. and Spicer, D.S., "The Kelvin-Helmholtz instability in photospheric flows – effects on coronal heating and structure," *Astrophys. J.* **403**, 769 (1993).

Kumar, A. and Rust, D.M., "Interplanetary magnetic clouds, helicity conservation, and current-core flux ropes," *J. Geophys. Res.* **101**, 15667 (1996).

Kusano K., Suzuki, Y. and Nishikawa, K., "A solar flare triggering mechanism based on the Woltjer-Taylor minimum energy principle," *Astrophys. J.* **441**, 942 (1995).

Longcope, D.W., Fisher, G.H. and Pevtsov, A.A., "Flux tube twist arising from helical turbulence: the Σ effect," *Astrophys. J.* **507**, 417 (1998).

Low, B.C., "Magnetohydrodynamic processes in the solar corona - flares, coronal mass ejections, and magnetic helicity," *Phys. Plasmas* **1**, 1684 (1994).

Lou, Y.Q. and Rosner, R., "The damping of the Alfvén mode in stochastic astrophysical fluids," *Astrophys. J.* **309**, 874 (1986).

Lu, E.T. and Hamilton, R.J., "Avalanches and the distribution of solar flares," *Astrophys. J.* **380**, L89 (1991).

Martin, S.F., Marqette and Bilimoria, R., "The solar cycle pattern in the direction of the magnetic field along the long axes of polar filaments," in *The Solar Cycle* (Ed. K.L. Harvey), ASP Conf. Series **27** 53 (1992).

Matthaeus, W.H. and Montgomery, D., "Selective decay hypotheses at high mechanical and magnetic Reynolds numbers," *Ann. N.Y. Acad. Sci.* **357**, 203 (1980).

Moffatt, H.K., "The degree of knottedness of tangled vortex lines," *J. Fluid Mech.* **35**, 117 (1969).

Moffatt, H.K., *Magnetic Field Generation in Electrically Conducting Fluids*, Cambridge University Press, Cambridge (1978).

Moffatt, H.K., "Magnetostatic equilibria and analogous Euler flows of arbitrarily complex topology. 1. Fundamentals," *J. Fluid Mech.* **159**, 359 (1985).

Moffatt, H.K. and Ricca, R.L., "Helicity and the Calŭgareanŭ invariant," *Proc. Royal Society London A* **439**, 411 (1992).

Parker, E.N., "Magnetic neutral sheets in evolving fields. II – Formation of the solar corona," *Astrophys. J.* **264**, 642 (1983).

Parnell, C.E. and Jupp, P.E., "Statistical analysis of the energy distribution of nanoflares in the quiet Sun," *Astrophys. J.* **529**, 554 (2000).

Pevtsov, A.A., Canfield, R.C. and Metcalf, T.R., "Latitudinal variation of helicity of photospheric fields," *Astrophys. J.* **440** L109 (1995).

Pouquet, A., Frisch, U. and Léorat, J., "Strong MHD helical turbulence and the nonlinear dynamo effect," *J. Fluid Mech.* **77**, 321 (1976).

Priest, E.R., van Ballegooijen, A.A. and Mackay, D.H., "Model for dextral and sinistral prominences," *Astrophys. J.* **460**, 530 (1996).

Ricca, R.L. and Berger, M.A., "Topological ideas and fluid mechanics," *Phys. Today* **49**(12), 24 (1996).

Rust, D.M., "Spawning and shedding helical magnetic fields in the solar atmosphere," *Geophys. Res. Lett.* **21**, 241 (1994).

Rust, D.M. and Kumar, A. "Helical magnetic fields in filaments," *Solar Phys.* **155**, 69 (1994).

Rust, D.M. and Kumar, A., "Evidence for helically kinked magnetic flux ropes in solar eruptions," *Astrophys. J.* **464**, L199 (1996).

Ruzmaikin, A., "Redistribution of magnetic helicity at the Sun," *Geophys. Res. Lett.* **23**, 2649 (1996).

Seehafer, N., "Electric current helicity in the solar atmosphere," *Solar Phys.* **125**, 219 (1990).

Seehafer, N., "Nature of the α effect in magnetohydro-dynamics," *Phys. Rev. E* **53**, 1283 (1996).

Shimizu, T., "Energetics and occurence rate of active-region transient brightenings and implications for the heating of the active-region corona," *Publ. Astron. Soc. Japan* **47**, 251 (1995).

Similon, P.L. and Sudan, R.N., "Energy dissipation of Alfvén wave packets deformed by irregular magnetic fields in solar coronal arches," *Astrophys. J.* **336**, 442 (1989).

Solanki, S.K., "Small scale solar magnetic fields – an overview," *Space Sci. Rev.* **93**, 1 (1993).

Stribling, T., Matthaeus, W.H. and Ghosh, S., "Nonlinear decay of magnetic helicity in MHD turbulence with a mean magnetic field," *J. Geophys. Res.* **99**, 2567 (1994).

Sturrock, P.A. and Uchida, Y., "Coronal heating by stochastic magnetic pumping," *Astrophys. J.* **246**, 331 (1981).

Taylor, J.B., "Relaxation and magnetic reconnection in plasmas," *Rev. Mod. Phys.* **58**, 741 (1986).

van Ballegooijen, A.A., "Electric currents in the solar corona and the existence of magnetostatic equilibrium," *Astrophys. J.* **298**, 421 (1985).

van Ballegooijen A.A. and Martens, P.C.H., "Magnetic fields in quiescent prominences," *Astrophys. J.* **361**, 283 (1990).

van Ballegooijen, A.A. Cartledge, N.P. and Priest, E.R., "Magnetic flux transport and the formation of filament channels on the sun," in *New Perspectives on Solar Prominences* (Eds.) D. Webb, D.M. Rust and B. Schmieder, IAU Coll. 167 ASP 150 265 (1998).

Witten, E. "Quantum field theory and the Jone polynomial," *Comm. Math. Phys.* **121**, 351 (1989).

Zhao, X.P. and Hoeksema, J.T., "Unique determination of model coronal magnetic fields using photospheric observations," *Solar Phys.* **143** (1993).

Name index

Subject index

Printed and bound by CPI Group (UK) Ltd, Croydon, CR0 4YY

23/10/2024

01778249-0004